The Management of Estates and their Resources in the Egyptian Old Kingdom

Joyce Swinton

BAR International Series 2392
2012

Published in 2016 by
BAR Publishing, Oxford

BAR International Series 2392

The Management of Estates and their Resources in the Egyptian Old Kingdom

ISBN 978 1 4073 0984 2

BAR Publishing is the trading name of British Archaeological Reports (Oxford) Ltd. British Archaeological Reports was first incorporated in 1974 to publish the BAR Series, International and British. In 1992 Hadrian Books Ltd became part of the BAR group. This volume was originally published by Archaeopress in conjunction with British Archaeological Reports (Oxford) Ltd / Hadrian Books Ltd, the Series principal publisher, in 2012. This present volume is published by BAR Publishing, 2016.

Printed in England

BAR
PUBLISHING

BAR titles are available from:

BAR Publishing
122 Banbury Rd, Oxford, OX2 7BP, UK
EMAIL info@barpublishing.com
PHONE +44 (0)1865 310431
FAX +44 (0)1865 316916
www.barpublishing.com

For Stuart

ACKNOWLEDGEMENTS

As a student in the Egyptology department of Macquarie University I have been privileged to work in an atmosphere of intellectual generosity, enthusiasm and personal support. I would like to express my indebtedness and gratitude to all members of the department. In particular thanks are due to the following for their interest and support and for the time they have given to encouraging and helping me in so many ways over the years of this study: Susanne Binder, Linda Evans, Beth Thompson and Alex Woods.

I would like to take this opportunity to thank my supervisor, Professor Naguib Kanawati, for years of intellectual enjoyment, encouragement, patience and good humour. His personal guidance has always been greatly appreciated. Like so many others, I am a beneficiary of his vision for the development of Egyptology in this country.

I wish to express my gratitude to Leonie Donovan, to whom I owe a great debt for the generosity with which she has given her time and technological expertise to the production of both a dissertation and the present publication. I could not have produced a volume for publication without her help, which amounted to many, many long hours of work. Thank you, Leonie.

Through the years of research and writing my husband, Stuart, has consistently provided support and understanding of his wife's constant abstraction to another time and place. This says much about a man's generous character – and a woman's good fortune.

Sydney, Australia

TABLE OF CONTENTS

TABLES

THE DATA and THE NILE VALLEY

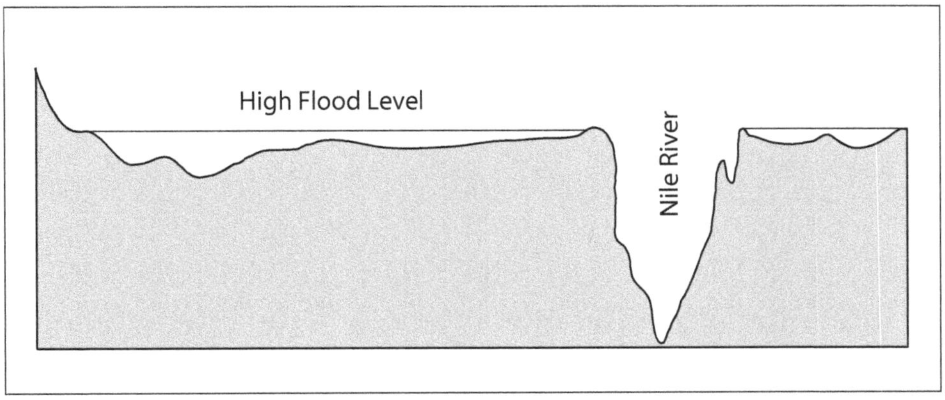

Figure 1: THIS CROSS SECTION OF THE NILE VALLEY AT SAQQARA SHOWS THE CONVEX NATURE OF THE FLOODPLAIN AND BACK SWAMPS ON THE OUTER MARGINS DURING A PARTICULARLY HIGH FLOOD LEVEL. After Hassan F.A. (1997).

The data for this study are the product of Egyptian society's adaptation to the hydrological dynamic of the river Nile and to the distinctive topography of the valley through which the river flows.

Configuration of the Nile valley

When the Nile flooded freely, marshland occurred not only in the Delta but also along much of the Egyptian valley. During an inundation, the topography of the Nile flood plain allowed the floodwater to spill laterally either side of the river into successive flood basins. As the Nile rose (mid August in southern Egypt), its waters found overflow channels and spread out to fill the flood basins. This situation moved northward between August and November. When the floodwater had drained, there emerged tracts of prime land for crops, seasonal pastureland for the grazing of livestock, and renewed, isolated marshy back swamps along the valley margins and the fringes of streams and oxbow lakes.[1] The process of flooding and drainage occurred on a much greater scale in the Delta. The Nile valley thus offered different kinds of terrain for exploitation: permanent swamps of the Delta and bordering the river, channels and stretches of standing water, seasonally watered and fertilised land for agrarian farming and grazing and "islands of habitation" that were only a few metres above the flood level but remained dry year round apart from times of excessive flooding. These features allowed Egyptians to maintain permanent living along the river valley, develop a simple basin irrigation system and to exploit the wetland resources, pasturage, arable land and even the desert fringe.

This riverine environment had particular characteristics. Bounded on both east and west by desert and barren cliffs, the basically convex landform of the valley caused a system of backwater swamps to develop on the outer margins of the river plain. Beyond the banks of the river and the areas of naturally formed levees resulting from the inundation, the birkets on the outer edges of the floodplain were fed initially by the annual flood and maintained by a reservoir of groundwater that gravity caused to seep outward.[2] This typography was not static; constant changes resulted from the varying levels of the annual inundation. Moreover, the width of the valley, and thus of the usable land, varies from one section of the river to another. Where the floodplain was relatively narrow, for example south of El Kab, there would have been few great estates.[3] Apart from the Delta, the part of Egypt with the broadest and most fertile valley floor lay in the middle provinces of Upper Egypt.[4] Although there are no direct pictorial or inscriptional references to the annual flood in the data from Old Kingdom tombs, these features underlay both the broad patterns and the detailed activities reflected in the funerary art decorating the walls of tomb chapels. This floodplain, on which society depended for its existence, shaped the Egyptian worldview as well as the country's economy and the organization of society.

1 Butzer (1959b) 78ff.

2 Hassan. F. A. (1997) 51 – 74.

3 Jacquet-Gordon (1962) 106-08, fig. F.

4 Kanawati *Reforms* (1980) 5-11, fig. 1-3.

INTRODUCTION

CHAPTER 1 AIMS and OBJECTIVES

1.1 Assets and resources of the ruling class

The economic prosperity of Old Kingdom Egypt depended on efficient exploitation of resources provided by the River Nile and valley. Many hundreds of decorated tombs mirror this dependence on a narrow strip of fertility that cut through barren rocky cliffs and deserts, providing funerary iconography and inscriptions that reflect a society in which peasant workers provided the labour needed to exploit the riverine environment, while their powerful and privileged rulers enjoyed the rewards.

The subject of this study is an examination of the resources at the disposal of the elite class of officials who administered the state on behalf of the crown. Their assets included one or more rural estates either owned outright or held in usufruct and/or enjoyed according to a land-owning system referred to as the *pr ḏt* (estate),[5] and all that the estate produced: a workforce if in some way bound to the estate, buildings, means of transport, household and personal effects. The resources available to these officials were the products of the estate: livestock, annually grown field crops and what could be procured from the desert margins, waterways and marshlands. Their assets and resources contributed to officials' status and authority and provided the crown with an elite class of administrators available for state service. Consequently, although officials were bureaucrats in the service of the crown and recipients of a redistributive system, the resources that gave them the freedom to undertake administrative duties depended on effective supervision of a rural estate, a dependent workforce experienced in pastoral and agrarian farming and hunting and, of course, efficient management of assets and resources in accommodation with the geography of the Nile valley.[6] The necessity for this accommodation introduced a duality into Egyptians' relationship with their natural environment, which is reflected in their religious beliefs and their art. While they confronted its power and occasional vagaries with awe, expressed by reverence for and fear of gods of the natural world, Egyptians also experienced the mundane realities of working with the land and usually enjoyed its material bounty.

1.2 The nature of the investigation

The aim of this study is to investigate funerary images and inscriptional material that may throw light on the economic basis of high officials and on the value that they attached to the different resources at their disposal. Using the data this way is beset with problems and uncertainties, many of which stem from the fact that we do not sufficiently understand the worldview or life experiences of the Old Kingdom to be able always to draw reliable evidence and valid information from funerary scenes. As a result, we either have to accept what is portrayed in the scenes at face value or decide on a method of analysis and interpretation. Both approaches rest on an essentially subjective ideological basis. Face value acceptance will sometimes leave unanswered questions or limit what can be inferred from scenes, and does not always encourage critical inquiry. On the other hand, in developing a method of inquiry to comprehend this ancient society, 90 per cent illiterate and totally dependent on the power of people, animals, wind and water, the researcher has to draw on his/her own worldview and technology-dependent, literate life experience. The acknowledgement that scenes and inscriptions may not always have been intended to be read literally takes the researcher into a range of complex responses that sometimes require a process of 'decoding' an image or an entire scene. There is not just one key that will 'unlock' the information contained in all scenes and images. Some scenes may have a particular funerary or afterlife significance; others may mirror knowledge of the natural world or merely contain mundane information about everyday life. Each type of scene needs an appropriate method of analysis. In some, the key to understanding lies in identifying the funerary implication; in others, images have to be recognised as the artist's 'aspective' portrayal of his observation of the natural world expressed in bas-relief or on a flat surface. Scenes contain information about 'daily life' that would be clear to the tomb owner's contemporaries but are obscure for literate, technology-dependent people of today who have no direct experience of how a simple farming community meets the demands of daily life. These factors have to be constantly borne in mind when examining the iconographic and inscriptional material relating to rural estates.

Whether Egypt's great estates were the property of the crown, temples or individuals, their management was the responsibility of high officials. As servants of the crown responsible for the administration of the state, they would have entrusted day-by-day operation of their estates to

[5] The nature of the *pr ḏt* referred to in so many Old Kingdom tombs has been investigated by J.J. Perepelkin in *Privateigentum in der Vorstellung der Ägypter des Alten Reichs* (1986).

[6] There is no evidence that officials owned mines and quarries or enjoyed profits from them: Bloxam, (2006). Officials' funerary monuments depict scenes of craftsmen at work making jewellery and household items, but these appear to be for domestic consumption, and scenes portraying the buying and selling of goods do not indicate that the tomb owner engaged in trade.

subordinates, as *Mrjj-Rꜥ-mrjj-Ptḥ-ꜥnḫ/Nḫbw* notes in his autobiography.[7] Investigation into the economic power base of the class of elite officials involves analysis of the composition and management of the resources available to this class, assessing the value placed upon these resources and identifying and analysing any trends in the portrayal of estate management that change over time.

A number of related issues are excluded from this study. The detailed practices and technology of resource production are not investigated partly because they have been intensively examined elsewhere,[8] but essentially because they are not strictly germane to the questions raised in this study. Two other more pertinent issues, however, have had to be omitted. Both the land held by the tomb owner and his workforce may be regarded as resources but the interpretation of the relevant data is contentious, making their investigation too large to be included in the present study.

1.3 Land ownership

While officials derived their resources from different kinds of terrain - annually inundated arable land, the riverine environment and the scrubland of the desert fringe - it is difficult to establish the legal arrangements giving officials entitlement to the land that was exploited on their behalf.[9] Statements of a general nature claiming what might amount to personal ownership of estates are frequently made in tomb inscriptions,[10] but extant legal records relating to this question are rare and often difficult to interpret. From these records it may be deduced that land could have been inherited, purchased, bestowed by the crown, perhaps as a stipend in return for carrying out official duties or held in usufruct. In addition, officials administered crown estates[11] and as priests, managed temple lands, which afforded them entitlement to a share of the produce of these lands.[12] Funerary inscriptions make it clear that officials at least enjoyed the use of landed properties and were able in some circumstances to hand on these rights to their heirs.[13] If data is drawn from the entire period of the Old Kingdom, there are clear records of no more than a few specific instances of the various kinds of landholding mentioned above. Statements about the nature of Old Kingdom landholding, therefore, have to be general as the evidence is drawn from a very small number of individual inscriptions that are geographically and chronologically scattered.[14] This is also true of the workforce at the officials' disposal.

1.4 The rural workforce

As in the case of landholding, there is insufficient evidence relating to the legal and the practical relations between the tomb owner and the people portrayed working on his land to support precise descriptions of this relationship. The farmhands shown labouring in the fields and marshes in private tombs are described as working in the *pr-ḏt* (or in teams of the *pr-ḏt*) of the tomb owner, but whether they were bound to the land in some form of serfdom, or peasants free to offer and withdraw their labour is uncertain. This is in part due to the difficulty of interpreting the meanings of the terms used to denote workers. Depicted labourers are variously referred to as *mrt* ('servant, menial'),[15] *nswtjw* ('the king's people')[16] or *jzwt* ('team').[17] A further group of unlabelled field-hands, whom Brovarski refers to as 'marsh dwellers', are sometimes associated with pastoral activities.[18] They are depicted as emaciated figures with unkempt hair, caring for the herd of cattle and sometimes leading an animal in a parade of livestock. A higher level of estate workers labelled as 'overseer', for example *jmj-r3 mḏt,*[19] *jmj-r3 ṯzwt,*[20] *jmj-r3 wḥꜥw*[21] ('overseer of the stalls', 'overseer of the gang of workmen', 'overseer of fowling'), are depicted supervising activities such as cattle husbandry, leading a line of cattle presented to the tomb owner or directing the team deploying the fowling clapnet, while a more senior figure presenting the document containing the record of accounts is sometimes titled *jmj-r3 pr* ('overseer of the house/estate').[22]

1.5 Location of portrayed estates

Knowledge of the locations of the individual estates owned or administered by high officials of the court is limited, as few tomb owners felt it necessary to record this information. Frequent references to 'settlements and estates of Lower and Upper Egypt' indicate that throughout the Old Kingdom high officials held or managed land both in the Delta and south of Memphis,[23] and it may be assumed that provincial officials both 'owned' land and, if they held appropriate titles, administered crown and temple estates in their

7 *Nḫbw* recounts that he administered the property of his brother for whom he worked. He also managed his brother's funerary estate, Cairo Text (JE 44608).
8 See below 'Studies' pp. 3 – 5.
9 Ward (1984) 63-64.
10 See TABLE 17: TOMB OWNER VIEWING THE WORK AND PRODUCE OF HIS ESTATE and Chapter 13: Administration of the Estate.
11 Moreno García (1999) passim.
12 Weeks (1983), the family of *Nj-k3-ꜥnḫ* was rewarded by the crown with land and revenue from the temple of Hathor at Tehneh. In late Dynasty 5 *Pn-mrw* was awarded a similar endowment by the crown: Ward (1984) 66. *Nj-k3-ꜥnḫ* also administered the agricultural installations of the crown as he was *jmj-r3 pr n ḥwt-ꜥ3t* and *jmj-r3 njwwt m3wt*: Moreno Garcia (2006) 102.
13 For the most recent publication of this see Thompson, Elizabeth, M., *The Old Kingdom Tombs in Middle Egypt*, vol. 1: *The Tombs of Nika-ankh I, Nika-ankh II and Ka-hep* (2012) forthcoming.
14 Eyre (1994) 107-109. See also Texts 106 (*K3.j-m-nfrt?*), 108 (*Mtn*), 110 (*Nj-k3-ꜥnḫ*), 111 (*Nj-k3w-rꜥ*), 112 (*Pn-mrw*), *1154* (*Ṯntj*) in Strudwick (2005) 189-191, 192-194, 195-199, 200-201, 202-203.
15 Translated by Perepelkin (1986) p. 25 and n. 6, 164ff and nn. 1, 2 as 'leibeigenes Gesinde'. See Jones I (2001) no. 1656, p. 443-444.
16 Translated by Helck as 'Frei-Bauern' in Allam (1994) 11. See Jones I (2001) no. 1828, p. 489.
17 Jones I (2001), no. 1275-7, p. 343.
18 Brovarski (2000) 44.
19 Jones I (2001) no. 873, p. 147.
20 Jones I (2001) no. 992, p. 276.
21 Jones I (2001) no. 421, p. 104.
22 Jones I (2001) no. 461, p. 114.
23 See TABLE 17: TOMB OWNER VIEWING THE WORK AND PRODUCE OF HIS ESTATE. See also *Mṯn* Urk. I 1-7; Strudwick (2005) 192-194.

province.[24] It is not possible to judge how much of Egypt's fertile land was administered by Memphite-based officials and how much was locally owned and controlled. Moreover, as discussed in Chapter 15, it is likely that the proportion of land under the control of officials appointed by the crown changed over time.[25]

1.6 Economic history of the Old Kingdom

The method of examining selected iconography and inscriptions of private tombs has been chosen because the reconstruction and analysis of any aspect of a society's economy calls for large amounts of quantitative data; in the case of the Old Kingdom the pictorial programmes of tomb chapels are one of the few areas of research to provide data in quantity. Furthermore, it may be hypothesised that the great estates were a vital factor in the economy of the Old Kingdom state, essential to provide sufficient taxation in kind to support the construction of the crown's great building projects, as well as being the source of officials' wealth and status.[26] Thus the supplies such as livestock and cereals to feed the construction workforce were both drawn from crown estates and paid as taxes by private estates. It is probable that the seasonal workforce, amounting to many thousands in some reigns, was also recruited from these sources.[27] This process may have helped to generate a dynamic impulse with social and political implications for the state. A society's economic features and changes are dynamic factors, essential to analysis of the cause/consequence nexus that is the core of history.[28] Although the methodology for researching Egyptian economic history is still being debated,[29] it may be possible in some periods of time to identify a causal relationship between economic development and the historical dynamic. The need for resources to sustain building programmes, for example, may have led to the creation of the office of *jmj-r3 Šm3* ('overseer of Upper Egypt') in the second half of Dynasty 5 as a means of tapping into under-exploited resources of the Upper Egyptian provinces. The ensuing changes made to the mode of operation of this office in Dynasty 6 suggest that in the later Old Kingdom the crown faced continuing difficulties in raising sufficient resources to support its projects.[30]

A major economic situation such as this would have affected the elite echelon of officials as well as the central administration and may be expected to have significant social and political consequences. To maintain a harmonious balance within the upper reaches of the country's administration, officials' resources would have had to be commensurate with the expectations prompted by their position within the administrative hierarchy. Yet faced with a financial crisis that emerged in late Dynasty 5, the crown would not have been able to satisfy the ambitions of the enlarged class of elite officials. Interpreting the data on which this hypothesis is based and testing it by comparison with new evidence from recent excavations and the reinterpretation of previously discovered data, is a task awaiting economic historians.[31]

1.7 Studies

The economic history of ancient Egypt is a comparatively new branch of Egyptology, post-dating by over a century the study of aspects such as religion and kingship. In the case of the Old Kingdom, the need to establish an acceptable method of investigation that sheds light on the economic record may in part explain the long delay. Scholarly debate continues over what amounts to appropriate data from which to draw evidence,[32] while the over-all lack of data has compounded the problems. Most Old Kingdom evidence reflects the interests and preoccupations of a small elite class, while the activities and preoccupations of the majority have left material invaluable for the investigation of technologies but less useful to the economic historian seeking an understanding of the demand/supply patterns of society as a whole. In the case of Old Kingdom Egypt even for the elite, sources of data from which to reconstruct the economic system are sparse and limited.[33]

Apart from studies of the scenes of 'daily life' in Old Kingdom tomb chapels, in particular by Montet and Vandier[34] who concentrated on description and understanding of technologies, the economic history of ancient Egypt was neglected by scholars for many years. In 1975 in reference to the New Kingdom where economic data was much more available, Janssen noted that the economic history of ancient Egypt was 'virtually non-existent' and what passed as 'economic history'

24 *K3r* of Edfu, Urk. I, 251-5; Strudwick (2005) 342-344; *Jbj* of Deir el-Gebrawi, Urk. I, 145; Strudwick (2005) 364-365.

25 Moreno Garcia (1999) 36-39 discusses the distinction between the great crown estates (*ḥwt ˁ3t*) and smaller units (*ḥwt*).

26 Kees (1932) 85ff. 92, 112; Goedicke (1956) *MIO* 1-10; Martin-Pardey (1976) 152ff; Kanawati (1980) *Reforms*, passim. Inked inscriptions on vases found in the Step Pyramid substructure provide early evidence of the great estates contributing provisions to the crown. Lacau–Lauer (1965) 10-11 [12], 48 [94], 49 [98], 56-57 [131], 57 [133, 134], 74-75 [194].

27 The discovery of a virtual town providing for a community of 18,000 workers underlines the crown's need for a constant supply of provisions: Hawass (1996) 53-67.

28 Eyre refers to this concept as 'the economy (being) a primary context of history': Eyre (2008) 88-90.

29 The issues revolve around the questions of macro- and micro-analysis and whether methods should focus on general theory or data collection. See below 'Studies' pp 3 – 5

30 Weeks (1983) p. 6 refers to Egypt in late Dynasty 5 and early Dynasty 6 as 'a state in financial crisis'.

31 With respect to recently discovered monuments in the Teti cemetery, the assembling of relevant data has been started. See Kanawati (2003) passim.

32 Whether, for example, data from tombs of officials is acceptable as evidence. For a discussion of these issues, see Chapter 2 Interpretation of funerary material.

33 Moreno Garcia (2001) 414. There is some evidence denoting a 'national' economy, for example regular cattle counts, a crown-imposed system of taxation and economic relations with neighbouring societies under central control. S. Allam (1988) 52-72; Helck (1975) passim.

34 Montet (1925); Vandier vols 1-5 (1952).

amounted either to vague theories or to the description of technologies.[35] Recognizing that farming was the predominant economic activity, Janssen assigned the socio-economic structure of pharaonic Egypt to the class of 'peasant societies'. Consequently, he insisted that the basic economic facts that were needed had to include detailed knowledge of the extent and productivity of Egyptian cultivation, the structure of the system of irrigation, the pattern and development of settlement and the relation of cities to countryside. Applied to the available evidence for the Old Kingdom, this amounts to an impossible 'wish-list'.[36] Documentary sources of the Old Kingdom that would provide the depth of economic history envisaged by Janssen are largely lost, although surviving papyri such as the Abusir and Gebelein collections and the Balat tablets suggest they did once exist.[37] Otherwise, there are a few legal inscriptions in tombs and royal decrees with a provenance spread over time and place,[38] thus making questionable the drawing of historical generalisations. For example, as so few legal inscriptions are inscribed in tombs, we cannot judge whether they were included because they record a typical or an atypical situation.[39]

In the 1970s, however, egyptologists began studying available Old Kingdom data as a means of understanding the operation of the economy. After Helck published his pioneering work on the economic history of pharaonic Egypt in 1975, the need for the collection and analysis of Old Kingdom data began to be addressed.[40] Basic investigations into individual themes in the decorative programmes, culminating in interpretation of their economic ramifications, began to appear. Among the notable publications are Ghoneim's compilation of the literature on the economic significance of cattle, the works of Sahrhage, van Elsbergen and Brewer-Friedman on the fishing industry and Mahmoud's work on the economic importance of the fowling industry.[41] In particular, the last four studies base their findings on data drawn from as many sources as possible, providing valuable evidence for further investigation into the working of the economy.

Scholars such as Menu and Eyre have taken a somewhat different methodological approach by emphasising general theory, which they combine with available data.[42] Writing a few years after Janssen in an article specifically concerned with the Old Kingdom, Eyre deplored both the failure to use 'analogy or general theory in the study of Egypt' and the preference for 'collection and detailed analysis of basic data over general writing of any sort'.[43] Janssen and Eyre make equally cogent criticisms, but the writing of economic history requires the blending of available data with relevant general theory so that hypotheses and conclusions feed into the historical causal chain helping to explain social change and the historical dynamic of a society. For example, theories such as Wittfogel's that the requirements of a large-scale irrigation system produced Egypt's centralized, bureaucratic administration,[44] or Carneiro's argument that an interaction of limited agricultural land with the growth of population led to conflict and conquest,[45] need to be tested against the historical evidence. The focus of Eyre on broader economic issues such as labour, water management and land ownership, for which Old Kingdom data are particularly sparse and difficult to interpret, leads to interpretations that appear to be more reflective of Middle and New Kingdom data.[46]

Moreno Garcia, whose specialty is investigation of the rural milieu of the Old Kingdom, has taken yet another path.[47] His methodology, eschewing all pictorial data, limited his analysis of economic features such as rural estates and their products and workforce.[48] This is particularly true of his study of the *ḥw.t*, which he interprets as an estate of the crown, as a factor in the Old Kingdom economy. While Moreno Garcia's self-imposed limitations did not allow him to investigate the work and management of a rural estate, his focus on administrative documents has contributed to an understanding of the interaction of institutions, peasants and the royal taxes and of increasing rural indebtedness,[49] classes of rural

35 J.J. Janssen 'Economic history during the New Kingdom', *SAK* 3, 1975, p. 128.
36 J.J. Janssen 'Economic history during the New Kingdom', *SAK* 3, 1975, p. 132.
37 Posener-Krieger–de Cenival (1968).
38 Posener-Krieger, Paule (1976) p. 479; Baer (1966) 1-9; Goedicke, *Königliche Dokumente aus dem alten Reich* (1967).
39 The written language was still developing in the Old Kingdom. The use of an abbreviated orthography and a standardised phraseology sometimes limited the capacity of writings to express precisely nuanced statements. Edel (1944) 1-90. This is particularly true of the Coptos decrees, whose clarity is further clouded by broken statements and lacunae. See Hayes 'Royal Decrees from the Temple of Min at Coptos' in *JEA* 32 (1946) pp. 3-23 and also Goedicke (1967) 128-137.
40 Helck (1975).
41 Ghoneim (1977); Brewer–Friedman (1989); Sahrhage (1998); Van Elsbergen (1997); Mahmoud (1991).
42 Menu (1982); Menu (1985); Menu (1994); Eyre (1987) 5-47, 167-221; Eyre (1994) 107-133; Eyre *JEA* 80 (1994) 57-80; Eyre (1995) 175-189; Eyre (1999) 33-60.
43 Eyre (1987) 5-47, 167-221. More recently Eyre has clarified his position, "In the current political-intellectual climate the interpretative traps lie on one side in the clichés of environmental determinism, and on the other in the paradigm of cultural evolution. These are both approaches that can easily fall into a form of historical retrodiction—hindsight as explanation—that glides too easily over proper evidence-based analysis in the attempt to provide explanations in socio-economic history. Despite the difficulties of quantification, and to draw conclusions drawn from micro-economic analysis of economic reality at a local level, this seems the best approach to avoid anachronism, and to advance understanding of socio-economic behaviour in pharaonic Egypt" Eyre (2008).
44 Wittfogel (1957).
45 Carneiro (1970) 733-8.
46 In his study of land ownership in the period from 2500 to 1000 B.C. Ward refers to the difficulties of interpreting extant legal documents, noting that they are sparse even for such a long period of time, truncated summaries of much larger collections (now lost) and for the scholar, bedevilled by difficulties of understanding terminology, Ward (1984) 63.
47 Moreno Garcia (1999).
48 While Moreno Garcia explained why he rejects the use of pictorial material in *Hwt et le milieu égyptien du IIIe millénaire*, his reasons seem to apply to all his work.
49 Moreno Garcia (2001) 411-450.

workers,[50] the territorial aspect of the agricultural administration of the crown[51], and the cattle industry.[52]
A further aspect necessary to the understanding of the rural economy, which has also received much recent attention, is the geography of the Nile valley. An appreciation of the topography and hydrology provides information crucial to an understanding of floodwater management, while soil and climate studies provide further basic information needed to understand the agriculture of this environment. For example, the climate of the Nile valley has evolved in the Holocene era providing intervening wet and dry phases. The last dry phase beginning between 5000 and 4000 B.P., and therefore, experienced by Old Kingdom farmers, helps to explain changes in the exploitation of estate resources.[53]
In an important study, Karl Butzer brought together the results of analysis of the geomorphology and hydrology of the Nile floodplain, climatic trends and what was known of the patterns of settlement to throw light on socio-economic developments,[54] This investigation has been further developed by Rushdi Said and Fekri Hassan; but while the scientific aspects of their geoarchaeological investigations reveal close connections between climate change and variations in Nile floods,[55] the hypothesis that these factors were a major cause of economic and political developments in the Old Kingdom needs a greater degree of supporting historical evidence with a closer chronological correlation.

Ultimately, studies in the economy of the Old Kingdom have to be investigated within a chronological framework, based on historical evidence. The Old Kingdom covers some 300 to 400 years, during which the economy as well as the polity was subject to change. Apart from the scientifically focused investigations mentioned above, economic studies that attempt to delineate changing trends are rare, partly because egyptologists tend to overlook the usefulness of statistics in identifying change. Helck examined the economic effects of royal building programmes, but did not apply statistical analyses to support his conclusions regarding the economic dynamic that the programmes unleashed.[56]
A study by Kanawati used a statistical method to investigate trends in the wealth disposal of the bureaucracy, which could have provided evidence for Helck's hypotheses.[57] This investigation, however, has still far to run. In examining the conditions of the bureaucracy, Weeks referred to the 'financial crisis' of late Dynasty 5 but did not attempt to analyse its causes or consequences.[58] Work over the last fifteen years in the Unis and Teti cemeteries at Saqqara may now throw light on the financial problems of the central administration in late Dynasty 5 to Dynasty 6.[59]

For Old Kingdom studies, only the iconography and associated inscriptions in private tombs offer a sufficient quantity of data for the reliable analysis of economic factors. Furthermore, the identification of changing trends that may be the cause or consequence of the historical dynamic requires data spread over time. This too, is supplied by private tombs. Today Old Kingdom economic historians have available data emerging from more recent findings such as the Teti cemetery tomb reports and from other branches of Egyptology, re-interpretations of 'older' data, general theories and analogies from anthropology and ongoing attempts to correlate geological, topographical and meteorological conclusions with the historical evidence.

50 Moreno García (2000) 123-139; Moreno García *JEA* 84 (1998) 71-83.
51 Moreno García (1996) 117-138; Moreno García (1998) 38-55.
52 Moreno García *ZÄS* 126 (1999) 116-131; Moreno García *RdE* 50 (1999) 241-257.
53 Kozlowski–Ginter (1993) 327-336.
54 Butzer (1976). See also Said (1993).
55 Hassan. F. A. (2005).
56 Helck (1975) Chs 3-20.
57 Kanawati (1977).
58 Weeks (1983) 6.
59 In *Conspiracies in the Egyptian Palace: Unis to Pepi I*, Naguib Kanawati provides data for economic analysis. (Kanawati 2003).

CHAPTER 2 RESEARCH METHODOLOGY

To judge from their funerary inscriptions, the leaders of Old Kingdom Egypt were bureaucrats who measured the success of their life's work in ponderous lists of administrative, courtly and priestly titles. It is likely that as nobles, the source of their wealth was derived from inheritance while, as servants of the crown, they received rewards for administrative service. Establishing this, however, depends on data that rarely offer direct and exact information on the nature and origins of these resources. In general, the data indicate that an official's degree of wealth was derived from the agricultural exploitation of the land so that his income was drawn from the resources of his own property or that held according to a system of usufruct provided by the crown or from the diversion of the production of temple estates as a stipend in kind. It is even more difficult to ascertain whether there existed a class of farmers who owned smaller quantities of agricultural land or whether the class of farm labourers included owners of small plots for cultivation. Consequently, it is not known how much of Egypt's productive land was in the hands of the three big powers: the crown, the temples and the class of elite officials, but it seems probable that these officials administered much of this land either as their own property or as servants of the crown.

2.1 Research methods

The research methods set out here are designed to investigate the available data in order to establish the resources that provided the class of elite officials with their wealth, power and status. The data for this study are drawn from the iconography and inscriptions of private tombs, Memphite and provincial, of the period from the early Fourth to the Eighth Dynasty. As the study has both a chronological and geographic basis, only scenes that have clear provenance and can be dated within a limited period of time have been consulted. Altogether 191 tombs containing scenes of estate activities have been used. These monuments have been grouped into six chronological periods: early Dynasty 4, mid to late Dynasty 4, early to mid Dynasty 5, late Dynasty 5, early Dynasty 6 and mid Dynasty 6 to Dynasty 8. The relationship of time periods to reigns is as follows:

Early Dynasty 4	Snefru, Khufu
Mid to late Dynasty 4	Djedefre, Khafre, Menkaure, Shepseskaf
Early to mid-Dynasty 5	Userkaf, Sahure, Neferirkare, Shepseskare, Raneferef, Niuserre
Late Dynasty 5	Menkauhor, Djedkare/Isesi, Unis
Early Dynasty 6	Teti, Pepi I
Mid-Dynasty 6 to Dynasty 8	Merenre, Pepi II, rulers in Dynasty 8

The principle of separating the above periods of time is based essentially on significant changes in the decorative programmes of tombs. [60] The use of these broad chronological groupings has generally made it unnecessary to address the vexed question of the dating of the private tombs of the Old Kingdom. In most cases a dating 'consensus' has been drawn from academic works dealing with the dating of each tomb.[61]

2.2 Data preparation

The data required for this study cover all themes and associated inscriptions relating to rural estates, as well as images of livestock in other situations such as being transported by boat, being hunted or brought back from the hunt, or on wall thicknesses being led towards a figure of the tomb owner and presumably part of a parade of animals. No two portrayals of the same theme in tombs of officials are identical. To meet this problem of infinite variation in detail, the basic methodology has been to order individual scenes and vignettes into groups of thematic similarity for purposes of comparison, so that significant aspects may be identified and trends established. To this end, the data have been arranged in tables, which pinpoint the elements to be analysed and allow them to be compared:[62]

Step 1:
A master list (TABLE 1: DATA DIGEST) of monuments (tombs and separate blocks from tombs) was drawn up. Only monuments with provenance and acceptable dating were included. These were separated into the time periods mentioned above, based on the preferred date for each tomb given in the Data Digest.

Step 2:
The data, line drawings where available and checked against photographs, were organised into resource categories:

(a) according to the type of resource: desert (wild animals); pastoral (domesticated livestock), agrarian (barley, wheat and flax), marshland (fishing, trapping wildfowl, gathering papyrus, building

[60] Based on the length of reigns suggested by Beckerath. See Beckerath, *Chronologie* (1997).

[61] See TABLE 1: DATA DIGEST. In the dates given in the Tables, 'early', 'mid' and 'late' are abbreviated to 'E', 'M' and 'L'.

[62] In his address at the Tenth International Congress of Egyptologists Eyre stressed the need for 'proper evidence-based analysis' involving quantification based on historical data: Christopher J. Eyre (2008) 88-90. This approach would also apply to the analysis of broad theories such as Wittfogel's argument that the dependence on large-scale irrigation on countries such as China and Egypt led to the rise of despotic, centralized states that were highly bureaucratized: Wittfogel (1957).

papyrus skiffs), the tomb owner facing a vertical inscription and panel of registers portraying agricultural activities;

(b) each bank of resource data was subdivided chronologically;

(c) the various themes in which each resource was portrayed were identified. Tomb by tomb, each type of resource was then sub-categorised according to these themes.

Step 3:
As described in *Step 1*, a basic chronological table was constructed for each type of resource with additional columns for entries of the sub-categories of themes. Scenes were entered in appropriate tables. Associated inscriptions were included if relevant to the aims of the study.

Step 4:
Once the resource tables had been constructed, scenes depicting each resource were studied according to time periods. Appropriate questions had to be framed. An inquiry that is focussed on the portrayal of the produce of rural estates as *resources* in funerary iconology is neither an art history project nor a study of farming technology, although findings from both may contribute valuable insights. Relevant questions had to revolve around resource management, the ultimate uses of each resource and the value placed on each. The issue of 'value' inevitably required a degree of subjective judgement.

2.3 Resource categories

The categories of resources were grouped under four heads: animals of the desert, domesticates of the pastoral environment, agricultural crops and marshland products. A further category has been devoted to estate management and the relationship of tomb owner to estate activities:

The desert fringe
- Desert animals (mainly ungulates but also striped hyena): sub-categories: animal parade; hunting scenes; returning from the hunt[63].

Pastoral
- Cattle: sub-categories: animal parade; husbandry.[64]
- Sheep: sub-categories: animal parade; planting field crops.
- Goats: sub-categories: animal parade; husbandry.
- Donkeys: sub-categories: animal parade; harvest; other work.

Agrarian (field crops)
- Soil preparation and planting: sub-categories: ploughing; hoeing; broadcasting seed; use of a flock of sheep.
- Cereals: sub-categories: nature of crop (barley or wheat); harvesting; dealing with the harvested material; storage of grain.
- Flax: sub-categories: harvesting; bundling plants.

Marshland produce
- Fishing: sub-categories: fish species; fishing methods; gutting and drying fish.
- Trapping wildfowl: sub-categories: bird species; operating the clapnet; transport of birds; fowl-yards.
- Papyrus: sub-categories: gathering papyrus plants; construction and use of papyrus skiffs; papyrus as cattle fodder.

Relationship of tomb owner to estate activities
- Depiction of tomb owner.
- The inscription column.
- Composition of associated panel of registers.

2.4 Interpretation of funerary material

As Weeks has noted, our understanding of the worldview of Old Kingdom society is limited. As a result we fail to recognise the intended meanings represented by an image or scene in Egyptian art because we cannot identify its 'significant attributes'.[65] Many debated issues relating to the use of funerary iconography as historical evidence stem from this failure. Furthermore, as the data used in this study are drawn from a funerary context, any findings are open to the objection that they are based on evidence that is invalid for an investigation of actual conditions. Tomb owners did not spell out their reasons for decorating their tombs so all interpretations relating to the purpose of funerary iconography are ultimately hypotheses. Consequently, the issue resolves into a question of which hypothesis best meets the conditions of the data. Do 'daily life' scenes represent actual historical conditions and the activities undertaken by the tomb owner during his lifetime or are they an idealised expectation of his 'after-life', or are the images merely stereotypical symbols expressing the ideals, beliefs and 'decorum' of an elite social class?

The hypothesis that the 'daily life' scenes in tombs portray the 'after-life' expected by the tomb owner does not in itself vitiate the material as historical evidence for life on earth. As a member of the most privileged and powerful class in Egyptian society, the tomb owner would be unlikely to envisage better or different circumstances than those he enjoyed in life. It can therefore be argued that such scenes reflect the best of the tomb owner's life, which he hopes to replicate in the 'next world'.[66]

[63] The presence of other animals such as lions is noted but not included as a resource.

[64] Slaughter scenes are not included in this study as they were considered to relate to the issue of ritual offerings rather than resources.

[65] Weeks (1979) 59-63.

[66] Sørensen (1987) 112-113; Dodson–Ikram (2008) 12-13.

There are, however, questions that a simple interpretation of Egyptian funerary art as a reflection of 'daily life' does not answer. As Roth points out, there are significant omissions from the scenes of daily employment on an estate.[67] Many activities, such as the trapping of waterfowl and the construction of papyrus skiffs are presented in one or two stereotyped scenes rather than as a sequential record of the full operation. Do these omissions support the contention that the pictorial programmes of chapels, apparently showing the conduct of rural estates, should be interpreted as stereotypical symbols perhaps of beliefs relating both to this world and the next?

Analysis of the distribution of 'daily life' themes in tomb chapels reveals that the great majority of relevant scenes depict farming, hunting and related activities carried out on rural estates.[68] Market scenes are comparatively rare and workshop portrayals show production for use in the tomb owner's household or estate. Only a relatively small number of (usually) multi-roomed tombs include scenes of commerce, and it is doubtful whether these are intended to represent trade for the profit of the tomb owner. In the case of smaller chapels with limited wall space, the 'daily life' aspect of decorative programmes is restricted to brief images representing activities of the tomb owner's estate. Although every Old Kingdom funerary programme has an individual layout with a unique variety of detail, the pictorial concepts denoting the main themes are repetitive. This may give a superficial impression that the choice and presentation of themes and their elements are largely stereotyped metaphors; but this need not have been the case. The character of the Nile valley meant that rural estates produced a similar range of resources using the same technology, so depictions of their activities would inevitably be similar. Furthermore, the craftsmen who decorated the tombs worked on a finite wall space. The felicity with which the best craftsmen packed a complex series of activities into a single register perhaps conceals the degree of difficulty involved in creating the composition. Craftsmen had to portray on a given limited space activities representing a progression of actions. Consequently they have frequently confined the presentation of each theme to one or two typical and immediately recognisable aspects. For example, to have portrayed every stage of trapping birds by clapnet, from the construction and setting up of the net to taking all the trapped fowl out of the net and placing them in baskets, would have required much more wall space than just the scene of the team of men springing the trap; yet this single scene sufficiently conveys the idea of the whole operation. It is likely that once a master craftsman had devised a scene that solved the problem of providing adequate communication within a limited space, his creation would have been copied and so become part of the basic repertoire representing the activities on estates. Our failure to appreciate the limitations and restrictions under which craftsmen conceived and executed a chapel's artistic programme mistakenly leads to the impression that a scene's repetition from one tomb to another results from the requirement to convey stereotyped symbols.

If the argument above is accepted, to Roth's apt question 'why are activities depicted in such abbreviated form?' may be added 'why in a few instances is so much detail provided?'. The theme of harvesting cereals, for example, is presented in a series of detailed activities. Depicted stages include at least twelve separate activities from reaping to filling granaries. No tomb includes every single aspect although a substantial number present six or more of these stages. One particular aspect, the herd of donkeys being driven back to the field to collect a load of sheaves, is certainly a motif that could be omitted without losing meaningful continuity. Surely this extended pictorial detail conveys a particular message that careful and speedy management of the grain harvest was of supreme importance on an estate or perhaps became so in the second half of Dynasty 5, when scenes providing details of harvesting began to proliferate?[69]

The fundamental objection to the use of funerary scenes as evidence of 'daily life' is that the images are metaphorical symbols reflecting a range of beliefs rather than the activities and circumstances of life on earth. To use an example given by van Walsem,[70] do the scenes of fighting bulls that first appear in late Dynasty 5 to early Dynasty 6 express the tomb owner's interest in bullfights or was the theme a metaphor for the challenges the deceased owner had to face, as Galan argues?[71] Van Walsem has questioned Galan's hypothesis by noting that the evidence for the symbolic and metaphorical interpretation of these Old Kingdom images is drawn anachronistically from a much later time period.[72] He further makes a cogent point that if such a theme symbolises important concepts that relate to vital aspects of the deceased's belief system and after-life as interpreted by Galan, it is surprising that they are not included in all tombs. Themes that are generally accepted as relating to after-life beliefs such as the scene of the offering table and the false door, features needed for the well-being of the ka, are consistent inclusions in all tombs, whereas few tombs present every theme of 'daily life' and some themes only appear infrequently.

Not all scholars accept the hypothesis that the images of 'daily life' are the symbolic expression of concepts

67 Roth, *IBAES* VI (2006) 243-244.

68 The comparative rarity of workshop scenes may be gauged by the recorded numbers of scenes in Porter–Moss (1974) vol. 3, Pt 1, pp. 355-356, 904-905. There are, for example, 78 entries under 'Agriculture' but only 23 entries for a combination of metalworkers, carpenters and stonemasons. Scenes of commerce are even less frequent.

69 See TABLE 12: INSCRIPTIONS – TOMB OWNER VIEWING PLANTING AND HARVESTING OF FIELD CROPS'

70 van Walsem (1998) 1206.

71 Galan (1994) 81-96. Galan's view is opposed to that of Kanawati, *BACE* 2 (1991) 51-58, who interprets the bullfighting scenes as merely portraying the tomb owner's interest in such activities. Lloyd (1978) 609-13 makes the case that the bullfight symbolises the conflict of Horus and Seth (between order and chaos).

72 van Walsem (1998)1206.

unconnected with everyday life on earth,[73] and thus are not an authentic or valid portrayal of the tomb owner's life.[74] However, this contention has to be responded to in order to support the validity of using funerary pictorial material as evidence of the world of the living. Moreno Garcia further challenges the credibility of the pictorial material as an authentic portrayal of life on earth by interpreting the composition of scenes as shaped, even distorted, by ideological considerations and rules of decorum that we may not even recognise.[75] Noting that the scenes are stereotyped and never correspond to events mentioned in the autobiographies, he rejects the conclusion that this 'standardized' iconography represents the realities of an individual's life. According to Moreno Garcia, Bianchi and Baines support this perspective by interpreting Egyptian art as the expression of the values and the ideas of the elite, which has given the art a canonical character that conforms to criteria developed to represent the values of this small but powerful segment of the population rather than reflecting the realities of a particular official's life.[76] Moreno Garcia's explanation for the frequent inclusion of scenes of the rural estates is that the concept of the tomb owner's estate is used as an ideological code denoting a timeless space, impossible to locate geographically, that includes all elements characteristic of the Egyptian landscape (desert, valley and river milieux).[77] Therefore, according to Moreno Garcia, the portrayed estate is a symbol of Egypt in miniature, an adaptation by officials of the royal ideology relating to the maintenance of Maat, rather than the representation of aspects of the life of the individual.

2.5 The multi-functional tomb

The heavy expense involved in constructing a great mastaba or rock-cut tomb and providing it with a decorative programme suggests that owners expected a lot from their tomb.[78] To the excavation of the concealed burial chamber with its massive stone sarcophagus was added the lengthy and costly process of mummification to preserve the body of the deceased, as required for the continued existence of the ka. This made the tomb a place of 'transformation' enabling the deceased to enjoy an after-life.

In addition, the tomb chapel was the focal point for the cult of the deceased. Officials went to lengths to make arrangements for the practice of their cult.[79] Consequently the living were expected, indeed encouraged, to visit the tomb after the death of the tomb owner. The *ḥtp dj nswt* formula, usually inscribed on the entrance and the false door, inform the visitor of the good standing of the deceased with king and gods; lists of titles, repeated at different places in the tomb chapel, define his successful career in the service of the crown. The 'appeal to the passer-by' and the threat to desecrators of the tomb assume continued visitations after the burial of the deceased. The regular visitors, personally close to the tomb owner during his lifetime, would be family and ka servants carrying out the ritual practices of the cult of the deceased.[80]

2.6 Multi-functional decoration

The tomb chapel was decorated, where possible sculpted in fine stone, with a pictorial programme that had more than one function. 'After-life' ritual is represented by the false door, usually situated above the burial chamber, thus providing access to the chapel for the ka of the deceased. Offering table scenes in all likelihood represented the participation of the ka of the deceased in the cultic celebration. Both the short ('thousands') offering list accompanying this scene and the great offering lists specify an ongoing supply of offerings, clearly referring to the after-life enjoyed by the deceased. Other cult elements such as the scene of the eldest son of the deceased presenting a lotus flower to his father may be included to stress the eternal existence of the ka.[81] Other images of the deceased at the entrance to the tomb communicate a welcome to visitors. Within the chapel the tomb owner usually presents his immediate family to the viewer.

Taken as a whole, the decorative programmes serve a number of functions vital to the deceased. That the images of 'daily life' portray aspects of the tomb owner's life on earth in order to provide both the magic needed to equip his afterlife and to convey information about the tomb owner to contemporary visitors to the tomb is the interpretation that best meets the essential conditions and features of scenes of estate activities. Pictorial programmes and motif details are never identical suggesting that individual circumstances rather than symbol and decorum influenced the choice of themes and detail. Symbolic and metaphorical meanings rarely seem to apply to these images, consequently making it necessary to cite supporting evidence anachronistically

73 E. Feucht in Luft (1992) 157-69; S. Quirke, *JEA* 80 (1994) p. 222; van Walsem in C.J. Eyre (1998) 1205-1213.

74 Besides the theme of bullfights, the tomb owner fishing and fowling in the marshes: LÄ pp. 147, 239, 1052; and, herding calves and sheep: Munro (1993) 95-136, have also been interpreted symbolically.

75 Moreno Garcia (1999) 63.

76 Moreno Garcia (1999) 64-65; Bianchi (1995) 2533-2554; Bianchi (1997) 34-48; Baines (1994) 67-94; Baines (1988) 192-214; Baines (1989) 131-149; Baines (1990) 1-23; Baines (1991) 123-200.

77 Moreno Garcia interprets associated inscriptions located in both Upper and Lower Egypt as underlining the concept of geographical unity: Moreno Garcia (1999) 71.

78 Roth (1994) 227-238. Strudwick (2005) 41 notes that most of the texts detailing payment for work on the tomb appear in the tombs of lower officials. This suggests that the less elevated official struggled to pay for his tomb. *Ttj-snb/Jrj*, whose highest title was *jmj-r3 zt ḫntj-š pr-ˁ3*, recounted that he and his family did the work within his tomb with their own hands: El-Khouli–Kanawati (1988) 9-10, pl. 3-4.

79 The invariable inclusion of the false door and the depiction of the ka of the deceased at the offering table in every tomb is significant support for this hypothesis, as are the cult items found in and around tomb chapels. *Nj-k3-ˁnḫ* of Tehneh detailed the arrangements for his cult on the walls of his tomb: Frazer (1903) 122ff, pl. III-V; Goedicke (1970) 131 ff, pl. 14 and Strudwick (2005) 195ff.

80 The inscription of *Nj-k3-ˁnḫ* exemplifies the lengths an official might go to in order to ensure his cult was practiced: Frazer (1903) 122ff, pl. III-V and Goedicke (1970) 131ff, pl. 14.

81 *Špss-k3.f-ˁnḫ, Jj-mrjj, Nfr-b3w-Ptḥ:* Weeks (1994) fig. 54, 43, 10, 20.

from later periods if such an interpretation is proposed. Occasionally a young son accompanying his tomb-owner father to view estate activities is depicted at a different age from one scene to another, which strongly gives the impression of historical reality and the passage of time in the tomb owner's life.[82]

Other features support the hypothesis that estate scenes are intended to reflect the individual's life on earth. All rural estates would have produced similar resources using the same methods. Yet individual tombs do omit themes, while other themes only appear in a few tombs. For example, only the multi-roomed tombs of the most powerful officials, such as *Ṯjj, K3-gm-nj* and *Mrrw-k3*, include clapnet to fowl-yard aspects of bird trapping or the force-feeding of hyenas, suggesting their desire to emphasise their status, wealth and lavish life-style. Tombs of lesser officials omit such scenes, perhaps implying that the choice of themes was shaped according to the relative wealth of the tomb owner. Chronological tables reveal that changes in farming methods are reflected in the pictorial programmes of later tombs.[83] The variations in individuals' choice of themes and details imply varied circumstances of life rather than a commitment to an elite decorum or a conceptual programme of beliefs expressed in symbols. Furthermore, while scenes invariably include a range of personnel who may be given their title and sometimes a personal name, minor individual historical characters such as a brother are occasionally portrayed in a scene of 'daily life'.[84]

The problematic aspects of the decorative programme may be explained within the hypothesis that scenes of 'daily life' represent selected aspects of the tomb owner's life on earth. For the maintenance of his cult and good memory, the deceased needed to portray a record of achievements rather than chronicle the vicissitudes of his working life. The 'daily life' elements of the pictorial programme documented the deceased's private, non-official life on earth. This may have accorded with the deceased's wish to remind visitors that their inheritance and payment as ka servants were a result of his efficient management of the estate and to have his life commemorated and his identity preserved and remembered with respect, thus ensuring the maintenance of his cult.

2.7 Service to Maat

Although the concepts of 'doing' or 'speaking Maat' are not specific in tombs until late in Dynasty 5, earlier biographical statements suggest that service to Maat was a pervasive ethical code by which all officials, from the king down, were expected to live and were judged.[85] Old Kingdom autobiographies do not explore the extension of the philosophy of Maat that appears in later writings,[86] but frequent references to 'doing Maat' imply a looking back over the life of the deceased rather than a reference to his after-life.[87] The king was ultimately responsible for maintaining Maat, the ordered world; yet in the Old Kingdom the threat to Egypt's orderly world was not external but came from social difficulties and conflict and from untamed nature, an inundation that was too high or too low or the failure to control the encroachments of the wilderness and manage its gifts. Concerned to portray his life as an autonomous individual, an official did not include scenes representing his service to the crown; for example, expedition leaders did not portray themselves leading expeditions and no *jmj-r3 k3t nbt nt nswt* ('overseer of all the works of the king') depicted himself supervising a great construction. In their official capacity on earth these tomb owners were agents of the crown implementing the king's service. In their tomb they sought to portray their individual service to Maat and did so as paterfamilias and administrator of an estate[88]. The official's efficient administration of the estate provided for family, retainers and estate workers and his settlements (*njwwt*), and yielded the capacity to aid the needy. The earlier autobiographies emphasise the tomb owner's personal ethical performance as a social leader and man of substance, rather than as an officer of the administration. Even in Dynasty 6, when tomb owners begin referring to their administrative duties in their autobiographies, they still retain statements of their individual commitment to Maat.[89]

Thus it is possible to explain problematic aspects of the decorative programme within the hypothesis that scenes of 'daily life' represent the non-official aspects of the tomb owner's life on earth. For the maintenance of his cult and good memory, it was the deceased's devotion to the code of Maat and his success in carrying out its dictates that needed to be portrayed.

2.8 Reading Egyptian art

When the Egyptian craftsman sculpted or painted a scene, his aim was to present a collection of associated ideas on behalf of the tomb owner rather than to express his own vision. Much of the decorative programmes in tomb chapels was intended to convey knowledge about the deceased to contemporary visitors. Modern visitors and scholars, however, have to acquire an understanding of the 'language' of this communication. Heinrich Schäfer has shown how the characteristic depiction of images in

[82] See Moussa–Altenmüller (1977) figs. 8, 12, 13, 18, 21 and 22.

[83] For example, TABLE 6: PORTRAYAL OF SHEEP and TABLE 7: MANAGEMENT OF DOMESTIC GOATS.

[84] *Mrrw-k3/Mrj* includes a vignette of his brother reclining on a skiff in a fishing scene: Duell I (1938) pl. 43-4. Scenes in the chapels of *Ḫnj/Špšj-pw-Mnw*: Kanawati II (1981) fig. 17, 18 and *K3(.j)-ḥp/Ṯtj-jkr*, Kanawati I (1980) fig. 8 include depictions of the artist and his brother.

[85] Lichtheim (1992) pp. 9-19.

[86] The bringing of all moral values such as justice, the maintenance of an orderly world and even of the cosmic order into the concept of 'Maat': Lichtheim (1992) pp. 20-101.

[87] Lichtheim (1992) 11.

[88] Lichtheim summarises the individual's motivation to 'do Maat' 'because it was good" and because "the god desires it". It was the principle of right order by which the gods lived, and which men recognized as needful on earth, and incumbent upon them. Lichtheim (1992) 19.

[89] Kloth (2002) 128-220.

Egyptian iconography may be read.[90] Freedom from the rules of artistic perspective allowed the Egyptian artist to employ devices that encapsulated a series of ideas in a single scene or image.

Illustrations of work on rural estates, however, present other 'reading' difficulties for the modern viewer with little first-hand experience of pre-industrial farming. Knowledge of estate management systems has to be inferred from one or two complex scenes representing an entire process and from a few terse, occasionally enigmatic inscriptions. Scenes may sometimes appear to be a pictorial code that requires deciphering in order to comprehend the conveyed information. The solution to the problem of reading a communication in pictorial form may be the recognition that an image has departed from 'reality' in order to convey a message. Evans, for example, has demonstrated that the scenes of physical hand feeding of cattle cannot be true to fact, as cattle cannot be safely fed in this manner.[91] The depiction of a farm labourer squatting in front of a recumbent animal and pushing food into its mouth, however, effectively conveys the concept of deliberately fattening selected animals (*Figures 7 and 14*). The abbreviated pictures of complex but normal activities – many carried out over a period of time – would have been immediately understood by contemporaries, especially as the devised scene became an accepted solution to the problem of pictorial communication confined to a limited space.

Even when the problems of 'reading' the pictorial data have been overcome, any interpretation of the purpose(s) of the tomb's decorative programme remains a hypothesis. The most adequate explanation is that the decorative programme, like the tomb itself, was multifunctional. Certain elements, such as the offering table scene and the presentations of offering bearers met the after-life needs of the deceased. Other elements, which were expanded over time, were included not only to provide an appropriate 'after-life', but also to inform visitors of the successful life-on-earth of the tomb owner and to keep his/her memory alive.

90 Schäfer (1986) 91, passim.
91 Evans (2010) 56-57; Swinton–Evans (2012). Sameh Shafik disagrees with this interpretation: Shafik (2010) 65-75.

SECTION ONE

RESOURCES OF THE DESERT

Figure 2: HUNTING THE ANIMALS OF THE DESERT, TOMB OF MERERUKA, SAQQARA. Kanawati–Abder-Raziq (2004) pl.46.

Rock drawings and paintings scattered across the land from the Eastern Desert to the borders of Libya and Nubia bear testimony to the relationship of Egyptians with the wild animals of the Nile valley. This evidence reaches back to Palaeolithic times with data in the form of rock carvings and ceremonial palettes [92] and knife handles engraved with depictions of animals from the Predynastic era, [93] all of which bear witness to the importance of wild animals in early Egyptian society. Egyptologists have identified a variety of elements in Egyptians' response to the wild creatures that inhabited the Nile valley and surrounding regions in pharaonic times and earlier. Accepting the requirement of food and the desire for magical protection from dangerous animals,[94] they have variously interpreted hunting as a reflection of moral victory over the constant threat of chaos[95] or as an entertainment for the elite.[96] The focus here, however, is on desert animals as a resource for high officials.

By Dynasty 4, climate change was bringing increased desiccation to the lush grasslands that in predynastic times had bordered the Nile flood plain.[97] As the greater aridity began to modify habitats, the responses of mammalian species varied, with the largest animals such as elephants and giraffes having disappeared from the region. The new environment is reflected in wall art in scenes of hunting desert animals where the ground, coloured yellow or red, is depicted as uneven and sparsely vegetated.

The Old Kingdom elite depicted surviving mammalian species in their tombs. These were mainly small creatures, hardy antelopes of the species that could survive with very little water, wild auroch, and a few predators such as lion, leopard and striped hyena. While the importance attached to domestic cattle may be inferred from evidence such as autobiographies of the provincial magnates of the later Old Kingdom,[98] these inscriptions contain no specific references to desert animals. Consequently, the value that may be attributed to the latter has to rely on deductions from wall scenes and their associated captions. The degree of prominence given to motifs such as the presentation of individual animals before the tomb owner, animal husbandry and the desert hunt is therefore used to evaluate desert animals as a resource, on the principle that themes and subjects portrayed in a funerary context were of abiding significance for the tomb owner and were not chosen at random or capriciously and were not merely copies of images from royal monuments or other private tombs.

92 Hierakonpolis Palette – Asselberghs (1961) pls. 70-71, figs. 127-8); Hunters Palette – Spencer (1980) pl. 63, no. 575; Four Dogs Palette – Asselberghs (1961) pls. 72-73, figs. 129-30.

93 Pitt Rivers knife handle – Bénédite (1918) fig. 1; Brooklyn knife handle – The Brooklyn Museum, Churcher (1984) p. 155f; Gebel el Tarif knife handle – Morgan (1896) fig. 136.

94 Dangerous animals and evil spirits were carved on curved pieces of ivory used to protect the sleeping and the dead: Hayes I (1953) p. 249.

95 Westendorf (1968) p.122f.

96 Montet (1958) p.76f.

97 Butzer (1976) 26-29.

98 Moreno Garcia, *RdE* 50 (1999).

CHAPTER 3 DESERT ANIMALS

3.1 Focus of enquiry: desert animals as an economic resource

TABLE 2: MANAGEMENT OF DESERT ANIMALS
TABLE 3: FEATURES OF DESERT HUNTING SCENES (a), (b), (c).

The data:
The iconography of desert animals and associated inscriptions in scenes of:

- the parade of animals
- hunting in the desert
- desert animals within an enclosure
- desert animals being brought back from the hunt
- funerary ritual: scenes of offerings, offering lists, ritual slaughter.

Issues:

- Meaning and significance of scenes portraying relationship between people and desert animals such as the parade of livestock and hunting.
- Whether Egyptians attempted to breed and domesticate desert animals.
- Increase of violence portrayed in the hunting scenes of late Dynasty 5 to Dynasty 6:
  - methods of depicting violence,
  - the tombs portraying the change,
  - significance of the change.

3.2 Distribution of data

The term 'desert animals' is applied to the species that successfully adapted to the changing conditions of the Holocene epoch. Although driven back from long stretches of the river banks where human societies developed, certain species were sufficiently adapted to inhabit the semi-arid landscape lying between the pastures closer to the Nile river and the parched, barren land and cliffs bounding the river valley. In the tombs chosen for this study these desert animals appear in the following frequencies:

Early Dynasty 4 to mid Dynasty 5:	in 26 tombs
Second half of Dynasty 5:	in 39 tombs
Early Dynasty 6 (Teti to Pepy I):	in 13 tombs
Mid Dynasty 6 to Dynasty 8:	in 12 tombs

3.2.1 Desert animals most frequently depicted in presentation scenes

Scimitar horned oryx: *Oryx dammah*
(*mȝ-ḥḏ:* Wb 2, 11,4-9).
The scimitar horned oryx is a large member of the species with a heavy body and long horns that curve slightly back from the animal's head. It lives in small herds in savannah and semi-desert plains and can survive for weeks without water. While there is no specific breeding season, winter rains that influence conception and a gestation period of eight to nine months lead to a concentration of births in winter and spring. Calves are concealed for up to six weeks for security from predators[99] (*Figure 3*). *Depicted in 74 tombs*

Nubian ibex *Capra ibex nubiana*
(*njȝ.w:* Wb 2, 202.1-4, alt. *nrȝ.w:* Wb 2, 280.3; *jȝm.t:* Wb I, 79, 1-2 Steinbock).
A small member of the *Caprini* subfamily, the ibex is a goat-antelope with long horns that arch up and back from its head in a great curve down towards its back. Males have a distinctive beard under their chin. It is a very agile animal well adapted to the steep cliffs and dry rocky wadis, where it lives in small herds. Ibex consume a wide variety of vegetation and require water for survival. As the mating season for the species is late autumn with a gestation period of five months, calves are born in spring (*Figure 4*). *Depicted in 53 tombs*

Dorcas gazelle *Gazella dorcas*
(*gḥs*: Wb 5, 191.1-8 Gazelle)
The Dorcas is a small gazelle with lyre-shaped horns. Its relatively long legs give it a running speed up to 80 km an hour. It is adapted to semi-arid conditions and can survive for long periods without water providing vegetation is accessible. Its mating season is autumn with a gestation period of six months so that calves are born in spring. Calves are concealed up to six weeks for security from predators[100] (*Figure 5*). *Depicted in 52 tombs*

Addax *Addax nasomaculatus* (*nwḏ.w:* Wb 2, 226.15-16 Mendes-Antelope) The addax, which can be tamed in captivity, lives in herds of 5 – 20 animals. The most desert-adapted of the antelopes, the addax can survive without water providing it has green food, so herds migrate across semi desert regions according to available pasture. Breeding occurs throughout the year but peaks in winter and spring. Calves are concealed up to the first six weeks for security from predators[101] (*Figure 6*).
Depicted in 33 tombs

---

99 Osborn–Osbornová (1998) 161-65; Walther, F. R. "Roan and Sable Antelopes" in *Grzimek's Encyclopedia* (1990) Volume 5, 437-447.

100 Osborn–Osbornová (1998) 175-77; Walther, F. R. "Gazelles and related species" in *Grzimek's Encyclopedia* (1990) Volume 5, 462-484.

101 Osborn–Osbornová (1998) 158-160; Walther, F. R. "Roan and Sable Antelopes" in *Grzimek's Encyclopedia* (1990) Volume 5, 437-447.

Striped hyena *Hyaena hyaena*
(*ḥṯt* Wb 3, 203. 16-17).
The striped hyena is one of the smaller hyenas. It is nomadic but occasionally gathers into small family groups. It is a nocturnal hunter, a solitary scavenger, carrion eater and hunter of small mammals. Its habitat is semi-desert and it can survive without water for long periods, although prefers to drink every night. It does not have a breeding season and cubs are born helpless and remain dependent on parental care up to twelve months. Although considered a shy animal they may be tamed[102] (*Figure 7*). *Depicted in 25 tombs*

3.2.2 Desert animals depicted less frequently in presentation scenes

Hartebeest *Alcelaphus buselaphus*
(*šsꜣ:* Wb 4, 543.5-6; *hbn:* Wb 2, 487.6)[103] *in 12 tombs*
Soemmering's gazelle *Gazella soemmeringi*
(*gsꜣ:* Wb 5, 206.2)[104] *in 4 tombs*
Fallow (dama) deer *Dama mesopotamica*
(*hnn:* Wb 2, 495.21)[105] *in 4 tombs*
Kob *Kobus (Adenota) kob*
(*ḏbn.w:* Wb 5, 568.2, *šs*: Wb 4, 543.5-6)[106] *in 3 tombs*
Barbary goat
Ammotragus lervia (*jbꜣw:* Wb 1, 62.18)[107] *in 2 tombs*

While the species portrayed in animal processions are limited to the above, a much greater variety of desert animals appears in scenes of the desert hunt and of animals kept in an enclosure: ungulates (auroch, duiker?) and small animals (cape hare, caracal, genet, golden jackal, hedgehog, ichneumon, jerboa, porcupine, red fox, shrew). Large predators (lions and leopards) must have found their way into enclosures as they are shown attacking ungulates. In the parade of livestock scenes, desert ungulates and the striped hyena appear in line with cattle.[108] Although they are frequently depicted in Old Kingdom chapels, desert animals are not portrayed in the great variety of themes in which domestic cattle are represented; they rarely figure in the diverse themes of animal husbandry that represent the management of domestic cattle. The only images of desert animals being attended by herders, apart from being led in parades, are rare depictions of hand feeding or of animals being transported by land and by boat.[109]

3.3 Economic value of desert animals

Did desert animals represent economic value for Old Kingdom officials? If so, did the animals retain their value throughout the period? These animals, mainly ungulates, are depicted in a few major themes: the parade, usually before the tomb owner, the desert hunt and herded within an enclosure.[110] Subsidiary transport themes such as the calves of desert animals brought back from the hunt or animals on transport boats are occasionally included. However, once widespread domestication of certain plants, in particular grains and flax, and animals (cattle, donkeys, goats and sheep) had been achieved, the continued economic importance of desert animals to Egyptian society is questionable. While desert ungulates provided meat, fat and by-products such as leather, bone, horn and gut, most of these products could be amply supplied by domesticates. On the other hand, owning herds of animals that were well adapted to the semi-arid conditions increasingly affecting pasturage might have had an economic advantage. In addition, there is the consideration that these desert animals continued to be maintained and slaughtered for status and/or ritual reasons. Evidence of these issues, however, has to be linked to the further questions of whether the hunting scenes represent attempts to capture desert animals alive or merely to kill them, and whether Egyptians had breeding and domestication programmes and pastoral capacity to maintain herds of desert ungulates. These issues will be investigated by analysis of available data.

3.4 Desert animals in the 'Parade of Animals'
TABLE 2: MANAGEMENT OF DESERT ANIMALS

3.4.1 Historical development of images and themes

Dynasty 4 to early Dynasty 5
Desert animals are depicted in scenes of parade in the following tombs:

*Rꜥ-ḥtp**	Medum	IVE
*Nfr-mꜣꜥt**	Medum	IVE
*Mṯn**	Saqqara	IVE
Nfr-mꜣꜥt	Giza	IV.E
Nfr	Giza	IV.M
Mr-jb.j	Giza	IV.L
Mrs-ꜥnḫ III	Giza	IV.L
Nj-kꜣw-Rꜥ	Giza	IV.L
Nb.j-m-ꜣḫtj	Giza	IV.L
Dbḥnj	Giza	IV.L
Pr-sn	Giza	V.E

102 Wagner, Aaron Parker, *Behavioral Ecology of the Striped Hyena (Hyaena hyaena)*, 2006, Doctoral dissertation, Montana State University; Osborn–Osbornová (1998) 97-104.

103 Osborn–Osbornová (1998) 171-73; Walther, F. R. "Hartebeests", in *Grzimek's Encyclopedia* (1990) Volume 5, 418-436.

104 Osborn–Osbornová (1998) 179-80; Walther, F. R. "Gazelles and related species" in *Grzimek's Encyclopedia* (1990) Volume 5, 462-484.

105 Osborn–Osbornová (1998) 152-55; Houlihan (1987). Houlihan concludes that there is insufficient evidence to be certain which specie of deer is depicted in these scenes but it is most likely to be *Dama mesopotamica*.

106 Osborn–Osbornová (1998) 170-71; Walther "Reedbucks, waterbucks and impalas", in *Grzimek's Encyclopedia* (1990) Volume 5, 448-461.

107 Osborn–Osbornová (1998) 189-192; Haas, G. "Barbary sheep (Genus *Ammotragus*)" in *Grzimek's Encyclopedia* (1990) Volume 5, 538-540.

108 There are also rare appearances of dama deer in animal parade scenes: *Mꜣ-nfr, Sḫm-ꜥnḫ-Ptḥ, Ṯjj; Mrjj-Ttj/Mrj*, see TABLE 2: MANAGEMENT OF DESERT ANIMALS.

109 *Ptḥ-ḥtp* (LS 31), *Ṯjj, Jnw-Mnw, Mrrw-kꜣ, Kꜣ-gmnj*, see TABLE 2: MANAGEMENT OF DESERT ANIMALS.

110 There are just a few exceptions. In the tombs of *Ptḥ-ḥtp* LD II 102b, 104 and *Mrrj* Davies et al (1984) pl. 6, a recumbent addax is represented being hand fed. In the tomb of *Mrrw-kꜣ* a row of tethered ungulates are eating or drinking from containers: Kanawati (2011) pl. 78.

Nswt-nfr	Giza	V.E
Snb	Giza	V.E
Sš3t-ḥtp/Htj	Giza	V.E
K3-pw-nswt/K3j	Giza	V.E

* Chapels with hunting scenes

The species depicted in these tombs:

Ungulates

addax, scimitar horned oryx, kob, dorcas gazelle, Nubian ibex, fallow (dama) deer, duiker?

Carnivores

leopard, striped hyena

Scenes depicting desert animals appear in most of the chapels of early Dynasty 4 used in this study.[111] Although the images of animals walking or standing are usually limited to one or two animals per register,[112] the manner of their portrayal (led by a herder[113] or walking unattended in line with a domesticated ox)[114] suggests they are being paraded before the tomb owner. In the chapel of *Rˁ-ḥtp* for example, the image of the oryx led by a herder is situated immediately beneath the feet of a large figure of the tomb owner who is viewing boat building.[115] The associated caption reads, 'Taking possession of gifts" (*jtj nḏ.t-ḥr*). In the chamber of *Jtt*, wife of *Nfr-m3ˁt*[116] the column inscription, *jnjt nḏ.t-ḥr n pr* [*ḏt*] ('Bringing the gifts/tribute of the [estate]') is associated with the two lowest registers of animals walking forward in line with their herders, also suggesting a parade of animals. Although the registers are located immediately beneath a desert hunt, the lines of animals probably represent a procession before the tomb owner, as it is unlikely that the domesticated ox in the scene would be included among the trophies of a hunt.

Animals associated with the semi-arid regions (all ungulates except for four depictions of striped hyenas)[117] appear in parade and offering procession scenes in eleven of the chapels dated from the second half of Dynasty 4 to early Dynasty 5. In these scenes, desert animals are shown either sharing a register with stall-reared cattle, or on a narrow register of their own above or beneath registers of domestic cattle.[118] The parade of animals is limited to one or two animals, possibly owing to lack of wall space, with the tombs of *Dbḥnj* and *Mrs-ˁnḫ* III as the only exceptions, perhaps due to their royal status.[119] The tombs of this period have L-shaped chapels with themes of funerary offerings and family relations as the major focus in their limited wall space. As a result, motifs of 'daily life' are only briefly treated on registers too narrow to contain depictions of more than one or two animals.[120]

Mid Dynasty 5

Desert animals are depicted in scenes of parade in the following tombs:

GIZA	
Jj-mrjj	*Sšm-nfr* II
Rˁ-wr I	*K3.j-m-nfrt**
*Ḫntj-k3w.s**	
SAQQARA	
Wr-jr.n-Ptḥ	*Nj-ˁnḫ-H̱nmw/H̱nmw-ḥtp**
ABUSIR	
*Ptḥ-špss**	

* Chapels with hunting scenes

Depicted species:

Ungulates

dorcas gazelle, Nubian ibex, scimitar horned oryx

Large carnivores

hyena

From mid Dynasty 5 on, particularly at Saqqara, officials were constructing tombs with greater interior wall space than was offered by the early L-shaped chapels of Giza.[121] The additional wall surface allowed for the portrayal of expanded parades of animals with a greater number of desert ungulates. The selections of depicted animals now include two or more of the species of addax, Dorcas gazelle, scimitar horned oryx, Nubian ibex, and striped hyena. In the extended tomb of *Jj-mrjj*, for example, three desert animals, a hyena, an ungulate with budding horns and a Dorcas gazelle occupy the top register above the doorway in the west wall of the first chamber, while on the east wall of the third chamber the tomb owner inspects a line of desert animals: an ungulate of unknown species, oryx and Nubian ibex.[122] In the east chamber of *Ḫntj-k3w.s* a procession of four animals (a bull? oryx, hyena and addax) and their herders occupies the lower register spanning much of the west wall.[123]

111 See TABLE 2: MANAGEMENT OF DESERT ANIMALS.

112 On the two lowest registers of paraded desert animals on the north side of the chamber of *Jtt* in the tomb of *Nfr-M3ˁt* three animals are depicted on each register: Petrie (1892) pl. 27.

113 Petrie (1892) pls. 11, 27.

114 Petrie (1892) pls. 24, 27.

115 Petrie (1892) pl. 11.

116 Petrie (1892) pl. 27.

117 One hyena is depicted being led on a leash at the end of a procession of figures representing estates in *Nj-k3w-Rˁ*: LD. II, pl.15. Hyenas are depicted in an animal parade in the tombs of *Mrs-ˁnḫ* III, *Pr-sn*, *Ḫntj-k3w.s*, *K3.j-dw3*, *Jj-mrjj*, *Nj-ˁnḫ-H̱nmw* and *H̱nmw-ḥtp*, *Nḫt-k3.j*, *Nfr*, *Sḫm-ˁnḫ-Ptḥ*, *Sšm-nfr* IV, *K3pj*, *Ṯjj*, *Mrw/Bbj* and *K3-gmnj/Mnj*. These are noted in TABLE 2: MANAGEMENT OF DESERT ANIMALS.

118 The only exception is on the north door jamb of the chapel of *Mrs-ˁnḫ* III, where a hyena and an oryx are depicted on separate registers, being presented to the tomb owner: Dunham–Simpson (1974) fig. 3a.

119 However, the extended scene of the parade of animals in the tomb of *Mrs-ˁnḫ* III contains no desert animals: Dunham–Simpson (1974) fig. 4.

120 The parade of animals has limited portrayal of one or two beasts on brief registers in the chapels of *Nj-k3w-Rˁ* LD II 57; *Mrs-ˁnḫ* III Dunham–Simpson (1974) fig. 4; *Sš3t-ḥtp/Htj* Junker II (1934) figs. 28, 30; *Nb.j-m-3ḫtj* LD II 14; *Mr-jb.j* LD II 19, 22a; *Pr-sn* LD II 83b, LD *Erg*. pl. 8; *Nfr* Reisner (1942) fig. 242; *Nswt-nfr* Junker III (1938) fig. 27; and *K3-pw-nswt/K3.j* Junker III (1938) fig. 18.

121 One of the earliest tombs with extended wall space is that of *Jj-mrjj* of Giza: Weeks (1994) 4-6.

122 Weeks (1994) fig.29, 31; LD II 50, 54.

123 Junker VII (1944) fig. 31. The animals on this register are an ox, oryx, hyena and addax. Although the hyena is a carnivore, it is sometimes depicted with ungulates in the parade of animals. Hyenas in Old Kingdom scenes are identified as the East African striped hyena (*Hyaena hyaena*), a smaller and less aggressive animal than the more common spotted hyena. Striped hyenas

Other indications of the value attached to ownership of desert animals by officials are the entire panels devoted to the motif. In the tomb of *Nj-ꜥnḫ-H̱nmw/H̱nmw-ḥtp* for example, *Nj-ꜥnḫ-H̱nmw* and his wife inspect a panel of at least five different species of desert ungulates, as do *H̱nmw-ḥtp* and his wife on the east wall of the vestibule.[124]

Many of these paraded desert animals are described as *'rn'*, usually translated as 'young'. The word *'rn'*, however, has a more encompassing and yet more precise meaning when it is applied to domestic cattle as it is reserved for prized, stall reared specimens, hand fed and fattened for slaughter.

Late Dynasty 5

Desert animals are depicted in scenes of parade in the following tombs:

GIZA	
Ꜣsn	*Sḫm-ꜥnḫ-Ptḥ*
Nj-mꜢꜥt-Rꜥ	*Sḫm-kꜢ*
Nfr (I)	*Sšm-nfr* III
Nfr-bꜢw-Ptḥ	*Sšm-nfr* IV
Nfr-ḫw(wj)	*Štwj*
Nḫt-kꜢ.j	*KꜢ.j-m-ꜥnḫ*
Rꜥ-wr II	*KꜢ.j-dwꜢ*
Sndm-jb/Mḥj	*KꜢpj*
SAQQARA	
Ꜣḫtj-ḥtp (PM 633)	*Nbt*
Ꜣḫtj-ḥtp (PM 599)	*Nfr-jrt-n.f*
Jj-nfrt	*Rꜥ-špss*
Jḥjj (r/u *Jdwt*)	*Ḥtp-ḥr-Ꜣḫtj*
Pḥn-wj-kꜢ.j	*Ḥtp-kꜢ*
Ptḥ-ḥtp (LS 31)	*Sḫm-ꜥnḫ-Ptḥ*
Ptḥ-ḥtp/Ṯfj	*KꜢ.j-m-nfrt*
MꜢ-nfr	*Ṯjj*
ABUSIR	
H̱krt-nbtj	
DESHASHA	
Jntj	
EL-HAWAWISH	
Ḥm-Mnw	

Depicted species:

Ungulates

addax, auroch, Barbary sheep, dorcas gazelle, hartebeest, kob, Nubian ibex, scimitar horned oryx, Soemmering's gazelle

Large carnivores

lion, leopard, striped hyena

The tombs of the second half of Dynasty 5 make clear the value attributed to desert animals, as they contain registers depicting the presentation of lines of these animals each led and controlled by one or two herders. Ten tombs portray four or more desert animals occupying a single register.[125] The greatest number of desert animals depicted on a single register appear on the top register of the south wall of the Saqqara chapel of *Sḫm-ꜥnḫ-Ptḥ* where eight animals (Nubian ibex, scimitar oryx, addax, dorcas gazelle, hartebeest, deer, striped hyena and Barbary sheep) parade in front of the tomb owner. Although this is not usually the case, none of the herders in this scene has to struggle to control his animal[126] and only the first two men turn back to hold the animal's muzzle.[127] The animals are a mixture of male, female and castrates.[128] In his pillared hall, where *Ptḥ-ḥtp* (LS 31) portrays his animals being hand-fed, an equal number of registers are devoted to desert ungulates as to domestic cattle.[129] *Ptḥ-ḥtp/Ṯfj* (PM 600-4)) combines the desert hunt and the presentation of desert animal in registers that span the long west wall of his chapel. On the left are two registers of desert hunting scenes, while on the right is a corresponding register of live animals being brought back from the hunt, and immediately below is a register of five animals of different ungulate species being presented to the tomb owner.[130]

The most highly developed and detailed iconography of this theme occurs in the tomb of *Ṯjj*.[131] In the Inner Hall four extended registers present a variety of desert ungulates paraded before the tomb owner with images depicting the different ways of presenting desert animals: tethered, herders turning back to control the animal by holding its horns and muzzle or just its horns, rope attached to the horns, guiding or controlling an animal from the rear, animals balking or walking unattended, in some instances accompanied by their young. Calves are carried in the arms of a herder or over his shoulders, and herders present very young animals in boxes slung on yokes suspended from their shoulders. Only two registers have survived, but the remains of a third register above show that it too contained images of paraded desert animals.[132]

---

appear in animal parades sometimes led on a rope and sometimes merely guided by a herder. The striped hyena can be tamed to accept humans and makes a surprisingly loyal pet: Kruuk (1976); Ikram (2001).

124 Moussa–Altenmüller (1977) fig. 18, 19, pl. 38, 40.

125 *Ꜣsn, Pḥn-wj-kꜢ.j, Jḥjj* (r/u *Sšsšt/Jdwt*), *Ptḥ-ḥtp* (II) (*Ṯfj*), *Wr-nw, Nfr, Rꜥ-wr* II, *Sḫm-ꜥnḫ-Ptḥ* (Giza), *Sḫm-kꜢj*, and *Ṯjj*. See TABLE 2: MANAGEMENT OF DESERT ANIMALS.

126 The feature of a herder struggling to control the animal he leads in a parade is much more frequently depicted for desert animals than for domestic cattle.

127 Evans (2006) 70 describes how ungulates may be controlled by grasping the animal's ear or horn with one hand, and pinching the septum nasi with the other hand.

128 The absence of udders or scrotums suggest a castrated male. The animals are always depicted in profile with the rear leg nearest the viewer stretched backwards, which allows for the depiction of udders or scrotum. Consequently, where the preputial sac is indicated on an animal without a defined scrotum, the implication must be that this is the depiction of a male castrate. Compare the first three desert animals (male, castrate, female) on the top register of the south wall of *Sḫm-ꜥnḫ-Ptḥ:* Simpson (1976) pl. D.

129 LD II 102a, 104d.

130 Paget–Pirie (1898) pl. 32, 33.

131 Wild III (1966) pl. 165, 166.

132 Wild III (1966) pl. 166. The position of the feet of herders and hooves above the images of a Dorcas gazelle and an addax, both females walking with young, suggest this was a depiction of a herder dealing with a balking ungulate.

Early Dynasty 6

Desert animals are depicted in scenes of parade in the following tombs:

GIZA
St-k3.j
SAQQARA
Jnw-Mnw *Nj-k3w-Jzzj*
Mrjj-Ttj/Mrj *Ḥzj*
Mrrj *Sʿnḫ-wj-Ptḥ*
Mrrw-k3
DEIR EL-GEBRAWI
Hnḳw/Ḫttj (I)

Depicted species:
Ungulates
addax, deer, dorcas gazelle, hartebeest, kob, Nubian ibex, scimitar horned oryx
Large carnivore
striped hyena

Desert ungulates continue to be shown paraded by their herders in the great Saqqara tombs in the Teti cemetery of early Dynasty 6, but the occasional obstreperous reactions of the animals are no longer depicted. There are, for example, no vignettes of ungulates actually balking, with both front legs stretched in front of their body in an attitude of refusing to move forward, which previously are quite common.[133] However, although herders no longer struggle for control, the distinction between the ease of leading domestic cattle and the need to be more watchful of paraded desert ungulates is maintained. Herders walk ahead or beside domestic cattle leading their animal by a rope attached to collar or muzzle. A second herder at the rear may guide or push an ox but rarely is any further need for control indicated. Leading a desert ungulate, on the other hand, is depicted as requiring more intervention. Even if the animals do not balk, herders continue to turn back to prevent them bolting by holding their horns and muzzle. Of 26 vignettes of desert animals being led on parade fifteen depict the animal being grasped by horns and muzzle and twelve animals are attended by two herders. Only four animals are shown with just one herder leading his charge by rope. In the tomb *qf Mrjj-Ttj/Mrj*, for example, on the West wall of room C1 six different species of desert ungulates are presented all except the ibex being controlled by horns and muzzle.[134]

Scenes of hand-feeding by herders occur in the tombs of *K3-gm-nj* (hyena), *Mrrw-k3* (hyenas and ungulates) (*Figures 7*), *Jnw-Mnw* (hyenas) and *Mrrj* (hyena and ungulates).[135] Just like cattle, the desert ungulates are portrayed in a recumbent posture as they are hand-fed, but the hyenas are depicted having been thrown on their back with their legs tied or held by one herder while another introduces food into the animal's mouth.[136] Hand-fed, recumbent cattle are labelled *jw3* or *rn jw3*, indicating that they are being specially fed for slaughter,[137] so it is highly likely that this is the destiny of the desert animals singled out for similar treatment. The bottom registers on the east and west walls of Room 5 of the tomb of *K3-gm-nj* portray rows of slaughtered and trussed domestic cattle and desert ungulates (dorcas gazelle, Nubian ibex and oryx), perhaps in preparation for the funerary feast.[138] While scenes of hand-feeding and slaughter do not offer evidence for the maintenance of entire herds, they suggest that in the ritual/funerary context desert ungulates had significant value.

Mid Dynasty 6 to Dynasty 8

Desert animals are depicted in scenes of parade in the following tombs:

Wr-nw	Saqqara	VI.M
Ppj-ʿnḫ-ḥrj-jb	Meir	VI.M
Ḥzjj-Mnw (G42)	el-Hawawish	VI.M
Ppj-ʿnḫ/Ḥnj-km	Meir	VI.L
Nj-ʿnḫ-Ppjj-km	Meir	VI.L
Ḫnj/Špsj-pw-Mnw	el-Hawawish	VI.L
K3(.j)-ḥp/Ṯtj-jḳr	el-Hawawish	VI.L
Ḏʿw	Deir el Gebrawi	VI.L
ʿnḫ-tj.fj	Moaʿlla	VIII

Depicted species:
Ungulates
gazelle, oryx

In this later period with one exception, the tomb of *Wr-nw*,[139] the chapels providing images of desert animals are provincial, all located in Upper Egypt. Here, there appears to be a distinction in the depiction of desert animals between the tombs of el-Hawawish (Upper Egypt 9) and those of other nomes. The customary scenes of lines of cattle and desert animals being paraded before the tomb owner continue to appear in the el-Hawawish tombs of *Ḫnj/Špsj-pw-Mnw*, *K3(.j)-ḥp/Ṯtj-jḳr* and *Ḥzjj-Mnw* (G42)[140]. In the scenes of *Ḫnj/Špsj-pw-Mnw* and *K3(.j)-ḥp/Ṯtj-jḳr*, both *ḥrj-tp ʿ3*, herd numbers are given. On the east wall of *Ḫnj/Špsj-pw-Mnw*'s chapel the tomb owner views four registers; the two upper registers devoted to butchery and the two lower registers to the presentation of cattle and desert animals with a gazelle labelled *r[n n] gḥs ḏbʿ* ('young gazelles ten thousand') and an oryx labelled *rn n m3-ḥḏ ḏbʿ* ('young oryx ten thousand'). All the other paraded animals are oxen and each is labelled *rn n jw3 ḏbʿ* ('ten thousand young oxen').[141] The only preserved part of the vertical inscription describing *Ḫnj/Špsj-pw-Mnw*'s interest in the

[133] *Nfr-m3ʿt* (G 7060), *Mr-jb.j/K3-pw-nswt*, *Mrs-ʿnḫ* III, *K3-pw-nswt/K3j*, *Jj-mrjj*, *Nj-ʿnḫ-Ḫnmw / Ḫnmw-ḥtp*, *J3sn*, *Nbt* (PM 624), *Rʿ-šps*s, *Ṯjj*. See TABLE 2: MANAGEMENT OF DESERT ANIMALS.

[134] Kanawati–Abder-Raziq (2004) pl. 46.

[135] Respectively, von Bissing (1905) pl. 1-13 and Wreszinski (1923) pl. 87; Duell II (1938) pl. 152; Kanawati–Abder-Raziq (2006) pl. 47; Davies et al I (1984) pl. 6(A).

[136] These scenes may represent the idea of intentionally fattening the animals rather than portraying a physical reality.

[137] See p. 34 for discussion of the meaning of the word *jw3*.

[138] Harpur (2006) Context Drawings 20 and 22.

[139] Davies et al (1984) pl. 31.

[140] Respectively, Kanawati II (1981) fig. 26; Kanawati I (1980) fig.11, 15; Kanawati VII (1987) fig. 6, 3c.

[141] Kanawati II (1981) fig. 26.

scenes presented on the registers reads *m33 jw3w* ... ('Viewing the young oxen ...'). The east wall of the chapel of *K3(.j)-ḥp/Ṯtj-jḳr* contains a comparable set of registers, although the figure of the tomb owner and accompanying column of inscription have not survived. The numbers given for the herds are strikingly smaller than the blanket 'ten thousand' given for each of the animals depicted for *Ḫnj*. On the third register the inscription above the first ox is '3014' and above the gazelle that follows *'rn n gḥs* 1023'. On the fourth register above the oryx is the inscription *'rn n m3-ḥḏ* 1000', and *'rn n jw3* 2012' accompanies the ox that follows.[142]

At Meir and Deir el-Gebrawi on the other hand, the treatment of desert animals paraded before the tomb owner is more cursory. In the Meir tombs of the provincial administrators, *Ppj-ꜥnḫ-ḥrj-jb* and *Ppj-ꜥnḫ/Ḥnj-km*, the scenes of animals presented for the cattle count have no images of desert animals.[143] In the chapel of *Ppj-ꜥnḫ-ḥrj-jb* the only representations of desert animals are a gazelle and an oryx being led as part of a line of offerings beneath the tomb owner seated at the offering table,[144] and in *Nj-ꜥnḫ-Ppjj/Sbk-Ḥtp/Ḥpj-km*'s chapel a single oryx takes its place in a register of live offerings.[145] Unlike the el-Hawawish nomarchs, the Meir administrators do not provide herd numbers for desert animals, although they give numbers for herds of domestic cattle.

In the tomb of *Jbj* at Deir el Gebrawi the only extant scene with desert animals is that of a desert hunt,[146] while in the tomb of his son and grandson, *Ḏꜥw* and *Ḏꜥw/Šm3j*, in a comparable position to that in *Jbj*'s chapel (the western side of the north wall) the scenes of desert animals are divided into an upper register depicting a hunt, with the register beneath providing a portrayal of two oryx being paraded before the tomb owner.[147] Although the register beneath and the vertical inscription indicate a cattle census, no numbers are given for the oryx, which may not have been included in the count.

3.5 The desert hunt

TABLE 3: FEATURES OF DESERT HUNTING SCENES (a), (b), (c).

3.5.1 Historical development of images and themes

Hunting the animals of the desert fringes continued long after the larger wild mammals had disappeared from the Nile valley. After the introduction of domesticates such as cattle, sheep and goats, the economic need to catch wild animals would have been less urgent, but wall scenes, ostraka and ornaments make it clear that Egyptians continued to be vitally interested in desert animals and to hunt them throughout pharaonic times. Old Kingdom portrayals and images of the hunt have few associated captions. Consequently, whether officials included hunting scenes in their tombs to record economic achievement or sporting prowess, or to depict in symbolic imagery resistance to and victory over 'chaos', or whether there was a combination of such aims, can only be inferred from the prominence given to these images. The desert hunt theme spans the entire Old Kingdom, although there is a long time gap, from mid Dynasty 4 to mid Dynasty 5, when hunting scenes were either not included in funerary chapels or have not survived.

The motif of the hunt contains a number of sub-themes. These are the portrayal of:

- lions and other wild predators attacking ungulates (*Figure 8*)
- hunting dogs, working singly or in packs, attacking ungulates and smaller animals (*Figures 8 and 9*)
- ungulates being lassoed by hunters
- ungulates and smaller animals fleeing
- desert animals engaged in natural activities such as copulating or giving birth
- the young of desert ungulates crouching in tussocks of vegetation (*Figures 8*)
- hunters returning with the spoils of the hunt.

If the earliest Dynasty 4 scenes are compared with those of the second half of Dynasty 5 and later, new emphases and moods become apparent in the later period with scenes containing vignettes both of animals behaving naturally as though unthreatened and, at the same time, of a greater degree of violence portrayed in the hunt.

Dynasty 4 to early Dynasty 5

Desert hunt scenes are only depicted in the following early Dynasty 4 tombs:

SAQQARA
Mṯn
MEDUM
Nfr-m3ꜥt *Rꜥ-ḥtp*

Tombs dated from the second half of Dynasty 4 to early Dynasty 5 do not contain scenes of the desert hunt.

Animals appearing in desert hunt scenes of early Dynasty 4 tombs:
Ungulates
addax, dorcas gazelle, kob, Nubian ibex, scimitar horned oryx, Soemmering's gazelle
Small mammals
Cape hare, red fox, hedgehog, ichneumon, jerboa
Large carnivores
leopard

Few depictions of desert hunting have survived from this period, which spans more than a hundred years. Scenes in the cruciform chapels at Medum of *Nfr-m3ꜥt*.[148] his wife,

142 Kanawati (1980) fig. 14.
143 Respectively, Blackman (1924) pl. 16 and Blackman–Apted (1953) pl. 32.
144 Blackman (1924) pl. 12 (Main room, west wall).
145 Blackman–Apted (1953) pl. 11.
146 Davies *Gebrâwi* I (1902) pl.7, 11, 12; Kanawati *Gebrawi* II (2007) pl. 52.
147 Davies II (1902) pl. 9.
148 Petrie (1892) pls. 17 and 27.

Ỉtt and *Rꜥ-ḥtp*,[149] and a narrow central panel on the southern wall of the cruciform chapel of *Mṯn* at Saqqara[150], all dated to the very beginning of Dynasty 4, are all that remain. However, in these early Dynasty 4 tombs, hunting in the desert is prominently included although depicted with comparatively little violence.[151] The only life-threatening attacks portrayed are those of hunting dogs with jaws around the neck of a cape hare and another small animal.[152] Elsewhere, single dogs bite at the tails of foxes and the hind legs of ungulates.[153] The hounds may have been trained to nip at the legs of the ungulates to drive them into prepared enclosures rather than kill them on the spot, so it is possible that the dogs were allowed to kill the smaller mammals for food. Even the dog attacking the cape hare is controlled on a long leash. Hunters are armed with boomerang shaped throw-sticks, weapons that would only have brought down small game. In *Mṯn*'s chapel the scenes are confined to a panel of five narrow registers; on the three lowest registers a hound bites the hind leg or tail of a fleeing desert animal. On each of the upper two registers, two animals are depicted in flight.[154] The purpose of the hunt, to drive animals into enclosed areas, may be inferred from the hunters' weapons, the nature of the attacks by the hunting dogs and scenes of desert animals presented to the tomb owner. No fencing is depicted in these early scenes, but the vertical statement in front of *Rꜥ-ḥtp* indicates that a fenced enclosure was used.[155] On the two upper registers foxes and ungulates are shown being driven forward by hunting dogs nipping at them from the rear. The vertical statement in front of the tomb owner reads *mꜣꜣ ꜥḥt* ('Viewing the trapping').[156] The determinative for 'trapping', Gardiner T24, indicates netting or fencing.

The tomb owner is once depicted taking part in the hunt when he is shown holding the leashes of two dogs that attack their prey.[157] Hunters are depicted armed with a stick and a boomerang-shaped weapon.[158] The greatest degree of violence portrayed shows single hunting dogs attacking smaller animals by the neck. It is possible that these dogs, used to chase animals into a fenced enclosure, were rewarded by being allowed to kill smaller animals such as the cape hare for their own consumption.

The only surviving desert hunt scene from mid Dynasty 4 until the second half of Dynasty 5 is that of a royal hunt from the mortuary temple of Sahure. This depicts the king, standing outside a fence, using a bow and arrow to shoot desert animals that are trapped within an enclosure.[159]

Mid Dynasty 5
Hunting scenes appear in the following two tombs:

SAQQARA
Nj-ꜥnḫ-H̱nmw/H̱nmw-ḥtp *Rꜥ-m-kꜣ.j*

The depicted animals are:
Ungulates
addax, dorcas gazelle, Nubian ibex, scimitar horned oryx
Large carnivores
striped hyena, lion
Small mammals
Cape hare, red fox

Late Dynasty 5
Hunting scenes appear in the following tombs:

Nj-mꜣꜥt-Rꜥ	Giza
Sšm-nfr IV	Giza
ꜣḫtj-ḥtp (D 64)	Saqqara
Pḥn-wj-kꜣ.j	Saqqara
Ptḥ-ḥtp (II) (*Ṯfj*)	Saqqara
Ṯjj	Saqqara

The depicted animals are:
Ungulates
addax, auroch, Barbary goat, dorcas gazelle, hartebeest, Nubian ibex, scimitar horned oryx, Soemmering's gazelle, wild donkey
Large carnivores
hyena, jackal, leopard, lion
Small mammals
Cape hare, hedgehog, gerboa, porcupine

Pairs of animals shown mating in desert hunt scenes of late Dynasty 5:

Ungulates:		
Wild donkey	*Nj-mꜣꜥt-rꜥ*	Giza
Addax	*Nj-mꜣꜥt-rꜥ*	Giza
Oryx	*Nj-mꜣꜥt-rꜥ*	Giza
Hartebeest	*Pḥn-wj-kꜣ.j*	Saqqara
Dorcas gazelle	*Sšm-nfr* IV	Giza
Other desert animals:		
Striped hyena	*Nj-mꜣꜥt-rꜥ*	Giza
Leopard	*Nj-mꜣꜥt-rꜥ*	Giza
-do-	*Ptḥ-ḥtp* II/*Ṯfj*	Saqqara
Jackal	*Nj-mꜣꜥt-rꜥ*	Giza
Caracal	*Nj-mꜣꜥt-rꜥ*	Giza
Striped hyena	*Nj-mꜣꜥt-rꜥ*	Giza
Red fox	*Nj-mꜣꜥt-rꜥ*	Giza
-do-	*Nj-ꜥnḫ-H̱nmw/ H̱nmw-ḥtp*	Saqqara
-do-	*Ptḥ-ḥtp* II/*Ṯfj*	Saqqara

149 Petrie (1892) pls. 17 and 27.
150 LD II pl.6. In the Old Kingdom running animals were depicted with front feet together and extended forward and rear feet together with their legs extended backward: Osborn–Osbornová (1998) 9.
151 *Rꜥ-ḥtp* and *Nfr-mꜣꜥt* Osborn–Osbornová (1998) pls. 9, 17, 27 and *Mṯn* LD II pls.3, 4, 6.
152 Petrie (1892) pl. 17.
153 Petrie (1892) pl. 9; Harpur (2001) fig. 92.
154 LD II pl.6.
155 Petrie (1892) pl. 9. The motif of desert animals within a fenced enclosure continues to appear. Sometimes the 'hunt' is located within the enclosure, which suggests that the actions do not portray a true hunt but rather the selecting of captive animals for slaughter and, perhaps, for training and feeding hunting dogs.
156 Montet (1924) pp. 89-90.
157 Petrie (1892) pl. 27. Apart from this early image, tomb owners are never portrayed within the hunting scene.
158 Petrie (1892) pl. 17.
159 Borchardt (1913) pl. 17.

Weasel	*Nj-m3ʿt-rʿ*	Giza
Hedgehog	*Nj-m3ʿt-rʿ*	Giza
-do-	*Ṯjj*	Saqqara
Shrew	*Nj-m3ʿt-rʿ*	Giza

The motif of hunting desert animals re-appears in eight tombs of the second half of Dynasty 5, the earliest, in the tombs of *Nj-ʿnḫ/H̱nmw* and *H̱nmw-ḥtp* and *Rʿ-m-k3.j*, dating from the end of the reign of Neuserre to early Djedkare. These hunting scenes include three new features.[160] Firstly, in the tombs of *Nj-ʿnḫ/H̱nmw* and *H̱nmw-ḥtp* and *Ṯjj* [161] rather than showing animals being herded into a fenced enclosure, the hunt appears to take place within it. This new element may mirror the hunting scene in the temple of Sahure, in which the king is portrayed outside of a fenced enclosure shooting animals within.[162] It is not possible to judge whether the fencing, shown as parallel bands either side of the registers containing hunting scenes, represents a temporary holding area into which the animals have been previously driven for the purpose of hunting, or a more permanent animal compound. The second innovation is the introduction of images of wild animals behaving naturally as though under no threat; depictions of pairs of animals copulating and females feeding their young, are included in the hunting scenes.[163] The third new feature is a greater degree of violence than previously depicted, first noted by Montet,[164] with young ungulate calves hiding in tussocks (*Figure 8*). The earliest extant examples of these 'new' features occur in the scene from Sahure's mortuary temple,[165] where on an upper register a pair of leopards mate and a gazelle feeds its calf while on two lower registers hunting dogs tear at the neck of desert ungulates and a lion grasps the muzzle of an auroch in its jaws. The motif of predatory lions occurs in seven scenes in officials' chapels of this period[166] (*Figure 8*).

Early Dynasty 6
Hunting scenes appear in the following tombs:

Jnw-Mnw	Saqqara
Mrjj-Ttj/Mrj	Saqqara
Mrrw-k3	Saqqara
Sʿnḫ-wj-Ptḥ	Saqqara
Mrw/Bbj	Sheikh Said

The species are:
Ungulates
addax, auroch, Dorcas gazelle, hartebeest, Nubian ibex, scimitar horned, oryx
Large carnivores
hyena, lion
Small mammals
cape hare, hedgehog, jerboa

Mid to Late Dynasty 6
Hunting scenes appear in the following tombs:

Jbj	Deir el-Gebrawi	VI.M
Mrw/Jjjj	Naga ed Der	VI.M
Ḥzjj-Mnw/Zzj	el Hawawish	VI.L
Ḫnj/Špsj-pw-Mnw	el-Hawawish	VI.L
Ḏʿw	Deir el-Gebrawi	VI.L

The species are:
Ungulates
auroch, Dorcas gazelle, hartebeest, Nubian ibex scimitar horned oryx
Large carnivores
cheetah or lioness, lion
Small mammals
hedgehog, jackal, red fox

Violent motifs continue to be portrayed in scenes from late Dynasty 5 to Dynasty 6. Indeed the degree of portrayed violence increases in Dynasty 6 tombs with dogs shown fastening their jaws on the necks of ungulates and packs of hunting dogs tearing at a fallen animal. These changes may reflect an altered attitude to desert animals as a resource for officials or an advance in the technique of artists to portray more complex images, which may have included a merging of episodes representing the animals in different time periods, before and during the hunt. The combination of images may also represent a merging of distances, as desert animals behaving naturally usually occupy a register separate to that of the scenes of the actual hunt:[167]

Beginning with the tombs of *Nj-ʿnḫ/H̱nmw/H̱nmw-ḥtp* and *Rʿ-m-k3.j* [168] hunters are depicted more closely involved in the attack on the desert animals. They are shown lassoing and clubbing the animals. In the hunt scenes of *Pḥn-wj-k3.j* and *Nj-ʿnḫ/H̱nmw/H̱nmw-ḥtp* a hunter even grasps an ungulate with his bare hands, an ibex in the case of *Pḥn-wj-k3.j* and a gazelle on the case of *Nj-ʿnḫ/H̱nmw/H̱nmw-ḥtp*. As well, hunters are depicted controlling and directing the hunting dogs by pointing at prey.[169]

3.6 Returning from the hunt

The young of many ungulate species are portrayed being carried by hunters, either in their arms or on their shoulders or, if very small, in boxes slung from shoulder yokes; these images are sometimes associated with men returning from the hunt, but may also be included in animal parades or represent the presentation of an offering. In the case of the party returning from the hunt, these images may relate to two hunting motifs: young ungulate calves crouching on sub-registers, presumably

[160] For an analysis of Old Kingdom desert hunting scenes see TABLE 3 (a), (b) and (c): FEATURES OF DESERT HUNTING SCENES.
[161] Moussa–Altenmüller (1977) pl. 38, 40; Wild (1953) pl. 127.
[162] Mortuary Temple, South passage, South wall, Berlin 21783, Borchardt (1913) pl. 17.
[163] *Pḥn-wj-k3.j* LD II 46; *Nj-m3ʿt-Rʿ* Roth (1995) pl.189.
[164] Montet (1924) 84-85.
[165] Borchardt (1913) pl. 17.
[166] See TABLE 3: FEATURES OF DESERT HUNTING SCENES (a) Animals attacked by hunting dogs or lions.
[167] *Pḥn-wj-k3.j* LD II, 46; *Nj-m3ʿt-Rʿ* Roth (1995) pl. 189.
[168] Hayes (1953) fig. 56.
[169] See TABLE 3: FEATURES OF DESERT HUNTING SCENES, (c) Hunters.

hidden by vegetation,[170] and adult animals being captured or killed during the hunt.[171] The very young calves of addax, gazelle and oryx instinctively hide immobile when endangered and thus could be easily captured. It is doubtful whether mature animals were ever depicted being brought back from the hunt. Occasionally images of mature animals being led towards the figure of the tomb owner are located immediately beneath or adjacent to hunting scenes, but these animals appear to have become accustomed to walking beside a herder, which would not be true of recently captured wild animals.[172]

Ptḥ-ḥtp (II) (*Ṯfj*) presents a most unusual scene of two of the most feared predators, a lion and a leopard captive in stout cages, being dragged by four hunters.[173] The caged animals lead the homeward procession, while to the rear of three porters carrying the usual ungulate calves, another hunter leads two animals on leashes. The inclusions of the caged lion and leopard and the leashed animals are difficult to explain. The reason for bringing home two powerful predators and the specific species of the leashed animals, which have the sloping backs and pricked ears of hyenas, are unclear. Hartmann suggested that they are wild dogs[174], although this does not explain why they are leashed or their acceptance of being leashed. If they are hyenas captured during the hunt, it is also unlikely that they would have accepted being controlled on a lead. It is unlikely that hyenas were used as hunting dogs taken to the hunt to bring down ungulates but it is possible that their role was to sniff out hidden calves. This is suggested by the controlling leash, which would prevent them tearing the young animals to pieces, and their position immediately following the herders carrying captured calves in the scene of 'returning from the hunt'.

3.7 Breeding and domestication

True domestication of a species is the result of genetic modification brought about by selective breeding over many generations. Taming, on the other hand, merely involves accustoming individual animals to contact with and handling by humans.

In Old Kingdom scenes of desert animals there are no images of any species being subject to controlled breeding. While scenes of cattle husbandry present humans controlling the copulation of bulls and cows and assisting in the birth of calves, there are no such images for desert animals. When desert ungulates are pictured copulating, they appear to be acting naturally and beyond human control. On the other hand, scenes show desert ungulates submitting to being led in animal parades, although they are typically portrayed responding less willingly than cattle. Herders are frequently shown taking measures to control the desert animals in their charge in ways that are never portrayed for cattle. A comparison of images of herders leading desert animals and domesticated cattle in the parade of animals from the tomb of *Ṯjj* makes this clear. The pictorial evidence suggests that the desert ungulates portrayed in the parade were, like the cattle which are frequently labelled '*jw3*', individuals selected from a small number of animals of a variety of species that had been brought back to the estate as calves to be reared for ritual slaughter. Despite the enormous size of herds of desert ungulates claimed by a few tomb owners[175], it is unlikely that estates could have maintained large numbers of these animals year after year. Such a practice would not have been viable in view of the demands on pasturage and fodder throughout the year and the difficulty of moving thousands of essentially wild animals from one grazing ground to another.

3.8 Summary

- Desert animals are depicted in themes of the parade of animals, the desert hunt and herded within an enclosure. The most frequently depicted species are scimitar horned oryx, Nubian ibex, Dorcas gazelle, addax and striped hyena.
- While desert animals appear in scenes throughout the Old Kingdom Period, their portrayal becomes cursory in the provinces of Upper Egypt in Dynasty 6. El-Hawawish tombs, in which large herds of desert animals are cited, are an exception.
- Images of desert animals in the 'parade of animals' scenes multiply in Dynasty 5 when officials begin building tombs with more than one room. Occasionally the scene occupies an entire panel of registers.
- Scenes of hand feeding desert animals are rare, only appearing in the tombs of magnates of later Dynasty 5 and early Dynasty 6. These are the only portrayals of husbandry relating to desert animals.
- Although the theme of the desert hunt spans the entire Old Kingdom, hunting scenes do not appear in funerary chapels from mid Dynasty 4 to mid Dynasty 5.
- In the earliest tombs the main aims of the hunt appears to have been either the capture of live ungulates or to drive them into a fenced enclosure. Hunters are armed with throw sticks, which would not have brought down the larger animals, or lassoes,

170 See Evans (2006) 131-2 for an explanation of this behaviour. It is not always possible to identify the species of the calves as they are sometimes depicted with 'nub' horns. In depictions of both young and mature desert ungulates the shape of the horns is the best indication of species of ungulate.

171 See TABLE 2: MANAGEMENT OF DESERT ANIMALS (Care: Young Carried by Herder) for frequency of this theme.

172 These scenes are classified as 'Hunters returning from the desert with game' by the *Oxford Expedition to Egypt:* Scene-details, Table 2.2.

173 Paget–Pirie (1898) pl. 33.

174 Dümichen (1869). Appendix of notes by R. Hartmann (cited in Paget–Pirie (1898) p. 28).

175 Notably two nomarchs of Upper Egypt 9, *Ḫnj/Špsj-pw-Mnw* and *K3(.j)-ḥp/Ṯtj-jkr*.

which would have captured the animals without killing them.

- When the hunting theme reappears in mid Dynasty 5 new moods are apparent: a greater degree of violence in the chase is combined with vignettes of animals behaving naturally as though beyond range of the hunt.

- After early Dynasty 4, Old Kingdom tomb owners never appear personally taking part in the desert hunt. Dependants of the tomb owner, who may be depicted viewing the action, carry out the hunt.

- From early Dynasty 4 on there are suggestions that desert ungulates were herded into or kept in stockades but the purpose of the stockade is unclear. While scenes of hunters returning from the hunt show ungulate calves being brought back, it is unlikely that images of herders leading mature animals represent the same concept.

- Hunting scenes do not have captions apart from an occasional naming of an animal or a brief phrase such as 'lassoing an oryx' (*spḫ m3-ḥḏ*). Consequently, whether these scenes have any symbolic meaning beyond the mere recording of an actual hunt is a moot point and a matter of speculation rather than argument from the evidence.

- Throughout the period of the Old Kingdom certain desert ungulate species, in particular addax, dorcas gazelle, Nubian ibex and scimitar horned oryx are depicted as sufficiently accustomed to human contact to be paraded before the tomb owner.

- The evidence does not establish that these animals were truly domesticated or that they were maintained as a herd.[176] Images portray paraded desert ungulates requiring a greater degree of control than domesticated cattle. Representations of desert animals provide evidence that they were partly tamed but not that attempts were made to domesticate them.

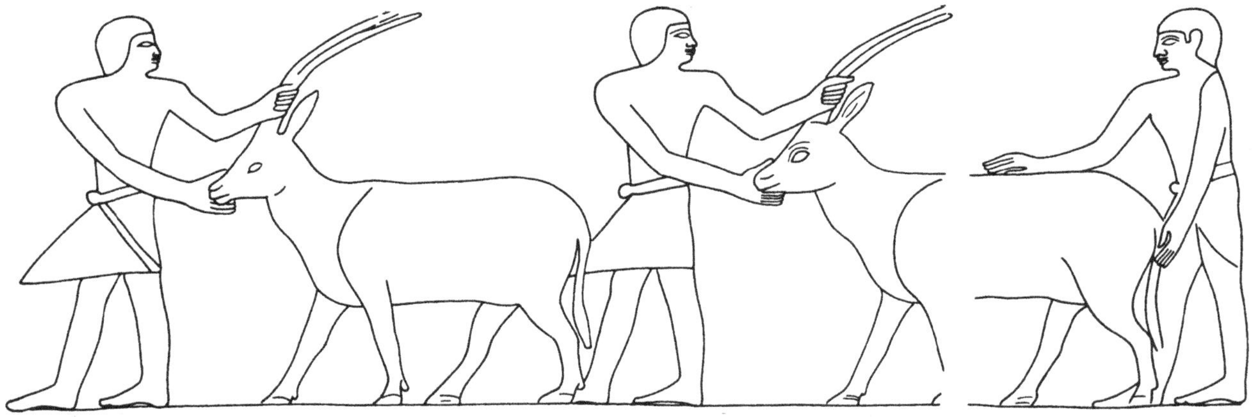

Figure 3: SCIMITAR HORNED ORYX, TOMB OF MERERUKA, SAQQARA. Kanawati–Abder-Raziq (2004) pl. 48.

[176] Only the el-Hawawish tombs of *Ḫnj/Špsj-pw-Mnw* and *K3(.j)-ḥp/Ṯtj-jḳr* provide herd numbers for desert ungulates. These numbers are suspect.

Figure 4: LEADING AN IBEX, TOMB OF MERERUKA, SAQQARA.
Kanawati–Abder-Raziq (2004) pl. 48.

Figure 5: TWO MEN PRESENTING A DORCAS GAZELLE, TOMB OF HESI, SAQQARA.
Kanawati–Abder-Raziq (1999) pl. 65(b).

Figure 6: LEADING AN ADDAX, TOMB OF MERERUKA, SAQQARA.
Kanawati–Abder-Raziq (2004) pl. 48.

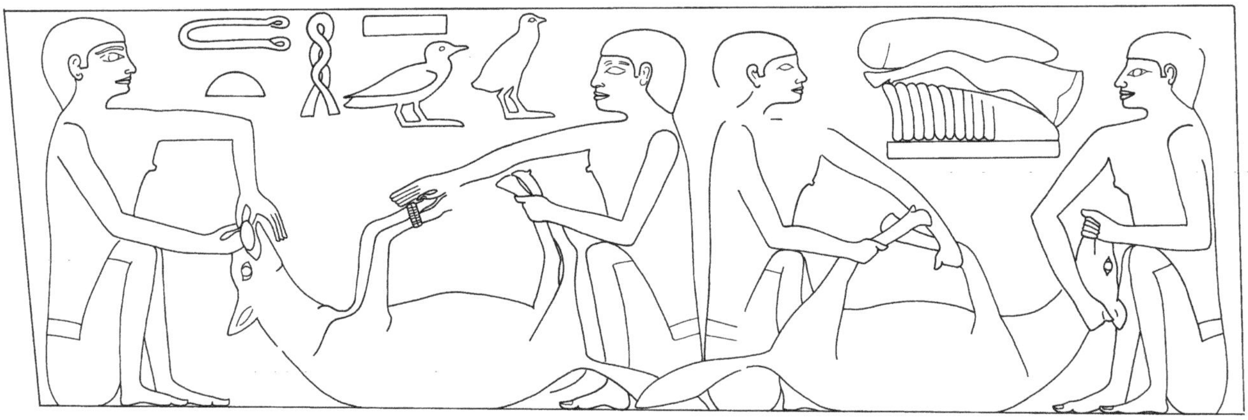

Figure 7: STRIPED HYENAS BEING FORCE FED, TOMB OF INUMIN, SAQQARA.
Kanawati (2006) pl. 47.

Figure 8: VIOLENCE DEPICTED IN AN EARLY DYNASTY 6 REPRESENTATION OF A DESERT HUNT, TOMB OF MERERUKA, SAQQARA.
Kanawati (2010) pl. 73.

Figure 9: HUNTING DOGS USED IN THE DESERT HUNT, TOMB OF INUMIN, SAQQARA.
Kanawati (2006) pl. 47.

SECTION TWO

THE PASTORAL ENVIRONMENT

Figure 10: Horned and polled cattle presented to the tomb owner, tomb of Mereruka, Saqqara. Kanawati (2011) pl. 74.

By the time of the Old Kingdom domesticated mammal species were a vital element of the Egyptian world. They were at the heart of the economy providing power and transport (heat from bodily excretions and haulage through their strength), foods from their body and an array of useful raw materials such as fat, hides, horns and bone. Ownership or management of herds and flocks had become a significant component of capital investment, bringing the individual wealth, power and social status. The scenes of 'daily life' on the walls of the tombs of ancient Egypt's elite reflect this in the attention and space they give to the presentation of domestic animals.

Scenes containing depictions of these animals are studied for a variety of reasons: to work out the technologies and detail of 'everyday' activities, which for many Egyptians would have been connected with farming and food production, to establish the precise species depicted and to investigate religious and after-life beliefs, as domesticated animals appear in 'ritual' contexts as well as 'daily life' themes. The aim in this study is to investigate the portrayed species as economic resources of value to the tomb owner and the rural estate depicted in his tomb.

The question of how much pasturage was available to support the herds and flocks is problematic in view of the enormous numbers cited particularly in Dynasty 6. Butzer has calculated that much of the Nile valley including the Delta was underpopulated in the Old Kingdom.[177] His judgement is that the Delta, which comprised approximately half Egypt's farmland, was not completely settled until the Ptolemaic period.[178] This provides a basis for the investigation of the 'exploitation' of Upper Egyptian provinces in Dynasty 6. Yet the simple basin irrigation of the Old Kingdom meant that the Nile valley only afforded one harvest a year, which implies that available pasturage would have been seasonal and short lived. As a consequence, the term 'pastoral environment', which normally refers to grass covered regions, has a more varied application in the context of the Nile valley and Delta. While grasslands may have survived year round in the Delta, the pasturage of the valley must have dried off in the hot summer season. Consequently, a study of domesticated livestock in the Old Kingdom has to encompass both a seasonal pattern of cattle management that includes the annual movement of herds and the use of the ecotone between the desert and the watered region for goats, sheep and donkeys.

177 Butzer (1976) 100, 108.

178 Butzer (1976) 83, 94, 96.

CHAPTER 4 DOMESTIC CATTLE

4.1 Egyptians and domesticated cattle

There is evidence of the domesticated *bos taurus* in both Lower and Upper Egypt from at least the fifth millennium B.C. onwards.[179] Early pictorial and archaeological traces suggest that a metaphorical significance, as well as material value, was attributed to the 'idea' of the animal, so potent yet controlled by man. The portrayal of the bull to represent the power of the king[180] and the ceremonial burial of cattle or parts of cattle bodies[181] may suggest that a symbolic value was accredited to cattle in Pre- and Early Dynastic society, although evidence for prehistoric cults of the bull is uncertain.[182] Cults of the Apis Bull and of goddesses portrayed as cows with human faces, however, had their origins in Early Dynastic times and remained vital aspects of Egyptian religious life throughout the Old Kingdom and beyond, as did the symbol of the bull to denote royal might.[183] Much of the surviving evidence relating to cattle in early Egyptian society appears to reflect these 'non-material' aspects of cattle culture;[184] yet domesticated herds were a valuable economic resource. By transforming plants into edible and usable products, domesticated cattle supplied many basic needs of their owners, while the integration of herds into the economic life of early Egypt profoundly affected the development of this society as well as enhancing the animals' 'non-economic' value.[185] Domestic herds provided communities with constant and reliable supplies of meat, fat, bone, sinew and hides, reducing the need to face the fortunes and dangers of hunting wild animals. When men became herders they adapted to the needs of the herds, which in the Nile valley entailed an annual regime of moving stock from one pasturage to another. Cattle management also helped to shape royal administrative practice, providing a further link between the economic and non-economic values attributed to cattle. The early use of a cattle count for dating purposes indicates that this census was a singularly important event.[186] Without the evidence from the private tombs of officials, however, the economic and social significance of cattle holdings and husbandry in the Old Kingdom would be much less clear, leaving our knowledge to lie mainly in the symbolic, religious and administrative spheres.

4.2 Focus of enquiry: cattle herds as an official's resource and the measure of his wealth

TABLE 5: LIVESTOCK NUMBERS – Cattle Variety

The data:

- Iconography of cattle parades
- Iconography of cattle husbandry
- Inscriptions: i. accompanying the iconography
  ii. from autobiographies.

Areas of research:

- Cattle varieties
- Management of herd
- Parade of animals.

In Old Kingdom tombs cattle are portrayed in the following themes:

- scenes of cattle husbandry, including herds being guided across a stretch of water and pairs of fighting bulls[187] (*Figures 18, 19 and 20*)
- paraded in scenes of presentation, frequently before the image of the tomb owner (*Figure 10*)
- working for man, pulling ploughs and threshing, and occasionally hauling heavy items (*Figures 11 and 12*)
- scenes of ritual butchery[188].

Domesticated cattle are the most frequently depicted animals in the chapels of Old Kingdom officials. In the 191 tombs cited in this study cattle appear in over 80 per cent of the chapels.[189] The value attributed to a herd may be gauged by their depiction as the major element in the parade of animals before the tomb owner, and in scenes of attentive husbandry, including controlled mating, giving birth, milking and handling of calves. Cattle are

179 Epstein I (1971) p. 213; Chenal-Vélardé I. (1997) 1.2; Wendorf-Schild propose a much earlier date (c. 11,000 B.C.) for Nabta Playa evidence: Wendorf-Schild (1994), 118-128; Wendorf-Schild (1998) 97-123; Wendorf-Schild et al. (eds.) (2001).

180 Reference: The Narmer Palette (Great Hierakonpolis Palette), Cairo J.E. 14716, C.G. 32169.

181 Brass (2003) 105-106.

182 Wengrow (2001) 91–104.

183 Kings attached 'Victorious ka-bull' at the beginning of their Horus name from Thutmose I onwards: von Beckerath (1984) 83ff and 225ff.

184 The interpretation of early cattle images is often subject to cliché and doubtful generalisation: Wengrow (2001) 91.

185 Badarian/Naqada I occupation sites typically have few traces of permanent constructions, yet layers of animal droppings combined with remains of enclosures reflect the seasonal movement of controlled herds: Butzer (1976), 14, 107. This evidence follows that of the more sedentary Mesolithic sites, suggesting a major change in the pattern of life: Haaland (1993) 47–86. Brass suggests that domesticated cattle were introduced into Nile valley society as 'a new religious and socio-economic element' in the Badarian era: Brass (2003) p. 108.

186 While issues surrounding the cattle census are much debated, there is no doubt that counts were made as early as the beginning of Dynasty 4: Helck/Eberhard (1975) 1036-9.

187 For analysis of scenes of cattle crossing water and fighting bulls see Chapter 5.

188 Ritual butchery is not investigated as the theme does not apply to economic issues.

189 Van Walsem points out the value of and inherent problems in using numbers as statistical data for the development of historical inference: van Walsem (2005) pp. 41-42.

the most frequently depicted animal being transported by boat, and appear regularly in scenes where they work for man.[190] Furthermore, cattle are the only domesticated animals to be shown in clearly defined varieties. They are frequently labelled *jw3* or *ng3.w* or, less often, *ḥr db*ʿ, *k3*, *ḥmt*, *jḫt*, *mn*ʿ*t* and *bḥs*.[191] The prominence given to representations of cattle may have been due to their symbolic and religious significance but as they most frequently appear in scenes portraying elements of 'daily life', it may be argued that it also reflects their economic and/or status value to tomb owners.

4.3 Types of cattle

Linnaeus identified three separate species of cattle: *Bos taurus*, the European cattle, including similar types from Africa and Asia; *Bos indicus*, the zebu; and the extinct *Bos primigenius*, the aurochs. Today these three varieties are considered to be subgroups of the same species. All Old Kingdom depictions of domesticated cattle are of the *Bos taurus* subspecies; there are no depictions of the zebu with its distinctive hump. The term 'breed' refers to a stock of animals or plants within a species with certain distinguishable characteristics that have been developed by deliberate selection and maintained by controlled propagation. Some scholars have identified differences in the characteristics of depicted cattle of the Old Kingdom. For example, it is claimed that the difference in body size and mass of the animals labelled *jw3* and those labelled *ng3.w* is evidence of separate breeds.[192] However, Epstein has pointed out that as all the horned types are of the primigenius cranial variety, they are of different horn types within a single breed.[193] Consequently, the issue resolves into the question of whether the depicted *polled* cattle were a separate, deliberately bred variety (*Figure 10*).

4.3.1 Polled cattle

The study of polled cattle is based on depictions in the following tombs:

*R*ʿ*-ḥtp*	Medum	IV.E
Mr-jb.j	Giza	IV.L
*Mrs-*ʿ*nḫ* III	Giza	IV.L
K3-pw-nswt/K3j	Giza	V.E
Jj-mrjj	Giza	V.M
*R*ʿ*-ḫ*ʿ*.f-*ʿ*nḫ*	Giza	V.M
Sšm-nfr II	Giza	V.M
K3.j-m-nfrt	Giza	V.M
Pḥn-wj-k3.j	Saqqara	V.L
Njwtj	Saqqara	V.L
*R*ʿ*-wr* II	Giza	V.L
*R*ʿ*-špss*	Saqqara	V.L
Sšm-nfr IV	Saqqara	V.L
Ṯjj	Saqqara	V.L
Mrjj-Ttj/Mrj	Saqqara	VI.E
Mrw/Bbj	Sheikh Said	VI.E
Mrrj	Saqqara	VI.E
Mrrw-k3	Saqqara	VI.E
Ḥzjj-Mnw	el-Hawawish	VI.E
Ḥm-r3/Jzj	el-Gebrawi	VI.E
*Sḫm-*ʿ*nḫ-Ptḥ*	Saqqara	VI.E
*Ppj-*ʿ*nḫ-ḥrj-jb*	Meir	VI.M
*Ppj-*ʿ*nḫ/Ḥnj-km*	Meir	VI.M

Brewer and Ghoneim classify the depicted polled cattle as belonging to a separate breed to the more commonly depicted horned variety. Brewer identifies four major physical types of cattle portrayed in the iconography: a long-horned variety of cattle that was tall and lean but heavily muscled (the variety most frequently depicted in the Old Kingdom); a short-horned variety, for which, according to Brewer, there is evidence from Dynasty 5; a hornless breed called *ḥr db*ʿ, also known from the Old Kingdom; and the zebu, introduced in the New Kingdom. Brewer states that these four varieties were sufficiently differentiated morphologically to be regarded as distinct breeds. He refers in particular to the fact that certain animals (the hornless variety) are portrayed with especially heavy bodies.[194] Where there are sufficient images of cattle in a single tomb to make a comparison, however, the data rarely show any difference in body mass or proportions between horned and hornless *jw3*.[195] Scholars, who have classified images of polled cattle with especially heavy body shape as a separate breed, may have relied on a few depictions from particularly well known tombs such as that of *Mrrw-k3*.[196] A more comprehensive analysis of the data shows that of the instances where a comparison can be made between a polled animal and horned cattle labelled *jw3*, over 85 percent of the scenes present polled animals with the same body mass and shape as the horned variety. Thus, if the polled cattle do represent a separate breed, the only consistent physical characteristics would be hornlessness and the domed head that accompanies the *naturally* polled condition[197] (*Figure 10*). Yet even the domed head is not sufficient evidence of a separate *breed*, as horned breeds occasionally produce polled individuals by chance. Whether the Egyptians actually succeeded in developing a separate breed of polled cattle is debatable. Cattle carry two genes that govern the horned condition. If only one of these is the polled gene; it is dominant and the animal will be polled, but the animal may pass on the other, horned gene rather than the polled gene to its offspring.[198] Consequently, producing a polled breed of cattle is a slow, trial-and-error process, with many

190 Cattle are also portrayed accompanying offering bearers and representations of the tomb owner's estates and in scenes of ritual slaughter. These motifs emphasise the variety of offerings brought to the deceased rather than the specific value of the cattle as a component of an official's resources.

191 For discussion of the meaning of these terms see p. 33ff.

192 Brewer et al (1994) 82 – 84.

193 Epstein vol. 1 (1971) 221.

194 Brewer et al (1994) 84; Ghoneim (1977) 76 – 80.

195 The two instances where polled animals are depicted with heavier bodies than those of other cattle in the scene: *Mrrw-k3* Duell (1938) pl. 152, Kanawati (2011) pl. 74 and *Ppj-*ʿ*nḫ/Ḥnj-km* Blackman–Apted (1953) pl. 41. Polled cattle with heavier bodies but without labels also appear in *Nj-*ʿ*nḫ-Ḫnmw*: Moussa–Altenmüller (1977) fig.17.

196 *Mrrw-k3* Duell (1938) pl.152; Kanawati (2011) pl. 74. See also *Nj-*ʿ*nḫ-Ḫnmw* Moussa-Altenmüller (1977) fig. 17.

197 Phillips-Coventry (May 2005) p. 5.

198 Phillips-Coventry (May 2005) p. 2 - 3.

setbacks, which may account for the frequent portrayals of polled cattle receiving special attention.

Depictions suggest that polled cattle were docile and especially valued, as they are often pictured being led together in groups of two or three and are portrayed as half grown or with a decorative blanket over their back. In the tomb of *Rˁ-ḥˁ.f-ˁnḫ*, for example, the figure '220' is inscribed above a group of polled animals.[199] It may be questioned whether the figure '220' is intended to apply to a herd that is specifically polled. Yet the occurrence of polled animals in so many tombs seems too great to be the result of random births. Physical dehorning is an unlikely explanation for the frequency of the depiction of hornless animals; it involves significant risk of infection, which would not have been easily controlled and would have posed a serious threat to valued livestock. The distinct dome to the top of the head showing that an animal was born polled is clearly depicted in at least 21 tombs.[200] The frequency with which small groups of polled animals are depicted suggests that attempts were made to breed this variety in Dynasty 5, perhaps with only limited success as polled specimens are depicted being treated with especial care.

The interest in breeding polled cattle appears to have waned in the second half of Dynasty 6 when the major preoccupation became the production of increased herd sizes. Provincial tombs of the second half of Dynasty 6 give less prominence to polled animals. Furthermore, polled bulls would have had no part in bullfights, which were a marked feature of interest among nomarchs and administrators of Upper Egypt at this time.

FINDING

As polled cattle are safer for herdsmen to handle and unlikely to do each other injury, in those times as now there would have been a strong motive to maintain as many polled cattle as possible and to breed from them. However, there is no conclusive evidence that polled cattle had become a distinct breed.

4.3.2 Short-horned cattle

According to Brewer, there is some evidence from Dynasty 5 of the distinctive breed of short-horned cattle, which become familiar in the Hyksos period.[201] Brewer does not cite his evidence, but it must be drawn from only a few tombs such as that of *Ptḥ-ḥtp* II (*Ṯfj*).[202] In this tomb, two groups of cattle have comparatively short crescent-shaped horns, while the two oxen on the register below have longer lyre-shaped horns, although of different sizes. These few images, however, may portray a random outcome, as herds naturally produce animals with varying sizes and shapes of horns.[203]

FINDING

While consistent depictions of cattle with much shortened horns are clear evidence of a breed of shorthorns in the New Kingdom, the data is too limited and uncertain to accept the claim that such a breed was introduced in the Old Kingdom.

4.3.3 Types of cattle identified by captions

In Old Kingdom tomb chapels, cattle are the only domesticated mammal to be shown throughout their life cycle being cared for by farmhands and depicted with a variety of markings, colorations, body shapes, sizes and horn arrangements, as well as being identified by a range of descriptive labels: *jwꜣ, rn jwꜣ, ngꜣ.w, ḥr ḏbˁ*.[204] The terms *ḥmt, jḥt, mnˁt, mrw, kꜣ* and *bḥs* are also applied to individual depictions of cattle but are clear descriptions of gender and age. The animals so described are usually given appropriate physical features, scrotum to indicate a bull and udders for a cow.[205] In husbandry scenes these animals are frequently gender differentiated by their activities; bulls are shown sniffing at or mounting cows or fighting each other, and images of cows show them giving birth, accompanied by or suckling calves, and being milked by herders. As images of animals engaged in gender differentiated activities or given a gender label are almost invariably depicted with appropriate sexual organs, it is logical to assume that animals portrayed lacking both of these identifying markers are male castrates (oxen). Moreover, images of animals that appear to be oxen are regularly portrayed in scenes where male castrates would normally be expected: the parade of livestock and as working animals.

The meanings of the first group of labels (*jwꜣ, ngꜣ.w* and the descriptor *ḥr ḏbˁ*) are less obvious. Scholars have used these terms to arrive at a variety of interpretations and classifications. They have also discerned different functions and body shapes for cattle labelled *jwꜣ* and *ngꜣ.w*. Of the four major physical types of cattle that he identifies, Brewer considers that the tall and lean but heavily muscled long-horned breed called *ngꜣ.w* was the variety most frequently depicted in the Old Kingdom and that there was also a hornless breed called *ḥr ḏbˁ*.[206]

Ghoneim, like Brewer, accepts that breeding programmes were carried out in the Old Kingdom. He classifies the depicted cattle into three main breeds: longhorn,

199 LD II pl. 9.
200 Phillips-Coventry (May 2005) 5.
201 Epstein (1971) op. cit. 288; Brewer et al (1994) 84.
202 Paget–Pirie (1898) pl. 31.
203 Dyson notes that Indian herds often include animals with horns of different sizes and advises the need for ample evidence to establish a distinct breed. Dyson (1955) 661-673.

204 In the earliest Dynasty 4 chapels captions such as *jwꜣ* or *ngꜣ.w* are not used to specify different types of cattle). *Rˁ-ḥtp*, for example, portrays an ox labelled *rn* ('young' or 'prime') with *mḏ.wt pr n ḏt* ('stalls of the estate') inscribed between the herder leading the animal and the tomb owner's staff.
205 Occasionally an animal whose gender is clearly established by its activity, such as a cow giving birth or a bull mating a cow, is portrayed without its defining sexual organs. Examples: Mating scenes – Simpson (1976) pl. C; Munro (1993) pl. 13; Kanawati–Abder-Raziq (2006) pl. 49(a); Hassan V (1944) fig. 123; Birth scenes – Hassan VI (1950) fig. 80; Simpson (1976) pl. C; Simpson (1992) pl. C; Munro (1993) pl. 13; Smith, (1978) fig. 226[c]; Paget/Pirie (1898) pl. 31; Kanawati-Abder-Raziq (2003) pl. 54; Simpson (1980) fig. 22.
206 *ḥr* (Wb 3, 131.1-132.23: mittels 'by means of'); *ḏbˁ* (Wb 5, 562.11-565.9: 'finger').

shorthorn and hornless, but is reluctant to describe cattle labelled *jwꜣ* and *ngꜣ.w* as distinct breeds. Instead, Ghoneim interprets these labels as distinguishing between functions, the heavier bodied *jwꜣ* cattle being specially reared and fattened for meat production and the *ngꜣ.w* trained as work animals.[207] Boessneck also identifies the animal labelled *jwꜣ* or *rn jwꜣ* as an ox deliberately fattened for slaughter.[208] Smith identified the *jwꜣ* as the 'bull' preferred for slaughter, describing it as 'exceedingly fat with pendulous belly' standing 'relatively low on its haunches; whereas the '*ng*' (*ngꜣ.w*) was the 'native wild bull' of the Nile valley, 'depicted as lean and standing relatively high on its haunches', captured and tamed to be Egypt's 'draught animal par excellence'.[209]

FINDING

In almost every instance where it can be verified, images of cattle labelled *jwꜣ, rn jwꜣ, ngꜣ.w,* and *rn ngꜣ.w* are oxen, that is, castrated males, without scrotum or udders; only the preputial sheath of the male appears, frequently but not consistently.[210] In a preponderant number of instances cattle depicted in parade scenes or recumbent being hand-fed are oxen labelled *jwꜣ*. A few paraded oxen are described as *ngꜣ.w*; otherwise the label is confined to animals depicted in pastoral scenes of husbandry. Here they are described as *ngꜣ.w n mr*, but when portrayed recumbent, it is not possible to tell whether they are bulls or oxen.[211]

jwꜣ and *rn jwꜣ*

(*jwꜣ* Langhornrind; Ochse (allg.) Wb 1, 49.9-11)

An analysis of oxen labelled *jwꜣ* and *rn jwꜣ* (altogether 72 tombs) provides the following data. The majority of images with depictions of oxen labelled *jwꜣ* and *rn jwꜣ* occur in presentation scenes with cattle being paraded by herders. While there are also images of cattle labelled *jwꜣ* and *rn jwꜣ* in scenes other than the animal parade, all except one of these images (tomb of *Jḫjj*)[212] are either scenes of butchery or trussed carcasses of cattle or are of recumbent animals being hand fed. In each of the images of handfeeding, the animals are all labelled *jwꜣ* or *rn jwꜣ*, sometimes accompanied by the word *wšꜣ* ('fattening'). In addition, phrases such as 'belonging to the stable' or 'of the stable' (*n mḏt, jmj mḏt, m mḏt, ḥꜣt mḏt*), of which there are eight instances, only appear as a description of an ox labelled *jwꜣ*.[213] If the images of cattle labelled *jwꜣ* and *rn jwꜣ* are compared to images of cattle that are clearly bulls and cows in the same tomb, the *jwꜣ* animals appear to have a thicker and more ponderous body shape.[214] Mature polled cattle are always depicted with the heavier body shape and, if labelled, are called *jwꜣ* or *jwꜣt*. Inscriptions in the tombs of *Rꜥ-špss* and *ꜣḫtj-ḥtp* (D 64) indicate that *jwꜣ* cattle were not only slaughtered for the benefit of the tomb owner but also for festivals that were held in honour of a god such as Thoth.[215]

FINDING

The labels *jwꜣ* and *rn jwꜣ* indicate the animal's function. These are oxen (castrated males) selected to be stable reared, as descriptive phrases containing the word *mḏt* ('stable') imply, and fed a special diet. These animals were fattened for slaughter on special occasions, such as a funerary celebration and feast days and are the variety normally depicted in parades of livestock. The essential function of these animals was for slaughter and consumption in ritual ceremonies.

ngꜣ.w and *rn ngꜣ.w*

(*ngꜣ.w* Langhornrind Wb 2, 349.1-5)

Tombs with cattle labelled *ngꜣ.w*:

Mr-jb.j	Giza	IV.L
Nb.j-m-ꜣḫtj	Giza	IV.L
Snb	Giza	V.E
Sšꜣt-ḥtp/Htj	Giza	V.E
Kꜣ.j-m-nfrt (LG63)	Giza	V.E
Jj-mrjj	Giza	V.M
ꜣḫtj-ḥtp (D64)	Saqqara	V.L
Jꜣsn	Giza	V.L
Ḥzjj-Mnw	el-Hawawish	V.L
Ḥtp-ḥr-ꜣḫtj	Saqqara	V.L
Sḫm-ꜥnḫ-Ptḥ (G7152)	Giza	V.L
Kꜣ.j- m-ꜥnḫ	Giza	V.L
Jttj-Šdw	Deshasha	VI.E
Ppj-ꜥnḫ/Ḥnj-km	Meir	VI.L

Relatively few animals are labelled *ngꜣ.w* in Old Kingdom tombs. They either appear in scenes of cattle in parades or husbandry scenes while four figure in scenes of butchery. In four of the husbandry scenes the animal labelled *ngꜣ.w* is described as *n(j) mr* ('of' or 'belonging to the (cattle) pasture'[216]), and the only occasion when the cattle pictured crossing water are given a label, they are called *ngꜣ.w*.[217] The phrase *rn ngꜣ.w* is reserved for oxen depicted in parades and never used in husbandry scenes. The descriptor *rn* is rarely applied to cattle labelled *ngꜣ.w* and only appears in eight scenes. Although the *ngꜣ.w* are occasionally given a slimmer body shape than the *jwꜣ*,[218] in other parade scenes[219] the *ngꜣ.w* are depicted with body proportions identical to those of the *jwꜣ* cattle in the same scene.

207 Ghoneim (1977) 47-65.
208 Boessneck (1988) 68.
209 Smith (1969) 308.
210 The only exceptions occur in the tomb of *Ḥtp-ḥr-ꜣḫtj* where one animal with a scrotum is labelled *rn jwꜣ* and another with udders is labelled *jwꜣt*: Mohr (1943) fig. 4, p. 39.
211 Weeks (1994) fig. 34; Simpson (1980) fig. 30; *Kꜣ.j-m-ꜥnḫ*, Kanawati *Giza* I (2001) pl.36.
212 Saleh (1977) fig. 61, pl. 18.
213 *Mr-jb.j, Jḫjj r/u Jdwt/Sšsšt, Ptḥ-ḥtp, Ptḥ-ḥtp/Ṯfj* (II); *Rꜥ-špss, Ḥww-wr, Kꜣ.j-m-nfrt*.
214 Making comparisons of size and shape of animals across different tombs is unacceptable because variations may have been due to the individual style of different craftsmen.
215 *mꜣꜣ jwꜣ n ḏḥwtt inn m ḥwwt-kꜣ njwwt nt pr-ḏt m Tꜣ-Mḥw Šmꜥw*: *ꜣḫtj-ḥtp* Davies *Ptahhetep* II (1901) pl.21,; pls. 20, 21.
216 Wb 2, 97. 14-15.
217 *Rꜥ-špss* LD II pl. 60.
218 *ꜣḫtj-ḥttp* Davies *Ptahhetep* II (1901) pl.21.
219 *Kꜣ.j-m-nfrt* LD II 63; *Jj-mrjj* LD II 54, Weeks (1994) fig. 41; *Sḫm-ꜥnḫ-Ptḥ* Badawy (1976) fig. 29.

FINDING

The *ng3.w* had a number of functions in Egyptian rural industry. Being oxen, they were more docile, and although not always labelled as *ng3.w*, they most likely were the beasts of burden, pulling ploughs, statues and coffins, sometimes used on the threshing floor instead of teams of donkeys or to trample seed into newly ploughed land.[220] However, as their appearance and occasional label as *rn ng3.w* in parades suggests, they might also be fattened for slaughter. (See below for discussion of the significance of the term *rn*.) As working animals, they were considered part of the herd of cows and bulls that browsed in the upland fields and, presumably[221] for part of each year, in the lower lying marshlands, as the phrase *ng3.w n mr* indicates.[222] It is possible that the *ng3.w* depicted in parade scenes were chosen for sacrificial slaughter either because they had naturally grown fatter or because their days of service were over. They are never portrayed being hand-fed.

The descriptor *rn*

The descriptor *rn* is first used in the chapel of *Jtt* to describe a tethered dappled ox,[223] and the caption *rn jw3* first appears above the oxen being led towards the tomb owner in the chapels of *Nj-k3w-rʿ* and *Sš3t-ḥtp/Ḥtj*.[224] Usually translated as 'young' in egyptological studies and tomb reports, *rn* is frequently applied to mature animals in the parade of livestock or being handfed, but rarely to calves or half grown animals or to bulls or cows.[225] In the chapel of *Mr-jb.j*, for example, a mature lyre-horned ox standing alone has the caption *rn ng3.w*, although on the register below a smaller polled animal, perhaps half grown, and captioned *jw3 n mḏt ḥr-ḏbʿ* ('ox of the stall suckled on the finger'), is not given the description *rn*. The phrase *ḥr-ḏbʿ*, rather than *rn*, is more likely to be applied to half grown animals. Furthermore, the term *rn* is not used in connection with either working cattle or vignettes of husbandry. Consequently, the translation 'young' for *rn* may provide the wrong nuance.

FINDING

The label *rn* is reserved for cattle and desert animals portrayed as mature animals prepared for ritual sacrifice. 'Prime' (in the agricultural sense) may better convey the intended meaning of *rn*, as the labelled animals either appear in a presentation scene or recumbent being hand-fed or being butchered.

Cattle labelled *jw3.t*

(*jw3.t* Kuh, Wb 1, 49.13; *jw3.t* Rindvieh (coll.), Wb 1, 49.14)

The depictions of polled cattle cited below are drawn from the images of cattle that are given descriptive labels. There are many more images of polled cattle without labels but these do not relate to the question of the meaning of *jw3.t*.

Mrs-ʿnḫ III	C*	Giza	V.E
K3-pw-nswt/K3j	C*	Giza	V.E
Sšm-nfr II	C*	Giza	V.M
Pḥ.n-wj-k3.j	C*	Saqqara	V.L
Rʿ-wr II	C*	Giza	V.L
Sšm-nfr IV	C*	Giza	V.L

C* male castrate

In six parade scenes one or more of the animals depicted as polled oxen is labelled *jw3.t*, which in some tomb reports is translated as 'cow', although the animal has no udders.[226] Indeed, no animal labelled *jw3.t* is depicted as a cow; all appear to be polled oxen, as the Table above indicates..

FINDING

The use of the descriptor *jw3.t* applied to images of polled cattle has a collective rather than a female connotation.

The meaning of *ḥr(j) ḏbʿ*

(*ḥr ḏbʿ* hornloses, Rind Wb 3, 136.6)

The phrase *ḥr(j) ḏbʿ* translates literally as 'on' or 'by' or 'by means of the finger', suggesting reference to a method of rearing an animal. The great majority of cattle depicted without horns are not described as *ḥr(j) ḏbʿ*. Altogether the phrase *ḥr(j) ḏbʿ* is used in eleven scenes but only applies in eight instances to polled cattle, a variety depicted in 23 tombs. In two of these instances the associated caption couples *jw3* or *jw3.t* with *ḥr(j) ḏbʿ*; in *Mr-jb.j* the caption reads *jw3 n mḏt ḥr(j) ḏbʿ*,[227] and in *Sšm-nfr* II *jnt jw3.t ḥr(j) ḏbʿ*.[228] On the doorway into Room 2 of the tomb of *Rʿ-špss* (LS 16) hornless cattle are depicted on the three lowest registers of a narrow panel behind the standing figures of the tomb owner and his son.[229] However, only the three animals grouped on the third register from the top are described as *ḥr(j) ḏbʿ*. The three hornless animals on the fourth register and the single hornless animal being prepared for slaughter on the bottom register are labeled *rn jw3*. This difference in labeling makes it clear that not all polled cattle were called *ḥr(j) ḏbʿ*. The interpretation of the phrase as referring to cattle which have a 'pointed' head, originally proposed by Andersson, does not meet the fact that most cattle with a high rounded head (polled animals) are not labelled *ḥr(j) ḏbʿ*.[230] On the other hand, Goedicke's interpretation, that *ḥr(j) ḏbʿ* refers to a method of weaning a 'finger suckled' calf misses the point of the phrase.[231] The rare use of the term suggests that it applies to a situation or process that occurred infrequently. As it implies human intervention in the rearing of the animal, it

220 In *Ptḥ-ḥtp* (II) (*Ṯfj*) the cattle on the threshing floor are labelled *ng3.w*: Davies, *Ptahhetep* II (1901) pl. 8.

221 It is unclear whether the entire herd was driven to the marshland and back each year. It is feasible that that *jw3* cattle, fed by hand, were kept throughout the year in upland stables.

222 *mr* Viehweide ('cattle pasture') Wb 2.97.14-15.

223 Petrie (1892) pl. 24.

224 LD II pl. 15; Junker II (1934) fig. 28; *Seshathetep/Heti* Kanawati *Giza* II (2002) fig. 45.

225 Mature desert animals being led in animal parades may also be described as *rn*.

226 Dunham-Simpson (1974) fig. 4 and p. 11.

227 LD II pl.22a

228 Kanawati *Giza* II (2002) pl. 61

229 LD *Erg*. pl. 41.

230 Andersson (1912) 145-164.

231 Goedicke (1957) 159-161.

probably refers to an animal that had been fed by hand from birth, perhaps because it had lost its mother. Such an interpretation satisfies the implicit significance of the rarity of the phrase, the use of which indicates careful and successful husbandry.

FINDING

The phrase *ḥr(j) ḏbꜥ* does not refer to a breed of hornless cattle but to a method by which herders hand-reared calves that were removed from or had lost their mother.

4.4 Cattle husbandry

Management of the herds was inevitably determined by the context provided by the Nile valley: prevailing climatic conditions, the annual inundation and the topography of the valley. Not all the important husbandry practices are depicted in Old Kingdom tombs; many male calves were castrated. From time to time infectious diseases would have decimated herds and drought would have reduced cattle to walking skeletons, but methods of dealing with these aspects are never portrayed. In keeping with all other themes of 'daily life' depicted in private tombs, only images of success are presented. Every animal has a healthy, substantial body indicating that it is well fed. Bulls mount cows, which give birth to large, healthy calves.[232] Older, experienced herders then handle and inspect the calves, and cows are successfully milked.[233] Selected oxen are stable reared and fattened and the herd always returns safely from the marshland to higher ground after crossing waterways where dangerous crocodiles lurk. Herders directed by overseers, are depicted living with the herd in the field or marshlands, and their zeal and skills are made clear in husbandry vignettes.

4.4.1 Historical development of images and themes

Cattle husbandry is depicted in the following tombs:

Rꜥ-ḥtp	Medum	IV.E
Ṯntj	Giza	V.E
Nb.j-m-ꜣḫtj	Giza	V.E
Jj-mrjj	Giza	V.M
Ptḥ-špss	Abusir	V.M
Nfr/Kꜣ-ḥꜣ.j	Saqqara	V.M
Kꜣ.j-m-nfrt (PM 263)	Giza	V.M

Distribution of Husbandry Images (Figures 13, 14, 15, 16 and 18)	*Tombs*
Milking	3
Fattening recumbent oxen	3
Crossing/emerging from waterway	3
Cow suckling calf	2
Copulation	1
Birth of calf	1
Care of calves	1

In tombs of the Fourth and early Fifth Dynasties there are no extended images of cattle husbandry, although there is a scene of cattle crossing water in *Nb.j-m-ꜣḫtj*,[234] The consistent portrayal of the phases in the life cycle of cattle with herders and supervisors in attendance are first seen in mid Fifth Dynasty tombs. Only seven of the 33 tombs of this early period present features of cattle husbandry, and in only one tomb, that of *Jj-mrjj,* are a number of these features depicted. *Jj-mrjj*'s chapel is exceptional for its time. Although *Jj-mrjj* was not a member of the royal family or even an official of the highest rank, his chapel consists of a number of decorated rooms. Cattle are portrayed on five walls of this chapel, with an extensive scene of four registers depicting most aspects of cattle management on the south wall of the second chamber. At a time when *Jj-mrjj,*was building and extending his tomb, varied and extensive treatment of 'daily life' features was more likely to be seen in Saqqara.

Later Dynasty 5

Cattle husbandry is depicted in the following tombs:

GIZA	
Jꜣsn	*Snḏm-jb/Jntj*
Nfr-bꜣw-Ptḥ	*Kꜣ.j-m-nfrt* LG 63
Snḏm-jb/Mḥj	*Kꜣpj*
SAQQARA	
ꜣḫtj-ḥtp D64	*Nbt*
Jrj-n-kꜣ-Ptḥ	*Rꜥ-špss*
Jrw-kꜣ-Ptḥ	*Ḥtp-ḥr-ꜣḫtj*
Jḥjj (*r/u Jdwt*)	*Sḫm-ꜥnḫ-Ptḥ*
Ptḥ-ḥtp	*Kꜣ.j-m-ꜥnḫ*
Ftk-tꜣ	*Kꜣ.j-m-nfrt*
Mꜣ-nfr	*Ṯjj*
DESHASHA	
Jntj	
EL-HAWAWISH	
Ḥzjj-Mnw	
ZAWYET EL-MAIYETIN	
Ḫw-ns	

Distribution of Husbandry Images (Figures 13, 14, 15, 16, 18, 19 and 20)	*Tombs*
Milking	15
Care of calves	10
Fattening oxen	10
Birth of calf	9
Crossing/emerging from waterway	7
Copulation	5
Cow suckling calf	5
Bullfight	1

By the second half of Dynasty 5 all the main themes used in the portrayal of cattle husbandry had been developed. Bulls are depicted sniffing cows in oestrous as well as

232 Every birth is attended by a herder shown helping to release the calf's head from the mother, who is clearly in need of help. A second herder sometimes stands in front of the cow uttering words such as 'release' or 'relax'. Occasionally a master drover stands by to give advice. These vignettes suggest both the importance attached to successful cattle births and selective cattle breeding, with particularly large bulls chosen to father large calves.

233 The handling of the calves may have been to accustom them to human contact or to select those suitable for stable rearing and slaughter.

234 LD II 12.

copulating[235] and selective breeding is indicated. In both *3ḫtj-ḥtp* and *Sḫm-ʿnḫ-Ptḥ* a herder brandishes a stick attempting to control a bull mounting a calf.[236] In two scenes depicting instances of bulls engaged in anovaginal sniffing, a prelude to copulation, a herder is warned to separate the animals – *Sšsšt/Jdwt: [ḥnt jw3t wn] m wp[.k] k3 [p]n* ('[Tether the cow quickly] as you separate this bull!') and *Ṯtw/K3.j-nswt: jj mnjw pw ḥnt jw3t wn m wp.k k3 pn* ('O this herdsman, tether the cow quickly as you separate this bull!').[237] The birth of calves is always attended by at least one herder. A herder is depicted crouching to receive the calf as it drops from its mother, often with an overseer standing behind or in front of the cow to supervise and check on the condition of the cow. Occasionally a third figure in front of the cow stretches out a hand to control and soothe the animal.[238] Calves are portrayed suckling and being weaned.[239] The process of selecting calves for stable rearing and fattening for slaughter or to accustom them to human contact is depicted.[240] Milking receives close attention, showing the technique used to control the reluctant cow and its attendant calf.[241] Recumbent cattle, almost always labelled *jw3* or *rn jw3*, are depicted being fed by hand by their herder who sometimes appears to push the feed into the animal's mouth.[242] The only bull fighting scene occurs in the el-Hawawish tomb of *Ḥzjj-Mnw* (M22), dated by Kanawati to later Dynasty 5.[243] That this early attestation of the theme occurs in el-Hawawish is perhaps not surprising in the light of the interest in the feature manifested by the nomarchs of Upper Egypt Nome 9 later in Dynasty 6.[244]

Early Dynasty 6

Cattle husbandry is depicted in the following tombs:

GIZA

Jdw	*Ṯtw/K3.j-nswt*
Mrjj-Rʿ-nfr/Ḳ3r	

SAQQARA

Jnw-Mnw	*Mṯṯj*
ʿnḫ -m-ʿ-Ḥr	*Nj-k3w-Jzzj*
Mrw/Ttj-snb	*Ḥzj*
Mrrj	*K3-gmnj*
Mrrw-k3	

DESHASHA

Jttj-Šdw

EL-GEBRAWI

Ḥm-r3/Jzj (I)

Distribution of Husbandry Images (Figures 13, 14, 15, 16, 18, 19 and 20)	*Tombs*
Milking	8
Crossing/emerging from waterway	10
Care of calves	5
Fattening oxen	5
Birth of calf	4
Copulation	3
Bull fighting	1

While most features of cattle husbandry appear in at least one or two of the great multi-roomed tombs of the Teti Cemetery in early Dynasty 6, the significant loss of decoration makes it difficult to judge whether all the powerful officials at this time considered cattle husbandry important and if so, which features they wished to portray. However, ten of the tombs of this time include detailed scenes of the herd of cattle swimming across a waterway, which contrasts with the frequency of late Dynasty 5 tombs, when only eight out of 58 tombs contain this scene. The higher frequency of this portrayal in early Dynasty 6 may have been due to an increase in this activity.

Mid Dynasty 6 to Dynasty 8

Cattle husbandry is depicted in the following tombs:

Jbj	Deir el-Gebrawi	VI.M
Ppj-ʿnḫ-ḥrj-jb	Meir	VI.M
Ḫntj	el-Khokha	VI.M
Sšm-nfr/Jwfj	Saqqara	VI.M
Jḥjj	el-Khokha	VI.L
Ppj-ʿnḫ/Ḥnj-km	Meir	VI.L
Ḥzjj-Mnw/Zzj	el-Hawawish	VI.L
Ḫnj/Špsj-pw-Mnw	el-Hawawish	VI.L
K3(.j)-ḥp/Ṯtj-jkr	el-Hawawish	VI.L
Ḏʿw	Deir el-Gebrawi	VI.L
ʿnḫ-tj.fj	Moʿalla	VIII
W3hj	el-Hagarsa	VIII
Mrjj-ʿ3	el-Hagarsa	VIII
Sbk-ḥtp	Moʿalla	VIII
Gḥs3/Nbjj	el-Hawawish	VIII

Distribution of Husbandry Images (Figures 13, 14, 15, 16, 18, 19 and 20)	*Tombs*
Bull fight	8
Crossing/emerging from waterway	8
Copulation	5
Birth of calf	3
Cow suckling calf	3
Milking	3
Fattening oxen	?2

From the second half of Dynasty 6 onwards, depictions of cattle husbandry are drawn from the tombs of Upper Egyptian provincial administrators. Despite the heightened degree of interest shown in the size of herds

[235] Macramallah (1935) pl. 7; Kanawati–Abder-Raziq (2003) pl. 54.
[236] *3ḫtj-ḥtp* Ziegler (1993) pp. 130-31; *Sḫm-ʿnḫ-Ptḥ* Simpson (1976) pl. C.
[237] *Sšsšt/Jdwt* Macramallah (1935) pl. 7; *Ṯtw/K3.j-nswt* Simpson (1980) fig. 22.
[238] For example, *Jj-mrjj* Weeks (1994) fig. 34; *Nj-ʿnḫ-ḫnmw* and *Ḫnmw-ḥtp* Moussa/Altenmüller (1977) pl. 79.
[239] For example, *Jj-mrjj* Weeks (1994) fig. 34; *Ṯntj* LD II pl. 31 [a]; *Snḏm-jb/Jntj* LD II pl. 77, Davies *Ptahhetep II* (1901) pl. 32.
[240] Calves may be tethered or held by a herder from whom they try to pull away: *Ṯjj* Wild II (1953) pl. 124; *3ḫtj-ḥtp* Davies *Ptahhetep II* (1901) pl. 4, 11; *Ptḥ-ḥtp* (II) *Ṯfj* Paget/Pirie (1898) pl. 31-32.
[241] *Ṯjj* Wild II (1953) pl. 124; Davies *Ptahhetep* II (1901) pl. 17.
[242] Sometimes this scene is divided into two vignettes: in the first the herder encourages the animal to lie down by pulling on a rope attached to the its lower jaw, while the second vignette depicts the actual feeding: *Jj-mrjj* Weeks (1994) fig. 31. The purpose of this time consuming practice is occasionally made explicit with a one-word caption, *wš3* ('fattening'). Whether these scenes are intended to portray an actual method of fattening cattle is very doubtful. See Evans (2006) 129-134.
[243] Kanawati IV (1983) fig. 8, 11.
[244] See Chapter 5, pp. 51-52.

by these officials, detailed vignettes of many aspects of the cattle husbandry theme are limited. At Meir and Deir el-Gebrawi even the scene of cattle crossing a waterway has limited treatment, although the traditional portrayal with boat of herders accompanying the swimming herd is maintained at el-Hawawish and el-Hagarsa into Dynasty 8. On the other hand, there is a marked increase in the portrayal of pairs of bulls fighting each. At el-Hawawish, the nomarch *Ḫnj/Špsj-pw-Mnw* devoted two registers to the theme, while *K3(.j)-ḥp/Ṯtj-jkr* gave it an entire panel of registers.[245]

FINDING

Few images represent details of husbandry in the early Old Kingdom tombs; certainly there are no sustained scenes of cattle management until the tomb of *Jj-mrjj* dated early to mid Dynasty 5. Images and scenes portraying aspects of cattle management appear much more frequently in later Dynasty 5, in 23 tombs of this period. At this time birth and care of calves, milking and the fattening of oxen are the most frequently depicted features. Captions and images suggest that estates carried out a policy of controlled breeding, one aspect of which was mating cows with the largest and most powerful bulls. The birth of particularly large calves is reflected in the scenes where a second and sometimes a third senior herder supervises the birth and in captions suggesting the distress of the parturient cow or the size of the calf.[246]

The daily management of the herd generally receives less attention in Dynasty 6. If scenes of the herd crossing water and pairs of bulls fighting are excluded, other husbandry features only appear as brief vignettes in 19 of the 44 tombs. In the chapel of *Jbj*, for example, a bull is depicted mounting a cow and a second cow is being milked, but both vignettes are on registers in which a first animal is perhaps being presented to the tomb owner. The text above the ox on the lower register reads *jrt jrw mḏwt* ('making the inventory of the cattle stalls'), which suggests a cattle count rather than husbandry.[247]

4.5 Herders of the estate[248]

The big estates maintained a hierarchy of farm workers to care for the herds of cattle, and the expertise of the overseers and herders is implicitly acknowledged in tomb iconography portraying scenes of herders at work in a variety of activities. Husbandry scenes portray overseers who, leaning on a staff and often dressed in a kilt with a projecting fringe,[249] give directions to subordinates to separate animals about to copulate,[250] supervise and advise herders attending the birth of calves[251] and the milking of cows.[252] They are also depicted in charge of the herd as it swims in deep water or leaves the marshes for higher land. There is no evidence that horses had been introduced into Egypt at this time or that trained dogs were used to control and direct herds. Consequently, if the size of herds approached the figures cited in tombs, considerable numbers of herders would have been needed on a permanent basis to manage the herd and to travel with the animals on their annual movement to and from the marshland. Both Montet and Ghoneim were of the opinion that herders travelled and lived with the herds for considerable periods of time.[253]

The herders' skill and dedication are implicit in the portrayal of husbandry details and the accompanying captions, such as overseers giving advice to herders dealing with the birth of a calf. In the tomb of *Jttj-Šdw* the appreciation of the herders' expertise is explicit.[254] On the west wall of the chapel the column of inscription dividing the figure of the tomb owner from registers of cattle husbandry reads:[255]

> *m33 k3(w) ḥmt nt ḫt.f rḫt n k3(w).f n sḏm mdw nrw nw pr n ḏt.f*
>
> 'Viewing the bulls and cows of his property and the number of all his bulls, and hearing the speech of the herders of his estate.'

The term 'hearing (or 'listening to') the speech of the herders' suggests *Jttj-Šdw*'s approval. The vertical inscriptions in front of the tomb owner viewing scenes of activities imply that these sights are pleasing to him and are sometimes explicit in this respect with phrases such as "*Viewing the work of ... and every good thing ...*".[256] The reference to the speech of the herders therefore suggests that the words of the herders also please the tomb owner. An extended caption above a line of cattle on the third register facing the tomb owner is even more affirmative:

> *mk k3w.k msw k3w.k n [ḫt?].k ...3 ms sn ntj m sḫwt r sḫ3 n[.k?] bnrw nḥb nrw*
>
> 'Behold, your oxen and the offspring of your oxen, of your property, and [the master drover?] who delivered them, who was in the marshlands, to

245 Respectively, Kanawati II (1981) fig. 20; Kanawati I (1980) fig. 10.

246 Evans (2006) 474-482.

247 Kanawati *Gebrawi* II (2007) pls. 54, 73.

248 See Ghoneim (1977) 39-42.

249 Men wearing this style of kilt have a variety of 'overseer' titles. (See Brovarski (2001) 58, n.310)

250 *3ḫtj-ḥtp* Ziegler (1993) 80 (plate), 81; *Jrj-n-k3-Ptḥ* Moussa–Junge (1975) 41, pl. 13; *K3.j-m-nfrt* Simpson (1992) 19, pl. 22, pl. G; *Sḫm-ʿnḫ-Ptḥ* Simpson (1976) 8, pl. 6, pl. C; *Ppj-ʿnḫ-ḥrj-jb* Blackman *Meir* IV (1924) 38, pl. 14.

251 *J3sn* Simpson (1980) pls. 43 [b], 44 [b], fig. 30; *K3.j-m-ʿnḫ* Junker, *Gîza* IV (1940) pl. 2, 12 [lower], Kanawati *Giza* I (2001) pl. 13 [b], 36; *Jj-mrjj* LD. *Erg*. 7, Weeks (1994) fig. 34, pl. 17 [c], 19 [b]; *Sḫm-ʿnḫ-Ptḥ* Simpson (1976) pl. 6, pl. C; *K3.j-m-nfrt* Simpson (1992) pl. 22, pl. G; *Ṯjj* Wild II [1] (1953) pl. 92 [A] (part), 124.; W*ʿtt-ḫt-Ḥr* Kanawati–Abder-Raziq (2008) pl. 56; *3ḫtj-ḥtp*: Davies (1901) pl. 13; *Jḥjj* (r/u *Sšsšt/Jdwt*) Macramallah (1935) pl. 7, Kanawati–Abder-Raziq (2003) pl. 54; *Nbt* Munro (1993) pl. 13; *3ḫtj-ḥtp* Ziegler (1993) pp. 80-1 (plates) 81; *Nj-ʿnḫ/Ḫnmw* & *Ḫnmw-ḥtp* Moussa–Altenmüller (1977) pl. 77, 79 [a]; *Mṯtj* Ziegler (1990) pp. 129-30, 143.

252 *Ḫnmw-ḥt* Harpur (1985) fig. 8, Sharawi/Harpur (1988) 64-6 and n. 21; *Ptḥ-špss* Verner (1986) photo 71, pl. 44; *Nj-ʿnḫ/Ḫnmw* / *Ḫnmw-ḥtp* Moussa–Altenmüller (1977) pl. 76; *Jrj-n-k3-Ptḥ* Moussa–Junge (1975) pl. 13; *Ṯjj* Wild II [1] (1953) pl. 124; *Sḫm-ʿnḫ-Ptḥ* Simpson (1976) pl. 6, pl. C; *Ḫw-ns* LD II 106 [a]; *Nbt* Munro (1993) pl. 21; *3ḫtj-mrw-nzwt* Smith (1946) 208, fig. 79; *Jntj* Petrie (1898) 7, pl. 5; Kanawati–McFarlane (1993) 28, pls. 8, 33; *Nj-k3w-Jzzj* Kanawati–Abder-Raziq (2000) pl. 49.

253 Montet (1924) 110; Ghoneim (1977) 68.

254 Kanawati–McFarlane (1993) pl. 46.

255 Kanawati–McFarlane (1993) pl. 51.

256 *Snḏm-jb/Mḥj* Brovarski (2000) figs. 97, 98; LD *Erg*, 12.

recall/report to you the pleasant (matters) which the herders provided.'[257]

Herders are portrayed, wearing attire that ranges from a short, tight fitting wrap-around kilt to a garment that partially covers the lower half of the trunk or just a strip of material that is scarcely more than a belt around the waist. The men depicted wading through water or handling calves are sometimes shown naked. Most herders are clean-shaven, but a few bearded figures have been identified as 'marsh-dwellers' whose regular habitat was the low-lying marshlands.[258] The equipment and activities of the herders indicate that they travelled and lived with the herd[259] and the impression conveyed is that the herds were never left unattended. More than any other group of workers such as ploughmen, harvesters and herdsmen parading desert animals, many of those tending cattle are portrayed as older men who are partly bald, rarely wear a wig and sometimes have the sagging stomach muscles of the elderly. In particular, overseers are often portrayed as older men, which suggests that experience and practical knowledge were particularly valued in cattle management. The impression that the skills of herders were appreciated is reinforced by scenes in the tomb of *Ṯjj* where those with the delicate task of handling young calves, perhaps to judge which are to be stabled and hand-fed, are all portrayed as elderly.[260]

4.6 Cattle in the parade of animals (Figure 10)

In this study the terms 'parade of livestock' and 'parade of animals' denote a line of animals, each normally led or guided by a herder. The 'line' may be depicted on one or more registers and usually includes domesticated cattle. The animals are sometimes portrayed being presented to the tomb owner, whose majestic figure matches the height of the entire panel of registers containing the images of animals and other daily activities. This study also uses the term 'parade of animals' to denote just one or two animals depicted on a narrow register or the thickness of a doorway, as these are also directed towards a figure of the tomb owner. Occasionally, individual cattle in a line of animals are depicted without a herder or tethered to a hoop driven into the ground or being led in groups of two or three. Unlike many depictions of desert animals showing herders struggling for control, domestic cattle are rarely represented balking or giving their herder trouble. Cattle, mainly oxen, are led by a herder walking ahead of the animal and controlling it by holding in one hand a rope attached to the animal's collar or lower jaw. A second herder may follow the animal guiding it with one hand and holding the rope of a following animal. Sometimes, an ox walks forward in the line of animals without any human control, or a herder may lead three oxen holding their ropes in one hand.

4.6.1 Historical development of images and themes

Domesticated cattle, almost always oxen (castrated males),[261] are depicted in parades in the following number of tombs:

Numbers of tombs portraying cattle in a parade of animals:

Dynasty	*Memphite*	*Provincial*
IV.E	5	-
IV.L	6	-
V.E	9	2
V.M	13	1
V.L	34	2
VI.E	10	4
VI.M	-	8
VI.L	-	7
VIII	-	6

Dynasty 4 to early Dynasty 5

Cattle are depicted in parades in the following tombs:

Jtt	Medum	IV.E
Mṯn	Saqqara	IV.E
Nfr-m3ꜥt	Medum	IV.E
Rꜥ-ḥtp	Medum	IV.E
Nfr-m3ꜥt	Giza	IV.M
Sš3t-ḥtp/Htj	Giza	IV.M
Mr-jb.j	Giza	IV.L
Mrs-ꜥnḫ III	Giza	IV.L
Nj-k3w-rꜥ	Giza	IV.L
Nfr	Giza	IV.L
Dbḥnj	Giza	IV.L
Wḥm-k3	Giza	V.E
Pr-sn	Giza	V.E
Nswt-nfr	Giza	V.E
Ḥmt-Rꜥ	Giza	V.E
Snb	Giza	V.E
K3.j-nj-nswt I	Giza	V.E
Ṯntj	Giza	V.E
K3(.j)-ḫnt (A2)	el-Hammemiya	V.E
K3(.j)-ḫnt (A3)	el-Hammemiya	V.E

The animals are either led by a herder, walk unattended or are occasionally depicted tethered. Almost all are oxen (castrated males). They may appear as a single animal on a narrow register or in a line of animals approaching the figure of the tomb owner.

Despite the availability of wall space in the cruciform chapels of early Fourth Dynasty tombs, pride of place in the wall panels exhibiting scenes of 'daily life' is given to the hunting of desert animals and, to a lesser extent, to marshland activities. Cattle, either being presented to the

257 Kanawati–McFarlane (1993) pl. 51.
258 Brovarski (2000) p. 44 and note 104.
259 In the panel of registers of cattle husbandry in *Jj-mrjj*, the top register portrays herders working on a woven mat, making rope, baking bread and roasting fowl; all activities that would be needed of men preparing to spend day and night with the herd. LD *Erg*. 7; Weeks (1994) fig. 34, pls. 17 [c], 19 [b]. The caption above the third register of the west wall of the tomb of *Jttj-Šdw*, quoted above, states that the master-drover accompanied the herd to the marshland.
260 Wild *Ti* II (1953) pl. 124.

261 With the exception of the herd depicted in the tomb of *Snb* (Junker V (1941) fig. 19) and a bull in *K3-pw-nswt/K3j* (Junker III (1938) fig. 18) all appear to be oxen.

tomb owner or as working animals or in images of husbandry, are not the most prominent element.[262] Cattle depictions are of single animals perhaps intended to represent an entire herd. The chapels of *Jtt* of Medum and *Nfr-m3ʿt* of Giza contain the earliest extant portrayals of the parade of animals. *Jtt*'s scene, on two registers, appears to favour desert animals over cattle.[263] The scene in the chapel of *Nfr-m3ʿt* of Giza gives a similar impression. Here the animal parade occupies three registers, partly lost.[264] In *Sš3t-ḥtp/Ḥtj* the comparison is even more pronounced. The desert ungulates are controlled by a herder holding their horns and muzzle, while the ox, led on a rope by an elderly herder, follows passively.[265]

Perhaps because she was wife of a king, *Mrs-ʿnḫ* III's multi-roomed chapel contains the earliest extant parade of animals on one extended register. Each of the four cattle is distinctively depicted, although all appear to be oxen. The first two have lyre-shaped horns but different markings; the third animal is polled and the fourth, half-grown, is dappled and has a blanket over its back. Each animal is led by an individually depicted herder and described by a separate caption.[266]

Mid Dynasty 5

Cattle are depicted in the following tombs either led by a herder or walking unattended or tethered, either singly or in a line of animals:

GIZA	
Jj-mrjj	*Ḫntj-k3w.s*
Njwtj	*Sšm-nfr* II
Rʿ-ḫʿ.f-ʿnḫ	*K3.j-m-nfrt*
Rʿ-wr I	
SAQQARA	
Wr-jr.n-Ptḥ	*Nfr* and *K3-ḥ3.j*
ABUSIR	
Ptḥ-špss	

By mid Dynasty 5 Memphite officials were building funerary chapels with greater interior wall space,[267] which allowed development of the pictorial representation of themes dealing with their resources, a trend reflected in the few decorated provincial tombs of this time.[268] The earliest tomb to present this new trend is that of *Jj-mrjj* where a parade of animals appears on three walls in two separate rooms: the east wall of Room 1,[269] the west wall,[270] and the east wall of Room 3.[271] Above the animals in Room 3 is the statement *jnt jw3 m nḏ.t-ḥr r prt-ḫrw* ('Bringing *jw3*-oxen as gifts for invocation offerings)[272]. Yet another parade of cattle is depicted on the south wall of the second room. Although this panel of registers is devoted to husbandry, on the second register four animals are led forward; all are oxen described as *rn jw3*. While the last animal is badly damaged, the first three are distinctively portrayed. The first animal is lyre-horned, the second polled and the third has deformed horns that curl under.[273]

In this period (early to mid Dynasty 5) a number of features are introduced into the scene of the presentation of cattle or become more frequent. Polled cattle are depicted in *Wr-jr.n-Ptḥ*[274] and *K3-pw-nswt/K3j*;[275] with three polled animals led together by one herder in *Njwtj*, *Nfr* and *K3-ḥ3.j* and *Sšm-nfr* II.[276] Curiously, an animal with crescent shaped horns is depicted in both the chapels of *K3(.j)-ḥnt* (A2) and *K3(.j)-ḥnt* (A3) of el-Hammemiya,[277] although crescent shaped horns do not appear in the capital until the end of Dynasty 5. Overlapping lines of cattle are portrayed for the first time in the chapels of *Wḥm-k3*, *Snb* and *Rʿ-ḫʿ.f-ʿnḫ*, the latter two including a caption giving herd numbers. The oxen depicted being led by herders are labelled *jw3* or *rn jw3*, and the tethered animals *ng3.w*, perhaps suggesting that the *ng3.w* cattle were not usually trained to be led by a rope. On the north wall of *K3-pw-nswt/K3j* an obstreperous bull labelled *rn mrw* has to be controlled by a herder wielding a stick.[278] Of the four ungulates on this panel, three (two desert animals and the bull) are clearly difficult for their herders

262 These generalisations are drawn from a very limited number of tombs, those of *Rʿ-ḥtp* and *Nfr-m3ʿt* of Medum (Petrie (1892) and *Mtn* of Saqqara (LD II 3-7).

263 Petrie (1892) pl. 27. While the first animal in the upper register is lost, all the other animals appear to be desert ungulates apart from one dappled steer. Herders are depicted leading and controlling the desert animals, but the solitary ox walks on without herders controlling it.

264 *LD II* pl. 17b. One desert ungulate is tethered and the other is balking, but the ox on the register above gives no trouble to its herders.

265 Junker *Giza* II (1934) fig. 28; Kanawati *Giza* II (2002) fig. 45. The herder leading the ox is partly bald, wears the kilt of an overseer of herders and is labelled *ḥk3* ('chief').

266 The first two herders wear similar kilts and wigs, the second figure turns back towards his charge. The third animal is led by an emaciated, bearded and balding figure: over the first animal is *jtt jw3 sḫt* ('bringing an ox of the field') and over the second, *jtt jmj mḏt* ('bringing [one] that belongs to the stalls'). The third polled animal is described as *jw3t nt nḏ.t-ḥr* ('oxen of offering'), and the fourth animal, also led by a similar figure but wearing the kilt of an overseer of cattle, is captioned *jtt bḥs šdj ḥr ḏbʿ* ('bringing a calf suckled on the finger'): Dunham-Simpson, (1974) 10-11, fig. 4.

267 Dodson–Ikram (2008) 173-174.

268 At el-Hammamiya, for example, a father and son, both named *K3,j-ḫnt*, depict in their chapels a parade of cattle on three successive registers: El-Khouli–Kanawati (1990) *K3,j-ḫnt* (A2), pl. 45 and *K3,j-ḫnt* (A3), pl. 68.

269 Weeks (1994) fig. 27. A large figure of the tomb owner faces four registers, on the second of which a polled ox is, unusually, portrayed balking. On the register below a herder leads two oxen labelled *rn jw3*, behind which is another tethered ox labelled *rn ng3.w*.

270 Weeks (1994) fig. 31. A major figure of *Špss-k3.f-ʿnḫ*, standing in front of his son *Jj-mrjj*, faces four registers, with images of cattle on the top two. The first register presents three polled oxen led(?) by a single herder and captioned *jnt nḏ.t-ḥr jw3* ('bringing gifts of *jw3*-oxen'), while a tethered *rn ng3.w* and a [*rn*] *jw3* led by a herder occupy the second register,

271 Again the animal captioned *ng3.w* is tethered, while the two being led are described as *jw3*.

272 Weeks (1994) fig. 41. *Jj-mrjj* and his family view another panel of registers, the third of which is devoted to the presentation of cattle.

273 Weeks (1994) fig. 34.

274 James (1961) pl. 30.

275 Junker *Giza* III (1938) fig. 18.

276 Respectively, LD. *Erg*. pl. 30; Moussa–Altenmüller (1971) pl. 6; Kanawati *Giza* II (2002) pl. 61.

277 El-Khouli–Kanawati (1990) pl. 45, 68.

278 Junker *Giza* III (1938) fig. 18.

to control, while the fourth, a smaller polled animal, labelled *jwꜣt* (presumably the collective noun as the animal appears to be an ox) walks peacefully behind its herder.

Five tombs at the capital, which are dated to this period, provide labels for varieties of cattle. Cattle in parades are more often labelled *jwꜣ* than *ngꜣw*. Portrayed being led as a group in the chapel of *Sšm-nfr* II, three polled cattle are labelled *ḥr(j) ḏbꜥ*.[279] The distribution of distinguishing labels is as follows:

jwꜣ:	*Jj-mrjj, Wr-jr.n-Ptḥ, Njwtj, Sšm-nfr II, Kꜣ.j-m-nfrt*
ngꜣw:	*Jj-mrjj, Kꜣ.j-m-nfrt*
ḥr(j) ḏbꜥ:	*Sšm-nfr* II

Late Dynasty 5

Cattle are depicted in the following tombs either led by a herder or walking unattended or tethered, singly or in a line of animals:

GIZA	
Jꜣsn	*Sḫm-ꜥnḫ-Ptḥ*
Rdj	*Sḫm-kꜣj*
Nfr (1)	*Sšm-nfr* III
Nfr-bꜣw-Ptḥ	*Sšm-nfr* IV
Nfr-ḫw(wj)	*Štwj*
Nḫt-kꜣ.j	*Kꜣ.j- m-ꜥnḫ*
Rꜥ-wr II	*Kꜣ.j-dwꜣ*
Snḏm-jb/Mḥj	*Kꜣpj*
Sḥtpw	
SAQQARA	
ꜣḫtj-ḥtp (D 64)	*Nbt*
ꜣḫtj-ḥtp (Louvre)	*Nfr-jrt-n.f*
Jj-nfrt	*Rꜥ-m-kꜣ.j*
Jḥjj (r/u *Jdwt/Sšsšt*)	*Rꜥ-špss*
Wr-jr.n.j	*Ḥtp-ḥr-ꜣḫtj*
Pr-nb	*Ḥtp-kꜣ.j/Tp-kꜣ.j*
Pḥn-wj-kꜣ.j	*Ḫnwt*
Ptḥ-ḥtp	*Sḫm-ꜥnḫ-Ptḥ*
Ptḥ-ḥtp (II)/*Ṯfj*	*Kꜣ.j-m-nfrt*
Ptḥ-ḥtp/Jj-n-ꜥnḫ	*Kꜣ.j-m-rḥw*
Mꜣ-nfr	*Ṯjj*
DESHASHA	
Jntj	
EL-HAWAWISH	
Ḥzjj-Mnw	

Paraded cattle appear in 41 of the 73 late Dynasty 5 tombs used in this study. Registers with a line of animals have now become commonplace as lines with three or more cattle appear in 24 of the tombs. In addition, seventeen tombs represent paraded cattle on multiple registers, not always in the same panel of scenes. Most paraded cattle are oxen led by a herder and are depicted with neither udders nor testicles. Yet no two scenes of parades are identical, as the animals are depicted with a variety of distinguishing features, which may vary for each animal in a parade line. Eleven tombs show paraded cattle with at least five of twelve identified features:

• Leading rope fastened to lower jaw	33 tombs
• Rope or collar around neck	29 tombs
• Polled animal	21 tombs
• Leading rope fastened to lower jaw and collar	19 tombs
• Groups of two to five animals led by a single herder	14 tombs
• Tethered to stake or hoop in ground	11 tombs
• Animal with distorted horn	8 tombs
• Sash hanging from neck	8 tombs
• Animal struggling with herder	7 tombs
• Lotus or papyrus umbel attached to collar	5 tombs
• Blanket over animal's back	4 tombs
• Possible crescent shaped horns	2 tombs

Both polled and horned animals are shown being led by a single herder in groups of two to five animals, but the polled variety may have been more highly valued as only the smaller, presumably half grown polled animal is depicted with a blanket over its back. These animals have a rope collar and, occasionally, rope tied around their muzzle. Animals are perhaps marked out for impending ritual slaughter by a decorative sash that hangs from the neck, or a collar of 3 to 5 strands of rope finished with an attached lotus flower or papyrus umbel. In almost every depiction cattle are shown being led by a rope that attaches to their lower jaw. Although the rope is never portrayed going into the animal's mouth in the manner in which a bit goes into a horse's mouth, the most likely explanation is that these are stylised depictions and that the rope was passed through the animal's mouth.[280]One or more of the depicted animals in the parade may be labelled *jwꜣ, rn jwꜣ, ḥr ḏbꜥ, ngꜣ.w* or *rn ngꜣ.w*, The numbering of herds is not unknown but is a rare addition to the scene before mid Dynasty 6.[281]

Early Dynasty 6

Cattle are depicted in the following tombs either led by a herder or walking unattended or tethered, singly or in a line of animals:

SAQQARA	
Jnw-Mnw	*Nj-kꜣw-Jzzj*
Mrjj-Ttj/Mrj	*Nfr-sšm-Rꜥ*
Mrw/Ttj-snb	*Rmnj/Mrwj*
Mrrw-kꜣ	*Ḥzj*
DESHASHA	
Jttj-šdw	
GEBRAWI	
Hnḳw/Jj…f	
Hnḳw/H̱ttj	

Although 38 tombs are dated to early Dynasty 6, only thirteen tomb chapels provide scenes of a parade of animals, partly due to the loss of wall decoration as in the case of *Kꜣ-gm-nj* and *ꜥnḫ-m-ꜥ-Ḥr*. Perhaps owing to this loss of scenes, only two chapels, those of *Jnw-Mnw* and

279 Kanawati *Giza* II (2002) pl. 61.

280 Suggested by Linda Evans, 2nd January 2009.

281 Three Dynasty 5 tombs record herd numbers: *Rꜥ-ḥꜥ.f-ꜥnḫ, Nfr-bꜣw-Ptḥ* and *Rꜥ-wr* II.

Mrrw-kꜣ, have registers with more than two animals. Where scenes have survived, as in the tombs of *Mrrw-kꜣ*, and *Ḥzj*, the basic features of the parade of cattle depicted in late Dynasty 5 tombs do not change. However a singular image occurs on the west wall of the small chapel of *Rmnj/Mrwj*, recently discovered in the Teti cemetery at Saqqara. Facing the false door are images of herders leading three animals, two desert ungulates and an ox. All three animals have a decorated banner wrapped around their neck and the herder leading the ox, inscribed as *rn jwꜣ*, turns back to his charge holding it by horn and muzzle in the manner used to control fractious desert ungulates. No similar image has been found in any other tomb but it may have been included here to match the images above of the two desert ungulates.

Mid to late Dynasty 6

DEIR EL-GEBRAWI	
Jbj	*Ḏʿw*
MEIR	
Nj-ʿnḫ-Ppjj-km	*Ppjj-ʿnḫ/Ḥnj-km*
Ppjj-ʿnḫ-ḥrj-jb	
EL-HAWAWISH	
Jn-jt.f/Bḫnt	*Ḫnj/Špsj-pw-Mnw*
Bꜣwj (BA48)	*Ḥzjj-Mnw/Zzj*
Mmj	*Kꜣ-ḥp/Ṯtj-jkr*
Nbt	

Dynasty 8

EL-HAWAWISH	
Bꜣwj (BA14)	*Rḥw-r-ꜣw.sn*
EL-HAGARSA	
Wꜣhj	*Mrjj-ꜣʿ*
MOʿALLA	
ʿnḫ-tj.fj	

In the provincial tombs of Dynasty 6 different themes related to cattle are included on the same register. For example, in the chapel of *Jbj* a figure of the tomb owner faces a panel of 5 registers. On registers three and four the first animal appears part of a presentation scene, but is followed by husbandry vignettes. In a number of scenes a paraded animal may be depicted with crescent shaped horns. In *Jbj* an ox facing the seated figure of the tomb owner is portrayed with particularly large crescent-shaped horns[282] and the third paraded animal on the third register in *Ḫnj/Špsj-pw-Mnw* also appears to have crescent shaped horns.[283]

In the second half of Dynasty 6 a new tone develops in the captions accompanying parades. Inscriptions above groups of cattle citing very large herd numbers, rare in the late Fifth and early Sixth Dynasties, become almost a regular feature of the provincial parade scene in mid Sixth Dynasty.

[282] Kanawati *Gebrawi* II (2007) pl. 7.
[283] Kanawati II (1981) fig. 26.

4.7 Inscriptional material relating to cattle

Three kinds of inscriptions provide specific references to cattle herds and their management. Two of these accompany Old Kingdom scenes of 'everyday life': vertical columns of inscriptions separating the figure of the tomb owner from a panel of registers, and captions usually positioned above or beneath a scene. The vertical inscriptions, almost invariably introduced by *mꜣꜣ* ('viewing' or 'inspecting'), refer generally or specifically to the contents and activities depicted on the registers that face the large figure of the tomb owner. The lines of inscription included within the registers give the typical speech of working men and women, sometimes their names and titles or a brief description of the vignette. The third genre includes the autobiographies and personal statements inscribed in tombs.

4.7.1 Historical development

While the inscriptions of Dynasty 4 and the first half of Dynasty 5, introduced by *mꜣꜣ*, occasionally refer to specific resources, most are limited to general statements. Typical of these statements is:

mꜣꜣ sš [nḏ.t-ḥr jn]nt r prt-ḫrw m njwt nt pr ḏt
'Viewing the recording [of the gifts which are brought] as funerary offerings from the towns of (his) (funerary) estate.'[284]

The only vertical inscription with a specific reference to cattle occurs in the tomb of *Mr-jb.j*.[285] Towards the end of the Fifth Dynasty vertical inscriptions from four tombs[286] make more precise reference to cattle, such as:

(mꜣ)ꜣ jwꜣ ʿwt ḫꜣst ꜣpdw jnw n m njwwt pr-ḏt
'Viewing the cattle, desert animals and fowl brought by the towns of (his) estate.'[287]

mꜣꜣ jwꜣ n ḏḥwtt inn m ḥwwt-kꜣ njwwt nt pr-ḏt m Tꜣ-Mḥw Šmʿw
'Viewing the cattle of the Thoth festival brought from the properties of the ka and the estates belonging to the tomb of the North and the South country.'[288]

The vertical inscriptions in Memphite tombs of early Dynasty 6 retain the traditional general statements or make reference to a range of resources. Only two tombs give specific mention to cattle.[289]

In the tombs of the highest official in the Upper Egyptian provinces of Dynasty 6, however, there develops a marked trend to emphasise cattle. This applies to vertical inscriptions, captions and autobiographies. The first

[284] Weeks (1994) pl. 41.
[285] LD II pl. 22a. This inscription refers to cattle as invocations brought from the Residence (*pr-nswt*).
[286] These inscriptions occur in four tombs: *ꜣḫtj-ḥtp* Davies II (1901) pl. 29, 21; *Ptḥ-ḥtp* (II) (*Ṯfj*) Paget-Pirie–Griffith (1898) pl. 31, 33 (above registers five and six); *Mꜣ-nfr* LD II 68; *Nbt* Munro (1993) pl. 17, 19.
[287] *Sḫm-ʿnḫ-Ptḥ* Badawy (1976) fig. 20.
[288] *ꜣḫtj-ḥtp* Davies II (1901) pl. 20, 21.
[289] *Sꜣbw/Jbbj* Borchardt (1964) pl. 21; *Kꜣ-gm-nj* von Bissing I (1905) pl. 8-16.

intimation appears in the tomb of *Jttj/Šdw* of Deshasha. On the northern part of the west wall of this chapel a depiction of the tomb owner faces a panel of four registers all referring exclusively to cattle.[290]The inscription reads:

m33 k3 ḥmt nt ḫt.f rḫt n k3.f nb sḏm mdw nrw[291] *nw pr n ḏt.f*

'Viewing the bulls and cows of his property and the number of all his bulls and hearing the speech of the herdsmen of (his estate).'[292]

The horizontal inscription above the scene on the third register reads:

mk k3w.k msw k3w.k n [ḫt?].k … ms sn ntj m sḫwt r sḫ3 n[.k?] bnrw nḫb nrw

'Behold, your oxen and the offspring of your oxen, of your property, and [the master drover?] who delivered them, who was in the marshlands, reporting to you the pleasant (matters) that the herdsmen provided.'[293]

This entire wall is devoted to motifs that are emphasised as pleasures for *Jttj-Šdw*'s ka.[294] On the south side the scene of a flautist and harpist playing for the seated tomb owner is described:

'Behold, this entertainment(?) is for your ka.'[295]

The marked emphasis on cattle also appears at Deir el-Gebrawi early in the Sixth Dynasty. *Hnḳw/Jj…f*, who is dated to the first half of the reign of Pepy I[296] describes his administration of the province (Upper Egypt 12):

jw gr mḥ.n(.j) wḏbw.s m mnmnt mḫrw.s m ꜥwt jw gr ss3.n(.j) wnšw nw ḏw ḏrjwt nt pt m ḫ3w n ꜥwt j mr(.j) 3ḫ s ntj jm.s … 3ḫ(.j) n.s m z3w nw mnmnt grgwt nt wḥꜥw jw grg.n(.j) j3t.s nb m rmt mnmnt ꜥwt m bw m3ꜥ n ḏd(.j) grg jm

'Moreover I filled its (the province's) river banks with cattle and its pastures with small cattle. Furthermore, I satisfied the jackals of the mountain and the kites of the sky with the hides of small cattle, for I wished the man in it to benefit.'.....'I benefited it with stables for cattle and dwellings for trappers. I settled all its mounds with people, large and small cattle, and I do not tell lies therein.'[297]

The character of this increased interest in cattle in the inscriptions in the early Dynasty 6 provincial tombs is traditional: the portrayal of the tomb owner's oversight of good husbandry practices and the beneficial administration of the lands under his control. In the case of *Hnḳw/Jj…f*, cattle receive no more attention than other domesticated grazing animals. In mid Dynasty 6, however, the mood and emphasis of inscriptions relating to cattle change, as pride in the presentation of cattle and good husbandry give up their premier position to pride in the size of herds. On the west wall of his chapel a figure of *Jbj*, nomarch of U.E. 8 and 12, views five registers of livestock. The vertical inscription in front of *Jbj* reads:

(m)33 ꜥwt ḫ3st z3 nfr n k3w n k3 n

Viewing the desert animals and the beautiful herd of cattle for the ka of …[298]

Cattle, portrayed in vignettes of presentation and husbandry, occupy the third, fourth and half the fifth register.[299] On the third register a scribe writes on his scroll:

sꜥꜥ n k3w ꜥwt 32 400

'Statement of bulls (cattle) and small cattle, 32 400.'

In an autobiographical statement *Jbj* refers to the grant of 203 arouras of land, equipped with cattle and workers, with which the king rewarded him (Urk. I. 144-145):

ḫr mrt nt ḏt(.j) mḥ m k3w m ꜥnḫwt m ꜥ3w

'equipped with *mrt*-workers of my property and filled with bulls, goats and donkeys'.[300]

The emphasis on herd numbers appears in other provinces of this time. At Meir, *Ppj-ꜥnḫ-ḥrj-jb*, a contemporary of *Jbj*,[301] recorded an enormous count of cattle totalling 767 777, with the inscription, *'Exaction of the impost of cattle and goats due from the middle nomes.'*[302] The autobiography of *Ḳ3r* of Edfu, also a contemporary, reflects this emphasis by stressing the growth of the cattle industry at Edfu under his leadership:

jw rdj.n(.j) wn jw3w nw sp3t tn r-ḥr n jw3w mḏ3w(.j) m ḥ3t Šmꜥw mj ḳd.f n gmt.n(.j) js pw m-ꜥ ḥrj-tp wn m sp3t tn tp-ꜥwj n rs-tp(.j) n mnḫ ḫrp(.j) ḫt n ḫnw

'I made (it so) that the bulls of this nome were the best bulls and my cattle sheds were at the head of the whole of Upper Egypt. It is certainly not a thing which I found achieved by the chief who was in this nome before me, because I was vigilant and because effectiveness held sway when I directed matters for the Residence.'[303]

4.8 Herd numbers

TABLE 5: LIVESTOCK NUMBERS

Jbj, Ppj-ꜥnḫ-ḥrj-jb and *Ḳ3r* were newly appointed to their respective provinces by Merenre as resident administrators and their immediate successors maintained their emphasis on herd numbers. At Deir el-Gebrawi the tomb of *Jbj*'s son and grandson, *Ḏꜥw/Šm3j* and *Ḏꜥw*, recorded a herd of perhaps 24,000.[304] *Ppj-ꜥnḫ-ḥrj-jb*'s

290 In the top register a scribe may be recording the varieties and numbers of cattle, but only the varieties remain visible. Kanawati–McFarlane (1993) pl. 51.

291 For this transliteration see Montet (1925) 99ff and Kanawati–McFarlane (1993) 54.

292 Kanawati–McFarlane (1993) 51.

293 *mk k3w.k msw k3w.k n [ḫt?].k …3 mz sn ntj m sḫwt r sḫ3 n[.k?] bnrw nḫb nrw* (ibid, pl. 51).

294 The west wall of the chapel north and south of the shrine.

295 *m n3 bw nfr n k3.k* Kanawati–McFarlane (1993) pl. 50,

296 Kanawati *Gebrawi* I (2005) 12-20.

297 Kanawati *Gebrawi* I (2005). (Urk. I 77:10-12; 78:17-79:1).

298 Davies *Gebrâwi* II (1902), pls. 9, 11-22; Kanawati *Gebrawi* II (2007) pls. 52, 71.

299 The remaining registers contain scenes of a desert hunt and a herd of goats.

300 Davies *Gebrâwi* II (1902) pl. 7; Kanawati *Gebrawi* II (2007) pls. 54, 73.

301 For this dating see El-Khouli–Kanawati (1990) 11ff.

302 Blackman *Meir* IV (1924) pl. 16.

303 Urk. I. 254:3-11.

304 Davies (1902) pl. 9. The number of 24,000 is uncertain owing to a lacuna.

successor at Meir, *Ppj-ʿnḫ/Ḥnj-km*, recorded herd numbers on walls of two rooms in his chapel. On the east wall of Room B the number 1000 is written above an ox (*ngꜣ.w*) while beneath the seated figure of the tomb owner the caption '*counting the large and small cattle*' is written beside a figure who seems to be pointing across the intervening doorway to the registers of livestock. On the south wall of Room F the seated figure of *Ppj-ʿnḫ/Ḥnj-km* inspects four registers of herds of goats and cattle '*from his own settlements*' (*m njwwt.f*) with numbers inscribed above the animals.[305]

The tomb of another of *Jbj*'s contemporaries, *Kꜣ.j-ḥp/Ṯtj* who was appointed nomarch of U.E. Nome 9 by Merenre, has lost most of its decoration, but *Kꜣ.j-ḥp/Ṯtj*'s successors, *Kꜣ(.j)-ḥp/Ṯtj-jkr* and *Ḥnj/Špsj-pw-Mnw*, both record herd numbers. On the east wall of his chapel *Kꜣ(.j)-ḥp/Ṯtj-jkr*, whom Kanawati dates to mid Pepy II, recorded numbers for both cattle and desert ungulates.[306] In a comparable position in the chapel of *Ḥnj/Špsj-pw-Mnw* the number '10,000' is inscribed above the depictions of a gazelle and two separate oxen, while the same number is recorded above an oryx and an ox on the register beneath, making a total of 30,000 head of cattle and 20,000 desert ungulates.[307] All that remains of the vertical inscription in *Kꜣ(.j)-ḥp/Ṯtj-jkr*'s tomb is … *jrw m kꜣw* …., but Kanawati suggests that it would have originally read *'(Viewing the exaction) of the cattle tax of bulls …*', indicating that the cattle and goats depicted on the opposing third register represent an official census.[308]

It is not easy to judge whether the emphasis was on the increase of cattle or of livestock in general:

- *Ḥnkw/Jj□f* (II) and *Jbj* both refer to large and small cattle in their autobiographical statements.
- *Jbj* further refers to 'a statement of 32400 cattle and small cattle'.
- *Ppjj-ʿnḫ-ḥrj-jb* refers to 'exaction of the impost of cattle and goats due from the middle nomes'.
- *Ppjj-ʿnḫ/Ḥnj-km* refers to 'Viewing the exaction of the impost of cattle and all small cattle', and 'Viewing the cattle and small cattle from his towns'.
- *Kꜣ(.j)-ḥp/Ṯtj-jkr* refers to numbers of cattle and oryx.
- *Ḥnj/Špsj-pw-Mnw* refers to numbers of cattle and gazelle.

The above data suggest that the emphasis was on increasing the numbers of pastured livestock.[309] It also suggests that the reasons for increasing herds were more complex than the explanation of Moreno Garcia, who describes the emphasis by officials on the increase of their cattle herds as offering proof of their administrative competence and that it was the response to the crown's interest in promoting the numbers of provincial cattle in order to supply expeditions.[310]

The vertical inscriptions in front of the figure of the tomb owner do not single out cattle as the tomb owner's greatest 'resource' interest until late Dynasty 5, although by mid Dynasty 5 the iconography is beginning to give cattle pride of place. Captions accompanying cattle scenes, however, merely present the brief conversation of minor figures or tersely identify a particular animal or activity. While inscriptions in provincial tombs of early Dynasty 6 indicate a growing interest in cattle herds, the major change occurs later, in mid Dynasty 6, when a consistent stress on great herd numbers appears in the tombs of the most important provincial administrators of Upper Egypt. In some instances, as with *Ppj-ʿnḫ-ḥrj-jb* and his successor at Meir, *Ppj-ʿnḫ/Ḥnj-km,* and *Kꜣ(.j)-ḥp/Ṯtj-jkr* of Akhmim, certain numbers represent an official cattle census for taxation purposes rather than a count of the tomb owner's property. The interest in detailing herd sizes does not appear to continue in the smaller Upper Egyptian tombs of late Dynasty 6 and Dynasty 8.

It is difficult to accept the accuracy of the cited herd numbers, particularly as it is estimated that the total fertile land of the Nile valley only amounted to 16,100 square kilometers in the Old Kingdom.[311] While the middle nomes of Upper Egypt contained the most extensive tracts of fertile land, only a proportion would have been pasturage, and this growth would have been seasonal.[312] Consequently the numbers cited by *Ppj-ʿnḫ-ḥrj-jb* totalling over three quarters of a million head of cattle appear to be much inflated, although they refer to the official census and, probably, *Ppj-ʿnḫ-ḥrj-jb*'s position as *jmj-rꜣ Šmʿ m spꜣwt ḥrjwt-jb*.[313] In fact, all the cited herd numbers must have been inflated in view of the seasonal nature of natural pasturage and the consequent need either to hand feed or move herds amounting to tens of thousands of head of cattle for part of the year. Moreover, there are other objections to the acceptance of the given numbers. The figures vary enormously from those cited by one generation of tomb owners to the next. For example, at El Hawawish the total number cited in *Kꜣ(.j)-ḥp/Ṯtj-jkr*'s tomb are 3226 cattle and 2023 desert ungulates, [314] while his successor, *Ḥnj/Špsj-pw-Mnw*, cites a total of 30,000 head of cattle and 20,000 desert

305 The numbers of herds are inscribed on all four registers: '32,400 males' above the herd of goats on the top register; '500 *jwꜣ*', '1400 *ngꜣ.w*' and '300 cows' on the second register; on the third register '1400 *jwꜣ*', '1300 *ngꜣw*' and '500? cows'; on the lowest register '300 *jwꜣ*'. Blackman/Apted (1953) pl. 41.

306 On the third register the number 3014 is inscribed above the depiction of an ox and 1023 above a gazelle. On the register beneath 1200 appears above an oryx and 212 above an ox. Kanawati I (1980) fig. 14.

307 Kanawati II (1981) fig. 26.

308 Kanawati I (1980) fig. 14.

309 Although sheep are usually considered to be 'small cattle', they are not included in any scene of the parade of animals of this time; nor are they shown associated with ploughing. It is probable, therefore, that 'small cattle' references apply only to goats. See Chapter 6 'Small Cattle'.

310 Moreno García *RdE* 50 (1999) p. 241. In particular Moreno Garcia attributes *Kꜣr* of Edfu's claim that he caused the herd numbers in U.E. 2 to increase to the need to supply crown expeditions into the Eastern Desert.

311 This assumption is based by Allen on the work of Butzer (1976) 83: Allen (1997) 145, note 13.

312 See CHAPTER 5: TWO FEATURES OF CATTLE MANAGEMENT.

313 Blackman *Meir* IV (1924). Pl. 4A, 1.

314 Kanawati I (1980) fig. 14.

ungulates.[315] That 20 000 wild ungulates could have been held in enclosures and adequately fed when seasonal pasturage died out in the height of summer is not credible. Furthermore, the enormous herd numbers do not match the size of the parcels of land of a few hundred arouras (five to six hectares) that tomb owners proudly claim were royal grants.[316] Yet it is unnecessary to go as far as Moreno Garcia who argues that the fantastic numbers of cattle supposed to belong to the deceased undermine the historical value of pictorial programs of tombs.[317] The most likely reading of the herd numbers is that they represent an emphatic statement that the policy of increasing herds in Upper Egypt was being carried out with outstanding success.

4.9 Summary

- Cattle are portrayed in the themes of husbandry, parades of animals, working for man and ritual butchery.

- Depictions of cattle develop from one or two animals in Dynasty 4 to early Dynasty 5 tombs to lines of cattle, sometimes interspersed with other ungulates. In the second half of Dynasty 6 the size of cattle herds is emphasised with Upper Egyptian administrators citing enormous numbers.

Varieties of cattle

- The terms *jw3, rn jw3* and *ng3w* refer to function and do not denote separate breeds of cattle. The term *jw3t* is used as a collective noun.

- The phrase *ḥr(j) ḏbꜥ* does not specify a polled animal but is a descriptive term applied to cattle reared by hand.

- It is likely that attempts were made to breed polled cattle, which were highly valued.

- It is unlikely that polled cattle depicted in Old Kingdom iconography constitute a separate breed.

- Data portraying short-horned cattle is insufficient to provide reliable evidence for a distinct breed in the Old Kingdom.

Herd management

- Few images represent details of husbandry in the early Old Kingdom tombs. There are no sustained scenes of cattle management until the tomb of *Jj-mrjj* in mid Dynasty 5.

- Images and scenes portraying aspects of cattle management appear much more frequently in later Dynasty 5, by which time all the main themes of cattle husbandry portrayed had been developed. Birth and care of calves, milking and the fattening of oxen are the most frequently depicted features.

- The daily management of the herd generally receives less attention in Dynasty 6.

- The increasing frequency of scenes of cattle crossing a stretch of water in the Fifth Dynasty in the tombs of high officials suggest that this activity was a customary and important aspect of the management of cattle herds. The frequency of portrayal is maintained in Dynasty 6, although at Meir and Deir el-Gebrawi its depiction seems to lose the significance of the activity.

- This reduction of the scene, however, is not the general trend in Upper Egypt; the traditional representation is maintained in the tombs of the Upper Egypt 9.

- The big estates maintained a hierarchy of farm workers to care for the herds of cattle, and the expertise of the overseers and herders is implicitly acknowledged in tomb iconography portraying scenes of herders at work in a variety of activities.

- The vertical inscriptions in front of the figure of the tomb owner do not single out cattle as the tomb owner's greatest 'resource' interest until late Dynasty 5, although by mid Dynasty 5 the iconography of the accompanying registers is beginning to give cattle pride of place.

- While inscriptions in provincial tombs of early Dynasty 6 indicate a growing interest in the size of cattle herds, the major change occurs later, in mid Dynasty 6.

- In mid to late Dynasty 6 the emphasis on building up of great herd numbers appears and is reflected in the captions and autobiographies in the tombs of the most important provincial officials of Upper Egypt, whose appointments were part of a crown policy of locating trusted resident administrators in many of the Upper Egyptian provinces.

- It is difficult to accept the accuracy of the cited herd numbers, particularly in view of Butzer's assessment that the total fertile land of the Nile valley only amounted to 16,100 square kilometers in the Old Kingdom.[318]

[315] Kanawati II (1981) fig. 26.
[316] For example, *Jbj* Kanawati *Gebrawi* II (2007) 53, pl. 54.
[317] Moreno García (1999) 63.

[318] The issue of numbers is discussed in Chapter 14, pp. 132–3.

Figure 11: WORKING CATTLE: PULLING A PLOUGH, TOMB OF KAIEMANKH, GIZA.
Kanawati *Giza* I (2001) pl. 36.

Figure 12: WORKING CATTLE: THRESHING CEREAL CROPS, TOMB OF MERERUKA, SAQQARA.
Kanawati (2011) pl. 84.

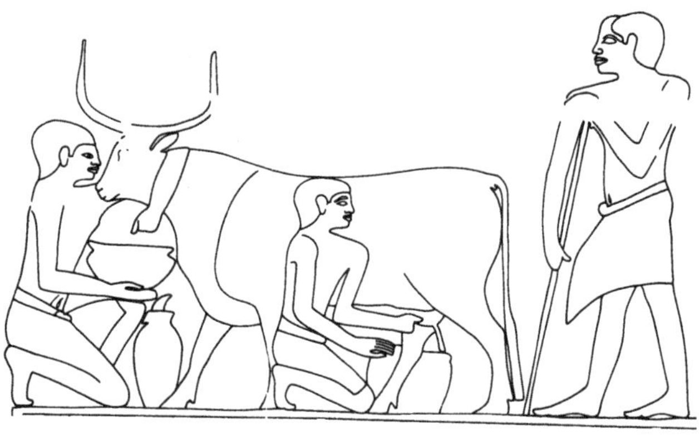

Figure 13: CATTLE HUSBANDRY: SUPERVISED MILKING, TOMB OF NIKAUISESI, SAQQARA.
Kanawati–Abder-Raziq (2000) pl. 49.

Figure 14: CATTLE HUSBANDRY: FATTENING CATTLE FOR RITUAL SLAUGHTER, TOMB OF MERERUKA, SAQQARA. Kanawati (2010) pl. 73.

Figure 15: CATTLE HUSBANDRY: CATTLE MATING, TOMB OF SHEDU, DESHASHA. Kanawati–McFarlane (1993) pl. 51.

Figure 16: CATTLE HUSBANDRY: COW SUCKLING CALF AND COW GIVING BIRTH, TOMB OF KHENI, EL-HAWAWISH. Kanawati II (1981) fig. 20, register 3.

Figure 17: CATTLE HUSBANDRY: OVERSEERS OF THE HERD, TOMB OF KHENI, EL-HAWAWISH. Kanawati II (1981) fig. 20, register 4.

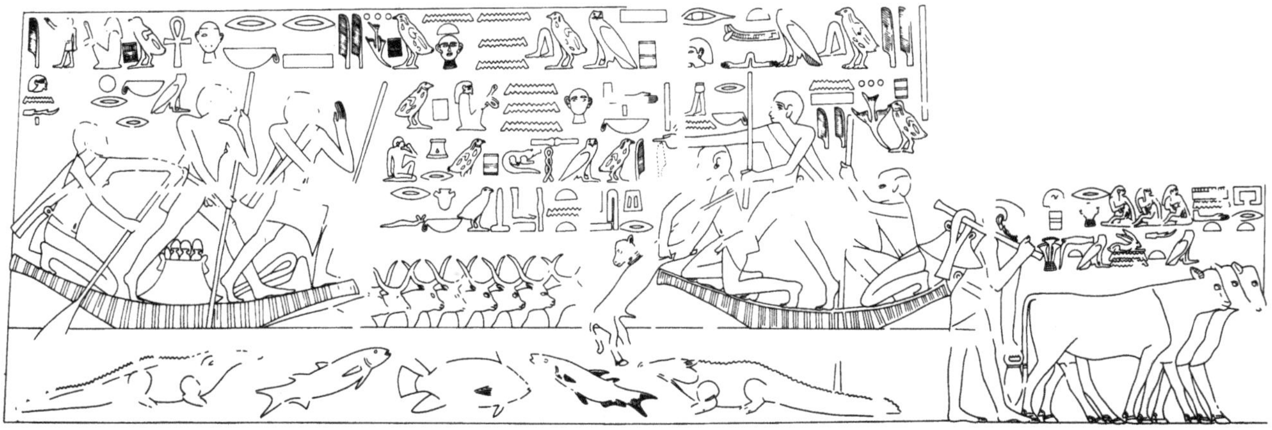

Figure 18: HERD OF CATTLE CROSSING AND EMERGING FROM A STRETCH OF WATER, TOMB OF HESI, SAQQARA.
Kanawati–Abder-Raziq (1999) pl. 55.

Figure 19: Fighting bulls, tomb of Kheni, el-Hawawish.
Kanawati II (1981) fig. 20.

Figure 20: TWO PHASES OF A BULLFIGHT, TOMB OF MERY, EL-HAGARSA.
Kanawati (1993) pl 43.

CHAPTER 5 TWO FEATURES OF CATTLE MANAGEMENT

5.1 Annual movement of cattle, Dynasties 4 to 8

TABLE 4: CATTLE CROSSING WATER

From late in the Fourth Dynasty certain tombs carry scenes of cattle swimming or wading through water or being led out of a stretch of water.[319] Only three of these scenes survive from the period of Dynasty 4 to mid Dynasty 5, although it is possible that the theme was included in the Dynasty 3 tomb of *Ḥzjj-Rˁ*.[320] There are two distinct variants of the theme of cattle in water. In the first variant cattle are depicted wading in shallow water or emerging from from a stretch of water. The animals are accompanied by a herder who either follows the herd (*Figure 18*, right), sometimes brandishing a stick to drive the cattle forward, or leads the herd out of the water by carrying a calf whose anxious mother follows, consequently leading the entire herd.[321] The second variant portrays cattle swimming through deep water, usually following a skiff. A calf is frequently depicted being towed through the water at the stern of the skiff and followed by the herd. In these latter scenes a submerged crocodile may sit on the bed of the channel. One or two skiffs of watchful herders accompany the herd (*Figure 18*, left). As the associated inscriptions do not reveal the precise purpose of the herd travelling through or out of the stretch of water, various interpretations of these scenes have been proposed. The most widely cited is an early interpretation that the scene portrays the herd being driven from the Delta marshlands southwards to the tomb owner's estate. While Montet originally accepted these inferences, first suggested by Maspero and Erman, in a later article he revised his interpretation deciding that the cattle are not depicted moving from one location to another but having been led into a stream to cool off during the heat of summer.[322] However the first variant, cattle wading in shallow water, is typically accompanied by inscriptions conveying the information that the herd is being led to higher ground (*r ḥrj-tp*):

> *prt m mḥjt r ḥrj-tp jn nrw ḥnˁ k3(w)t.sn k3w wrw*
> 'The coming out of the marshland to <u>the top</u> (the high ground) by the herders with their produce (and?) large cattle.' (The herd wades in shallow water.)[323]
>
> *prt ng3.w r ḥrj-tp*
> 'The coming out of the *ng3.w* (oxen) to the top (i.e. the high ground).' (The herd wades in shallow water.)[324]
>
> *prt m mḥjt*[325] *m pḥ r ḥrj-tp*
> 'The coming out the marshland and the channel to the top (the high ground).' (The herd wades in shallow water)[326]

The traditional interpretation of these scenes, that they represent the herd travelling south from the Delta to the tomb owner's estate, has been based on the translation of 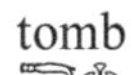 as *T3-mḥw* ('Lower Egypt/the Delta'). Moussa and Altenmüller, however, translate  as *mḥjt* ('marshland') rather than giving it the meaning of *T3-mḥw*, arguing that the concept 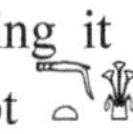 is being contrasted with *ḥrj-tp* ('the high ground'), so that in this context it is a topographical rather than a toponymical term.[327] It may have been feasible for cattle from estates located close to Memphis to be driven the comparatively short distance to the Delta marshlands during the months preceding the inundation and back to higher ground when the Delta began to flood; but it is unlikely that herds from much further south could have been driven to the marshy regions of Lower Egypt. In the inscriptions stating that the cattle are leaving the 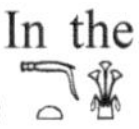 for the  both terms occur in the same sentence, while the destination of the cattle is always *ḥrj-tp* ('the top'/'high ground'), never *rsj* ('south') or *Šmˁw* ('Upper Egypt'). It is therefore unlikely that these scenes represent the movement of herds from the Delta southwards to their original pastures, a journey which for many herds would have taken days. Pasturage along such a route would not have supported numerous herds of cattle travelling over the same land at the same time of year. For an annual movement of cattle from marshlands to higher ground to have been feasible, the marshlands that the cattle left must have been in the vicinity of the estate that owned them. With perhaps an exception of some herds belonging to estates situated very close to the Delta, the movement of cattle must have taken place between marshy locations along banks of oxbow lakes and streams and canals meeting or flowing parallel to the Nile and the higher ground above flood level, where pasturage, if any, was maintained by precipitation. Butzer describes the Nile valley as a classic example of a convex floodplain characterised by 'natural levees that constitute the low-water channel banks' and having 'the lowest areas distant from the river and often

---

[319] The earliest surviving scene occurs in the tomb of *Nb.j-m-3ḫtj* (PM 230) but part of it has been lost due to the construction of a doorway: LD II, 12.
[320] Quibell (1913) 16, pl. 7.
[321] Evans (2006) 175 explains the nature of this behaviour, which Egyptian herders understood and used to move their herds.
[322] Montet (1954) 49. This interpretation has been more recently accepted by Ghoneim (1977) 30.
[323] Moussa–Altenmüller *Nianchchnum* (1977) pls. 76-7.
[324] *Rˁ-špss* LD II pl. 60 left.
[325] The Lepsius inscription has Gardiner I10 (*ḏt*) instead of Gardiner V23 (*mḥ*), which must be either a scribal error or a mistake in recording.
[326] *Ḥw-ns* LD II pl. 105b, right.
[327] Moussa–Altenmüller (1977) pl. 76-77, p. 155.

situated near the outer margins of the valley'.[328] This topography, which prevailed in dynastic times, would have required cattle to cross oxbow lakes or local streams to arrive at higher ground. Although these distances were not great, the herders would have had to travel and live with the cattle both in the marshlands and the higher pastures to prevent them straying, to protect them from wild animals and generally care for them.[329] Whether the herd is portrayed wading through shallow water or has to swim, it is portrayed accompanied by alert herders who use their knowledge of cattle behaviour to control and direct the progress of the herd.[330]

The conclusion that the Delta is not always indicated in the depicted movement of cattle is based on the examination of scenes in which the cattle wade through shallow water. The scenes in the great Memphite tombs of the end of the Fifth Dynasty and the early Sixth Dynasty of cattle swimming in deep water provide very little additional support for this conclusion because their associated captions rarely contain topographical or toponymical expressions; nor do they explain the purpose of the activity. Instead the inscriptions refer to the danger from a crocodile portrayed on the bed of the channel beneath the accompanying canoes, and the measures taken by the herders to protect the calf being towed by the leading canoe.[331] However, two provincial and two Memphite tombs provide additional support for the above conclusion.

On the east wall of the chapel of *Jttj/Šdw* of Deshasha, where the tomb owner is portrayed viewing registers of rural and marshland activities, there are two inscriptions referring to the herd crossing water. Above the herd the caption reads:

> *rdjt sḫn m ḥr-jb mw /////* [*ḏ3t?*] *n.f* [*k3w*]*.f r m33*
> 'Causing to remain motionless in the midst of the water. The [crossing?[332]] of his cattle to him for viewing.' (The cattle swim in deep water.)

A column of inscription in front of the figure of the tomb owner reads:

> [*m3*]*3 ///// sḫt ḏ3t n.f ////* [*k3*]*w.f*
> 'Viewing) ....(from) the marshland and the crossing of his [cattle] to him...'[333]

A similar arrangement of themes occurs in the chapel of *Ḥzj* in the Teti cemetery at Saqqara (*Figure 18*). On the bottom register of the east wall of the portico a scene of cattle in deep water is preceded by a portrayal of cows apparently emerging from the water; the line representing the water level continues across the legs of a bearded herder following the animals. The caption above the swimming cattle reads:

> *j nrw pw ˁnḫ ḥr.k r r šjj pw ntj ḥr mw jw m šp-tp tm jwjj n šjj pw*
> 'Oh this herdsman, may you be watchful against that aquatic which is in the water, who comes like a blind-of-head, and may they (the cattle) not come to that aquatic.'

A caption above three cows reads:

> *prt mnˁt r ḥrj-tp m-ḫt wnn m mḥt*
> 'Bringing the milking cows[334] to the top (the high ground) after being in the marshland'.[335]

Paired with the scene in the forecourt of his tomb of cattle wading through shallow water [second inscription, page 49 above and note 324) *Rˁ-špss* has a second scene of cattle swimming in deeper water, accompanied by two skiffs. The herder in the leading skiffs calls out:

> *nr ˁnḫ ḥr.k jr šjj pw ntj ḥr mw*
> 'Oh herdsman, may you be watchful against that aquatic which is in the water.'[336]

The warning probably applies to the calf being towed behind the first skiff.

This double representation is repeated in the tomb of *Ḫw-ns* of Zawyet el-Amwat where both themes are presented on a register beneath a scene of fighting boatmen. On the left of the register, the inscription above the herd in deep water reads:

> *nr pw ˁnḫ ḥr.k wrt jr šjj pw ntj ḥr mw jw.f m šp-tp*
> 'Oh herdsman, may you be very watchful against that aquatic which is in the water, and may he be blind-of-head.'[337]

On this register the swimming herd follows behind the skiff.

Just as the double statement in *Jttj/Šdw* refers to two aspects of the same activity (the safe return of the herd), so the pairs of scenes in the chapels of *Ḥzj, Rˁ-špss* and *Ḫw-ns* also represent two different stages of the same journey, which allows both the danger of the journey and its successful outcome to be emphatically portrayed.

In Dynasty 6 the theme of a herd of cattle swimming behind a skiff was only continued in certain provinces in Upper Egypt. It is given much less prominence in the tombs of the administrators of Meir and Deir el-Gebrawi, while at el-Hawawish and el-Hagarsa the established,

---

328 Butzer (1976) 16.

329 The birth of a calf, depicted in many tombs, is always attended by a herder.

330 Evans (2006) 173.

331 For example, *j nrw pw ˁnḫ ḥr.k r šjj pw ntj ḥr mw tm jw nw n šjj pw jw.f m šp tp ˁnḫ ḥr.k r.f wrt* 'Oh this herdsman, be watchful against that aquatic which is in the water. May these not go to that aquatic and may he be blind-of-head. Be extremely watchful against him': *ˁnḫ-m-ˁ-Ḥr* Kanawati–Hassan (1997) pl. 37; *nr pw ˁnḫ ḥr.k r šjj pw ntj ḥr mw jw.f m šp-tp jr tw r.f wrt* 'Oh herdsman, be watchful against that aquatic which is in the water, and may he be blind-of-head. Pay great attention to him': Kanawati–Abder-Raziq (2003) pl. 54.

332 Kanawati supplies 'crossing' from the associated vertical inscription.

333 *Jttj/Šdw*: Kanawati–McFarlane (1993) pl. 46; Petrie (1898) pl.15.

334 The three cows are polled while the swimming cattle that follow have lyre shaped horns. This order of animals (polled cows first) is reflected in many scenes of swimming cattle. See Kanawati–Abder-Raziq (1997) pl. 55.

335 Kanawati–Abder-Raziq (1997) p. 32 translates *ḥr-tp* as 'the land'.

336 *Rˁ-špss* LD II pl. 60 (right).

337 *Ḫw-ns* LD II pl. 105b (left).

traditional Memphite treatment of the theme is maintained with the herd depicted in an overlapping line, crossing a stretch of water. In the second half of the Sixth Dynasty only the high officials buried at el-Hawawish follow the early Dynasty 6 scene of the herd swimming in deep water accompanied by herders in boats and customary inscription.[338] At el-Hagarsa on the border of this province, the Memphite style of representation was continued into the 8th Dynasty in the tombs of *Mrjj-ˁ3* and *W3hj*.[339] In the tombs of Meir, the 'cattle in water' scene is attested only in the tomb of *Ppj-ˁnḫ/Ḥnj-km*, where the herd occupies the lower half of a divided register.[340] The cattle, depicted wading in fairly deep water, share the register with a papyrus thicket, a hippopotamus hunt, a presentation of marsh products to the tomb owner and, on a sub register above, farm workers cooking, eating and gathering papyrus. In the tomb of *Ḏˁw* of Deir el-Gebrawi the rationale behind the scene of cattle crossing water appears to have been almost lost.[341] An image of a calf facing a cow(?) and possibly a few more cattle is tucked beneath the prow of the skiff on which the tomb owner is harpooning fish, while a further image of cattle in water is included beneath a fishing net controlled by men standing on the boat above. Here a dappled calf faces its mother(?) while the rest of the herd appears to be moving in the opposite direction. In these tombs the theme of cattle crossing water is closely associated with, but subordinate to, the motifs of the tomb owner fishing and fowling from a canoe in the marshes and the hunting of the hippopotamus.

FINDING

The increasing frequency of scenes of cattle crossing a stretch of water in the Fifth Dynasty in the tombs of high officials suggest that this activity was a customary and important aspect of the management of cattle herds. The frequency of portrayal is maintained in Dynasty 6, although at Meir and Deir el-Gebrawi its depiction seems to lose the significance of the activity. At Meir it only appears in one tomb, that of *Ppj-ˁnḫ/Ḥnj-km*, where it is partly relegated to a sub-register.[342] In *Jbj*'s chapel at Deir el-Gebrawi its position, tucked beneath registers of returning boatmen and the stern of the tomb owner's skiff, gives the impression that it was added as a space-filler or an after-thought. In the tomb of *Ḏˁw* the small herd swimming beneath a fishing skiff is only just recognisable as representing the motif. This reduction of the scene, however, is not the general trend in Upper Egypt; the traditional representation is maintained in the tombs of the Upper Egypt Nome 9.

The portrayal of the crossing of a stretch of water did not necessarily entail a journey from the marshes of the Delta. Inscriptional evidence suggests that movement was from lower-lying marshland to higher pastures.

Apart from the later Dynasty 6 departure from the traditional representation noted at Meir and Deir el-Gebrawi, the scenes of swimming and wading cattle confirm and elaborate on the impression drawn from other scenes of husbandry that good management of the cattle herds was of especial importance to officials. They portray the watchfulness of herders taking their cattle across a crocodile infested channel and the successful conclusion of the journey. The scenes represent the care bestowed on the herds by experienced and skilled herders in a dangerous episode that occurred annually in the management of the herd.

5.2 Fighting bulls

Pairs of fighting bulls are portrayed in the following tombs:

Ḥzjj-Mnw	el-Hawawish	V.L
Jttj-Šdw	Deshasha	VI.E
Jbj	el-Gebrawi	VI.M
Ṯ3wtj	el-Qasr wa-'l-Saiyad	VI.M
Jdw/Snnj	el-Qasr wa-'l-Saiyad	VI.L
Ppj-ˁnḫ/Ḥnj-km	Meir	VI.L
Mrjj	el-Hagarsa	VI.L
Ḥzjj-Mnw/Zzj	el-Hawawish	VI.L
Ḫnj/Špsj-pw-Mnw	el-Hawawish	VI.L
K3(.j)-ḥp/Ṯtj-jkr	el-Hawawish	VI.L
Ḏˁw	el-Gebrawi	VI.L
W3hj	el-Hagarsa	VIII
Mrjj-ˁ3	el-Hagarsa	VIII
Ḫw-ns	Qubbet el-Hawa	VIII
Rḥw-r-3w.sn	el-Hawawish	VIII

The portrayal of a pair of bulls locked in combat, which first appears in the tomb of *Ḥzjj-Mnw* (M22) at el-Hawawish in late Dynasty 5,[343] is essentially a Sixth Dynasty motif.[344] It is included in *Jttj-Sdw*'s tomb at Deshasha in early Dynasty 6[345] and while it grows in popularity as the Dynasty progresses with more than one pair of bulls depicted fighting on the same register, in the Old Kingdom all attestations are found in the provinces of Upper Egypt.[346] As the theme develops at Deir el-Gebrawi and el-Hawawish, the later nomarchs (*Ḏˁw* and *K3(.j)-ḥp/Ṯtj-jkr*) allot it greater wall space[347]. In the tomb of *Mrjj* of el-Hagarsa the motif is presented in two stages showing the bulls clashing head to head and then one bull throwing the other to the ground[348] (*Figures 19 and 20*).

338 *Ḫnj/Špsj-pw-Mnw* Kanawati II (1981) fig. 22; *K3(.j)-ḥp/Ṯtj-jkr* Kanawati I (1980) fig. 12 and *Ghs3/Nbjj* Kanawati VII (1987) fig. 30.

339 Respectively, *Mrjj-ˁ3* pl. 37 and *W3hj* pl. 20 in Kanawati (1995).

340 Blackman–Apted *Meir* V (1953) pl. 39

341 *Ḏˁw* Davies *Gebrâwi* II (1902) pl. 5.

342 The herd follows two skiffs, with a calf tied to the stern of the second craft. The three men on the skiff seem to be ignoring the following cattle although a caption over their heads records the herder in the prow warning of a lurking crocodile. The two men in the first skiff are bringing a variety of offerings received by *Ḥnn-jt* (*Ppj-ˁnḫ/Ḥnj-km*'s son), while men on two skiffs behind the herd are harpooning hippopotami: *Ppjj-ˁnḫ/Ḥnj-km* Blackman–Apted *Meir* V (1953) pl. 30.

343 Traces of this scene are preserved in the tomb of *Ḥzjj-Mnw* (M22) at El-Hawawish: Kanawati IV(1983) fig. 8, 11.

344 Kanawati I (1991) 51-58.

345 Petrie (1898) pl. 18; Kanawati–McFarlane (1993) pl. 51.

346 All Old Kingdom bullfighting scenes are located in the middle provinces of Upper Egypt except for two scenes in the tomb of *Ḫw-ns* at Qubbet el-Hawa: Smith (1946) 227; Vandier V (1969) 60, fig. 35. 2.

347 The theme occupies two registers in the chapel of *Ḫnj/Špsj-pw-Mnw*: Kanawati II (1981) fig. 20, and an entire panel of registers in the tomb of his successor, *K3(.j)-ḥp/Ṯtj-jkr*: Kanawati I (1980) fig. 10.

348 Kanawati *el-Hagarsa* I (1993) pl. 43.

At el-Hagarsa and el-Hawawish the motif continues into the 8th Dynasty.[349] It reappears in the Middle and New Kingdoms.[350]

The bullfighting scene has been included as a husbandry motif because it is presented as an activity, organised and controlled by herders with the tomb owner depicted viewing the scene. In the scene in *K3(.j)-ḥp/Ṯtj-jkr*'s tomb, for example, a figure wearing the kilt of an overseer of farm workers and titled *jmj-r3 ṯzt* appears twice: on the top register the figure raises a stick presumably to control the bulls and on the register beneath he leans on a staff watching the combat, while the tomb owner stands behind.[351]

5.3 Summary

Annual movement of cattle

- From the late Fourth to the late Sixth Dynasty, certain tombs include scenes of cattle engaged in the seasonal fording of a stretch of water.

- The two variants of the theme are the herd wading in or emerging from shallow water and cattle swimming through deep water. In both types of scenes herders accompany the herd.

- The most widely cited interpretation is that the herd is being driven from the Delta marshlands southwards.

- However, inscriptions accompanying scenes of cattle wading in shallow water suggest they are being led from the marshes to higher ground, which is a more feasible explanation.

- A departure from the traditional representation of the scenes of cattle swimming and wading is noted at Meir and Deir el-Gebrawi in later Dynasty 6.

- The scenes represent a dangerous annual episode in the management of the herd and the skills and vigilance of herders.

Fighting bulls

- The motif of a pair of bulls locked in combat only appears in Upper Egyptian tombs from late Dynasty 5 to the end of the Old Kingdom.

- The motif becomes more popular as Dynasty 6 progresses with scenes being given more wall space and elaboration.

- As the fighting bulls are shown organised and controlled by herders, the theme is an aspect of cattle husbandry.

349 *W3hj* Kanawati (1995) pl. 23; *Rḥw-r-3w.sn* Kanawati VII (1987) fig. 15.
350 Galan (1994) 87 - 90 (Galan's list of later bullfighting scenes).
351 Kanawati I (1980) fig. 10; Galan (1994) 85.

CHAPTER 6 “SMALL CATTLE” (SHEEP and GOATS)

SHEEP

6.1 Focus of enquiry: (1) Domesticated sheep as an economic resource

TABLE 6: PORTRAYAL OF FLOCK OF SHEEP

The data:

- Iconography of sheep flocks and associated inscriptions
- Material evidence of sheep raising and mutton consumption from Kom el-Hisn and the Giza Plateau Mapping Project
- Related inscriptional evidence.

Areas of research:

- Analysis of the scenes portraying flocks of sheep
- The use of and value attributed to sheep.

Ovis aries (*sr* Wb. III, 414)
All the sheep portrayed in Old Kingdom scenes appear to have belonged to the same type, with long legs, a thin tail, pendant ears and twisted horns growing laterally from the head.[352] Males have pronounced manes. Variously coloured, with hairy rather than woolly coats, these animals would have been kept for meat, hides and milk, but probably not for their wool,[353] which grew beneath the hairy coat and was shed each spring in a thick mat.[354] Unlike goats, which naturally browse on the sparse flora of mountain ranges, sheep graze on lower uplands. In a region such as the Nile valley or delta, sheep may have been reared in areas drier than those needed for cattle, although they would not have had the capacity to thrive over such a varied range of environments as goats. The limited birth season of sheep produces more males than are needed for breeding. Surplus males are normally slaughtered for consumption.

6.2 The portrayal of sheep in Old Kingdom tombs

The analysis of scenes offers only three motifs:

- the majority of scenes show a flock of sheep either following or preceding the depiction of ploughing or hoeing
- occasionally a line of overlapping sheep is included with the portrayal of other domesticated animals
- a flock of sheep threshing the grain crop is a rare motif.

Sheep are portrayed in the following number of tombs used in this study:

24	Sheep trampling seed driven by herders with whips and/or sticks (*Figure 21*)
20	Sheep trampling seed led by a coaxer-sower (*Figure 21*)
6	In parade of animals
2	On threshing floor (*Figure 22)*

In the period up to the end of Dynasty 5 sheep are depicted in connection with soil preparation and planting in 30 tombs, all Memphite excepting the tombs of *Wr-jr.n.j* of Sheikh Said and *H̱w-ns* of Zawyet el-Maiyetin. These depictions present a small flock of sheep, driven forward by herders brandishing whips and sticks. In seventeen of these scenes the flock of sheep is portrayed in close association with ploughing or hoeing, either following or preceding these activities. The flock is being used to trample newly sown seed into the soil as related inscriptions suggest. The captions revolve around the words *skꜣ*[355] and *zḥ.t*[356] and extend in length from *skꜣ m zḥ.t* (‘cultivating with a flock of sheep’)[357] to *skꜣ m zḥ.t jn jzwt.f nt pr.f n ḏt* (‘cultivating with a flock of sheep by his crews of his estate’).[358] Evans suggests that the men, armed with whips and sticks, are running beside the flock to keep the animals herded and moving forward because sheep habitually bunch together when alarmed. This would have kept them in line with the sower scattering the seed.[359] It is probable that the image of a flock of sheep controlled by men with whips and sticks must have always been associated with sowing. In eleven instances, however, the flock of sheep is portrayed although the scene of soil preparation, either by plough or men with hoes, is located on a separate register or is omitted.

352 The earliest representation of this species is on the reverse of the Libyan Palette (JE 27434 = CG 14238), which dates from the Naqada III or Protodynastic Period of Egypt (circa 3,200 to 3,000 BC). The palette is unprovenanced but believed to be from Abydos.
353 Boessneck (1988) 72-3.
354 The wool bearing sheep with horns that curved down and backwards does not appear in Egyptian iconography until Dynasty 12: P. Newberry I, *Tomb of Khnemhotep* (1893) pl. 25. Furthermore, the animal embodiment of the creator god Khnum was depicted with the lateral horns of *ovis longipes palaeoegyptiacus* until Dynasty 12.

355 *skꜣ* to plough, to cultivate (Wb 4, 315-316.9)
356 *zḥ.t* "Schafherde (die Saat festtritt)" (Wb 3, 464.2)
357 *Mrs-ꜥnḫ* III, *Nj-ꜥnḫ-H̱nmw* and *H̱nmw-ḥtp, Ḥtp-ḥr-ꜣḫtj, Nfr-jrt-n.f, Wr-jr.n.j* and *Sḫm-kꜣj.* See TABLE 6: PORTRAYAL OF FLOCK OF SHEEP.
358 *Jj-mrjj* and *Nfr-bꜣw-Ptḥ.* See TABLE 6: PORTRAYAL OF FLOCK OF SHEEP.
359 Evans (2006) 142-48.

Only four scenes from this time period have been found in which the flock is not depicted treading seed.[360] In these tombs, dating from late Dynasty 4 to early Dynasty 6, a flock of sheep is included either in a parade of animals or is portrayed on the threshing floor.[361] Although many Memphite tombs of the second half of Dynasty 5 develop the motif of the 'parade' of animals and give it considerable prominence, sheep are rarely included.

6.2.1 Historical development of images and themes

Flocks of sheep depicted in the following tombs:

Late Dynasty 4 to early Dynasty 5

Mrs-ꜥnḫ III	Giza	IV.L
Nb.j-m-ꜣḫtj	Giza	IV.L
*Wḥm-kꜣ.j**	Giza	V.E
*Snb**	Giza	V.E

* portrayed either in the parade of animals or threshing cereal

Mid Dynasty 5

GIZA
Jj-mrjj
SAQQARA
Wr-jr.n-Ptḥ — *Nfr* and *Kꜣ-ḥꜣ.j*
Nj-ꜥnḫ-Ḫnmw and *Ḫnmw-ḥtp*
ABUSIR
Ptḥ-šps

Late Dynasty 5

GIZA

Jꜣšn	*Snḏm-jb/Jntj*
Jj-nfrt	*Snḏm-jb/Mḥj*
Nj-mꜣꜥt-Rꜥ	*Sḫm-kꜣj*
Nfr (1)	*Kꜣ.j-m-nfrt*
Nfr-bꜣw-Ptḥ	*Kꜣ-ḥj.f*
Sꜣ-jb	

SAQQARA

ꜣḫtj-ḥtp (Louvre)	*Ḥtp-ḥr-ꜣḫtj*
Jrj-n-kꜣ-Ptḥ	*Ḫnwt*
Wr-jr.n.j	*Sḫm-ꜥnḫ-Ptḥ*
Ptḥ-ḥtp/Jj-n-ꜥnḫ	*Sḫntjw ꜥnd Nfr-sšm-Ptḥ*
Nb-j-kꜣw-ḥr/Jdw	*Kꜣ.j-m-nfrt*
Nbt	*Kꜣ.j-m-ḥzt*
Nfr-jrt-n.f	*Ṯjj*

ZAWYET EL-MAIYETIN
Ḥw-ns

The portrayal of sheep in the tombs of Dynasties 4 and 5 receives similar treatment to that of donkeys in that they are almost always depicted as working animals used for agricultural purposes. It is likely that flocks of sheep were grazed on the poorer pastures of an estate and that rams were culled for slaughter, although such activities are never portrayed. Only their 'trampling' use, either in the field or on the threshing floor, is portrayed. Yet flocks of sheep had further economic value. Sahure records bringing 243,688 sheep from Libya.[362] At the Old Kingdom location of Kom el-Hisn in the Western Delta, a site known as Room 17 was devoted to cooking; the dung used as fuel was dominated by sheep and goat remains.[363] These findings were not typical of the site as a whole. Moens and Wetterstrom report that the study of charred material in general, from locations across the entire site produced few sheep and goat pellets in the dung used for fuel.[364] Cagle suggests that Room 17 was used to prepare the food of the elite of Kom el-Hisn, while charred remains found elsewhere were the product of cooking for the ordinary people who worked at the site. This suggests that while they were perhaps not sacrificed ceremonially, sheep and goats belonged to the diet of the upper classes.

In general, the portrayal of sheep is similar to that of donkeys; they are never depicted in scenes of husbandry. Sheep are typically portrayed as working animals with their nature and responses shown being used by man. In order to get the sheep to follow a given direction, the man sowing seed walks ahead of the flock. Frequently he holds out a hand to the leading animal, presumably enticing it forward with the offer of seed so that the flock will follow.[365] The flock is also driven forward by herders with raised whips, to which the animals respond by instinctively herding together, thus keeping them on line.[366] These two factors, tempting the leading sheep to follow the sower and frightening the animals to herd together, indicate the use of the instinctive responses of the species to get the animals to work for man.[367] Unlike scenes of paraded cattle and desert animals, which are depicted as fine specimens that herders proudly present to the tomb owner, or even like domesticated goats taken to browse under the care of a goatherd, sheep are nearly always represented in their 'working' capacity. As Evans describes, observant and creative tomb craftsmen departed from the formal depiction of the flock as a strictly layered line of images to depict the variety of responses of alarmed animals as they followed the sower of seed.[368] But this was an artistic device to make the scene more interesting for the viewer. While it portrays an understanding of animal behaviour, it reflects efficient work practices rather than the pride of ownership that representations of other livestock display.

Flocks of sheep depicted in the following tombs:

Dynasty 6 to Dynasty 8

Mrrw-kꜣ/Mrj	Saqqara	VI.E
Mḥw	Saqqara	VI.E
Ḥzj r/u *Sšm-nfr*	Saqqara	VI.E
*Ppj-ꜥnḫ/Ḥnj-km**	Meir	VI.L
*Ḏꜥw**	Gebrawi	VI.L

360 *Snb, Wḥm-kꜣ.j* and *Wr-jr.n-Ptḥ*. See TABLE 6: PORTRAYAL OF FLOCK OF SHEEP.

361 The tombs of *Snb, Wḥm-kꜣ.j, Jrj-n-kꜣ-Ptḥ* and *Rꜥ-ḫꜥ.f-ꜥnḫ*. See TABLE 6: PORTRAYAL OF FLOCK OF SHEEP.

362 Borchardt (1910-13) Bl. 1.

363 Cagle (2002).

364 Moens–Wetterstrom (1988) 167.

365 Fisher–Matthews "The social behaviour of sheep" in Keeling–Gonyou (2001) 218-220.

366 Hafez, E.S.E., Cairns, R.B., Hulet, C.V. and Scott, J.P. (1969) "The behaviour of sheep and goats" in Hafez (1969) 332.

367 Evans (2006) 151-2

368 Evans (2006) 153-4

With the exception of the above scenes, two associated with sowing seed[369] and two further scenes associated with the parade of animals,[370] sheep disappear from the record in Dynasty 6. In the tomb of *Ppjj-ʿnḫ/Ḥnj-km* of Meir a small flock of sheep follows a herd of donkeys being paraded before the seated tomb owner.[371] Although the figure of *Ppjj-ʿnḫ/Ḥnj-km* is separated from the animals by a doorway, the herders standing in front of the line of animals face the tomb owner on the other side of the doorway. The gestures of both figures make it clear that they are presenting their charges to *Ppjj-ʿnḫ/Ḥnj-km*. The caption above the sheep states '1000 females'. Above the doorway is an explanatory inscription reading, 'Viewing the exaction of the impost of bulls and all kinds of small cattle'. The statement describing the action of the tomb owner, which regularly begins with *m33,* is usually written as a column separating the tomb owner from the scenes he is inspecting.[372]

In the other late Dynasty 6 scene portraying sheep, the tomb owner, *Ḏʿw*, views herds of animals that also include cattle, goats and donkeys.[373] In this scene on the east wall of *Ḏʿw*'s chapel, a flock of sheep, some nibbling tall grass and accompanied by a herder, face the standing figure of the tomb owner. There is no explanatory inscription separating the tomb owner from his herds or associated with any of the animals. However, the similarity of the scenes in the two chapels of *Ppjj-ʿnḫ/Ḥnj-km* and *Ḏʿw* suggests that they both represent a count of the animals and that the inspection is taking place in the fields. The *Ppjj-ʿnḫ/Ḥnj-km* panel of scenes has the portrayal of goats browsing on the foliage of trees immediately below the register of sheep and donkeys, and on the three lower registers cattle are depicted walking through shrubs and grass. In the tomb of *Ḏʿw* the sheep, as well as the cattle on the register above, are standing in tall grass, with some of the animals feeding on the herbage; so this scene too probably represents *Ḏʿw* viewing the count of animals. In both tombs the panels of sheep, donkey, goat and cattle herds appear on the east walls.

The inclusion of sheep in the animal census in the tombs of *Ppjj-ʿnḫ/Ḥnj-km* and *Ḏʿw* suggests an interest in maintaining flocks but the reason for this is not clear, as the use of sheep to trample seed into the soil at the time of ploughing or on the threshing floor is not portrayed after early Dynasty 6.

DOMESTIC GOATS

6.3 Focus of enquiry: (2) Herds of domesticated goats as an economic resource

TABLE 7: MANAGEMENT OF DOMESTIC GOATS

The data:

- Scenes from 24 tombs, Dynasties 4 to 8
- Material remains, mainly from Kom el-Hisn and the Giza Plateau Mapping Project.
- Related inscriptional and artefact evidence.

Areas of research:

- The nature of the portrayal of domestic goats
- The geographic and chronological distribution of scenes with domestic goats.

Capra hircus (*wnḏw* Wb. I, 326)
According to evidence from the Zagros Mountains, *Capra hircus* may be the first herd animal to have been domesticated, approximately 10,000 to 11,000 years ago.[374] The domestic goat was introduced about 7000 B.P. into Egypt from Asia[375] where it had already been a domesticated species for two to four thousand years.[376] Old Kingdom depictions portray the animal with twisted horns, lop ears, beard and a coat of short hair. The species is adapted to rocky, even mountainous terrain and is a mixed feeder. Although it often prefers to graze on grass, it will also browse on weeds, fruit and leaves of shrubs and trees and even bark. When ground herbage is sparse, it has sufficient agility to climb into the canopy of trees to consume leaves on higher branches. Flexible feeding habits and ability to digest lower quality herbage enable the species to survive in environments that would not support cattle or sheep. Its browsing behaviour, efficient digestive system and ability to reduce its metabolism enable it to survive prolonged periods of severely limited food supply. When conditions improve, a herd of goats can rebuild its numbers, as it will reproduce speedily with females frequently giving birth to twins.[377]

6.4 The portrayal of goats in Old Kingdom mortuary chapels

Analysis of scenes with goats provides the following motifs:

- life cycle – males fighting, mating, giving birth, suckling young, kid accompanying female, slaughter (*Figure 23*)
- husbandry – herders attending, herders pulling down branches of a tree, herder protecting a kid being born from a predatory animal, feeding an animal by hand (*Figure 23*)
- goats eating – standing to eat grass, weeds and low shrubs, rearing on hind legs to eat lower leaves of a tree, standing on tree branches to eat leaves, standing

369 *Mḥw; Ḥzj* r/u *Sšm-nfr*. See TABLE 6: PORTRAYAL OF FLOCK OF SHEEP.
370 *Ppj-ʿnḫ/Ḥnj-km; Ḏʿw*. See TABLE 6: PORTRAYAL OF FLOCK OF SHEEP.
371 Blackman–Apted (1953) pl. 32.
372 *m33* is usually taken as the infinitive form of the verb 'to see' or 'to view' but in the context of the tomb owner viewing estate activities is best translated in English by the present participle.
373 Davies (1902) pl. 8.

374 Zeder–Hesse (2000).
375 Reed (1969) 372.
376 The dating of the original domestication of the species is disputed: Brewer et al (1994) 90.
377 McGregor (2000).

on upper sub-register to eat tree leaves, recumbent being hand fed (*Figure 23*)
- herders depicted standing or seated watching herd, skinning a slaughtered animal, pulling down branches or chopping down a tree, drinking from a jar, playing a pipe (*Figure 23*)
- presentation of herd to tomb owner.

6.4.1 Historical development of images and themes

Domesticated goats as a browsing herd are depicted in the following tombs:

Dynasty 4 to mid Dynasty 5

Nb.j-m-3ḫtj	Giza	IV.L
Sḫm-k3-Rʿ	Giza	V.E
Nj-ʿnḫ-H̱nmw and *H̱nmw-ḥtp*	Saqqara	V.M
Nfr and *K3-ḥ3.j*	Saqqara	V.M
Rʿ-ḥtp	el-Hammamiya	V.M

Domestic goats are portrayed as a herd grazing and browsing in five tombs dated from late Dynasty 4 to mid Dynasty 5.[378] The typical depiction of goats is as a small herd browsing among shrubs and trees accompanied by one to three herders. In two of these tombs, those of *Nb.j-m-3ḫtj* and *Sḫm-k3-Rʿ*, the scenes are very fragmented.[379] The scenes in *Nfr* and *K3-ḥ3.j* and *Nj-ʿnḫ-H̱nmw* and *H̱nmw-ḥtp* portray watchful husbandry with herders protecting their charges.[380] In the *Nfr* and *K3-ḥ3.j* scenes goats stand on their hind legs to nibble at leaves on the lower branches of trees.[381] In the scene in the chapel niche a supervisor leaning on his staff watches two herders knocking down tree branches for the herd, and at another tree a female suckles her kid as she rears up to eat the leaves. In an accompanying vignette a herder skins an animal carcass that hangs from a tree. The vertical inscription separating the registers from the figures of the tomb owner and his wife does not refer to the register containing the scenes of the herd of goats; it reads, 'Viewing fishermen, fowlers and scribes of his towns of Lower and Upper Egypt' (*m33 ḥ3mw ʿhw zšw n njwt.f nt mḥwj šmʿj*).[382]

The scene in *Nj-ʿnḫ-H̱nmw* and *H̱nmw-ḥtp* provides a more detailed account of goat husbandry. On the top register from right to left, vignettes present a herder knocking down branches of a tree so that two goats can feed on the foliage, another herder, supervised by an overseer, beating off a wild cat as it prepares to spring at a baby goat being born, and a third herder skinning the carcass of a slain goat. Further to the left, cuts of meat are being cooked in a large kettle, presumably as a meal for the herders and tree fellers pictured next to the kettle. On a sub-register above the birthing scene, there are three goats and a kid protected in a cage. The theme is not a major feature in either tomb. Images are confined to a single register and the vertical inscriptions separating the panel of registers from the figures of the tomb owner do not refer to the register containing the scenes of goat herding. For example, the *H̱nmw-ḥtp* inscription reads, *m33 šdt m š3bt š n dwjt ʿḥʿt pr dt* 'Viewing the construction of a *š3bt*-boat, papyrus gardens and the possessions of the estate'.[383]

Late Dynasty 5 to early Dynasty 6

3ḫtj-mrw-nswt	Giza	V.L
3ḫtj-ḥtp (Louvre)	Saqqara	V.L
Nb-k3w-ḥr/Jdw	Saqqara	V.L r/u VI.L
Ḥtp-ḥr-3ḫtj	Saqqara	V.L
Ḥw-ns	Zawyet el-Amwat	V.L
Sšm-nfr IV	Giza	V.L
Jttj/Šdw	Deshasha	VI.E
Mrrj	Saqqara	VI.E
Nj-ʿnḫ-Ppjj	Zawyet el-Maiyitin	VI.E

The theme of goat husbandry only appears in nine of the 127 chapels used in this study dated to the period covering the second half of Dynasty 5 and early Dynasty 6. The most frequently depicted motif for this species is of goats feeding on trees. Only the chapel of *Ḥtp-ḥr-3ḫtj* contains a scene of the herd grazing on herbage; on the upper sub-register four goats nibble at a variety of low growing plants.[384] When portraying the herd browsing on the upper canopy of trees, the craftsman uses one of two devices: either showing individual animals standing on branches[385] or, more frequently, on a sub-register that raises the depiction of the goat(s) level with the arboreal canopy.[386] Goats remaining at ground level but browsing on trees are portrayed rearing up on their hind legs to get at the lower leaves.[387] The other frequently included themes, females with young or giving birth or animals copulating, emphasise the expansion of herd numbers.[388] Herders are included in all but three scenes that are too fragmentary to provide an entire picture. The only scene that definitely diverges from the motif of the herd browsing are the depictions in the tomb of *Mrrw-k3* of four splendid, recumbent animals being hand fed just like cattle.[389] Each animal is labelled *rn wndw* ('young/ specially fattened[390] goat'). Although trees and, more rarely, low growing plants are depicted, no scene provides any idea of the nature of the terrain; all are pictured on flat register lines.

---

378 See TABLE 7: MANAGEMENT OF DOMESTIC GOATS

379 *Nb.j-m-3ḫtj* Hassan (1943) fig. 75; *Sḫm-k3-Rʿ* LD Erg. 36, Hassan (1943) fig. 57 and Harpur (1985) fig. 10.

380 *Nfr* and *K3-ḥ3.j* Moussa–Altenmüller (1971) pl. 1, 4, 18,19; *Nj-ʿnḫ-H̱nmw* and *H̱nmw-ḥtp* Moussa–Altenmüller (1977) fig. 18.

381 *Nfr* and *K3-ḥ3.j* Moussa–Altenmüller (1971) pl. 4.

382 *Nfr* and *K3-ḥ3.j* Moussa–Altenmüller (1971) pl. 3.

383 *Nj-ʿnḫ-H̱nmw* and *H̱nmw-ḥtp* Moussa–Altenmüller (1977) pl. 8.

384 *Ḥtp-ḥr-3ḫtj* Mohr (1943) fig. 21.

385 *3ḫtj-mrw-nswt* Smith (1946) fig. 239.

386 *3ḫtj-ḥtp* (Louvre) Ziegler (1993) pp. 44 (plates), 137-9 (drawing); *Nb-k3w-ḥr/Jdw* Hassan (1932) fig. 12-13; *Ḥtp-ḥr-3ḫtj* Mohr (1943) fig. 21; *Ḥw-ns* LD II, 108 and *Mrrj* Davies et al (1984) pl. 7.

387 *3ḫtj-mrw-nswt* Smith (1946) fig. 239; *3ḫtj-ḥtp* (Louvre) Ziegler (1993) pp. 44 (plates), 137-9 (drawing); *Nb-k3w-ḥr/Jdw* Hassan (1932) fig. 12-13; *Ḥw-ns* LD II 108 and *Jttj/Šdw* Petrie (1898) pl. 15 and Kanawati–McFarlane (1993) pl. 14, 46.

388 *Nb-k3w-ḥr/Jdw* Hassan (1932) fig. 12-13; *Ḥtp-ḥr-3ḫtj* Mohr (1943) fig. 21; *Ḥw-ns* LD II 108; *Sšm-nfr* IV Junker XI (1953) Abb. 93; *Jttj/Šdw* Petrie (1898) pl. 15 and Kanawati–McFarlane (1993) pl. 14, 46.

389 Duell II (1938) pl. 152, Room A13, north wall.

390 See Chapter 4, p. 35 for discussion of the meaning of *rn*.

Herders are depicted in at least six scenes[391] engaged in a variety of activities. In *Nb-k3w-ḥr*'s scene two workers are chopping down trees on which goats are browsing. It is unlikely that the information to be conveyed to the viewer is that the trees are being felled to provide the herd of goats with fodder. The animals are already in the canopy and the captions read, *sḳd* ('cutting') and *sḳdwt* ('cutting to shape'), which would refer to the associated activity of boat building. The implication, however, is that while the timber from the felled trees will be used to build a boat, the herd of goats are able to consume the foliage so that nothing goes to waste. In the Deshasha tomb of *Jttj/Šdw* an overseer ([*jmj-r*] *t̠zt nb r* ...'any master drover to ...') is referred to in a broken inscription.[392] Although an entire register is devoted to the depiction of goats with scenes of goats browsing in trees and a line of overlapping animals marching towards the standing figure of the tomb owner, the vertical inscription dividing *Jttj/Šdw* from the panel of registers that he is viewing concentrates on the marshland activities and does not refer to the herd. Although fragmented, the inscription can be read as [*m3*]*3* {*k3t*} *sḫt d3t n.f* [*k3*]*w.f* ... ('Viewing the work of the marshland and the crossing of his cattle (back) to him'). In six tombs a herder is portrayed skinning the carcass of a goat that hangs from a tree. Although this may indicate that goat meat was considered an inferior food fit for workers or that slaughtering a goat was a perquisite of goat herders, such conjectures do not fit in with Cagle's suggestion that Room 17, where the fuel was dominated by sheep and goat remains, was used to prepare the food of the elite of Kom el-Hisn, nor with the scene in the tomb of *Jbj* where the cooking of goat meat is supervised by the 'director of the dining hall'.[393]

Mid Dynasty 6 to Dynasty 8

Jbj	Gebrawi	VI.M
Jn-jt.f/Bḫnt	el-Hawawish	VI.M
Jḥjj	el-Khokha	VI.M
Ḥzjj-Mnw/Zzj (G42)	el-Hawawish	VI.M
Ppjj-ʿnḫ-ḥrj-jb	Meir	VI.M
Ḫntj	el-Khokha	VI.M
Ppjj-ʿnḫ/Ḥnj-km	Meir	VI.L
Ḫnj/Špsj-pw-Mnw	el-Hawawish	VI.L
K3.j-ḥp/Ṯtj-jkr	el-Hawawish	VI.L
Dʿw	Deir el-Gebrawi	VI.L
W3hj	el-Hagarsa	VIII
Mrjj-3ʿ	el-Hagarsa	VIII
Gḥs3/Nbjj (GA11)	el-Hawawish	VIII

All the scenes portraying goats in this time period are drawn from the central Upper Egyptian nomes (U.E. 9 to U.E. 14). Goats appear in thirteen of the thirty-one chapels and there is a notable development in their presentation; although the theme of the browsing herd is retained, it gives way in importance to that of the herd being paraded before the tomb owner and in some cases being shown as part of the cattle count. In seven instances the herd is depicted as part of the parade of animals before the tomb owner.[394] In the Meir tombs of *Ppjj-ʿnḫ-ḥrj-jb* and *Ppjj-ʿnḫ/Ḥnj-km* goats are part of the census and are specifically mentioned in the captions that identify the scenes that the tomb owner is 'viewing'. Six of the tombs present a herd of goats twice, in two distinct motifs: browsing among trees and the animal parade.

6.5 Goats in the parade of animals

6.5.1 Historical development of images and themes

Dynasty 4 to mid Dynasty 5

In this period only one tomb presents goats as part of the parade of animals. On the second lowest register, following a herd of cattle *Rʿ-ḫʿ.f-ʿnḫ* depicts a herd of goats as overlapping figures. Above the goats is written the number '2235'.[395]

Late Dynasty 5 to early Dynasty 6

Only two tombs, both provincial, present goats in the parade of animals. The scene in *Ḥzjj-Mnw* (M 22 of el-Hawawish) is too damaged to provide detail on how goats are portrayed.[396] The tomb of *Jttj-Šdw* of Deshasha is the first to present the two scenes of goat husbandry and a herd of goats in an animal parade, that become a feature in later Dynasty 6 and Dynasty 8.[397]

In the provincial tombs of this period the elements of goat husbandry are much reduced and in some cases disappear altogether. Even when the theme of goat husbandry is retained, as in the tomb of *Ppj-ʿnḫ-ḥrj-jb*, [398] it is associated with cattle rather than with tree felling and boat building, which is generally the association in earlier tombs. In all, goats are associated with cattle and other livestock in parades in nine tombs dated to the period between mid Dynasty 6 and Dynasty 8, while a number is provided for the herd of goats in three tombs: 32,400 for *Ppj-ʿnḫ/Ḥnj-km* of Meir, 12,000 for *Mrjj* and 21,000 for *W3hj*, both of el-Hagarsa. The positioning of the goat theme, together with the numbers suggests that goats were included in an animal census at this time. In the latest tombs the two themes of husbandry and parade of animals are combined on the same register suggesting that the concept of husbandry was giving way to the growing dominance of numbered herds.

6.6 Material evidence

Neither Kom el-Hisn nor the Giza Plateau Mapping Project offers direct evidence of resources enjoyed by

[391] Herders may have originally been included in *Jttj/Šdw*, but the central portion of the scene is lost: Petrie (1898) pl. 15; Kanawati–McFarlane (1993) pl. 46.

[392] Kanawati–McFarlane (1993) pl. 46, p. 50.

[393] Kanawati *Gebrawi* II (2007) pl. 50.

[394] This is also the case in the scene of *Jttj/Šdw*: Kanawati–McFarlane (1993) pl. 46, p. 50.

[395] LD II 9.

[396] *Ḥzjj-Mnw* Kanawati IV (1983) fig.12.

[397] *Ppj-ʿnḫ/Ḥnj-km* Blackman–Apted (1953) pl. 32; *Mrjj-ʿ3* Kanawati *el-Hagarsa* III (1995) pl. 37, 40; *W3hj* Kanawati (1995) *el-Hagarsa* III pl. 20, 22-3.

[398] *Ppj-ʿnḫ-ḥrj-jb* Blackman (1924) pl. 14.

high officials but they do provide a rough picture of the consumption of livestock at the Giza construction site in Dynasty 4 and the production and consumption of livestock at Kom el-Hisn, which may have been a typical Delta pastoral establishment. A central authority brought young, male sheep and goats in large numbers to Giza where they were slaughtered; the vast majority of faunal remains consist of sheep, goat and cattle bones. Kom el-Hisn is only four kilometres from the desert fringe, which would have allowed herders to exploit this transitional area with flocks of goats and sheep.[399] Yet at Kom el-Hisn pig rather than sheep or goat bones dominate the record, suggesting that sheep and goats were not slaughtered in any quantity *in situ* but were driven to the capital perhaps as provisions for a temple or construction site. Whether pastoral estates such as Kom el-Hisn were managed by high officials is not known, but goats were included in the official 'cattle census' represented in a number of tombs, particularly in Upper Egypt in Dynasty 6 and later, indicating that they were valued by the state as well as locally.

6.7 Summary

Sheep

- Sheep portrayed in Old Kingdom scenes have long legs, a thin tail and twisted horns growing laterally from the head. Their wool, which was shed in a dense mat, grew beneath a hairy coat. In the Nile valley and Delta sheep would have grazed on the drier upland areas.

- The majority of Old Kingdom scenes with sheep portray small flocks, driven by men with whips and sticks, trampling seed into prepared soil. Sheep are rarely included in other themes such as the parade of animals or threshing grain.

- Although sheep are typically portrayed trampling seed, findings from Kom el-Hisn and the Giza Plateau Mapping project provide evidence that sheep were consumed in considerable quantity by the elite classes and workers.

- All except two of the scenes of sheep trampling seed into the soil are Memphite, of which two appear in Dynasty 4 tombs and two in early Dynasty 6 tombs. The remainder occur in tombs dated to Dynasty 5.

- The two scenes portraying a flock of sheep trampling seed in Dynasty 6 are both Memphite and date to early in the dynasty. Two other mid-dynasty provincial scenes present a flock of sheep in what appears to have been a count of animals. Apart from these four depictions, sheep disappear from the pictorial record in the Sixth Dynasty.

- In general, the portrayal of sheep is similar to that of donkeys. They are never depicted in scenes of husbandry and rarely shown being proudly paraded before the tomb owner. As with the donkey, the depicted interest in sheep is of making use of the characteristics of the species.

Goats

- The two major themes are goat husbandry and the presentation of the herd of goats before the tomb owner.

- In Dynasty 5 and early Dynasty 6 the theme of husbandry dominates.

- These scenes concentrate in particular on the provision of adequate fodder for the herd from terrain that appears to provide little pasturage. Consequently the most frequent image is of goats nibbling the foliage of trees.

- Herders are assigned a prominent role in husbandry scenes. They are portrayed pulling down branches of trees so that the goats can browse on the foliage, separating fighting males and protecting the birth of kids from other animals. Young goats are portrayed in protective cages presumably erected by herders.

- Herders must have lived day and night with the herd as they are portrayed skinning the carcass of a goat hanging from a tree and cooking joints of meat.

- Until mid Dynasty 6 only three tombs contain scenes in which goats are portrayed as part of the parade of animals. In the later provincial tombs the herd of goats depicted in animal processions becomes more prominent, while husbandry becomes a minor theme often tacked on to the end of the paraded herd. In three tombs the herd of goats appears to be part of a census.

- Although the Giza Plateau Mapping Project offers no direct evidence of resources enjoyed by high officials, faunal remains show that in Dynasty 4 young, male goats were consumed in considerable quantity, presumably supplied by the state.

[399] Redding (1992) 99-107; Redding in Lehner (2002) 171-78.

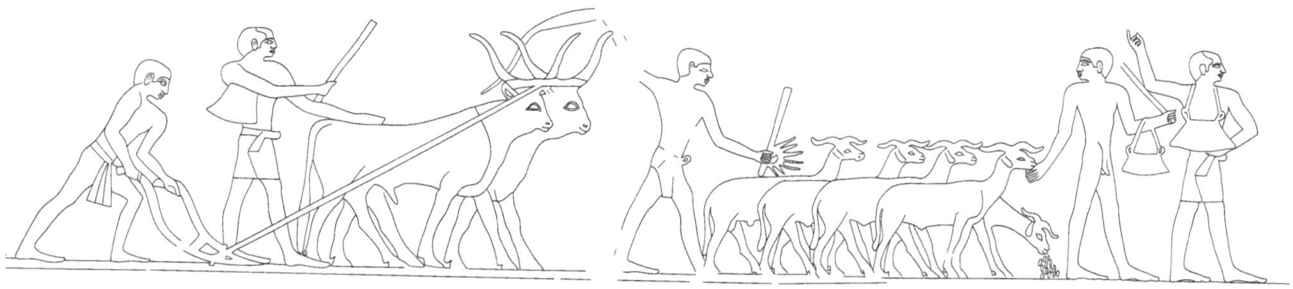

Figure 21: SHEEP TRAMPLING SEED INTO SOIL, TOMB OF NEFER AND KAHAY, SAQQARA.
Line drawing courtesy of the Australian Centre for Egyptology.

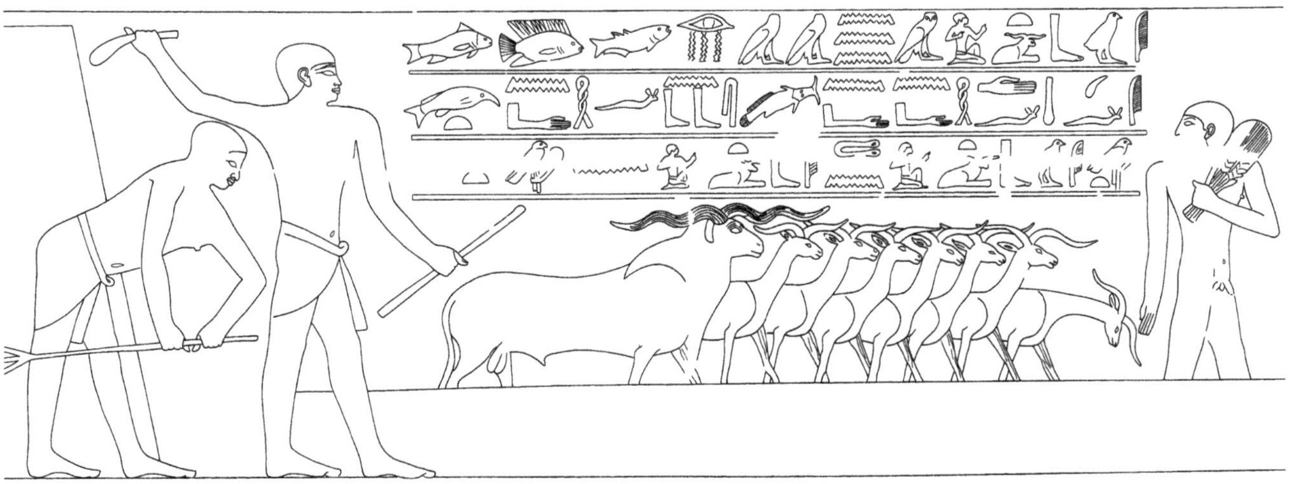

Figure 22: SHEEP ON THE THRESHING FLOOR, TOMB OF MERERUKA, SAQQARA.
Kanawati (2011) pl. 83.

Figure 23: HUSBANDRY OF A HERD OF GOATS, TOMB OF NEFER AND KAHAY, SAQQARA.
Line drawing courtesy of the Australian Centre for Egyptology.

CHAPTER 7 DONKEYS

7.1 Focus of enquiry: donkeys as an official's resource

TABLE 8: REPRESENTATION OF DONKEYS

The data:

- The iconography and associated inscriptions relating to the use of donkey herds on estates
- Other inscriptional data relating to the size of donkey herds.

Areas of research:

- Themes in which donkeys appear
- The use of donkeys on estates
- The value placed upon the ownership of a herd of donkeys.

Equus asinus (*ˁ3* Wb 1, 165.6-11)
Unlike the other domestic mammals, most of which were introduced into predynastic Egypt from the Middle East, *equus asinus* was probably indigenous to Africa. Mitochondrial DNA comparisons suggest that domestic donkeys were originally bred from the Nubian *equus africanus africanus* and the Somali *equus africanus somaliensis,* two subspecies of the wild African ass (*equus africanus*).[400] The earliest evidence of domesticated donkeys in Egypt, from El-Omari, dates to the fifth millennium B.C. although the first hard evidence for donkeys used as transport animals has come from the discovery of ten articulated donkey skeletons buried in the funerary complex of Abydos, dating to the Early Dynastic period around 3000 BC. Despite their formal burial, all the Abydos donkey skeletons exhibited a range of joint wear and other pathologies in their bones consistent with load carrying and typical microscopic fractures that arise in the vertebral bones due to overloading stress.[401]

Donkeys are herbivores that prefer grass, but with teeth adapted for tearing and chewing plant matter they also eat a range of shrubs and desert plants and can survive on a range of poor fodder, Having a higher thirst threshold than other equids, donkeys can go without water for up to three days and tolerate dry, hot conditions browsing or grazing on any available forage.[402] If donkeys are not provided with fodder, they require seven hours daily to graze or browse.

The domesticated animal can live up to 50 years, which gives it a working life of over 40 years. Donkeys in the service of humans have the advantage of being easy to maintain. Primarily used for the transport of goods, they are reported to be able to carry loads in excess of 200 kg for short distances.[403] Although the animal has a reputation for stubbornness, it is usually amenable to commands. Power has noted that in an Old Kingdom corpus of 550 images, donkeys are only portrayed once, in the tomb of *ˁnḫ -m-ˁ-Ḥr*, expressing antagonism to their herder.[404]

[400] Beja–Pereira (2004).
[401] Rossel (2008) 3715–3720.
[402] Grinder–Krausman–Hoffmann (2006) 1-9; Moehlman (1998) 132.

7.2 The portrayal of donkeys in Old Kingdom mortuary chapels

Apart from a comparatively small number of scenes of a herd of donkeys in a parade of animals,[405] the only theme in which donkeys are consistently portrayed is that of a 'beast of burden', usually involved in the cereal harvest or, vary rarely, carrying either the tomb owner or a crate of young desert ungulates. Depicted individually or in small herds, donkeys feature frequently in Old Kingdom iconography from Dynasty 5 on, when estate activities became major themes in the wall scenes of the tombs of officials. In these scenes donkeys appear working for man in two major themes: either as a means of transporting the produce of the estate[406] (*Figure 25*) or threshing the cereal harvest (*Figure 26*). Harvest transport scenes occasionally contain portrayals of individual donkeys that are not actively working but the context of these vignettes is always that the animal is being brought under control or waiting to be used.[407] Power interprets the portrayal of the herder holding the animal's foreleg and/or ear in these vignettes, as in the chapel of *Mrrw-k3,* as a method of controlling the animal[408] (*Figure 25*). Sometimes a small herd, usually under human control, is portrayed in the parade of animals,[409] but this theme curiously does not appear in the period from late Dynasty 5 to mid Dynasty 6.

7.2.1 Historical development of images and themes

Dynasty 4 to mid Dynasty 5
Donkey 'themes' are portrayed in the following number of tombs (*Figures 24, 25, 26*):

[403] Brewer (1994) 100.
[404] Power (2004) 134-35, fig. 8.
[405] See TABLE 8: REPRESENTATION OF DONKEYS.
[406] In three scenes donkeys are pictured carrying the tomb owner. See Moussa–Altenmüller (1977) pl. 42, 43; LD II 43 [a]; Hassan (1944) fig. 104; Martin (1979) pl. 33 [90].
[407] Hassan (1944) fig. 26; LD II, pl. 9; LD II, pl. 47 upper; LD II, 80 [a]; Junker XI (1953) fig. 75, pl. 20 [b], 21 [a-c]; Wild III (1966) pl. 153; Duell II (1938) pl. 169.
[408] Power (2004) 135; Duell II (1938) pl. 169.
[409] See TABLE 8: REPRESENTATION OF DONKEYS.

Transport of harvest:	Herd driven to field	3
	Laden donkeys	9
Threshing		4
Other transport		1
Animal parade		5

Donkeys featured in one or more of the above themes are depicted in the following tombs:

Late Dynasty 4 to early Dynasty 5

Nj-k3w-Rᶜ	Giza	IV.L
Wḥm-k3.j	Giza	V.E
Rᶜ-wr	Giza	V.E
Ḫnmw-ḥtp	Saqqara	V.E
Snb	Giza	V.E

Mid Dynasty 5

GIZA

Jj-mrjj *K3.j-m-nfrt*

Rᶜ-ḫᶜ.f-ᶜnḫ

SAQQARA

Wr-jr.n-Ptḥ

Nj-ᶜnḫ-Ḫnmw and *Ḫnmw-ḥtp*

Rᶜ-m-k3.j

In Dynasty 4, the only tomb with an extant scene containing donkeys is that of *Nj-k3w-Rᶜ*, where a single animal is led in a parade of animals and estates towards the tomb owner. By mid Dynasty 5, however, the two major themes associated with working donkeys are developed: transport of the cereal harvest and threshing the spikes of grain. The tomb of *Jj-mrjj* is perhaps the earliest with surviving, detailed harvest scenes. The north wall of the second chamber of his chapel presents the two main aspects of the transport of the harvest to the threshing floor: a line of donkeys driven towards the bales of cereal sheaves, and donkeys carrying their loads to the threshing floor. Each animal is driven to the pile of bales by a herder. The tomb of *Nj-ᶜnḫ-Ḫnmw* and *Ḫnmw-ḥtp* presents a similar scene.[410] The theme of loading bales on to the backs of the donkey herd has survived in six tombs of early to mid Dynasty 5 and the journey to the threshing floor is represented in nine tombs of the same period. In each case two or more workers load the animals and two figures accompany each laden animal. The more senior figure is portrayed brandishing a stick to drive the animal forward and the other, sometimes a smaller, younger figure, walks beside the animal to keep the load in place. The shape of the bales of sheaves varies from one tomb to another. Blankets are thrown over the laden animals to protect their backs. The donkeys being driven to the field to be loaded are depicted as a group with one figure of an animal partly superimposed on the animal in front, according to Schäfer a device to represent a forward motion.[411] In the scenes of loading the donkeys and leading them to the threshing floor each group consisting of the animal and its handlers is portrayed as an individual vignette, with the focus of interest on the safe transport of the load of sheaves.

[410] *Nj-ᶜnḫ-Ḫnmw* / *Ḫnmw-ḥtp*: Moussa–Altenmüller (1977) pl. 56-9.

[411] Schäfer (1986) 293-4.

Where donkeys are portrayed in the threshing scenes two herders, pictured at each side of the herd, use sticks to control the animals. Confusion appears to reign among the pack of donkeys on the threshing floor in the tomb of *Rᶜ-ḫᶜ.f-ᶜnḫ* and three men are needed to control the situation. One herder raises his stick to an animal intent on going in the wrong direction; another herder controls an animal by holding its foreleg and a third herder stands behind the pack with his arms stretched wide, presumably to keep the animals moving in the required direction. Two animals have lowered their heads to eat the cereal on the threshing floor.[412]

Although donkeys must have provided an ubiquitous form of transport on large estates, they are rarely depicted carrying anything except grain to the threshing floor. However, the richly decorated tomb of *Nj-ᶜnḫ-Ḫnmw* and *Ḫnmw-ḥtp* provides a rare vignette of a donkey laden with a crate of the young of desert ungulates, probably gazelle to judge from the horns of eight calves whose heads poke out of the crate. The register consists of an adult scimitar horned oryx guided by two herders, a laden donkey, an overseer offered a drink and two porters, the first carrying two crates of desert ungulate young and the second a small carcass and a bag. Above the men is the caption: *jwjt m jmnt* 'Coming from the West', which suggests that the images on this register represent the party returning with animals captured by a hunting expedition on which donkeys were taken as means of transport. On the other hand, the vertical inscription flanking the panel of registers is a generalised statement: *m33 nḏ.t-ḥr jnnt m njwwt.f nbt tzt jḫt bnrt r pr-ḏšr* 'Viewing the presentations which are brought from all his towns and the heaping up of sweet things at the *pr-ḏšr*' that has no specific reference to the register containing the image of the donkey. Two separate concepts, the return from the hunt and the tomb owner viewing activities on and presentations from his estate, are included in the same panel of registers.

In the tomb chapels of *Wḥm-k3.j* and *Rᶜ-ḫᶜ.f-ᶜnḫ* animal parade scenes demonstrate more clearly the donkey herd as valued livestock. In *Wḥm-k3.j*'s scene a line of donkeys follows lines of cattle and sheep on two sub-registers, with scribes making a record that is presented to the tomb owner. *Rᶜ-ḫᶜ.f-ᶜnḫ*'s scene has the number '860' inscribed above the last animal in line.

Second half Dynasty 5

Transport of harvest:	Herd arriving	23
	Laden donkeys	26
Threshing		25
Other transport		2
Animal parade		0

[412] LD II 9 (lower).

Donkeys featured in one or more of the above themes are depicted in the following tombs:

GIZA

3ḫtj-mrw-nswt	*S3-jb*
J3zn	*Snḏm-jb/Jntj*
Nj-m3ʿt-Rʿ	*Snḏm-jb/Mḥj*
Nj-ḥtp-Ptḥ	*Sḫm-ʿnḫ-Ptḥ*
Nfr (1)	*Sḫm-k3j*
Nfr-b3w-Ptḥ	*Sšm-nfr* IV
Rʿ-wr II	*K3-ḥj.f*
Ḫww-wr	*Kd-nfr*

SAQQARA

3ḫtj-ḥtp (Louvre)	*Ḥtp-ḥr-3ḫtj*
3ḫtj-ḥtp D64	*Ḥtp-k3.j/Tp-k3.j*
Jrj-n-k3-Ptḥ	*Ḫnwt*
Jḥjj (r/u *Jdwt*)	*Spd-ḥtp*
Pḥn-wj-k3.j	*Sḫm-ʿnḫ-Ptḥ*
Ptḥ-ḥtp	*K3.j-m-nfrt*
Nj-k3w-ḥr	*K3.j-m-rḥw*
Nfr-jrt-n.f	*Ṯjj*
Nfr-sšm-Ptḥ and *Sḫntjw*	

SHEIKH SAID

Wr-jr.n.j

ZAWYET EL-MAIYETIN

Ḫw-ns

The theme of transporting the cereal harvest continues to include the two motifs in which donkeys are represented. The first of these, donkeys being driven back from the threshing floor to collect further bales of cereals, maintains the portrayal of a line of layered, overlapped herders driving forward a line of layered, overlapped animals. Each animal must have had its own driver as drivers and donkeys are almost invariably depicted in equal number. The number of drivers and donkeys varies from four to eight. The small herds appear to be moving at a gallop with each driver holding a stick over his shoulders and the first driver sometimes waving two sticks. While the depicted numbers of animals are usually small, the scene in *3ḫtj-ḥtp* carries the caption *sḥ3ṯ* 2500 *ʿ3* ('driving back 2500 donkeys'),[413] Occasionally the first driver steps forward either to keep the leading animal or the entire pack moving.

Scenes of laden donkeys transporting bales of sheaves to the threshing floor have considerable detail and variety, although few tombs show the actual loading of the bales on to the animals' backs. In the list of tombs used in this study only that of *Ṯjj* (D22) portrays workers lifting a bale on to the back of a donkey.[414] This stage, however, is suggested in a number of tombs, either by depicting the returning pack of donkeys adjacent to, and facing men packing and tying bales,[415] or by a worker, with piles of sheaves behind him, struggling to control the lead donkey by holding its ear and foreleg.[416] Usually two farmhands accompany each animal on its journey to the threshing floor. As in the chapel of *3ḫtj-ḥtp* (D64), a man to the front may control his donkey by holding its head, while another walks beside or behind the animal to steady the bale of sheaves. In eight of the scenes the figure walking beside the donkeys is that of a young boy. In *Pḥn-wj-k3.j, Ḫw-ns, K3-ḥj.f* and *Ṯjj* foals accompany the donkey pack, perhaps as a means of training. Some scenes make it quite clear that the donkeys are carrying their load to the threshing floor. In the tombs of *Nfr-jrt-n.f, Nfr-b3w-Ptḥ* and *Rʿ-m-k3.j* the laden donkeys are depicted next to the stack of sheaves awaiting threshing.

Although cattle and sheep are depicted in seven tombs on the threshing floor, most of the threshing appears to have been carried out by a pack of donkeys. Besides being portrayed on a second threshing floor in the tombs where cattle or sheep are also shown to be threshing, donkeys appear in a further eighteen threshing scenes in this time period. Each scene has a slightly different arrangement of animals but the basic composition remains the same. Two men on either side of the pack brandish raised sticks to force the animals to keep moving in the required direction so that their trampling on the sheaves separates the kernels from the husks. Most of the pack appears to be moving but one or two animals lower their heads to the floor, presumably trying to eat some of the corn. One or two animals raise their heads towards a herder in an aggressive pose, with mouth open and ears laid back.[417] Many of the images of donkeys in the threshing scenes and harvest transport scenes are depicted with ears inclined backwards although not flattened against the neck. Evans interprets this as a sign of tiredness and judges that the portrayals of working donkeys generally represent the animals as being overworked.

Dynasty 6

Domesticated donkeys are portrayed in the following number of tombs dated to the first half of Dynasty 6:

Transport of harvest:	Herd arriving	4
	Laden donkeys	11
Threshing		9
Animal parade		0

The number of scenes extant from late Dynasty 6 to Dynasty 8:

Transport of harvest:	Herd arriving	0
	Laden donkeys	9
Threshing		2
Animal parade		3

Donkeys featured in one or more of the above themes are depicted in the following tombs:

Jn-Snfrw-jšt.f	Dahshur	VI.E
Jttj-Šdw	Deshasha	VI.E
ʿnḫ-m-ʿ-Ḥr	Saqqara	VI.E
Mrrw-k3/Mrjj	Saqqara	VI.E
Mḥw	Saqqara	VI.E
Mṯtj	Saqqara	VI.E
Nj-ʿnḫ-Ppjj	Zawyet el-Maiyetin	VI.E
Ḥm-r3/Jzj (I)	Deir el-Gebrawi	VI.E

413 Davies II (1902) pl. 8.
414 Wild III (1966) pl. 154.
415 van de Walle (1978) pl. 12.
416 Mohr (1943) fig. 50.

417 Evans (2006) 212-13.

Ḫnmw-ntj	Giza	VI.E
Jbj	Deir el-Gebrawi	VI.M
B3wj BA 49	el-Hawawish	VI.M
Ppj-ʿnḫ-ḥrj-jb	Meir	VI.M
Mmj	el-Hawawish	VI.M
Snfrw-ḥtp	Giza	VI.M
Ppj-ʿnḫ/Ḥnj-km	Meir	VI.L
Mḫw and *S3bnj*	Qubbet el-Hawa	VI.L
Ḥzjj-Mnw/Zzj	el-Hawawish	VI.L
Ḫnj/Špsj-pw-Mnw	el-Hawawish	VI.L
K3(.j)-ḥp/Ṯtj-jkr	el-Hawawish	VI.L
Ḏʿw	Deir el-Gebrawi	VI.L
ʿnḫ-tj.fj	Moʿalla	VIII
W3hj	el-Hagarsa	VIII
Mrjj-ʿ3	el-Hagarsa	VIII
Rḥw-r-3w.sn	el-Hawawish	VIII

Donkeys continue to be depicted in harvest scenes well into Dynasty 8, laden with sheaves that they carry to the threshing floor or working as small herds driven round in circles so that they trample on the sheaves to separate the grain from husks. But the scene of the herd being driven back to the fields by stick-wielding herders for further loads does not appear in tombs dating to the second half of Dynasty 6. The number of Dynasty 6 scenes showing the donkey pack arriving at the fields to be loaded is remarkably low.[418] The omission of this theme from provincial tombs of mid Dynasty 6 and later may be due to reduced wall space.

However, donkey herds remained an important resource and their value may have increased in Dynasty 6. There is clear evidence of breeding programmes in the reign of Pepy I, perhaps practiced throughout the Old Kingdom. At el-Hawawish a *sḥḏ ḥm-nṯr, ḥrj-tp nswt pr ʿ3* named *Ḳrrj* recorded that he had acquired a number of female donkeys from which he bred a herd of 200:

> *jw jr.n(.j) ḥmwt bnnt jw rdj.n.sn ʿ3 200 jr s nb sw3.tj.fj ḥr w3t tn*[419]

In the second half of the dynasty herds of donkeys begin to appear on registers of animals being presented for the census to the tomb owner. *Ppj-ʿnḫ/Ḥnj-km* of Meir and *Ḏʿw* of Deir el-Gebrawi include donkey herds among their parade of animals but do not give numbers of their herds as they do for other domestic animals.[420] In Dynasty 8 *Mrjj-ʿ3* of el-Hagarsa also portrays donkeys among the parade of domestic animals.[421]

The value of a herd of donkeys would have been enhanced by their use for a variety of burdensome tasks beyond the demands made of them on the estates. Relations with the oases had to be maintained.[422] The men who led expeditions in the Old Kingdom would have taken large herds of pack animals to transport food and water[423] and bring back the heavy loads produced in mines and quarries or bales of tribute and trade from surrounding tribes.[424] Donkeys were thus important to the economy in many ways, although in officials' tombs their value is only reflected in agricultural scenes. Apart from these scenes only a couple of vignettes reflect the multifarious services that donkeys provided: a container of baby gazelles loaded on the back of a donkey, presumably being brought from a desert hunt,[425] and an occasional scene of a tomb owner riding a donkey.[426]

7.3 Summary

- Donkeys almost invariably appear either as a secondary element in the sequence of scenes representing the harvesting of cereals or as a small herd in the parade of animals.

- In harvest scenes donkeys appear in two themes, both of which represent the animals working for man, either as a means of transporting the produce of the estate or threshing the cereal harvest.[427]

- It may be assumed that these were the animals' major functions on an estate belonging to an official.

- In harvest scenes the pack is shown being driven by herders waving sticks to collect bales of cereals, being led to the threshing floor laden with a large bale of cereals and being forced by herders using sticks to keep moving round on the threshing floor.

- The greater number of scenes, with donkeys, present the herd transporting bales of cereals to the threshing floor. This theme is maintained in the smaller provincial tombs of mid Dynasty 6 and later, when the portrayal of the donkey pack being driven to collect their loads is omitted.

- Unlike portrayals of cattle and goats, donkeys are not depicted in scenes indicating husbandry; they are not shown mating or giving birth and foals only appear walking beside their mother. Donkeys are never depicted grazing freely.[428]

418 See TABLE 8: REPRESENTATION OF DONKEYS.
419 Kanawati VI (1986) fig. 20b.
420 *Ppjj-ʿnḫ/Ḥnj-km*: Blackman *Meir* IV (1924) pl. 32; *Ḏʿw:* Davies *Gebrâwi* II (1902) pl.6.
421 *Mrjj-ʿ3*: Kanawati *el-Hagarsa* III (1995) pl. 35, 37, 39-40.
422 Kaper–Willems (2002) 79-94.
423 Frank Förster discusses the logistical arrangements used in the late Old Kingdom/First Intermediate Period for long distance desert travel with herds of donkeys: Förster (2007).
424 Harkhuf records that on his third expedition to Yam he followed the chief of Yam to the Land of the Libyan" and finally returned with a caravan of 300 donkeys carrying luxury goods: Urk I 125, 13-127, 17.
425 Moussa–Altenmüller (1977) pl. 42, 43.
426 Hassan V (1944) fig. 104; *Ḥww-wr* rides two donkeys: LD II 43 [a]. The other two Old Kingdom riding scenes are Moussa–Altenmüller (1977) pls. 42 and 43.
427 See TABLE 8: REPRESENTATION OF DONKEYS.
428 In order to get sufficient nourishment, donkeys either have to graze for six to seven hours a day but if required to work, they have to be supplied with additional fodder: Evans (2006) 196-197; Aganga–Letso– Oaganga (2000).

- The size of herds claimed by tomb owners seems out of proportion to the needs of an individual estate. Apart from their depiction in 'animal parade' scenes, donkeys always appear as working animals; yet inscribed numbers, presumably presenting the size of the herd, far exceed the numbers of animals that would have been needed on even a large estate.
- Inscriptional evidence, such as officials' autobiographies, suggests that donkeys were important to the economy in many ways, although in the iconography of officials' tombs their value is only reflected in agricultural scenes. Apart from these scenes only a couple of vignettes reflect the range of services that donkeys would have provided.

Figure 24: HERD OF DONKEYS DRIVEN TO THE FIELD TO BE LOADED WITH SHEAVES OF CEREALS, TOMB OF MERERUKA, SAQQARA. Kanawati (2011) pl. 81

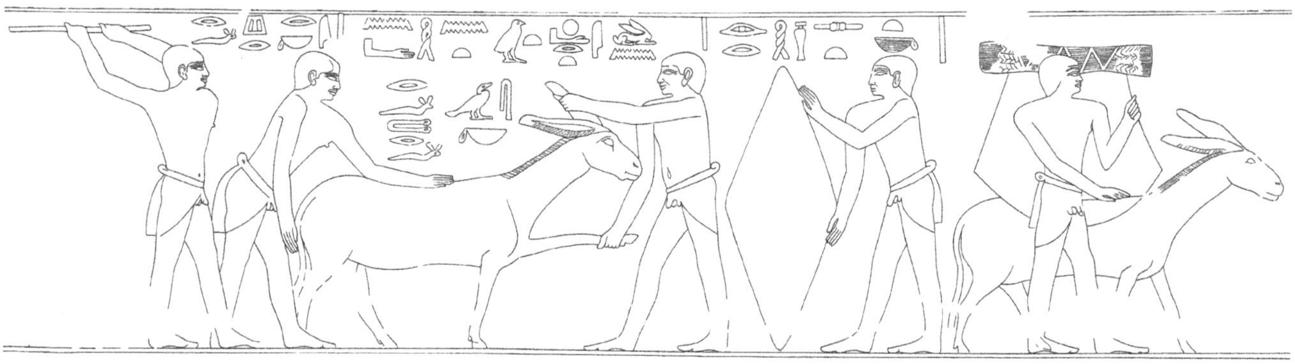

Figure 25: CONTROLLING A NERVOUS DONKEY AND A DONKEY CARRYING SHEAVES TO THE THRESHING FLOOR, TOMB OF MERERUKA, SAQQARA. Kanawati (2011) pl. 82

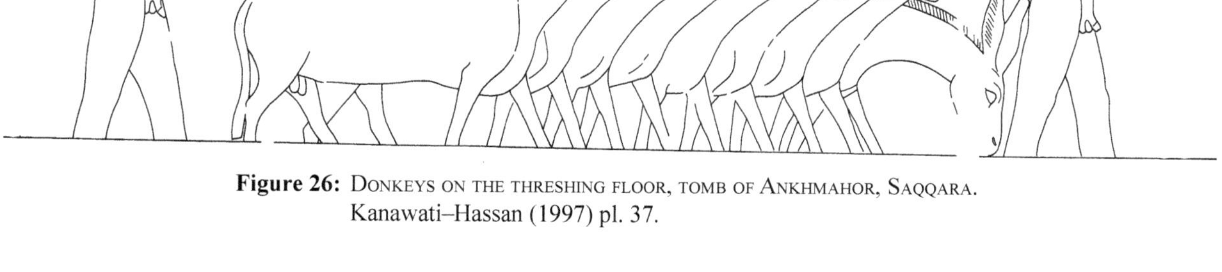

Figure 26: DONKEYS ON THE THRESHING FLOOR, TOMB OF ANKHMAHOR, SAQQARA. Kanawati–Hassan (1997) pl. 37.

SECTION THREE

THE AGRARIAN CONTEXT

This section investigates the production of field crops as factors in the economy of estates from which elite officials drew their resources. The consequences of the adoption of agrarian farming have been much debated by prehistorians. Gordon Childe proposed a general thesis that the introduction of agriculture was the precondition that enabled societies to produce a surplus of food, making possible the development of cities and a statewide administration.[429] This concept, essentially linked to the basic idea of a 'Neolithic Revolution', has been widely accepted although more recent research rejects Childe's hypothesis that there was a sharp break between hunter/gathering and agriculture.[430] In the case of ancient Egypt, although domestic crops were introduced between 6000 and 5000 B.C., archaeological evidence suggests there was a long transitional phase in which agricultural settlements coexisted with villages dependent on both foraging and farming, a situation that lasted into the third millennium B.C.[431]

Accepting Ermann's generalised view, that agriculture was the foundation of Egyptian civilization[432], J. H. Steward theorized that the need to construct and control a system of irrigation to support agriculture led to the formation of a state administration and bureaucracy.[433] R. C. Allen accepted this interpretation that the introduction of agriculture into the Nile valley triggered "a dynamic that led to a unified state in Egypt".[434] He claims that by producing a surplus of comparatively non-perishable crops and livestock and a *seasonal* surplus of labour, the adoption of agriculture made feasible both the exploitation of farmers by the state and a stratified society with an elite bureaucracy at its apex. He further maintains that the "main function of the Pharaonic state was to transfer a considerable fraction of the income produced by Egypt's farmers to an unproductive aristocracy." This, he argues, was due to the development of farming combined with the particular geography of Egypt, which restricted settlement to the Nile valley, thus allowing a central authority to impose demands and controls on the movement of the farming community.[435] Although he is not specific, Allen must be referring to the production of field crops because the farming of livestock is a year-round activity, which does not produce a seasonal labour surplus. Allen's acceptance that agrarian farming led to a unified Egyptian state may be challenged, but by the Old Kingdom the production of cereals surplus to the subsistence needs of a peasant population was required for the maintenance of a complex court and bureaucracy and the construction of royal works. Only large agricultural units such as the rural estates held or managed by elite officials could have provided in sufficient quantities the surpluses of food and (seasonal) workers needed to support the structure of the pharaonic state and its building programmes.

Agrarian farming was not an Egyptian discovery. The farming and consequent domestication of cereals first developed to the north-east of Egypt in the Fertile Crescent about nine and a half thousand years ago[436] and were introduced into Egypt one to two thousand years later.[437] By Dynasty 3, it is probable that all settlements up and down the Nile valley had adopted both animal husbandry and agrarian farming, although the evidence is patchy and scattered. The existence of early dynastic necropoleis at Helwan and Saqqara and large royal funerary structures make it clear that farming was well developed in the Memphite region. The massive royal building programmes of Dynasties 3 and 4 indicate both the regular if seasonal supply of surplus labour and the state's command of surplus food production with which to feed the workforce. This argues for the spread of farming down the Nile on a scale beyond that of subsistence farming and which could only have been carried out on large estates closely tied to the central authority.

Field crops of the Old Kingdom[438]

The only domesticated field crops depicted in Old Kingdom tombs are barley, wheat and flax, none of which originated in Egypt. In Egypt, barley, emmer wheat, and flax are winter crops; they are sown in late autumn – in Pharaonic times after the inundation had receded – and harvested in spring.[439]

---

429 Childe (1936) 74-117.

430 Binford (1962), (1972); Hodder (1989) 268-274; Hodder (1993) 253-257.

431 Krzyzaniak (1977) 85, 89, 90, 103, 132, 140; Martin et al. (1984) 196–197.

432 Ermann (1894, reprint 1971) 425.

433 Steward (1949) 1–27.

434 At the same time, Allen (1997) 135–154 appears to accept the paleopathologic conclusions of Cohen–Armelagos (1984) 586-594 that the adoption of agriculture in a number of hunter/gatherer societies is associated with reduced living standards and health.

435 Allen (1997) 135-154.

436 Redman (1978) 88–140; Moore (1985) 23, 39–43, 61.

437 The earliest evidence of barley and emmer wheat has been found at Merimda Beni Salami and the Fayum. See Wetterstrom (1993) 201; Hassan. F. A. (1988) 141; Arkell–Ucko (1965) 145-166; Trigger (1983) 17; Wenke–Long–Buck (1988) 29-51.

438 Old Kingdom funerary iconography shows a range of crops being grown on the estates of high officials, but the occasionally depicted fruit and vegetables would have been mainly perishable and probably intended for immediate consumption by the tomb owner and his family. Orchard scenes have been preserved in seventeen Old Kingdom tombs used in this study and gardening scenes in eight tombs. The crops in these scenes include fruits, leafy salad plants, legumes, roots and cucurbits.

439 While the present study does not centre on the technology of arable

farming, Siebels' research into the funerary iconography and associated inscriptions of Old Kingdom agriculture has been used in interpreting scenes and assessing the importance of the three portrayed field crops as a resource of officials: Siebels (2000) passim.

CHAPTER 8 FIELD CROPS – CEREALS

8.1 Focus of enquiry: cereal crops as economic resources for elite officials

TABLE 9: PRODUCTION OF FIELD CROPS
TABLE 10: GRAIN STORAGE, BREAD AND BEER PRODUCTION
TABLE 11: TITLES
TABLE 12: INSCRIPTIONS – TOMB OWNER VIEWING PLANTING AND HARVESTING OF FIELD CROPS

The data:

Iconography and associated inscriptions of:

- Soil preparation and planting
- Harvesting
- Dealing with the harvested crop
- Storage of the crop.

Other data:

- Funerary scenes: offerings, offering table scenes, offering lists
- Artefact finds.

Areas of research:

- Crop varieties
- Phases of field crop production
- Management of the phases
- Changes in presentation of themes and motifs over time and location.

8.1.1 Cereal crops represented in Old Kingdom tombs

Barley (*Hordeum* spp.)
(*jt* Wb 1, 142)

Barley is the most adaptable of all the cereals and can be grown in a variety of environments. While barley tolerates salinity, its resistance to dry heat makes it a particularly useful crop in near-desert environments; consequently it may have been more widely cultivated in Egypt than wheat.[440] However, although it has been established that the ratio of barley to wheat was 2:1 at Nagada Predynastic sites,[441] in the dynastic era the preponderance of barley is probable[442] but cannot be verified. The epigraphic evidence is unclear and confusing and in the past, finds of cereals have not been systematically analysed.[443]

Wheat: Emmer (*Triticum. dicoccum*)
(*bd.t* Wb 1, 486.14)

Emmer is an awned wheat with spikelets consisting of two well developed kernels. It originated from a mutation with no husk and probably dates from about 9000 B.P.[444] It is tall, growing to two metres, but its yield is low. Cultivated emmer has been found on sites across the Fertile Crescent from about 9000 years ago, and in Egypt from about 7000 years ago.[445]

Identifying the field crops

Emmer wheat and barley, considered to have been the two staple cereals in the prehistory of the Middle East, are the only Old Kingdom grain crops that can be identified in tomb scenes and can only be distinguished from each other by inscription. Depictions of ploughing do not allow the viewer to distinguish between the three crops (flax, barley and emmer wheat) and associated captions do not specify which seeds are being sown. Harvest scenes frequently show crops as a horizontal band across the landscape. The band may be divided by narrow vertical lines to suggest stalks, but the crop is hard to identify unless the flowering head of the flax plant or, in the case of grains, seed bearing spikes are depicted. Flax and grain crops can be distinguished from each other by the method of harvesting - the flax plants are pulled out of the ground by hand while the grain is cut with a short handled sickle – but it is much more difficult to distinguish barley from wheat unless a caption provides this information.

From the earliest harvest scenes in Dynasty 5 to Dynasty 8, crops ready for harvest are shown at a variety of heights: at waist height or less in 31 tombs; growing taller than waist high in 22 tombs while in nine tombs flax and grain are depicted growing at different heights. Lexicographic evidence from the scenes of cereal production indicates that Egyptians recognized different categories of both barley and wheat, but the principles on which the distinctions were made are unclear. The terms may represent different sub-species of barley and emmer, but as they are used as labels for the contents of granaries, they could refer to the quality of the crop or the intended use of the grain or even the place of origin of the seed crop.

The historical development of the iconography of cereal production has been divided into three themes:

- Preparation, which includes ploughing, hoeing, the scattering of seed and the use of a flock of sheep

440 Zohary–Hopf (1994) 55; Hassan F.A. (1997) 51-74.
441 Wetterstrom (1993) 165-226.
442 In 'viewing' inscriptions it is cited first in the list of crops being harvested. See Siebels (2000) 117.
443 For example, the Egyptian word *swt* may refer to a type of barley, although it is frequently read as 'wheat': Germer (1985) 14:210; Darby et al. (1977) 482, 490-1.
444 Harlan (1981); Zohary–Hopf (1993) Ch. 2.
445 The earliest Egyptian sites are in the Delta and the Fayum. See Wendorf et al (1970) 1168 and Hassan (1988) 151.

- Harvesting, which includes the reaping of barley and wheat, transport of the crops by a herd of donkeys, threshing, winnowing, sieving and stacking
- Storage, which includes measuring the grain, taking it to, and removing it from, the granaries.

The cultivation of flax is investigated separately. See Chapter 9: Field Crops – Flax.

8.2 Planting: ploughing, hoeing and seed

Figure 11

8.2.1 Historical development of images and themes

Between early Dynasty 4 and Dynasty 8 scenes representing the planting of seed appear in 61 of the tombs used in this study, eighteen of which date to Dynasty 6 or later.[446] Two forms of soil preparation are depicted: drilling furrows with ploughs pulled by oxen (occasionally by cows) and guided by one or two ploughmen, and hoeing by hand. The most consistently portrayed method of soil preparation is the turning over of the soil by ox-drawn ploughs. This method is depicted in tombs from early Dynasty 4 through to Dynasty 8.[447] In the list of tombs used in this study, the distribution of the theme of oxen pulling ploughs is as follows:

• Dynasty 4	4 chapels
• early to mid Dynasty 5	6 chapels
• late Dynasty 5	24 chapels
• early to mid Dynasty 6	6 chapels
• late Dynasty 6 to Dynasty 8	9 chapels

The four main components of this theme are:

- a ploughing team that usually consists of one man and yoked oxen pulling the plough, or two men, one of whom guides the plough, while the other drives the oxen forward (*Figure 11*)
- a team of men hoeing the soil with short handled hoes
- a sower casting seed grain from a bag hung from his shoulder
- a flock of sheep being driven by a gang of men with raised whips and led forward by a coaxer-sower offering handfuls of grain to the flock leader (*Figure 21*).

Mid Dynasty 5 to early Dynasty 6

Only six tombs, dating from mid Dynasty 5 to early Dynasty 6, contain all four features:

Jj-mrjj	Giza	V.M
Nfr-b3w-Ptḥ	Giza	V.L
Ḥtp-ḥr-3ḫtj	Saqqara	V.L
Sḫm-k3j	Giza	V.L
Mr.f-nb.f	Saqqara	VI.E
Ṯjj	Saqqara	V.L

The small quantity of tombs in this category is due to the limited number of scenes depicting farm labourers hoeing by hand. Otherwise, fifteen tombs contain three of the major elements, ploughing, casting seed and a flock of sheep, mainly dating to the period mid to late Dynasty 5. The ploughing element is retained to the end of the Old Kingdom. Other motifs that appear occasionally are an overseer leaning on his staff and the sower taking seed from a bag held by another field hand.

While the scenes of field hands working with hand-held hoes are rare and mainly confined to the second half of Dynasty 5,[448] the portrayal of an associated flock of sheep is included much more frequently. In the chapels of Dynasty 5 a flock of sheep accompanies the ploughing team in 29 seed planting scenes. This theme is usually interpreted as the use of the flock to trample the newly sown seed into the soil, but the relationship between the flock and the plough varies. The sheep may be depicted behind or ahead of the plough and the two elements, sheep and plough, may appear on two separate registers. This applies to the 'coaxer-sower'. The two different positions of the sower and flock of sheep vis-à-vis the plough may indicate two different techniques of sowing seed. When the sheep follow, the plough is used to create a furrow for the seed which is then covered by the trampling sheep; but when the flock is depicted in front of the plough, the sheep follow a sower trampling the seed into the soil, which is then turned over by the plough to cover the seed. The theme continues into early Dynasty 6 in the tombs of *Ḥsj* and *Mḥw* of Saqqara, the only two Sixth Dynasty tombs to depict sheep treading seed.[449]

Early Dynasty 4 to mid Dynasty 5

Ploughing is depicted in the following tombs:

*Jtt**	Medum	IV.E
Nfr-m3ʿt	Medum	IV.E
Rʿ-ḥtp	Medum	IV.E
Nb.j-m-3ḫtj#*	Saqqara	IV.L
Nswt-pw-nṯr#	Giza	V.E
Ḫnmw-ḥtp	Saqqara	V.E
Jj-mrjj#*	Giza	V.M
Jj-nfrt#*	Giza	V.M
Nj-ʿnḫ-Ḫnmw and *Ḫnmw-ḥtp*#*	Saqqara	V.M
Nfr ʿnd K3-ḥ3. j#*	Saqqara	V.M
Rʿ-ḥtp	el-Hammamiya	V.M
K3.j-m-nfrt#*	Giza	V.M
Dw3-Mnw	el-Hawawish	V.M

* Sower casting seed
\# Flock of sheep

446 See TABLE 9: PRODUCTION OF FIELD CROPS.

447 See TABLE 9: PRODUCTION OF FIELD CROPS.

448 The latest depiction of hand-held hoes appears in the early Dynasty 6 tomb of *Mr.f-nb.f*: Myśliwiec (2004) pls. 21, 66.

449 Siebels interprets a figure in front of a plough as a 'coaxer-sower' in scenes in *Mrrw-k3* Duell II (1938) pl. 168, *Ppj-ʿnḫ-ḥrj-jb* Blackman (1924) pl. 24 and *Ḥzjj-Mnw/Zzj* Kanawati (1987) fig. 4a. In *Mrrw-k3* the two figures in front of the plough wear the kilt of a supervisor. In *Ppj-ʿnḫ-ḥrj-jb* the equivalent figure is sowing seed. Blackman notes (p. 37) that the photograph of the scene (pl. 24, 2) shows seed falling from the sower's upraised hand. It is not likely that a coaxer-sower would be behind the sheep. The same comment may be made about the scene in *Ḥzjj-Mnw/Zzj*, where the figure in front of the plough is interpreted as a supervisor by Kanawati (p. 11).

The earliest scenes, on narrow registers in the Medum tombs of early Dynasty 4, are vignettes of ploughing and presumably represent the entire process of soil preparation and planting. The Dynasty 5 portrayal of the theme develops with the depiction of a team of men working the soil with hand-held hoes.

Late Dynasty 5
Ploughing is depicted in the following tombs:

GIZA	
Nfr (1)* #	*Sḫm-kꜣj*#*
Nfr-bꜣw-Ptḥ#*	*Kꜣ.j- m- ꜥnḫ*
*Ḫww-wr**	*Kꜣpj*
Snḏm-jb/Jntj#	*Kꜣ-ḥj.f*#*
Snḏm-jb/Mḥj#*	
SAQQARA	
Jrj-n-kꜣ-Ptḥ#*	*Rꜥ-špss#*
*Pḥn-wj-kꜣ.j**	*Ḥtp-ḥr-ꜣḫtj*#*
Nb-j-kꜣw-ḥr/Jdw#*	*Ḫnwt*#*
Nbt#*	*Sḫm-ꜥnḫ-Ptḥ*#*
Nfr-jrt-n.f#*	*Ṯjj*#*
Nfr-sšm-Ptḥ and *Sḫntjw*#*	
SHEIKH SAID	
Wr-jr.n.j#*	*Srf-kꜣ.j*
ZAWYET EL-MAIYETIN	
Ḫw-ns#*	

* Sower casting seed
\# Flock of sheep

Once the ploughing/sowing theme had been developed in the first half of Dynasty 5, there is little further artistic elaboration to the scene of soil preparation. In fact, the theme is reduced in detail in Dynasty 6 with the omission of the flock of sheep in all tombs except those of *Mḥw* and *Ḥzj*, both dated to early Dynasty 6. This reduction is particularly marked in the provincial tombs of Dynasty 6, where depictions of ploughing and sowing are maintained but all other motifs relating to planting are omitted.[450] This partial omission in the ploughing/sowing theme contrasts with the continued inclusion of multiple stages of the processing of the harvest in the same tombs but cannot be attributed to the lack of wall space in provincial rock-cut tombs.[451] In *Ppj-ꜥnḫ-ḥrj-jb*, for example, the tomb owner views six registers of livestock, three of which portray scenes of ploughing,[452] while *Ppj-ꜥnḫ/Ḥnj-km* is depicted viewing a long register portraying three sets of ploughing/sowing teams.[453] On a panel of four registers in the tomb of *Ḏꜥw* the two top registers each present three vignettes of ploughing teams, with the third register depicting the flax harvest and the fourth the cereal harvest.[454] These ploughing depictions are given little variation of detail, but their repetition may reveal heightened interest in the planting of crops.

8.3 Harvesting the cereal crops

Figure 27

8.3.1 Historical development of images and themes

Barley (*jt*) and emmer wheat (*bdt*) were winter crops, sown when the inundation subsided. Thus, scenes portraying the harvesting and storage of these crops represent activities taking place in the spring or early summer. Tombs begin to include detailed depictions of the harvest-storage process in Dynasty 5, although no tomb portrays every phase. The major activities were reaping, transport of sheaves, threshing, winnowing and storage in granaries, with additional minor themes and vignettes accompanying each stage.[455] A limited number of tombs portray separate fields with crops of either grain or flax, but the iconography does not permit the precise identification of cereal under cultivation.[456]

Reaping grain crops
The incidence of scenes used in the present list of tombs of reaping cereals and/or making bundles of sheaves is as follows:

• Dynasty 4	in 1 chapel
• early to mid Dynasty 5	in 9 chapels
• late Dynasty 5	in 32 chapels
• early to mid Dynasty 6	in 14 chapels
• late Dynasty 6 to Dynasty 8	in 8 chapels

Scenes of reaping cereal crops appear in the following tombs:

Dynasty 4 to early Dynasty 5

Nfr…w	Medum	IV.E
Rꜥ-wr	Giza	V.E
*Ḫnmw-ḥtp**	Saqqara	V.E

Scenes of workers engaged in cutting grain crops first appear in the Medum tomb of *Nfr….w*,[457] but are not attested again until Dynasty 5.[458]

Mid Dynasty 5

GIZA	
*Jj-mrjj**	*Rꜥ-ḫꜥ.f-ꜥnḫ**
Nj-mꜣꜥt-Rꜥ	*Ḫwfw-ḫꜣ.f* II*
SAQQARA	
Wr-jr.n-Ptḥ	*Rꜥ-m-kꜣ.j**
Nj-ꜥnḫ/Ḫnmw and *Ḫnmw-ḥtp*	
EL-HAMMAMIYA	
Rꜥ-ḥtp	

450 See TABLE 9: PRODUCTION OF FIELD CROPS.
451 There are fifteen ploughing scenes belonging to mid Dynasty 6 to Dynasty 8 but none includes a flock of sheep trampling seed. See TABLE 9: PRODUCTION OF FIELD CROPS.
452 Blackman (1924) pl. 14.
453 Blackman–Apted (1953) pl.30.
454 Davies II (1902) pl. 6.

455 For a complete listing of harvest scenes, see TABLE 9: PRODUCTION OF FIELD CROPS.
456 *ꜣḫtj-ḥtp* (Louvre) Ziegler (1993) 70-1, drawings 126, 129, 135-7; *Wr-jr.n.j* Davies *Sheik Said* (1901) 19-23, pls. viii, xvi; *Ḫw-ns* LD II 106 [b], 107.
457 Petrie (1892) pl. 28 [6].
458 The earliest Dynasty 5 chapels with scenes of reaping are those of *Rꜥ-wr* and *Ḫnmw-ḥtp*. See TABLE 9: PRODUCTION OF FIELD CROPS.

Late Dynasty 5

GIZA

Jꜣsn	*Snḏm-jb/Jntj*
Nj-mꜣꜥt-Rꜥ G2097	*Snḏm-jb/Mḥj*
Nfr (1)	*Sšm-nfr Tḥ**
Nfr-mḏr-Ḫwfw	*Kꜣ.j- m- ꜥnḫ*
*Ḫww-wr**	*Kꜣ-ḥj.f*
Sꜣ-jb	*Kd-nfr*

SAQQARA

ꜣḫtj-ḥtp (Louvre)	*Rꜥ-špss*
ꜣḫtj-ḥtp D64	*Ḥtp-ḥr-ꜣḫtj*
Jrj-n-kꜣ-Ptḥ	*Ḫnwt*
Pḥn-wj-kꜣ.j	*Spd-ḥtp*
Ptḥ-ḥtp I D62	*Sḫm-ꜥnḫ-Ptḥ*
Nfr-jrt-n.f	*Kꜣ.j-m-nfrt*
Nfr-sšm-Ptḥ and *Sḫntjw*	*Ṯjj*

SHEIKH SAID

*Wr-jr.n.j**	*Srf-kꜣ.j*

ZAWYET EL-MAIYETIN

*Ḫw-ns**

* Tying sheaves of cereals into bundles.

From mid Dynasty 5 on graphic details are added to the reaping scene in Dynasty 5 and may indicate a heightened degree of interest in cereal production.[459] Vignettes include overseers leaning on a staff as they supervise the reaping, shearers standing upright to take a drink or perhaps to ease their back,[460] a flute-player accompanying the reaping[461] and quail appearing among the crop.[462] The typical scene is of a row of harvesters against a backdrop of corn almost of the same height. Each harvester is leaning forward; with one hand he grasps a bunch of the growing corn which he cuts with a short handled sickle. As the dynasty progressed, the monotony of the line of identically depicted harvesters is broken up with one or two of the figures standing up straight, sickle under arm, usually holding up a sheaf of corn.[463]

The associated theme is the tying of sheaves into bundles, which are then stacked in preparation for transport to the threshing floor. A worker carrying an armful of sheaves to another who is tying them into bundles[464] or stacked bundles of sheaves awaiting transport may be depicted.[465]

Scenes of reaping cereal crops appear in the following tombs:

Early Dynasty 6

GIZA

Ḫnmw-ntj

SAQQARA

Mrw/Ttj-snb	*Mḥw*
Mr.f-nb.f	*Nj-ꜥnḫ-nswt*
Mrrw-kꜣ/Mrj	*Ḥzj*

DESHASHA

Jttj-Šdw

DEIR EL-GEBRAWI

Ḥm-rꜣ/Jzj (I)

Mid Dynasty 6

GIZA

Snfrw-ḥtp

DEIR EL-GEBRAWI

*Jbj**

MEIR

Ppj-ꜥnḫ-ḥrj-jb

EL-HAWAWISH

Bꜣwj BA48	*Nbt*
Mmj	*Kꜣ.j-ḥp/Ṯtj*

Late Dynasty 6

DEIR EL-GEBRAWI

*Ḏꜥw**

MEIR

Ppj-ꜥnḫ/Ḥnj-km

EL-HAWAWISH

Ḥzjj-Mnw/Zzj	*Kꜣ(.j)-ḥp/Ṯtj-jkr**
*Ḫnj/Špsj-pw-Mnw**	

QUBBET EL-HAWA

Mḫw and *Sꜣbnj*

Dynasty 8

EL-HAGARSA

Wꜣhj	*Mrjj-ꜥꜣ*

* Tying sheaves of cereals into bundles.

Transport of sheaves to the threshing floor

The incidence of scenes used in the present list of tombs of transporting bundles of sheaves from the field is as follows:

- early to mid Dynasty 5 — in 8 chapels
- late Dynasty 5 — in 35 chapels
- early to mid Dynasty 6 — in 11 chapels
- late Dynasty 6 to Dynasty 8 — in 8 chapels

This transport theme is presented in three separate motifs:

- A pack of donkeys is driven to the field to be loaded with sacks of sheaves. Behind the pack there is normally a line of drivers waving sticks to drive the donkeys forward[466] (*Figure 24*).
- Sacks of sheaves are loaded on the backs of donkeys. A worker may control a nervous animal by holding its

459 The crop is presented either as an uninterrupted band of crops acting as a backdrop to the farmhands {the earliest examples occur in *Rꜥ-ḥtp* El-Khouli–Kanawati (1990) pl. 73 [A]; *Ḫnmw-ḥtp* Petrie–Murray (1952) pl. 17 [4-5, 7] and Shaawi–Harpur *JEA* 74 (1988) fig. 6} or in clumps indicating the ongoing reaping process {the earliest examples occur in *Wr-jr.n-Ptḥ* Hall–Lambert (1922) pls. 10-11 and James (1961) pl. 29 [2]}. Farmhands are depicted bending down to cut the plants, either all oriented in the same direction as in *Ḫnmw-ḥtp* Petrie–Murray (1952) pl. 17 [4-5, 7] and Shaawi–Harpur *JEA* 74 (1988) fig. 6, or facing each other in pairs, presumably intended to be seen as working on different rows as in *Wr-jr.n.j* Davies *Sheik Said* (1901) pl. 16.

460 For example, *Nfr-sšm-Ptḥ* and *Sḫntjw* Moussa–Junge (1975) pl. 4 [b].

461 For example, *Ḥtp-ḥr-ꜣḫtj* Mohr (1943) fig. 48.

462 For example, *Ḥzj* Kanawati–Abder-Raziq (1999) pl. 9-10.

463 For example, *Ḥtp-ḥr-ꜣḫtj* Mohr (1943) fig. 47; *Ḫw-ns* LD II 106b.

464 *Ḫnmw-ḥtp* Petrie–Murray (1952) pl. 17 [4-5, 7]; Shaawi–Harpur, *JEA* 74 (1988) fig. 6.

465 *Jj-mrjj* LD II 51; Weeks (1994) fig. 39.

466 Among the earliest scenes are those in the tombs of *Jj-mrjj* LD II, 51 and Weeks (1994) fig. 39 and *Ḥtp-ḥr-ꜣḫtj* Mohr (1943) fig. 50.

foreleg and ear.[467] A blanket may be depicted over the donkey's back, presumably to protect it from the heavy load it was required to carry[468] (*Figure 25*).

- Laden donkeys are driven to the threshing floor. Each animal is accompanied by one or two men, one of whom holds the load in place,[469] or by a driver and a young boy.[470] A foal may accompany its mother.[471] The shape and form of the bundles of sheaves carried by the donkeys are portrayed with great variety[472] (*Figure 28*).

Threshing

Figure 26

The incidence of scenes of animals threshing cereals in the list of tombs used in this study is as follows:

- early to mid Dynasty 5 — in 3 chapels
- late Dynasty 5 — in 28 chapels
- early to mid Dynasty 6 — in 9 chapels
- late Dynasty 6 to Dynasty 8 — in 3 chapels

The scene of the threshing floor has two main stages and a number of variations. In the first stage workers build a stack of sheaves of barley or emmer wheat that awaits threshing. Men may be depicted throwing sheaves on to a trapezoid-shaped stack.[473] The major stage portrays one or two men controlling a herd of animals on the threshing floor. Donkeys are shown being driven around the threshing floor to trample on the sheaves,[474] Normally the animals are depicted as a small overlapped herd, with one or two animals lowering their head to eat the corn or resisting the command of their controller.[475] Occasionally, a flock of sheep or a herd of cattle are portrayed as the means of threshing the cereal harvest. The tomb of *Mrrw-k3* presents three images of the threshing floor, each with a different group of animals: cattle, donkeys and sheep. Two farmhands with sticks held above their heads control the cattle; one farmhand with a raised stick keeps the donkey pack moving, while the sheep are led by a figure in a similar position to a coaxer-sower and urged on at the rear by a farmhand wielding two sticks.[476]

Scenes of animals threshing cereals appear in the following tombs:

Early to mid Dynasty 5

GIZA

Rˁ-wr	*Rˁ-ḫˁ.f-ˁnḫ*

SAQQARA

Ḫnmw-ḥtp	*Rˁ-m-k3.j*

Late Dynasty 5

GIZA

3ḫtj-mrw-nswt	*Snḏm-jb/Jntj*
Nj-m3ˁt-Rˁ (G2097)	*Snḏm-jb/Mḥj*
Nj-ḥtp-Ptḥ	*Sḫm-ˁnḫ-Ptḥ*
Rˁ-wr II	*Sšm-nfr* IV
Ḫww-wr	*K3-ḥj.f*

SAQQARA

3ḫtj-ḥtp (Louvre)	*Ḥtp-k3.j/Tp-k3.j*
3ḫtj-ḥtp	*Spd-ḥtp*
Jrj-n-k3-Ptḥ	*Sḫm-ˁnḫ-Ptḥ*
Pḥn-wj-k3.j	*K3.j-m-nfrt*
Ptḥ-ḥtp I	*K3.j-m-rḥw*
Nfr-jrt-n.f	*Ṯjj*
Ḥtp-ḥr-3ḫtj	

SHEIKH SAID

Wr-jr.n.j

ZAWYET EL-MAIYETIN

Ḫw-ns

Early to mid Dynasty 6

Jttj-Šdw	Deshasha	VI.E
ˁnḫ -m- ˁ-Ḥr	Saqqara	VI.E
Mrrw-k3/Mrj	Saqqara	VI.E
Mḥw	Saqqara	VI.E
Nj-ˁnḫ-Ppjj	Zawyet el-Maiyitin	VI.E
Nj-ˁnḫ-nswt	Saqqara	VI.E
Jbj	Deir el-Gebrawi	VI.M
Ppj-ˁnḫ-ḥrj-jb	Meir	VI.M
Mmj	el-Hawawish	VI.M
Snfrw-ḥtp	Giza	VI.M

Late Dynasty 6 to Dynasty 8

Ppj-ˁnḫ/Ḥnj-km	Meir	VI.L
W3ḥj	el-Hagarsa	VIII
Mrjj-ˁ3	el-Hagarsa	VIII

Winnowing and stacking the grain[477]

In the present list of tombs the incidence of scenes of winnowing and stacking the grain is as follows:

- Dynasty 4 — in 1 chapel
- early to mid Dynasty 5 — in 4 chapels
- late Dynasty 5 — in 23 chapels
- early to mid Dynasty 6 — in 7 chapels

This phase consists of a number of activities. The threshed grain is first shown collected in a stack and then moved into place by a long handled fork and swept into a pile to be winnowed, usually by women. The scene in the tomb of *3ḫtj-ḥtp* (Louvre) depicts two women sieving two separate piles of grain by lifting up the grain with two sticks or paddles. The grain is portrayed cascading down suggesting that a current of air is blowing away the chaff.

---

467 *Ptḥ-ḥtp* I Murray (1905) 15, pl. 11 (labelled *Ptḥ-ḥtp* II).

468 *Nfr-sšm-Ptḥ* and *Sḫntjw* Moussa–Junge (1975) 21-2, pls. 4 [b], 5.

469 *Rˁ-m-k3.j* Hayes (1953) 100-101, fig. 57; *Wr-jr.n-Ptḥ* Hall–Lambert (1922) pls. 10-11 and James (1961) pl. 29.

470 *Jj-mrjj* LD II 51and Weeks (1994) fig. 39; *Nfr-b3w-Ptḥ* LD II 56 [a, a bis] and Weeks (1994) fig. 9.

471 *Ptḥ-ḥtp* I Murray (1905) 15, pl. 11 (labelled *Ptḥ-ḥtp* II).

472 Vandier *Manuel* VI (1952-1978) 128-9. Siebels (2000) 238-249 also discusses the varied forms of these bundles.

473 Depiction of the untying of the bundles of sheaves is rare. See *Ḥtp-ḥr-3ḫtj* Mohr (1943) 83, fig. 49.

474 Occasionally a flock of sheep is depicted on the threshing floor: *Jrj-n-k3-Ptḥ* Moussa–Junge (1975) pl. 9. Usually two men brandishing poles control the animals and keep them on the move. The donkeys may be depicted as an untidy pack wanting to move in all directions: *Rˁ-ḫˁ.f-ˁnḫ*: LD II, 9 [lower].

475 *Wr-jr.n.j* Davies *Sheik Said* (1901) pl. 16.

476 Kanawati (2011) pls. 81-84.

477 For the complete list of agricultural scenes used in this study see TABLE 9: PRODUCTION OF FIELD CROPS.

A male figure bends forward to sweep the grain into a heap.[478] The grain is then forked into a second stack to await removal to a granary. Each one of these activities is portrayed in the tomb of *Ṯjj*.[479] As both barley and emmer wheat are hulled cereals that retain an envelope of chaff around the grain after they have been threshed, they require a further stage of pounding and sieving to separate the chaff from the grain. This activity may be represented in the scene of *Ṯjj*, where one female figure appears to hold a sieve.[480] This phase did not complete the processing of the grain in readiness for use. The final preparation, which would have entailed separating the hull from the seed, was not executed until the grain was taken from the granary for grinding and cooking.[481]

Stages of the harvest

There are at least twelve sequential stages portrayed in the harvesting of grain, if the stack of sheaves awaiting transport to the threshing floor, a further stack awaiting threshing and a third stack of grain ready to be transported to the granary are counted as separate stages. Most of these stages first appear in private tombs in early to mid Dynasty 5, but the source of their graphic inspiration probably derives from artistic creations that date at least to the preceding dynasty. The tombs of Dynasty 4 and early Dynasty 5 include very few depictions associated with the harvesting of grain, although it is likely that most of the graphic concepts portraying these activities were available to tomb owners and the craftsmen they employed; a scene of workers cutting grain appears in the Medum tomb of *Nfr...w* and the process of winnowing is portrayed in the late Dynasty 4 tomb of *Mrs-ʿnḫ* III at Giza. In tombs dating from mid Dynasty 5 to early Dynasty 6, however, harvest scenes frequently present a number of stages in the process of dealing with the crop. The tomb of *Ḫnmw-ḥtp*, which probably belongs to mid Dynasty 5, contains five of the sequential elements of harvesting grain.[482] The tombs of *Jj-mrjj, Wr-jr.n-Ptḥ* and *Rʿ-ḫʿ.f-ʿnḫ* present six harvest elements, while seven appear in the tomb of *Nj-ʿnḫ-Ḫnmw* and *Ḫnmw-ḥtp*.[483] In the second half of Dynasty 5 five or more of these elements appear in eighteen tombs and nine tombs contain eight or nine elements. Five or more of these elements appear in two tombs in the first half of Dynasty 6 and three in tombs belonging to later Dynasty 6.[484]

The major phase of the transport of sheaves to the threshing floor is a concept represented by four distinct scenes: the arrival of a pack of donkeys, the process of loading sacks on to the backs of donkeys, the laden donkeys being led to the threshing floor and piles of sacks near the threshing floor. Few tombs present all these stages. The chapels with the greatest number of 'transport' elements occur in the later Dynasty 5 Saqqara tombs of *Nfr-jrt-n.f, Sḫm-ʿnḫ-Ptḥ, Kȝ.j-m-nfrt, Ṯjj* and the Giza tomb of *Snḏm-jb/Jntj*, and in early Dynasty 6 in the Saqqara tomb of *Mrrw-kȝ*.

The final appearances of these themes vary in time. The pack of donkeys driven to the fields to be loaded is last seen in early Dynasty 6 in the tombs of *Mrrw-kȝ, Mḥw* and *Mṯṯj*,[485] while the scene of laden donkeys being led to the threshing floor and the stacking of the sheaves lasts considerably longer. The stacking of sheaves last appears in the tomb of *Ḏʿw*, whereas laden donkeys appear in the Dynasty 8 tombs of *Mrjj-ʿȝ* of Hagarsa and *Rḥw-r-ȝw.sn* of el-Hawawish.[486]

Barley or emmer wheat?[487]

As with the flax crop, the growing cereal is typically depicted as a rectangle,[488] which is sometimes just an outline but may be filled with lines representing the stalks of the plants. In a small number of Dynasty 5 tombs the crop is depicted in bunches grasped by the reaper.[489] Where colour has survived, the cereal crop is painted in shades of yellow and brown. The depiction of the crops does not distinguish between wheat and barley; only an occasional caption provides this information. Consequently it is impossible to judge with certainty which cereal was more frequently grown. However, the tombs specifying barley (*jt*) outnumber those specifying emmer wheat (*bdt*) by 33 to 19.[490]

8.4 Storage of the grain

Figure 29

8.4.1 Historical development of images and themes

A sequence of scenes, sometimes located so that it accompanies the threshing or winnowing of the corn, shows the grain being stored in a granary. There are four main phases of this sequence:

- workers taking grain from the heaps/stacks
- workers carrying containers to granary
- granary workers receiving the containers
- workers taking grain from a granary.

After it had been winnowed and sieved, the grain was collected in heaps or in stacks from which workers took measured amounts to carry in containers to granaries, structures with rounded roofs usually depicted in a

[478] Ziegler (1993) 137.

[479] Wild III [2] (1966) pls. 136-139,151-155

[480] Wild III [2] (1966) pls. 136-139 [A, B], 151-155.

[481] Murray (2000) 527. This interpretation is accepted by Samuel (2000) 545.

[482] See TABLE 9: PRODUCTION OF FIELD CROPS.

[483] Moussa–Altenmüller (1977) pl. 54 [b], fig. 24; pl. 56-9.

[484] See TABLE 9: PRODUCTION OF FIELD CROPS.

[485] Respectively, Duell II (1938) pls. 168-70; Altenmüller (1998) pl. 22 [b]-25; Kaplony (c1976) pp. 22-4, Nr. 3.

[486] Stacking sheaves: Davies, *Gebrâwi* II (1902) pl. 6. Laden donkeys: Kanawati (1993) pls. 35, 37, 39-40 and Kanawati (1987) figs. 15, 16, 25-6.

[487] According to both Murray (2000) 511-13 and Samuel (2000) 545 there is no evidence of any other grain grown in Egypt in the Old Kingdom.

[488] The rectangle may be divided into separate stands of crops indicating the ongoing reaping process. Siebels (2000) 155 provides a comprehensive list of this feature.

[489] *ȝḫtj-ḥtp* (Louvre), *ȝḫtj-ḥtp* (D 64), *Jj-mrjj, Rʿ-ḫʿ.f-ʿnḫ, Sḫm-ʿnḫ-Ptḥ* (Saqqara), *Kȝ.j-m-nfrt, Ṯjj*. See Table 9: 'Production of Field Crops'.

[490] See TABLE 9: PRODUCTION OF FIELD CROPS.

row.[491] The farmhand carrying the grain may be pictured climbing steps up the side of the granary to pour the grain through an aperture in the top of the granary or he may be depicted handing over the containers to granary workers who pour the grain into the granary. A 'crier' is depicted announcing these operations, which were checked and recorded by scribes, who are also shown checking and recording deductions by workers taking grain from an opening at the bottom of the granary.[492] A more senior official, *jmj-rꜣ pr*, is often depicted supervising all the above operations.
The incidence of scenes relating to grain storage in the list of tombs used in this study is as follows

- late Dynasty 4 — in 1 chapel
- mid Dynasty 5 — in 4 chapels
- late Dynasty 5 — in 17 chapels
- early to mid Dynasty 6 — in 3 chapels
- late Dynasty 6 to Dynasty 8 — in 3 chapels

Scenes of grain storage appear in the following tombs:

Late Dynasty 4 to mid Dynasty 5

Mrs-ꜥnḫ III	Giza	IV.L
Jj-mrjj	Giza	V.M
Wr-jr.n-Ptḥ	Saqqara	V.M
Nj-ꜥnḫ/H̱nmw and *H̱nmw-ḥtp*	Saqqara	V.M
H̱wfw-ḥꜣ.f II	Giza	V.M

Late Dynasty 5

GIZA

ꜣḫtj-mrw-nswt	*Kꜣ.j- m- ꜥnḫ*
Nfr-bꜣw-Ptḥ	

SAQQARA

Jrj-n-kꜣ-Ptḥ	*Spd-ḥtp*
Pḥn-wj-kꜣ.j	*Sḫm-ꜥnḫ-Ptḥ*
Ptḥ-ḥtp(LS 31)	*Kꜣ.j-m-nfrt*
Nj-kꜣw-ḥr	*Kꜣ.j-m-rḥw*
H̱nwt	

ABUSIR

Tp-m-ꜥnḫ

ZAWYET EL-MAIYETIN

Ḫw-ns

SHEIKH SAID

Wr-jr.n.j

Early Dynasty 6 to Dynasty 8

Mḥw	Saqqara	VI.E
Nj-ꜥnḫ-nzwt	Saqqara	VI.E
Jbj	Deir el-Gebrawi	VI.M
Ḥzjj-Mnw/Zzj	el-Hawawish	VI.L
H̱nj/Špsj-pw-Mnw	el-Hawawish	VI.L
Wꜣhj	el-Hagarsa	VIII
Mrjj-ꜥꜣ	el-Hagarsa	VIII

No tomb presents all four phases of the 'grain-storage' sequence, although three phases appear in the tombs of *Nj-ꜥnḫ-H̱nmw* and *H̱nmw-ḥtp, Spd-ḥtp* and *Nj-kꜣw-ḥr.*[493] Of the sequence of 'granary' scenes the only surviving element from Dynasty 4 is in the tomb of *Mrs-ꜥnḫ* III where a worker is portrayed taking grain from a mound. The majority of scenes of this sequence occur in the Memphite tombs of mid to late Dynasty 5. Apart from the tombs of *Mḥw, H̱nj/Špsj-pw-Mnw, Wꜣhj* and *Mrjj-ꜥꜣ,* Dynasty 6 tombs only contain the scene of workers taking grain from a granary, in which case it is treated as the first phase of a bread or beer-making sequence.[494] It is perhaps surprising that other elements of the 'granary' sequence have not survived from any other Dynasty 6 Memphite tomb. While much of the decoration is missing from the great tombs of *Mrrw-kꜣ* and *ꜥnḫ-m-ꜥ-Ḥr,* sufficient survives from other tombs of early Dynasty 6, such as those of *Kꜣ-gmnj, Ḥzj, Sꜥnḫ-wj-Ptḥ, H̱ntj-kꜣ.j/Jḫḫj* and *Mṯtj,* to make the absence of these scenes a matter of note.

In the provinces it is doubtful that any 'granary' scenes date to Dynasty 5, as depictions in the tombs of *Ḫw-ns* and *Wr-jr.n.j*[495] represent workers adding winnowed grain to a stack, not taking grain to or removing it from a granary.[496] The Dynasty 6 provincial 'granary' scenes are also very limited, apart from a sequence of scenes in *H̱nj/Špsj-pw-Mnw*, where workers take containers of grain from a mound and a worker on the register below is probably adding grain to a silo,[497] and a row of silos depicted in the tomb of *Jttj-Šdw* of Deshasha.[498] While Old Kingdom granaries are normally portrayed resting on a platform, those in the Dynasty 8 tombs of *Mrjj-ꜥꜣ* and *Wꜣhj* at Hagarsa are depicted within a pillared canopy and appear to be raised much higher above ground level.[499] These two tombs also portray men ascending stairs leading to the top of the granaries. The canopy and stairs are features that may first appear at the end of the Old Kingdom; they continue into the Middle Kingdom.[500] As in the Dynasty 6 Memphite necropoleis, scenes in

491 Siebels (2000) 404.

492 See TABLE 10: GRAIN STORAGE, BREAD AND BEER PRODUCTION

493 Respectively, Moussa–Altenmüller (1977) pl. 54, a, b, fig. 24; Harpur (1986) figs. 2, 6; Quibell (1909) pl. 62 [2].

494 See James (1953) pl. 9.

495 Respectively, LD II 107 and Davies (1891) pl. 16, in both cases register 3 from the top. These scenes are more likely to represent adding sheaves to a stack in readiness for threshing, which appears on the register below.

496 Despite their listing as such in the Oxford Expedition to Egypt Database.

497 Kanawati II (1981) fig. 17.

498 Petrie (1898) pl. 23; Kanawati–McFarlane (1993) pls. 15, 47. There may have been a scene of workers taking grain to the granaries, but it is completely lost.

499 Respectively, Kanawati (1995) 34, pls. 7 [b], 36; Kanawati (1995) 18, pls. 5 [b], 6 [left], 28, 31.

500 For granaries with stairs, see Vandier (1950) fig. 56; Jéquier (1929) figs. 44, 51, 83, 88, 117, 136 (the tombs of *Snj. Pnw, Mḥj, Sbḳw* and *Ḏgm/Ppjj-mr*). Harpur (1987) 263 notes that the depiction of men climbing stairs to pour grain into a granary only occurs at provincial sites and probably only at the very end of the Old Kingdom. For a Middle Kingdom representation see Davies–Gardiner (1920) pl. 15. Siebels (2000) 405, referring to the Dynasty 5 tomb of *Kꜣ.j-m-rḥw*, questions whether the pillared canopy sheltering the granaries was confined to the end of the Old Kingdom and later. This feature together with the raising of the floor of the granary may suggest that there was greater need to protect the contents of granaries towards the end of the Old Kingdom.

provincial tombs of this period are mainly confined to depictions of workers taking grain out of granaries.

Each phase of the operation may be accompanied by an official who checks or records the quantity of grain being handled:

- *nḫt ḫrw*
  This is the 'crier' who calls out quantities for the benefit of scribes who make the record. In granary scenes this refers to grain being taken from a mound,[501] In *Nj-ꜥnḫ-H̱nmw* and *H̱nmw-ḥtp* an official with this title is linked to the granary as *nḫt ḫrw n šnwt* (crier of the granary) in a baking scene, while the crier depicted in the granary scene is labelled, *nḫt ḫrw pr ḏt.* A similar situation pertains in *Nj-kꜣw-Jzzj,* where a *nḫt ḫrw n šnwt* is depicted in a register of bread and beer making, while an untitled crier named *Ḳd-ns* keeps a tally of the grain being taken from a granary.[502] This official is depicted in eleven tombs.

- *zš*
  A scribe appears in every 'granary' sequence where sufficient decoration has survived.[503] Scribes are depicted recording the extraction of grain from a mound to be transported to a granary, the extraction of grain from the granary and the presentation of accounts. In *Jj-mrjj, Nfr-bꜣw-Ptḥ,* and *Sḫm-ꜥnḫ-Ptḥ*[504] a scribe is designated *zš šnwt* ('scribe of the granary'), which in the view of Siebels indicates the particular administrative section of the estate to which he is attached.[505]

- *ḫrp ḫꜣ(w)* ('controller of grain measurers')
  This official is depicted in the scene of taking grain from the granary in preparation for bread and beer making in the tomb of *Nj-kꜣw-Jzzj.*[506]

- *jmj-r pr* ('overseer of the house')
  This is an important estate official who figures in a variety of themes and in many chapels. In this theme he was responsible for the final collection of records and their presentation to the tomb owner.

- *jrj ḫt* ('property custodian')
  In the tomb of *Nj-kꜣw-ḥr* he supervises workers taking grain to the granary.[507]

The grain-storage theme is sometimes depicted in association with a further theme usually called 'the rendering of accounts', in which a number of men are brought before scribes. The men, often depicted as elderly, balding and with sagging stomach muscles, are being presented to scribes. They may be given the title of *ḥḳꜣ ḥwt* ('estate manager/estate chief') or *ḥḳꜣ njwwt* ('chief of the towns/settlements') and are shown as supplicants. Two captions accompanying Dynasty 5 scenes read:

ḥsb ꜥwt jn ḏꜣt nt pr ḏt ('Examining the accounts of the farmers by the administration of the estate'.)[508]
ḥsb ḥkꜣw jn ḏꜣḏꜣt nt pr ḏt ('Examining the chiefs by the assessors of the estate'.)[509]

The above Dynasty 5 captions and the scenes they accompany represent examination and record-making; there is no implication of punishment for misdeeds or negligence. However, equivalent scenes from Dynasty 6 present portrayals of physical punishment of the headmen brought before the scribes[510] (*Figure 30*).

8.5 Inscriptional data

Four genres of inscriptions relate to cereals: captions accompanying depictions, vertical columns of inscriptions usually separating a large figure of the tomb owner from a panel of registers, references to bread and beer in offering lists connected to ritual, and statements regarding the payment of bread and beer to workers.

Captions associated with scenes of agriculture
Captions that refer to ploughing and sowing are usually terse and provide information evident from the iconography. They consist of one or just a few words such as *stjt* ('scattering') or *skꜣ m hb* ('cultivating with a plough') or *wꜣḥ hb* ('applying the plough').[511] Even the slightly longer statements provide little additional enlightenment. *Jj-mrjj*, for example, describes the work of hoeing by hand as *bꜣ jn jswt.f nt pr.f n ḏt* ('hoeing by his gang of his estate') and the flock of sheep being driven across the soil as *skꜣ m sḫt jn jswt.f nt pr.f n ḏt* ('ploughing with the sheep by his gang of his estate'),[512] thus informing the viewer that the work is being carried out by fieldhands who belong to his estate. Nor do the captions become more informative over time. Although later Dynasty 5 and Dynasty 6 scenes may give more of the mundane speech of the workers, they are no more enlightening than those of Dynasty 4,[513] with ploughing scenes retaining brief descriptions such as *skꜣ m hb*, and

501 In *Nj-ꜥnḫ-H̱nmw / H̱nmw-ḥtp*, a crier is portrayed for each of the two vignettes of a worker taking grain from a mound: Moussa–Altenmüller (1977) pl. 54, a, b, fig. 24.
502 Kanawati–Abder-Raziq (2000) pls. 13, 48, p. 37.
503 This includes the scene of the presentation of accounts and punishments.
504 Respectively, Weeks (1994) figs. 39-40, pls. 23-26; LD I, 56a, a [bis] and Weeks (1994) figs. 9-11; Simpson (1976) pl. D.
505 Siebels (2000) p. 401.
506 Kanawati–Abder-Raziq (2000) pls. 13, 48, p. 37.
507 Quibell (1909) pl. 62.2.

508 *ꜣḫtj-ḥtp* (Louvre) Ziegler (1993) 150-151.
509 *Kꜣ.j-m-nfrt* Simpson (1992) 18, pl. 21.
510 Duell I (1938) pls. 34 [B], 36-7, 38; James (1953) pl. 9.
511 Siebels (2000) 74-81, 105-107 provides lists of the attestation of the captions associating with tilling and planting and discusses their usage.
512 Weeks (1994) fig. 39.
513 On the west wall of Room I of *Nfr-mꜣꜥt*'s chapel the caption accompanying the two ploughing teams is *stjt* ('Scattering'): Petrie (1892) pl. 18 [right]; Harpur (2001) fig. 75, pl. 6. In the chapel of his wife, *Jtt*, a somewhat expanded caption reads, *skꜣ m hb* ('Cultivating with a plough') and over a figure of a sower a caption may read *s<tt?>* ('Sowing'): Petrie (1892), pl. 28 [4] and Harpur (2001) fig. 81, pl. 22.

w3ḥ hb.[514] Consequently, it is not possible to deduce from either the iconography of tillage and sowing or associated captions which crop, flax, barley or emmer wheat, is being sown or the extent of land under cultivation. Only the scene in the Saqqara tomb of *Mḥw* provides more information, but the translation of *stjt jt bdt in ḥk3 ḥwt*, as 'scattering *barley and emmer wheat* by the manager of the estate' requires the inference that the scene represents two separate plantings (on different parcels of land) that are merged together, perhaps to save wall space or prevent costly repetition of what would be identical scenes to the viewer.

Captions accompanying scenes of harvesting and the processing and storing of the cereal crop provide some information as to which crop is being reaped and, very occasionally, the size of the harvest. But these inscriptions are random, fail to give a clear indication which crop was more frequently grown and are not related to the size of land under cultivation.[515] Consequently, there is no indication of the yield per hectare.

Typical of the inscriptions are:

Jj-mrjj: 3zḫ jt bdt m njwwt.f 'Reaping barley and emmer wheat in his towns.'
Sḫm-k3.j: ḥwj mḥꜥ jn jzwt (nt pr-) ḏt.f 'Gathering flax by the gangs of his estate.'
Sndm-jb/Mḥj: dm3 mḥꜥ 'Binding flax.'
Ḥww-wr: 3zḫ m njwwt.f nt pr-ḏ[t] 'Reaping in his towns of the estate.'[516]

Two provincial tombs of late Dynasty 5 distinguish reaping gangs by description, but the significance of the distinction among the gangs is uncertain:[517]

Ḥw-ns of Zawyet el-Maiyetin states:
3zḫ jt jn jzwt pr-ḏt 'reaping barley by the gangs of the estate.'
3zḫ jt jn jz wr 'Reaping barley by the great gang.'
3zḫ bdt jn jzwt pr-ḏt 'Reaping emmer wheat by the gangs of the estate.'
3zḫ bdt jn ḥmw nzwt 'Reaping emmer wheat by the servants of the king.'[518]

Wr-jr.n.j of Sheikh Said states:
3zḫ jt jn jzwt pr-ḏ[t] 'reaping barley by the gangs of the estate.'
3zḫ jt jn[519] *jz wr* 'Reaping barley by the great gang.'
3zḫ bdt jn jz wr 'Reaping emmer wheat by the great gang.'
3zḫ bdt jn ḥmw nswt 'Reaping emmer wheat by the servants of the king.'[520]

According to these inscriptions there appears to have been three different teams at work on the harvest.

Two other groups of captions referring to the harvest are of interest as they indicate the need for speed in gathering the crops. These are in the form of orders to field hands and use the verb *wnj* ('to hurry') such as, *wn ṯw jr.k ṯ3w srf-jb mjw jr[.j] ntj(w) ḥnꜥ* ('Hurry yourself, industrious one. Come to me, comrades.').[521] A further variety of captions introduces the concept of 'the one who says he will act in time' (*jšst pw ḏd jrj.f m-ṯr*). This, too, suggests the need for speed in harvesting before the crop over-ripened. The impression is that the men selected as reapers gave assurances that they would accomplish the task competently within the required time. The phrase *ṯ3w (ṯ3j) srf-jb* ('industrious one') occurs in a number of these statements referring to the matter of time.[522] Captions associated with scenes of transporting cereals, threshing, winnowing and sweeping are either commands or statements referring to the activity they accompany.

The vertical column of inscriptions separating the figure of the tomb owner from the panel of scenes that he views may also refer to agriculture, but only when all or a majority of registers in the panel represent the theme of planting and harvesting crops.[523] The inscriptions vary from a terse (*m33*) *sk3* '(Viewing) the ploughing' to statements that specify a number of the relevant activities. The most detailed is that of *K3.j-m-nfrt* (D 23), which reads, *m33 sk3 jt ḥj mḥꜥw 3sḫ šd ... ḥj ḫ3ḫ3 jꜥb* ('Viewing the ploughing, gathering of flax, reaping, loading (donkeys), treading (of the threshing floors), winnowing, heaping up'). The statement of *Ṯjj* is the most precise as it accurately reflects the harvest activities represented on the registers facing the tomb owner: *m33 ḥwj mḥꜥ 3zḫ jt bdt šdjt wbs m spwt m njwwt nt pr-ḏt* ('Viewing the pulling of flax, the reaping of barley and emmer wheat, the transporting, the heaping on the threshing floors from (by?) the towns of the estate').

514 Siebels (2000) discusses these inscriptions in detail. See pp. 74-81, 89-90, 105-107.
515 See TABLE 9: PRODUCTION OF FIELD CROPS.
516 Respectively, LD II 51; Weeks (1994) fig. 39, pls. 23 [a, b], 24 [a, b]; Simpson (1980) pls. iii [c-e], iv [a-e], fig. 4; LD, *Erg.* 14; Brovarski (2001) pls. 115-16, figs. 114-15; LD II 43 [a]; Hassan (1944) fig. 104.
517 Both *Ḥw-ns* and *Wr-jr.n.j* present scenes of the flax harvest but without captions identifying the teams of workers. The possible significance of specifying different gangs of workers is discussed in Vandier *Manuel* VI (1978) 111-12; Siebels (2000) 181, n. 142-44.
518 LD II 106 [b]
519 Although the glyphs read '*jt*', Davies (1901) 21 translates them as '*jn*' to make better sense of the statement and to conform to other statements with the same import.
520 Davies (1901) pls. 8, 16.
521 *Nj-ꜥnḫ-Ḫnmw / Ḫnmw-ḥtp*: Moussa–Altenmüller (1977) pl. 56-9. The verb *wnj* is also used in *Ṯjj* Wild III [2] (1966) pls. 136-139 [A, B], 151-155 and *Mrrw-k3* Duell II (1938) pls. 168-70; Kanawati (2011) pl. 81-2, 84.
522 *Nj-ꜥnḫ-Ḫnmw / Ḫnmw-ḥtp* Moussa–Altenmüller (1977) pl. 56-9, *Ṯjj* Wild III [2] (1966), pls. 136-139 [A, B], 151-155; *K3-ḥj.f* Junker VI (1943) fig. 42-3, 45-7, pl. 10[b], 12-14 [a, b]; *Sḫm-ꜥnḫ-Ptḥ* Simpson (1976) and *Mrrw-k3* Duell II (1938) pls. 168-70 and Kanawati (2011) pl. 81-2, 84.
523 See TABLE 12: INSCRIPTIONS – TOMB OWNER VIEWING PLANTING AND HARVESTING OF FIELD CROPS. Inscriptions and panels devoted to the theme of agriculture occur in the tombs of *Jj-mrjj, 3ḫtj-ḥtp* (Louvre), *3ḫtj-ḥtp* (D64), *Jrj-n-k3-Ptḥ, Wr-jr.n.j, Nfr-jrt-n.f, Ḥtp-ḥr-3ḫtj, Ḥw-ns, Sḫm-ꜥnḫ-Ptḥ* (D 41), *K3.j-m-nfrt* (LG 63), *K3.j-m-nfrt* (D63), *K3-ḥj.f, Ṯjj, Ppj-ꜥnḫ-ḥrj-jb* (Meir), *Jbj ꜥnd Ḏꜥw* (Deir el-Gebrawi).

When the panel of registers is not restricted to agriculture but consists of a variety of resources, the vertical inscription either has a generalised import or identifies what was probably the tomb owner's paramount interest. In *Nfr* and *K3-ḥ3.j* for example, the tomb owner views scenes of husbandry, marshland activities, wine making and agriculture, while the inscription reads, *m33 ḥ3mw ꜥḥw zšw n njw(w)t.f nt T3-mḥw Šmꜥw* ('Viewing fishermen, fowlers and scribes of his town(s) of Lower and Upper Egypt').[524] The inscription of Queen *Nbt*, who views registers of agriculture, husbandry and offering bearers bringing wine and a range of offerings, reads, *(m)33 jrp rḏ.t m stp-z3 n* ... ('Viewing the wine, brought from the palace for ..'). On the other hand, Queen *Ḫnwt*, who views scenes of the grain harvest and transport by boat, has a general inscription that reads, *(m33) jrr(.w) m ḥwwt.s nwwt.s n(t) pr-ḏt* ('[Viewing what is] brought from her estates and her towns of the estate').

To summarise, the vertical inscriptions separating the figure of the tomb owner that refer to field crops give equal or more emphasis to grain than to flax, but only refer specifically to field crops when the field-crop theme occupies a majority of registers in the panel opposite.[525]

Ritual inscriptions

Bread and beer, the major products of the grain harvest, appear regularly in both the long and short forms of funerary offering lists. The glyphs for bread (*t*) and beer (*ḥnḳt*) appear almost without fail in the short offering list situated beneath or beside the portrayal of the tomb owner's ka seated at his offering table. In the great offering lists, many different kinds of foods made from grain are listed, and beer is usually specified. For example, in his great offering list *Sbk-nfr* of el-Hagarsa includes thirteen different kinds of bread and three kinds of cake.[526] A further class of inscriptions relating to offerings also highlights the ritual significance of bread. For example, on the south wall of Room 9 in the tomb of *Ḫntj-k3.j/Jḫḫj* the captions associated with bearers bringing variously shaped loaves of bread read, *t wꜥb pw n Jnpw jw [n Jḫḫj] jm3ḫw* ('This is pure bread of Anubis; it is [for] the revered [Ikhekhi].') and *t wꜥb pw n Wsjr jw n Jḫḫj mry nṯr* ('This is pure bread of Osiris; it is for Ikhekhi, beloved of the God.').[527]

Autobiographical inscriptions

Cereals, usually in the form of bread and beer, are sometimes attested as constituting part of the wages of workmen. In her discussion of this practice Roth notes that it is difficult to determine whether this type of payment was the general rule as the recorded instances are rare.[528] Roth cites five cases: those of *Rmnw-k3.j/Jmj, Mṯtj, 3ḫtj-mḥw, Ḥtp-ḥr-3ḫtj* and an anonymous false door.[529] All appear to date to the period from late Dynasty 5 to early Dynasty 6. *Rmnw-k3.j/Jmj, Mṯtj, 3ḫtj-mḥw* and *Ḥtp-ḥr-3ḫtj* all announce that they also paid their tomb workers with linen but *Ḥtp-ḥr-3ḫtj* informs the reader that he provided his tomb workers in addition with 'great quantities of barley' as well as bread and beer. Presumably part of the daily payment of bread and beer would have been consumed on the job, but a portion could have been taken home. Perhaps the workers, skilled tomb builders and craftsmen, would have been brought from some distance and consequently were provided with living quarters and rations on site.

8.6 Other evidence: data for the usage of the grain crop

While funerary iconography provides the evidence for the production of grains in the Old Kingdom, evidence for the utilization of these crops and the extent to which they were employed can be drawn from a variety of sources. Grains had essentially four uses: for bread, cake and beer, as payment for services and goods and to meet taxation demands.

There is considerable evidence for the baking of bread and brewing of beer in the funerary iconography of the Old Kingdom, but these scenes may be interpreted as producing beer and bread for consumption on the estate.[530] Artefact findings, however, provide clear evidence that bread and beer were the staple food and drink for rich and poor. There is archaeological evidence for the production of bread from the predynastic and early dynastic periods onwards. At two predynastic sites, Abydos and Hierakonpolis, residues from excavated vats contain evidence of emmer wheat probably connected with beer production.[531] By Dynasty 4 the archaeological data for the consumption of bread and beer provide evidence of the dietary patterns of ordinary workers.[532] The findings of the Giza Plateau Mapping Project provide massive evidence for the consumption of bread and beer by the workers as well as the supervisors and their families in the time of Menkaure.[533] Bread moulds and white carinated bowls appear in enormous numbers in the galleries, which probably housed the workforces employed on the Giza Plateau.

8.7 Summary

- The only cereals that can be identified in Old Kingdom tomb scenes are barley (*Hordeum* spp.) (*jt*) and emmer wheat (*Triticum. dicoccum*) (*bdt*). These are only identified by inscription because the

524 For this and the following references see TABLE 12: INSCRIPTIONS: TOMB OWNER VIEWING PLANTING AND HARVESTING OF FIELD CROPS.

525 See TABLE 12: INSCRIPTIONS: TOMB OWNER VIEWING PLANTING AND HARVESTING OF FIELD CROPS.

526 Barta (1963) 47-50, 73-5, 84-6; Kanawati (1993) 31-2, pl. 7b, 28.

527 James (1953) 59, pl. 31.

528 Roth, in Silverman (c1994) 236-38.

529 Respectively, Hassan II (1936) fig. 206; Goedicke (1958) pl. 2; Edel (1953) 327-33 and (1958) 17-18; Sethe (1903) 50.6-7; Sethe (1903) 226.1-2.

530 Bread baking scenes are included in 33 tombs used in this study, and beer brewing scenes in eleven tombs.

531 Samuel (2000) 539-541; Peet–Loat (1913) 1-7; Geller (1992b) 104-112.

532 Murray (2007b) 163-69, 267-58.

533 Murray (2005) 1, 6-9.

iconography makes no distinction between the two crops.

- The inscriptional data suggests that more barley than emmer wheat was planted but this cannot be verified as the basic data is scattered sporadically and recorded amounts may be unreliable.

- Cereal production may be divided into three themes: preparation and planting, harvesting and processing of the harvest, and storage. The portrayal of each theme has a different time span, with the first having the longest, from early Dynasty 4 to Dynasty 8.

- The components of each of the three themes are images representing a sequence of actions. In particular, the second theme, harvesting, is portrayed in approximately twelve sequenced stages. No tomb shows every stage, although 23 tombs contain at least five harvesting stages. All themes, together with all their components, are most frequently attested in the second half of Dynasty 5, which corresponds with the chronological distribution of tombs used in this study.

- A number of scenic components have a shortened time span. The image of workers tilling the soil with hand held sickles is confined to six tombs of the period mid Dynasty 5 to early Dynasty 6. The depiction of a flock of sheep used to trample seed into the soil first appears in late Dynasty 4 and is not attested after early Dynasty 6. The depiction of a herd of donkeys being driven to the field where harvesting is taking place extends from mid Dynasty 5 to early Dynasty 6, and that of women winnowing and sieving the threshed grain from late Dynasty 4 to early Dynasty 6.

- Four classes of inscriptions relate to field crops. Captions suggest that harvesters were selected for speed and competence. 32 tombs contain scenes of the tomb owner viewing panels with at least one register of crop production. Seventeen of these have the tomb owner viewing a panel with crop production scenes on every extant register and a column of inscription stating that the tomb owner is viewing the work of planting and harvesting field crops. The three kinds of ritual offering inscriptions invariably feature bread and beer, while autobiographic statements sometimes include claims by the tomb owner to have paid workers in linen, bread and beer.

- The depicted workforce for planting and harvesting is largely anonymous and lacks descriptive titles, except in grain storage scenes where a variety of minor officials who are depicted checking and recording the quantity of grain being handled, are given titles.

- Ample archaeological findings in the form of bread moulds, vats, and beer containers from many Old Kingdom sites, combined with the inclusion of bread and beer in offering lists, provide evidence that bread and beer, both the produce of cereals, were the staple food and drink of all classes.

Figure 27: REAPING THE CROP, TOMB OF HESI, SAQQARA.
Kanawati–Abder-Raziq (1999) pl. 52.

Figure 28: TRANSPORTING THE SHEAVES TO THE THRESHING FLOOR, TOMB OF MERERUKA, SAQQARA.
Kanawati (2011) pl. 81.

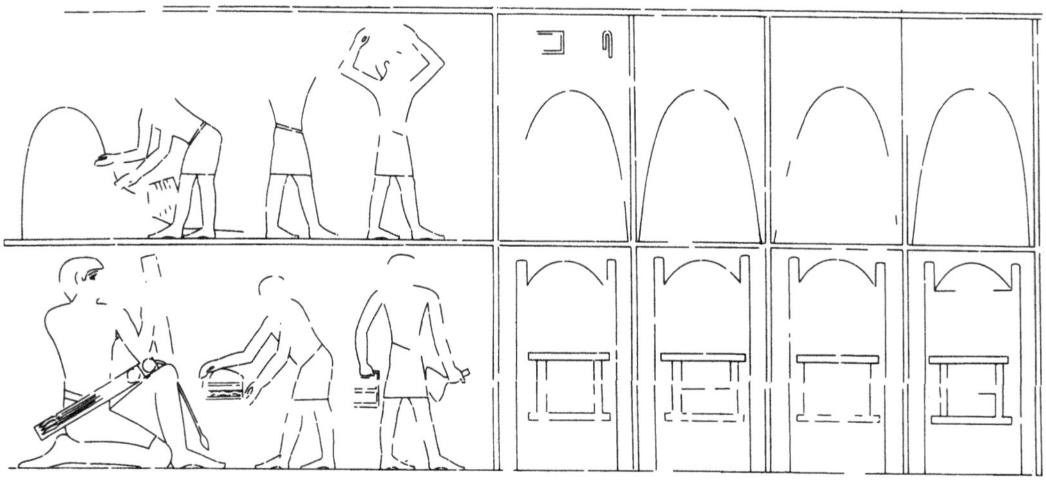

Figure 29: WORKERS TAKING GRAIN TO THE GRANARIES, TOMB OF KHENI, EL-HAWAWISH.
Kanawati II (1981) fig. 17.

Figure 30: HEADMEN RENDERING THEIR ACCOUNTS AND BEING PUNISHED, TOMB OF MERERUKA, SAQQARA.
Kanawati (2010) pl. 77, 78.

CHAPTER 9 FIELD CROPS – FLAX

9.1 Focus of enquiry: evaluation of flax as an economic resource

TABLE 9: PRODUCTION OF FIELD CROPS
TABLE 11: TITLES

The data:
The iconography and associated inscriptions of the two portrayed phases of flax production:
- Soil preparation and planting
- Harvesting.

Data from other sources mainly relating to linen:
- Material remains
- Offerings and offering table scene
- Slab stelae.

Areas of research:
- Changes in presentation of theme over time and location
- Uses of, and demand for flax products.

9.1.1 Flax crops represented in Old Kingdom tombs

Flax (*Linum usitatissimum*)
(*ḥmꜥ; mḥj; mḥꜥw; mꜥḥw* Hannig, 540).
Linum usitatissimum is a herbaceous annual plant that grows to a height of about one metre, with a main stem that branches to produce pale blue flowers. It thrives best in moist soils and temperate regions. In climates such as Egypt's, flax is grown as a winter crop and the damp, loamy soil and mild humid conditions of the Delta following the inundation provided the best environment. Although flax is not native to Egypt and may have been introduced from the Levant,[534] evidence of its use, from the Fayum region c. B.P. 7000, dates to Predynastic times.[535]

Linum usitatissimum has a number of uses. The long fibres from the bark of the plant's multi-layered stem are spun into linen yarn of varied fineness and strength. Its seeds may be crushed, producing about 40 per cent oil, used in ancient Egypt for cooking and medicinal purposes.[536] The crushed residue of the seeds may have been used to feed livestock.

Soil preparation and sowing seed
As discussed in Chapter 8, the earliest scenes relating to the theme of preparation and sowing of field crops, from the Medum tombs of *Nfr-mꜣꜥt* and *Rꜥ-ḥtp*, merely depict the action of ploughing.[537] There is no data to indicate whether these scenes illustrate preparation for the planting of flax or cereals; nor do later scenes that include the sowing stage distinguish between the planting of the three crops, flax, barley and wheat.[538]

The flax harvest
To achieve the best quality linen, the flax plant has to be harvested while still young and green when the fibres are at their peak condition.[539] As the quality of the flax fibre depends on the age of the harvested plant, it is likely that the flax harvest normally preceded that of the cereal crop.

There are two elements in the sequence of Old Kingdom scenes representing the flax harvest: men harvesting a field of flax by pulling the plants out of the soil and an associated portrayal of seated or squatting men tying flax plants into bundles. Neither motif appears in Dynasty 4 tombs. Other data[540] however, show that linen had the same social and economic value in the early Old Kingdom as in later times when the flax harvest was depicted,[541] so it is possible that the appropriate imagery had not yet been developed.

9.2 Harvesting flax plants

Figure 31

9.2.1 Historical development of images and themes

Scenes of the flax harvest first appear in the early Dynasty 5 tomb of *Rꜥ-wr*,[542] and the theme is included in 40 tombs used in this study. Unlike the harvest of grain, which is represented in tombs to the end of the Old Kingdom, the flax harvest does not appear in the tombs of Dynasty 8. The latest tombs to include the portrayal of pulling flax plants out of the soil are dated to the second

534 Germer (1985) 101.
535 Caton-Thompson–Gardner (1934) 40-46, 49, 51, 90.
536 Vogelsang-Eastwood in Nicholson and Shaw (2000) 294.

537 Petrie (1892) pl. 12, 18,
538 See Chapter 8.2, pp. 70-71 for an historical description of this theme. Vogelsang-Eastwood distinguishes between the sowing of flax and cereals by the arm action of the depicted sower; overarm for cereals and underarm for flax. Her illustrated example, from the tomb of *Wr-jr.n.j* Davies (1901) pl. 16, does not convince. The first sower with overarm action does not relate to the flock of sheep, which follows the second plough. The second sower, immediately in front of the flock, may be using an underarm action in order to entice the sheep forward. See Nicholson and Shaw (2000) 270.
539 Later harvesting produces a coarse fibre only useful for the production of rough cloth and materials such as rope for which strength is the most important criterion: Vogelsang-Eastwood in Nicholson and Shaw (2000) 270.
540 See below, pp. 83-84.
541 The prominence and complex organization of linen lists on Giza slab stelae suggest that linen was a particularly valued item, further emphasised by the fact that the stelae were the only decorated element of the tomb: Manuelian (2003) Ch. 1, 4d and Appendix 3.
542 Hassan I (1932) pl. 3 [3].

half of the reign of Pepy II, when the most securely dated instances are those of *Ppj-ꜥnẖ/Ḥnj-km*, *Ḏꜥw*, and *Mẖw* and *Sꜣbnj*.[543]

Scenes of pulling up flax appear in the list of tombs used in this study as follows:

Early to mid Dynasty 5

Rꜥ-wr	V.E
H̱nmw-ḥtp	V.E
Wr-jr.n-Ptḥ	V.M
Ptḥ-špss	V.M
Nj-ꜥnẖ-H̱nmw and *H̱nmw-ḥtp*	V.M

Late Dynasty 5

ꜣḫtj-ḥtp (Louvre)	*Ḥtp-ḥr-ꜣḫtj*
Jrj-n-kꜣ-Ptḥ	*Ḥtp-kꜣ.j*
Wr-jr.n.j	*Tp-kꜣ.j*
Pḥn-wj-kꜣ.j	*H̱w-ns*
Ptḥ-ḥtp I	*H̱nwt*
Ptḥ-ḥtp/Jj-n-ꜥnẖ	*Sndm-jb/Jntj*
Nj-mꜣꜥt-Rꜥ	*Sndm-jb/Mḥj*
Nb-j-kꜣw-ḥr/Jdw	*Sḫm-ꜥnẖ-Ptḥ*
Nfr (1)	*Sḫm-kꜣ.j*
Nfr-jrt-n.f	*Kꜣ.j-m-nfrt*
Nfr-sšm-Ptḥ and *Sḫntjw*	*Kꜣpj*
Rꜥ-špss	*Kꜣ-ḥj.f*
Ḥzjj-Mnw	*Ṯjj*

Early Dynasty 6

Mrw/Ttj-snb	*Nj-ꜥnẖ-nswt*
Mrrw-kꜣ	*Ḥzj*
Mḥw	

Mid Dynasty 6

Jbj	*Ppj-ꜥnẖ-ḥrj-jb*

Late Dynasty 6

Ppj-ꜥnẖ/Ḥnj-km	*Ḏꜥw*
Mẖw and *Sꜣbnj*	

The harvesting process depicts two to six men harvesting the plant against a backdrop of a field of growing flax.[544] The harvesters pull the plants out of the ground rather than cutting with a sickle in order to preserve the full length of the fibres unmarked.[545] The crop is frequently depicted as a rectangle that is either blank, possibly from loss of paint work, and marked only in outline or filled with vertical lines to represent the stems of the plants, which are occasionally portrayed with flowering heads. In the few tombs retaining colour, the stems of the plant are painted green and the flowering heads blue, indicating that the harvesting is taking place while the plants are immature and thus intended to provide linen of fine quality.[546] Although the flax plant actually grows to about a metre in height, the depicted height of the crop varies from the harvesters' knee height to level with their chest. A certain proportion of the plants would have been allowed to mature and produce seeds to be crushed for their oil or used in the following year's planting.

Scenes of the flax harvest depict harvesters in a variety of postures quite different from those of men cutting the cereal harvest with sickles.[547] Flax harvesters stoop or squat to grasp a bunch of stalks with both hands, straighten up to pull the plant out of the soil or stand erect holding up a bundle of plants. The men shown standing erect appear to be inspecting the plants they hold, perhaps to check whether any harm has been done to the stems and consequently the fibres of the plant, or to clean off the soil. The variety of postures in which the harvesters are normally depicted give the impression that pulling the plants out of the soil requires exertion and care.[548]

When captions accompany the depictions, they are usually quite terse. Those in the scene of *ꜣḫtj-ḥtp* (Louvre) describe the individual actions involved in harvesting the flax plant.[549] To the right a supervisor orders the line of labourers to hurry (*sḫs*); two men pulling plants out of the ground are described as 'tearing' or 'pulling' (*wnpt*); the next man holding up a bunch of plants is described as 'equalising' (*stwt*); the next worker picking up two bunches of plants advises 'hurry' (*sḫs*) while the following worker is described 'smoothing' or 'shaping' (*ꜥbb*) the bundle of plants he holds in his hand. Other phrases used to describe the actions of the flax harvesters are *ḥwj mḥꜥ* ('pulling up flax') and *ṯꜣt* ('pulled' or 'stripping').[550] The daylong activities of pulling, bending and stretching upright required of the flax harvesters would have been strenuous and tiring.

9.3 Dealing with the harvested flax plants

Figure 32

9.3.1 Historical development of images and themes

The second element portraying seated or squatting men tying flax plants into bundles is less frequently depicted than the actual harvesting of the plant, but the distribution over time, confined to Dynasties 5 and 6, is similar.

In the list of tombs used in this study, scenes of tying flax plants appear in:

Early to mid Dynasty 5

Wr-jr.n-Ptḥ	*Nj-ꜥnẖ-H̱nmw* and *H̱nmw-ḥtp*

[543] Respectively, Blackman–Apted (1953) pls. 22 [3], 58 [1]; Davies *Gebrâwi* II (1902) pl. 6; de Morgan (1894) p. 145 [d].

[544] See Vandier *Manuel* VI (1978) 69–73 and fig. 44.

[545] Siebels (2000) 143.

[546] Traces of pale blue on the flowering heads of the plants remain in *Jrj-n-kꜣ-Ptḥ*, *wr-jr.n.j*, *Nfr* (1), *Ḥtp-ḥr-ꜣḫtj*, *Kꜣ.j-m-ꜥnẖ* and *Ṯjj*: Siebels (2000) 134.

[547] Harpur (1987)165 discusses these postures.

[548] This feature does not appear in all scenes. In the tomb of *Ḥzj* the two teams of harvesters facing left and right are all depicted with virtually the same posture, standing and bent slightly forward either grasping the stalks or with their arms wrapped around them. It is possible that this presentation was adopted because the crop is depicted having grown almost to head height: Kanawati–Abder-Raziq (1999) pl. 52.

[549] Ziegler (1993) 126, 129.

[550] Refer TABLE 9: PRODUCTION OF FIELD CROPS – *ꜣḫtj-ḥtp*, *Wr-jr.n.j*, *Ḥzjj-Mnw*, *H̱w-ns*, *Sḫm-ꜥnẖ-Ptḥ*, *Sḫm-kꜣ.j*.

Later Dynasty 5

3ḫtj-ḥtp (Louvre)	*Snḏm-jb/Mḥj*
Wr-jr.n.j	*Sḫm-ᶜnḫ-Ptḥ*
Nb-j-k3w-ḥr/Jdw	*Sḫm-k3.j*
Nfr (1)	*K3.j-m-nfrt*
Nfr-jrt-n.f	*K3pj*
Nfr-sšm-Ptḥ and *Sḫntjw*	*K3-ḥj.f*
Ḥtp-ḥr-3ḫtj	*Ṯjj*
Ḫw-ns	

Early Dynasty 6

Mrw/Ttj-snb	*Nj-ᶜnḫ-nswt*
Mrrw-k3	*Ḥzj*
MḥW	

Mid Dynasty 6

Jbj

Late Dynasty 6

Ppj-ᶜnḫ/Ḥnj-km	*Ḏᶜw*

The associated scene of men seated or squatting on the ground tying flax plants into bundles usually adjoins that of the actual harvesting and the two motifs are sometimes linked by a figure delivering plants from the field.[551] The squatting worker who is handed a bunch of plants is described as 'bundling the flax' (*dm3 mḥᶜ*). Included in this scene are depictions of tied bundles awaiting collection. The care taken in handling the plants is further evidenced in the scenes of tying the harvested flax into bundles. The bundles are held vertically as they are taken, by hand, to those tying the bundles. The tying process is located adjacent or close to the field of flax in 23 tombs, suggesting that the plants had to be handled as little as possible, and the experienced workers tying the plants are depicted using hands and feet to form the bundles. The onward transport of the bundles for processing is never portrayed.[552]

9.4 Inscriptional data

Nine vertical inscriptions,[553] separating the figure of the tomb owner from the panel of activities that he inspects, include references to the flax harvest, but only as one of a number of harvesting activities. The description of *K3.j-m-nfrt* of Saqqara emphasises the grain harvest with, *m33 sk3 ḥj mḥᶜw 3sḫ šd ... ḥj ḫ3ḫ3 jᶜb* ('Viewing the cultivating of grain, pulling up flax, reaping loading (donkeys), striking, winnowing, heaping'), as does the inscription of *Ṯjj*, *m33 ḥwj mḥᶜ 3sḫ jt bdt šdjt wbz m spwt m njwwt nt pr-ḏt* ('Viewing the pulling up of flax, the reaping of barley and emmer wheat, the transporting (by donkey), the heaping of the threshing floors, from the towns of the estate'). The vertical inscriptions of *Jrj-n-k3-Ptḥ* and *Sḫm-ᶜnḫ-Ptḥ* of Saqqara show the same emphasis on grain: *m33 sk3 wḥᶜ mḥᶜ.w* <*3zḫ*> *šdjt* ('Viewing ploughing, pulling up flax, cutting with a sickle, transporting') and *m33 k3t sḫt sk3 3sḫ ḥwj mḥᶜw šdt jᶜ3w ḥwj jᶜ3w spwt ḫ3ḫ3* ('Viewing the work of the fields, ploughing, reaping, gathering flax, loading donkeys, donkeys treading the threshing floors, winnowing'). A small number of extended captions include: *ḥwj mḥᶜ jn jzwt ḏt.f* ('pulling up flax by the crews of his estate') and *mᶜ n ᶜt smᶜ* ('flax for the *smᶜ*-rope').[554]

9.5 Other evidence

Apart from the lone instance in the tomb of *Ftk-t3* of Abusir[555] no other scene relating to the treatment of the harvested flax appears in Old Kingdom chapels, perhaps because the remaining processes took place at a location that was not strictly part of the estate,[556] or because craftsmen had not yet developed a suitable pictorial scheme to represent the complex, lengthy and labour-intensive process to which flax fibres were subjected.[557] There are, however, a small number of scenes of weavers receiving payment for their services.[558] For the eventual product, linen, there is a considerable body of evidence that includes slab stelae, offering scenes in which rolls of linen and linen garments are presented to the tomb owner and linen chests, scenes depicting bags made of linen, scenes and inscriptions illustrating the use of linen as a medium of trade and as payment for services. In addition, there is a body of archaeological evidence of uses for linen from the Old Kingdom and previous dynasties consisting of garments, sails for boats, lengths of unused linen cloth and mummy bandages.

Vogelsand-Eastwood categorises the uses of linen into 'clothing', 'household', 'outside', 'economic', 'ritual' and 'funerary' but draws on all periods of ancient Egyptian history for her evidence.[559] The data offered here is confined to the Old Kingdom and earlier periods.

---

551 For example, *Ḫnmw-ḥtp*, Petrie and Murray (1952) pl. 17 [4-5, 7] and Sharawi–Harpur (1988) fig. 6; *3ḫtj-ḥtp* (Louvre) Ziegler (1993) 70-1 (plates) 71-4, 126, 129, 135-7.

552 The only Old Kingdom tomb to include scenes of processing flax plants is that of *Ftk-t3*. See LD Erg. 40[c]; Bárta *Abusir V* (2001) figs. 3.11, 3.18.

553 See TABLE 9: PRODUCTION OF FIELD CROPS – *Ḥtp-ḥr-3ḫtj, Jrj-n-k3-Ptḥ, Jbj , Ḏᶜw, Wr-jr.n.j, Ḫw-ns, K3.j-m-nfrt, Ṯjj, Ppj-ᶜnḫ-ḥrj-jb.*

554 *K3pj* Roth (1995) pls. 41d, 42a-b, 158-159..

555 Barta (2001) 88.

556 Production of linen as a cottage industry, similar to that referred to in the Hekanakhte Papers in the Middle Kingdom, may have already been in existence. 'Account VII' provides information regarding payment to a woman named Sitnebsekhtu for the production of linen on her property. The bundles of flax had been delivered to Sitnebsekhtu as raw materials, James (1962). A loose block found in the tomb of *Ftk-t3* of Abusir with a surviving line of inscription reads *pr-jrjwt nt ḥr*(*t*) 2430 ('the house of weavers, of items: 2430'): Barta (2001) 88. Barta does not give an opinion whether the tomb of *Ftk-t3* is the true provenance of the block.

557 This included retting, drying, breaking, scutching and combing the fibres before the final spinning and weaving.

558 *Nb.j-m-3ḫtj* LD Erg. 34 [a], Junker V (1941) fig. 11, Hassan IV (1943) fig. 82. *Snb* Junker *Giza V* (1941) fig. 8, pl. 5 [a]. *Ptḥ-špss* Vachala (2004) 262-3 [E99 (1897)]; *3ḫtj-ḥtp* (Louvre) *Giza V* (1941) fig. 9, Ziegler (1993) 60-2 (plates) 116-19 (drawings) 121-2 (texts). *Ptḥ-ḥtp* LS 31 LD II 103 [a], *Giza V* (1941) fig. 10. *Sḫm-k3.j, Sšm-nfr* IV Junker *Giza XI* (1953) fig. 65.

559 Vogelsang-Eastwood in Nicholson and Shaw (2000) 286.

For clothing

Apart from bead-netting dresses, archaeological evidence from the Old Kingdom and earlier periods amounts to a few linen garments.[560] Pictorial data and statues, however, provide a mass of evidence showing that men and women wore a range of linen garments.[561] Dresses, kilts and loincloths are painted white and are therefore intended to represent cloth rather than leather. As there is no Old Kingdom evidence of garments made out of cloth other than linen, it is virtually certain that the depicted clothing is the product of flax.

Household uses

Vogelsand-Eastwood lists cushions, curtains, bed linen, towels, lamp wicks and seals as items made from linen. No Old Kingdom relics of these items have been found, but cushions on the chairs of the tomb owner and his wife are portrayed in Old Kingdom scenes. These chairs are furnished with a cushion from Dynasty 5 onwards.[562] Bed making scenes appear in thirteen tombs of this study,[563] although no lengths of linen designed as bedding have been found dating to the Old Kingdom.[564]

Uses outside the home

Flax fibres were the strongest plant fibres available in ancient Egypt, so it is not surprising that linen was used for a variety of manufacturing and transport purposes.[565] Of items depicted in Old Kingdom scenes, Vogelsand-Eastwood lists bags and sacks,[566] cloths for straining liquids such as wine and oils,[567] cloths covering the backs of animals[568] and sails and awnings for boats.[569] All these items are depicted with varying degrees of frequency in the tombs of Old Kingdom officials. Linen bags are frequently pictured situated on the skiff in the scenes of the tomb owner taking a 'pleasure cruise' or fishing and fowling.[570] Traders in market place scenes are depicted with small linen sacks tied behind their shoulders.[571]

Economic usage

As a medium of exchange in barter transactions and to pay wages linen has considerable utility. It is durable and transportable; pieces of material can be easily measured and checked for quality. Roth cites a number of tomb inscriptions in which it is either clearly stated or may be inferred that the tomb owner claims to have paid the workers constructing his tomb with linen.[572] Roth notes, however, that from the small number of citations it is not possible to judge whether payment with linen was the norm.

Portrayals of quantities of linen being packed in wooden chests suggest either the storage of supplies of the material or a mode of transport. The importance of this procedure is exemplified in the tomb of *Ftk-tꜣ* of Abusir, where the associated inscription (*sẖp(j)t r pr-ḥḏ*) makes it clear that transport of the material to the *pr-ḥḏ* is intended, presumably as a form of taxation.[573] The *pr-ḥḏ* was the department responsible for the collection and redistribution of taxes in kind. Helck noted that the *pr-ḥḏ* presented linen as funerary equipment to officials.[574] At all periods of the Old Kingdom important officials recorded titles relating to linen, while holders of lesser titles attest to a hierarchy of scribes and officials connected to the supervision of linen production.[575]

Ritual and funerary uses

Wrapping for the bodies of Old Kingdom officials undoubtedly created the greatest funerary demand for linen,[576] but there were other funerary uses such as the wrapping of viscera in linen and the depositing of balls of natron in linen bags with deceased.

9.6 Summary

- While the seeds of the flax plant (*Linum usitatissimum*) were crushed to produce oil, the main use of the plant was the production of linen of a range of qualities.

560 From Tarkhan, a sheet of good quality linen dated to the First Dynasty (UC 28670); and a short dress with pleats (UC 28614Ai); dated to the period First Dynasty to the Old Kingdom. From Deshasha, two dresses found in a coffin dated to the Old Kingdom (UC 31182 and 31183). From Gebelein, a Sixth Dynasty pleated dress with sleeves (Schiaparelli no. 14087).

561 Every Old Kingdom scene and statue depicts men and women wearing linen garment, the style of garment depending on the status of the person. Officials wear tight-fitting or flared kilts and working-men brief kilts or, more usually, loincloths. Female tomb owners and female relatives of officials are portrayed wearing dresses and female workers simple loincloths. The few examples of dresses found would have hung loosely on the wearer and do not correspond to the portrayed tight-fitting dresses. The difference may be accounted for by the desire to portray the essentially female form and consequent female qualities of the women of high status.

562 For the list of chairs with cushions see Cherpion (1989) 147-154, Critères 2-6.

563 See for example, *Wr-jr.n-Ptḥ* James (1961) pl. 29.2 and *Mrrw-kꜣ/Mrj* in Duell I (1938) pl. 93, Kanawati (2010) pl. 98-99.

564 Vogelsang-Eastwood presents piled lengths of linen from the tomb of *Ḫꜣ* (TT8) as an example of bed linen (*Turin, Museo Egizio, Inv. Suppl.* 8629-36): Nicholson and Shaw (2000) 291, fig. 11.20.

565 Wicker (1997) 112-113.

566 For a study of sacks holding sheaves of cereals see Siebels (2000) 239-249.

567 See Nicholson and Shaw (2000) 588-590.

568 For cloths protecting the backs of donkeys see Siebels (2000) 251. See Chapter 7, p. 62 for cloths laid over the backs of donkeys.

569 The composition of sails and awnings is confirmed by the wrappings of a mummy in the Musée des Beaux Arts Lyon, which when laid out were found to have come from a boat's sail: Goyon–Josset (1988) 129-132 (cited by Vogelsand-Eastwood in Nicholson and Shaw [2000] 292).

570 See Woods (2008) 60, 63, 142 and note 354, 145 note 378, 182 and note 289,184, 204.

571 *Ftk-tꜣ*, LD II 96.

572 Roth (1994) 232-237: *Nfr-ẖw-w(j)* G 2098, *ꜥnḫ-m-ꜥ-Rꜥ* G 7837 and 7843, *Nj-mꜣꜥt-Rꜥ* PM 282-4, *Rmnw-kꜣ(.j)* PM 261, *Mṯṯj* PM 646-8, G2375, *Ḥtp-ḥr ꜣḫtj* D 60. Roth infers the payment of linen in the inscriptions of *Nfr-ẖw-w(j)* and *ꜥnḫ-m-ꜥ-Rꜥ* from the use of the term *ḏbꜣ* and *ḏbꜣw* for 'payment/s', determined by the 'cloth' glyph in *Nfr-ẖw-w(j)*'s tomb.

573 LD II 96; Barta (2000) 88-93.

574 Helck (1975) 97.

575 See TABLE 11: TITLES – Titles Relating to Linen.

576 Linen wrapping around the bones of the owner of Medum Mastaba 17, dated to early Dynasty 4, suggests that this aspect of mummification was practised at least from the beginning of the Old Kingdom: Petrie (1892) 14-15.-

- In the funerary iconography there is no distinction between the sowing of cereals and flax. However, the iconography clearly distinguishes the harvesting flax from that of cereals by the separate methods used in each process.

- A chronological study of the two pictured elements of the flax harvest, pulling the plants out of the ground and tying them in bundles, provides a life span for the flax harvest scenes from early Dynasty 5 to late Dynasty 6. In these scenes harvesters are depicted in a variety of postures, suggesting that the work was heavy labour. The men tying the bundles are shown either seated or squatting. The iconography of both themes suggests that great care was taken in handling the plants.

- In the Old Kingdom there is only one scene showing the production of linen, although weavers are sometimes depicted being paid for their services. The considerable evidence, pictorial, inscriptional and archaeological, of the variety of uses to which linen was put consists of garments, household needs, for bags and sacks, sails and awnings for boats, as a medium of exchange, payment for services and as a form of taxation. Ritual uses included wrappings for bodies, viscera and balls of natron.

Figure 31: HARVESTING FLAX, TOMB OF HESI, SAQQARA.
Kanawati–Abder-Raziq (1999) pl. 52.

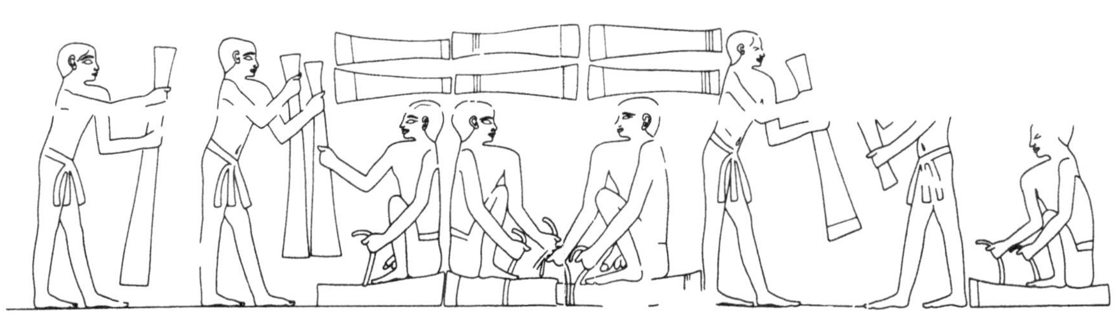

Figure 32: TYPING FLAX PLANTS INTO BUNDLES, TOMB OF HESI, SAQQARA.
Kanawati–Abder-Raziq (1999) pl. 52.

SECTION FOUR

RESOURCES OF THE MARSHLANDS

Figure 33: Representation of the marshland, tomb of Mereruka, Saqqara. Kanawati (2010) pl. 68.

The aim of this section is to identify the wetland resources exploited by officials and to investigate the value of this exploitation. High officials devoted considerable wall space in their tomb chapels to marshland activities - many registers, even entire panels of decoration - making it clear that they found the products of the marshes of great utility and that they required their workforce to capitalise on these resources for the benefit of the estate.

The wetlands and the configuration of the Nile valley in Egypt

All wetlands have many common features but no two are identical. Consequently, the following overview of the nature and distribution of the Egyptian wetlands in ancient times is intended to provide background to understanding how its resources were exploited.

In the pharaonic era marshland occurred extensively in the Delta and intermittently along much of the Egyptian valley. During an inundation the floodwater, spilling laterally from the river, filled successive flood basins on either side of the river[577] renewing the marshy back swamps along the valley margins and the fringes of oxbow lakes, providing habitats for freshwater fish and waterfowl in channels and lakes, as well as along the river. How far the floodwaters spread depended on the width of the valley and the level of the inundation. Where

[577] Butzer (1976) 17-18.

the Nile braided into major distributaries and many minor channels of the Delta, extensive tracts of wetlands consisting of papyrus swamps and interposed grazing land were created.[578] The Nile valley thus provided Egyptians with the swamps of the Delta, marshes bordering the river, subsidiary channels and backwater swamps that lay well away from the river, enabling Egyptians living along the river valley, as well as in communities close to the Delta, to exploit wetland resources for much of the year.

The produce of the wetlands, notably birds, fish and sedges, especially papyrus, were not Old Kingdom officials' most highly valued assets. Their funerary iconography and inscriptions, including in Dynasty 6 their autobiographies, provide evidence that these wealthy and powerful administrators prized their herds of cattle above all other resources. However, the marshlands offered both a plentiful supply of birds and fish (cheap sources of good quality protein) that enabled a well-endowed official to provide for a large household with numerous dependants and workers, and material for the manufacture of a multitude of items of everyday use. The most valuable raw material of the marshland was papyrus, which furnished Egyptians with the means for the construction of lightweight craft and the manufacture of cordage, many items essential for the exploitation of all other wetland resource, and fodder for their herds.

While it is not known whether estate owners claimed stretches of marshland and riverbank as part of the estate, tomb scenes make it clear that they freely exploited these regions.

[578] Excavations at Kom el-Hisn provide evidence of specialised cattle farming in the pastureland of the Delta that lay above the level of the inundation: Moens–Wetterstrom (1988)

CHAPTER 10 FISH

10.1 Focus of enquiry: fish and fishing as an economic resource for a great estate

TABLE 13: FISH AND FISHING – METHODS AND MANAGEMENT

The data:

- The iconography of fish and fishing by estate workers
- Captions associated with the iconography.

Areas of research:

- Fish species most frequently depicted being fished by workers.
- Methods of fishing: techniques of catching fish in quantity.
- Processing caught fish: gutting and drying fish; extracting roe.
- Transporting the catch.

Not all the themes in which fish are portrayed represent fish as an official's resource. Scenes of the deceased spearing fish or depictions of fish swimming freely in a stretch of water do not indicate a resource. This includes themes that may portray the landing of a catch for immediate consumption or an activity in which the actual catching of fish is a subsidiary element or to be interpreted as symbolising a statement of belief. For example, it cannot be inferred that scenes of the deceased spearing fish[579] or depictions of fish swimming freely beneath a boat indicate the acquisition of a resource that is to be stored and used at will in the future. Consequently, this study concentrates on scenes of fishermen making catches of fish and workers and bearers handling the catch.

Scenes of organised fishing, especially with a dragnet, and the associated theme of gutting and drying fish suggest that fish were caught in quantity because they were a significant product of the estate. Fish either have to be consumed, processed for preservation or transferred to holding ponds very soon after being landed, as dead fish quickly putrefy, particularly in hot weather. Consequently, only a catch that has been processed or kept alive in a holding pond can be regarded as a storable resource. The employment of teams of fishermen and the portrayal of the gutting process, the preliminary step in drying the fish and the only means of preservation used by ancient Egyptians,[580] indicate that the catch of fish was not intended for immediate consumption and could be used for trade or to reward the deceased's dependants and workers, as well as providing a storable supply of food for the household. Any trade, however, would probably have been small-scale, local commerce.[581] Transportation scenes consist of porters carrying fish and there are no portrayals of fish being transported any distance by ship. Whether the theme of market trading in fish represents a resource of any significance for tomb owners is doubtful.

Yet fishing was important to the life of the estate. To judge by the number of scenes, fishing with a dragnet (seine), of which 78 scenes are included in this study, was the most frequently used method of catching fish in quantity. Marsh scenes also include other fishing techniques: individual fishermen fishing from a boat using a hand-held net or a line with hooks, or funnel-shaped and rounded traps made of basketware. Although these methods do not represent catching fish on such a large and organised scale as by dragnet, they are considered here because they present fishing on behalf of the tomb owner.

10.2 Depiction of fish species in fishing industry scenes[582]

For the inhabitants of the Old Kingdom the waters of the Nile valley included the main river, a number of auxiliary channels, oxbow lakes and the distributaries of the Delta, all fed by the annual inundation. Much of the shoreline of these bodies of waters would have been fringed by marshland vegetation. This pattern of waterways provided diverse habitats: partly saline in the Delta, water of varying depths, flowing river water, comparatively still water in oxbow lakes, brackish muddy beds rich in vegetation and decomposing organic matter near the shore and clearer sandy beds. Consequently the valley provided environments suited to a great variety of freshwater fish. Scenes from Old Kingdom tombs provide evidence that many fish species, from the huge Nile perch to shoals of mullet and to various species of catfish were harvested in abundance from the waters of the Nile

579 This particular theme may represent the tomb owner's prowess and enjoyment of sport in the marshlands or symbolise a religious belief, but the scene never suggests catching fish in quantity or preserving the catch.

580 There is no evidence that ancient Egyptians smoked fish as a means of preservation: Ikram (1995) 155.

581 Dried fish would only have had limited value as a resource. Just as in the modern experience, the product of dried fish in Ancient Egypt would not have survived for long, as it would have been infested by blowfly larvae during the drying process, and would have attracted dermestid beetles, which would, "within a few months leave nothing but skins and skeletons": Van Neer (1989) 55–56.

582 This excludes the scene of the deceased spearing fish, but includes all other scenes of catching, drying and transporting fish by workmen.

valley.[583] The species of most fish depictions can be identified from the shape of body and fins. Over 26 species are portrayed, either in nets, traps, hooked on a line, gutted or being transported.[584] Mullet appears to have been the most popular fish. It figures in more tombs than any other species and in over 80 per cent of gutting scenes.[585] Other frequently depicted species are *Tilapia, Synodontis schall*, *Citharinus*, *Tetrodon fahaqa* and *Lates niloticus*. The majority of depicted species (15 out of 26) appear in at least fifteen tombs of the list used in this study.

The kinds of fish most frequently depicted in Old Kingdom fishing 'industry' scenes:

Species and number of tombs in which depicted[586]

Mugil sp.	57+
Tilapia sp	46+
Synodontis schall	41+
Citharinus	34+
Tetrodon fahaqa	32+
Lates niloticus	30+

Species depicted in 15–30 tombs:

Anguilla vulgaris	*Mormyrus caschivo*
Clarias lazera	*Mormyrus niloticus*
Gnathonemus cyprinoides	*Schilbe*
Labeo niloticus	*Synodontis batensoda*
Malopterurus electricus	

Mugil depicted in 57+ tombs
Although coastal fish, grey mullets (species *M. cephalus, M. capito* and *M. auratus*) enter the river Nile and are found as far south as Aswan. Many *mugil* species move between saltwater and the freshwater environments of rivers. The adult fish feed mainly on algae while inhabiting fresh waters. Average body length of the fish is 35 cm. The frequent depiction of *mugil sp*[587] in Old Kingdom tombs suggests that the fish was highly valued as well as being easy to catch. Mullets swim in schools and lose some of their wariness when they prepare to migrate out to sea to spawn.[588]

Tilapia depicted in 46+ tombs
Although a number of species of the *Tilapia*, particularly *T. nilotica* and *T. zillii*, frequent the Nile, these species cannot be differentiated in Old Kingdom art. *Tilapia* prefer shallow swampy water with plentiful vegetation, as they feed on the diotoms growing on the plants. *T. nilotica* grows to a body length of about 46 cm. and *zillii* to about 29 cm.[589]

Synodontis schall depicted in 41+ tombs
This species is an omnivore that feeds on insect nymph, larvae, eggs and detritus. It tends to inhabit the air-water interface, possibly to replace gas lost from the swimbladder as it adjusts to pressure changes by swallowing or releasing gas bubbles. It can grow to a body length of about 37 cm.[590]

Citharinus spp. depicted in 34+ tombs
Called the moon fish of the Nile, *Citharinus* has a thick body with a pronounced adipose fin, making it easy to identify in depiction. *Citharinus citharus* grows to approximately 60 cm in length and the smaller species, *Citharinus latus* to about 42 cm. From the size of the adipose fin, most Old Kingdom portrayals appear to be *Citharinus citharus*. The genus feeds on algae near the surface of the water and, at the bed of rivers, on Crustacea and diatoms.[591]

Tetrodon fahaqa depicted in 32+ tombs
This is a freshwater species of the genus *Tetraodon*, It lives on the muddy bed of shallow waters, feeding on molluscs and other denizens of this environment, and consequently has powerful jaws. *Tetrodon fahaqa* is a highly poisonous puffer fish, identifiable in depiction by its rounded, globular shape, which is achieved by inflating its stomach with air. It grows to about 43 cm in length.[592]

Lates niloticus depicted in 30+ tombs
The Nile perch is one of the largest freshwater species growing to nearly two metres and 200 kg. It is a fierce predator. Adults are found in all freshwater habitats but juveniles are restricted to shallow near-shore waters.[593]

Elements of the theme of fishing
The major motifs depicted in this theme:

- the process of catching the fish, whether by dragnet, hand net, funnel trap, basket trap or line and hook (*Figures 34, 35, 36*).
- the gutting and drying of the catch and extraction of roe (*Figure 37*)
- transporting the catch, either to the gutting area or to be presented to the tomb owner (*Figure 38*)
- scenes of registration of the catch and workmen cooking fish appear in very few tombs.[594]

10.3 Methods of catching fish

Apart from images of single fishermen fishing with a hook and line or laying small wicker traps, other methods of fishing show teams of men who either work in unison

583 At least nineteen different species are depicted in the tomb of *Tjj*; seventeen in the tomb of *Mrrw-k3* and eleven in the tomb of *K3-gm-nj*.
584 The number of different species portrayed in Old Kingdom tombs is inexact because a number of depictions of fish cannot be identified with certainty. See van Elsbergen (1997) 133-307.
585 van Elsbergen (1997) Ch 3.
586 The plus sign (+) indicates the depiction of fish that are of uncertain species. Identification draws on Gamer–Wallert (1970) 140-41 pls. 1–7 and on Boulenger (1907) vols. I, II.
587 It is not usually possibly to distinguish species.
588 Brewer–Friedman (1989) 72, Gamer–Wallert (1970) 14, Boulenger (1907) 427-429.
589 Brewer–Friedman (1989) 74-79, Gamer–Wallert (1970) 13.
590 Brewer–Friedman (1989) 67-68, Gamer–Wallert (1970) 12.
591 Brewer–Friedman (1989) 56, Gamer–Wallert (1970) 8.
592 Brewer–Friedman (1989) 80-81, Gamer-Wallert (1970) 14-15.
593 Brewer–Friedman (1989) 74-75.
594 *Mḥw* Altenmüller (1998) 137-9, pl. 34 [a, b], 36 [a, b], 37 [a, b] and *K3-gmnj/Mmj* von Bissing I (1905) 14-15, 20, pl. 18 - 19.

or cooperate in taking a large catch of fish. The dragnet (seine) scene, which appears in many more tombs than does any other method of catching fish, presents the catching of fish in the largest quantities.[595] Teams of men, divided into two groups, are depicted standing on the shore of a body of water as they work together to haul in the two ends of a long net laden with fish. This activity represents the final phase of the fishing process and usually shows many different species of fish trapped in the net. The earlier stages showing one end of the net stretched out from land to the middle of the water only appear in a few tombs[596]. Each end of the net was attached to a stout rope. One end was taken out by boat into the middle of the channel or lake, with the other end held on the bank by a team of men. The lower edge of the net, with sinkers attached, was allowed to drop vertically into the water and then the boat was rowed back to shore dragging its end of the net and trapping fish. When the boat reached the shore, the fisherman on the boat must have either jumped ashore or handed his end of the rope to one of the gang waiting on land, as most scenes show two gangs standing on the water's edge pulling the ropes on either end of the net. As the net was hauled in and the fish removed, it was rolled up for a new catch.[597] Van Elsbergen implies that this method of fishing was adapted to deep water such as the Nile and its main branches.[598] However, the scene in the tomb of Jbj shows the dragnet being towed through a body of water with vegetation, which suggests that the lower edge of the net reached almost to the floor of the channel.[599] Furthermore, a number of fish species depicted caught in the dragnet, such as the *Tilapia* and *Clarias*, prefer a habitat of shallow water.[600] So the dragnet may have had a wider usage than van Elsbergen assumes.

Other means of catching fish in quantity include the funnel shaped traps ('weirs') made of basketry, which are depicted in two sizes. The smaller traps, which must have been used in shallow water, were laid in rows on the floor of the lake or channel.[601] The much larger weir trap, intended for deeper water, was laid and retrieved from boats.[602] The interior of the weir was so designed that once a fish had swum into the trap it was unable to get out. When the filled trap was taken up, it was upended over a container, the narrow end untied and the trapped fish tipped into a sack-like container.[603]

Methods of catching fish in smaller quantities are also represented in Old Kingdom tombs. Fish are shown being caught by fishing line and hook and occasionally in shallow water by wicker baskets that were upended and placed over a fish.[604] Angling with line to which were attached a number of hooks is more frequently depicted and must have been a method for deeper water, as the angler always fishes from a boat. Depictions of fishing with a bag shaped hand net show the fisherman in a boat either leaning forward to tow the net through the water or lifting the net full of fish.[605] The triangular mouth of the net is held rigid by being attached to two rods. From the depiction of the size of fish caught in the net it seems that the method was used to catch smaller fish.

10.4 Fishing scenes

10.4.1 Historical development of images and themes

Distribution of scenes in tombs of the Old Kingdom:

Dynasty	Dragnet	Hand-held net/wicker trap	Line & Hook
IV.E-L	4	0	0
V.E	4	0	0
V.M	9	2	1
V.L	35	9	11
VI.E	15	4	9
VI.M	4	1	1
VI.L	6	5	4
VIII	2	0	0

Dynasty 4

The only fishing technique portrayed in this period is the use of the dragnet, which appears in the tombs of *Rꜥ-ḥtp, Nfr…w* and *Nfr-mꜣꜥt,* all dated to the beginning of the dynasty.[606] These are the earliest scenes of fishing by dragnet. They show teams of men, standing at the water's edge, hauling in the net of fish by pulling on the rope attached to each end of the net. The scene in *Rꜥ-ḥtp* shows the net with a row of sinkers attached to the lower side of the net and a row of floaters attached to the upper side, an arrangement to keep the net vertical as it is dragged through the water, but only three men needed to haul the net onto the shore.[607] No other methods of fishing appear in the tombs dated to Dynasty 4.

Early to Mid Dynasty 5

Dragnet scenes are depicted in the following tombs:

ꜣḫtj-ḥtp	Giza	V.E
Pr-sn	Giza	V.E
Sḫm-kꜣ-Rꜥ	Giza	V.E
Kꜣ.j-ꜥpr	Saqqara	V.E
Jj-mrjj	Saqqara	V.M
Nj-ꜥnḫ-Ḫnmw and *Ḫnmw-ḥtp*	Saqqara	V.M
Nj-kꜣw-Ḥwt-ḥr	Giza	V.M

595 See TABLE 13: FISH AND FISHING – METHODS AND MANAGEMENT.
596 See *Ḥwfw-ḫꜣ.f* II Simpson (1978) fig. 47, *Kꜣ.j-m-ḥzt* McFarlane (2003) pl. 44.
597 The construction of the dragnet and its manipulation is dealt with in detail by Montet (1924) 32-36, Vandier (1969) 559-601, Van Elsbergen (1997) 9.
598 Van Elsbergen (1997) 125.
599 Davies *Gebrâwi* I (1902) 12-13, pl. 4; Kanawati *Gebrawi* II (2007) pl. 45.
600 Brewer–Friedman (1989) 4.
601 Edel–Wenig (1974) pl. 16, Moussa–Altenmüller (1977) fig. 12.
602 Wild II (1953) pl. 111.
603 Wild II (1953) pl. 111, Duell I (1938) pl. 43.
604 Moussa–Altenmüller (1977) fig. 12.
605 Duell I (1938) pl. 43, Kanawati (2010) pl. 79: the scene in *Mrrw-kꜣ* shows both stages of fishing by hand net.
606 See TABLE 13: FISH AND FISHING – METHODS AND MANAGEMENT.
607 Petrie (1892) pl.11.

Nfr and *K3-ḥ3.j*	Saqqara	V.M
Rʿ-ḫʿ.f-ʿnḫ	Giza	V.M
Ḫwfw-ḥ3.f II	Giza	V.M

Fishing with hand-held net or wicker traps:

Ptḥ-špss	Abusir	V.M
Nj-ʿnḫ/Ḫnmw and *Ḫnmw-ḥtp*	Saqqara	V.M

Fishing with line and hook(s) from a boat:

Nj-ʿnḫ-Ḫnmw and *Ḫnmw-ḥtp*	Saqqara	V.M

By Dynasty 5, the gangs of men hauling the ropes attached to the dragnet begin to be portrayed in livelier and more varied poses illustrating the strain involved in pulling such a long net filled with fish to shore. That this form of fishing took place in the marshlands is confirmed by the position of the scene in tombs. In *Jj-mrjj,* the scene is located between a register of catching birds by clapnet and two registers of constructing a papyrus boat and fighting boatmen.[608] The dragnet is positioned either above or below the clapnet in the tombs of *Sḫm-k3-Rʿ, Rʿ-ḫʿ.f-ʿnḫ* and *Nj-ʿnḫ-ḫnmw* and *Ḫnmw-ḥtp.*[609] The figure of a supervisor is introduced in mid Dynasty 5. The scenes in *Rʿ-ḫʿ.f-ʿnḫ* and *Nj-ʿnḫ-ḫnmw* and *Ḫnmw-ḥtp* include an overseer standing between the two gangs giving orders to the haulers.[610] While the supervisor in *Nj-ʿnḫ-ḫnmw* and *Ḫnmw-ḥtp* is given the title, *jmj-r3 whʿt*, and is dressed in the kilt with the protruding front panel of an overseer, in the tomb of *Rʿ-ḫʿ.f-ʿnḫ* he leans on a staff wearing a regular knee-length kilt. In the scene of *Sḫm-k3-Rʿ*, the supervisor, in a less typical position, stands to the right behind the gang pulling the rope. In all dragnet scenes the haulers are depicted either wearing only a waistband tied in a bow or naked. In this time period the hauling teams begin to be equipped with bands worn over one shoulder and attached to the rope. This arrangement presumably enabled them to use more of their body strength in hauling up the net.[611]

By mid Dynasty 5 other methods of fishing appear. In the tomb of *Nj-ʿnḫ-ḫnmw* and *Ḫnmw-ḥtp*, a further 'aquatic' scene located immediately beneath the dragnet scene contains images of fishing with a wicker 'basket' trap, with three small weirs set in a row, a much larger weir trap manipulated by men on two boats and of a fisherman wielding a hand net. On either side of the dragnet is a vignette of a fisherman angling from a boat. These registers show the full range of fishing methods used by ordinary fishermen portrayed in the Old Kingdom.[612]

Second half of Dynasty 5

Dragnet scenes are depicted in the following tombs:

GIZA

3ḫtj-mrw-nswt	*Sḫm-k3j*
Jj-nfrt;	*Sšm-nfr* IV
Jttj	*K3.j-m-nfrt* II
Nj-m3ʿt-Rʿ	*K3pj*
Ḫww-wr	*K3-hj.f*
Snḏm-jb/Mḥj	

SAQQARA

3ḫtj-ḥtp (Louvre)	*Nfr-jrt-n.f*
Jj-nfrt	*Ḥtp-ḥr-3ḫtj*
Jrj-n-k3-Ptḥ	*Jntj*
Jrw-k3-Ptḥ	*Ḥtp-k3.j/Tp-k3.j*
Jdwt/Sšsšt (r/u *Jḥjj*)	*Ḫnwt*
Pḥn-wj-k3.j	*K3.j-m-nfrt*
Ptḥ-ḥtp/Jj-n-ʿnḫ	*K3.j-m-rḥw*
Nj-k3w-ḥr	*K3.j-m-ḥzt*
Nb-j-k3w-ḥr/Jdw	*Ṯjj*

ABUSIR

Ftk-t3	*Ṯp-m-ʿnḫ*

DESHASHA

Jntj

SHEIK SAID

Wr-jr.n.j	*Srf-k3.j*

EL-HAWAWISH

Ḥzjj-Mnw

ZAWYET EL-AMWAT

Ḫw-ns

Fishing with wicker traps:

SAQQARA

Ptḥ-ḥtp/Ṯfj	*Ṯjj*

Fishing with hand held net:

GIZA

Nj-m3ʿt-Rʿ	*Snḏm-jb/Mḥj*

SAQQARA

3ḫtj-ḥtp	*K3.j-m-nfrt*
Jdwt/Sšsšt (*Jḥjj* r/u)	*K3.j-m-rḥw*

Fishing with line and hook(s) from a boat:

GIZA

Nj-m3ʿt-Rʿ	*Sšm-nfr* IV
Snḏm-jb/Jntj	

SAQQARA

3ḫtj-ḥtp	*Ḥtp-ḥr-3ḫtj*
Jj-nfrt	*Ḫnwt*
Jdwt/Sšsšt (*Jḥjj* r/u)	*K3.j-m-nfrt*
Ptḥ-ḥtp/Ṯfj	*Ṯjj*

EL-HAWAWISH

Ḥzjj-Mnw

Dragnet scenes from the second half of Dynasty 5 continue to portray the details first appearing in the middle of the dynasty, although most tombs contain images of only the final stage of the action with two gangs on the shore pulling in the net teeming with fish. Only two tombs, those of *Jntj* of Deshasha and *K3.j-m-ḥzt* of Saqqara, show the earlier phase when one end of the

608 Weeks (1994) fig. 40, second chamber, north wall, east.

609 *Sḫm-k3-Rʿ* LD II 42 [a, lower]; Hassan *Giza* IV (1943) pp. 111, 114, fig. 58; *Rʿ-ḫʿ.f-ʿnḫ* LD II 9 [lower] and *Nj-ʿnḫ-ḫnmw* / *Ḫnmw-ḥtp* Moussa–Altenmüller (1977) pl. 31, 36 [b] (part), fig. 12.

610 *Rʿ-ḫʿ.f-ʿnḫ* LD II 9; *Nj-ʿnḫ-ḫnmw* / *Ḫnmw-ḥtp* Moussa–Altenmüller (1977) fig. 12.

611 *Nj-ʿnḫ-ḫnmw* / *Ḫnmw-ḥtp* Moussa–Altenmüller (1977) fig. 12, *Jj-mrjj* Weeks (1994) fig. 40 and Simpson (1978) fig. 47.

612 Van Elsbergen (1997) Ch 1.

net is laid by boat.[613] The supervisor of the dragnet may be depicted standing either between the two gangs or to one side, directing the hauling of the net or holding (checking?) a fish. A new elaboration to the scene shows the ends of the rope neatly rolled up into balls.[614]

Other methods of fishing are more frequently depicted in the second half of the dynasty, although only four images of wicker traps (weirs) have been found in Dynasty 5 tombs. Scenes from the tombs of *Nj-ꜥnḫ-ẖnmw* and *H̱nmw-ḥtp* and *Ṯjj* show that the long weirs needed to be handled by men operating from two boats.[615] Images of a fisherman on a boat wielding hand held nets to scoop masses of small fish out of the water become more frequent, while fishermen on boats are portrayed holding lines that have a number of hooks. These less frequently depicted methods of fishing are not always located together on the same or associated registers or even on the same wall as the dragnet scene, as they are in the tomb of *Nj-ꜥnḫ-ẖnmw* and *H̱nmw-ḥtp* where all the methods of fishing in quantity are grouped together on two registers, one beneath the other. In the tomb of *Jdwt/Sšsšt*, for example, there are remains of vignettes of fishermen on a boat using a hand net and angling on the south wall of Room I, more remains of the same methods of fishing as well as a dragnet scene on the north wall of the same room, while on the lowest register of the west wall of Room III the imposing figure of the princess standing on a canoe is flanked on either side by images of fishermen using a hand net, wielding a harpoon (in front of *Jdwt*) and angling with a line which has four hooks (behind *Jdwt*).[616]

Early to Mid Dynasty 6

Dragnet scenes are depicted in the following tombs:

Jnw-Mnw	Saqqara	VI.E
Wꜣtt-ḫt-ḥr/Sšsšt	Saqqara	VI.E
Wnjs-ꜥnḫ	el-Khokha	VI.E
Mrw/Ttj-snb	Saqqara	VI.E
Mr.f-nb.f	Saqqara	VI.E
Mrrw-kꜣ	Saqqara	VI.E
Mḥw	Saqqara	VI.E
Nj-ꜥnḫ-Ppjj	Zawyet el-Amwat	VI.E
Nj-kꜣw-Jzzj	Saqqara	VI.E
Hnḳw/Jj…f (II)	Deir el-Gebrawi	VI.E
Ḥzj	Saqqara	VI.E
Ḫwꜣw	Naga ed-Der	VI.E
Ṯmrrjj	Naga ed-Der	VI.E
Kꜣ-gmnj	Saqqara	VI.E
Jbj	Deir el-Gebrawi	VI.M
Jdw I	Dendera	VI.M
Ppj-ꜥnḫ-ḥrj-jb	Meir	VI.M
Ḫntj	el-Khokha	VI.M

Fishing with wicker traps:

Mrrw-kꜣ	Saqqara	VI.E
Mḥw	Saqqara	VI.E
Ḥzj	Saqqara	VI.E
Kꜣ-gmnj	Saqqara	VI.E
Jbj	Deir el-Gebrawi	VI.M

Fishing with hand held net:

Mrrw-kꜣ	Saqqara	VI.E
Ḥzj	Saqqara	VI.E
Kꜣ-gmnj	Saqqara	VI.E
Jbj	Deir el-Gebrawi	VI.M

Fishing with line and hook(s) from a boat:

Wḏꜣ-ḥꜣ-Ttj/Ššj/ Nfr-sšm-Ptḥ	Saqqara	VI.E
Mḥw	Saqqara	VI.E
Mrjj-Rꜥ-mrjj-Ptḥ-ꜥnḫ/ Nḫbw	Giza	VI.E
Mrrj	Saqqara	VI.E
Mrrw-kꜣ	Saqqara	VI.E
Nj-ꜥnḫ-Ppjj	Zawiet el-Amwat	VI.E
Nj-kꜣw-Jzzj	Saqqara	VI.E
Ḥzj	Saqqara	VI.E
Kꜣ-gmnj	Saqqara	VI.E
Jbj	Deir el-Gebrawi	VI.M

In the early Dynasty 6 tombs the dragnet scenes appear at their most elaborate. In *Mrrw-kꜣ* there are two separate scenes of men hauling in a dragnet.[617] In the scene on the north wall of Chamber A6, two teams one with twelve fishermen and the other with eleven, haul in a net laden with 31 fish of eighteen different species. Each man wears a waistband and a shoulder strap attached to the rope, and is depicted in a different stance showing how each braced himself as he hauled in the net. There is no overseer standing between the two teams. The men pulling in the dragnet on the east wall of Chamber A4 are depicted rather differently; the two teams have nine men each and they wear short kilts. An overseer, wearing a longer kilt with protruding front panel, stands between the two teams leaning on his staff. In the tomb of *Jnw-Mnw* the teams (four men in each) and the supervisor have adorned themselves with lotus flower headdresses.[618] To judge by their balding pates, at least five of the haulers in *Ḥzj*'s dragnet scene are depicted as elderly, which seems surprising in view of the exertion needed to drag the laden net to shore.[619] Perhaps the liveliest scene is that of *Nj-kꜣ-Jzzj* where the two teams of five men each (over half of them balding) kneel, squat and lean to left and right.[620] At either end of the scene a kneeling man takes the same species of fish out of the net. This is a *Synodontis batensoda,* which has a reputation for jumping. None of the early Dynasty 6 Saqqara scenes of dragnet fishing contains any significant new development, merely elaboration of detail, which suggests that catching fish in quantity was important to these powerful officials. Provincial scenes of this date also simply portray the same basic elements, although

613 *Jntj* Petrie (1898) pl. 5, Kanawati–McFarlane (1993) pl. 33.
614 *Pḥn-wj-kꜣ.j* LD II pl. 46, *ꜣḫtj-ḥtp* (Louvre) Ziegler (1993) 133-4, 140, 142-3 (drawing).
615 Moussa–Altenmüller (1977) fig. 12, Épron–Daumas I (1939) pl. 9.
616 Kanawati–Abder-Raziq (2003) pl. 53 (d), (f), 54.
617 Duell I (1938) pl. 43, 55, Kanawati (2011) pls. 79, 84.
618 Kanawati (2006) pl. 48.
619 Kanawati–Abder-Raziq (1999) pl. 55.
620 Kanawati–Abder-Raziq (2000) pl. 47.

with less artistic facility.[621] Dragnet scenes continue in the provincial chapels from mid Dynasty 6 to Dynasty 8. They appear in tombs located in (nomes) Upper Egypt 4, 8, 9, 12, 14 and 16.

Other fishing techniques

Other means of catching fish in quantity include the funnel shaped traps ('weirs') made of basketry, which are depicted in two sizes. The smaller traps, which must have been used in shallow water, were laid in rows on the floor of the lake or channel.[622] The much larger weir trap, intended for deeper water, was laid and retrieved from boats.[623] Apart from vignettes of fishermen angling with line and hooks which continue into late Dynasty 6, other fishing techniques only appear in the second half of Dynasty 5 and early Dynasty 6.

Methods of catching fish in smaller quantities show the use of line and hook and occasionally in shallow water wicker baskets that were upended and placed over a fish.[624] Angling with line to which were attached a number of hooks is more frequently depicted and must have been a method for deeper water, as the angler always fishes from a skiff. Depictions of fishing with a bag shaped hand net show the fisherman in a boat either leaning forward to tow the net through the water or lifting the net full of fish.[625] From the depiction of the size of fish caught in the net it seems that the method was used to catch smaller fish. Fishing with hand nets only appears in a few tombs and seems to represent catching smaller fish[626] which may have swum in shoals either in shallow water or near the surface of the water. In the tomb of *Mrrw-k3* four older fishermen, working on papyrus skiffs, scoop up masses of small fish. Van Elsbergen suggests that the hand net was possibly used either in waters with papyrus thickets[627] or in pools that were gradually draining after the inundation or in channels with a high velocity flow of water, but certainly where the water was clear enough to sight the fish.[628]

Late Dynasty 6 to Dynasty 8

Dragnet scenes are depicted in the following tombs:

Ppj-ꜥnḫ/Ḥnj-km	Meir	VI.L
Nj-ꜥnḫ-Ppjj-km	Meir	VI.L
K3(.j)-ḥp/Ṯtj-jkr	el-Hawawish	VI.L
Ṯtj-ꜥ3	el-Hawawish	VI.L
Ḫnj/Špsj-pw-Mnw	el-Hawawish	VI.L
Ḏꜥw	Deir el-Gebrawi	VI.L
Mrjj-ꜥ3	el-Hagarsa	VIII
Gḥs3/Nbjj	el-Hawawish	VIII

Fishing with hand-held net or wicker traps:

Ppj-ꜥnḫ/Ḥnj-km	Meir	VI.L
Ḫnj/Špsj-pw-Mnw	el-Hawawish	VI.L
K3(.j)-ḥp/Ṯtj-jkr	el-Hawawish	VI.L
Ḏꜥw	Deir el-Gebrawi	VI.L

Fishing with line and hook(s) from a boat:

Ppj-ꜥnḫ/Ḥnj-km	Meir	VI.L
Ḫnj/Špsj-pw-Mnw	el-Hawawish	VI.L
K3(.j)-ḥp/Ṯtj-jkr	el-Hawawish	VI.L
Ḏꜥw	Deir el-Gebrawi	VI.L

Scenes of dragnet fishing continue to be included in tomb decoration into Dynasty 8 but with no significant developments to the main elements of the scene. While the scenes in the earlier tombs of the great provincial magnates such as *Ppj-ꜥnḫ/Ḥnj-km* of Meir and *K3(.j)-ḥp/Ṯtj-jkr* and *Ḫnj/Špsj-pw-Mnw* of El Hawawish are well crafted,[629] those in later tombs demonstrate less artistic ability.[630] Again, only the earlier tombs contain depictions of other methods of fishing.

The scenes of fishing with a dragnet do not show any critical change over time. Two teams of men haul on either end of the rope attached to the net, in which is portrayed a variety of fish species. The numbers of men depicted pulling in the net of fish and the fish depicted in the net increase until they reach a maximum in the scenes of high officials in the Teti cemetery in early Dynasty 6. Elaborating details are added, usually in tombs of late Dynasty 5 and early Dynasty 6, although some first appear in the tomb of *Nj-ꜥnḫ-Ḫnmw* and *Ḫnmw-ḥtp*, dated to mid Dynasty 5. These include the strap attached to the rope and worn over the shoulder of the haulers, the ends of the rope curled into a neat ball and a supervisor standing between the two teams. The scene showing the dragnet stretched out into the water with one end held by a fisherman on a boat is confined to a few tombs of mid to late Dynasty 5.

Scenes of fishing with baskets or weirs appear in mid Dynasty 5 but these images, which expand the fishing theme, are confined to the tombs of the highest officials and those closest to the king. Fishing with hand held nets or line and hook(s) is not always associated with the methods of fishing in quantity; these methods of fishing occasionally appear in other 'aquatic' scenes such as that on west wall of Room III of Jdwt where the princess stands on a papyrus skiff.[631]

10.5 Scenes of fish gutting and drying

The number of extant scenes of this theme suggest that tomb owners considered gutting and drying fish to get rid of the fish's body wastes an important part of the operation of catching fish in large quantities. The figures

621 See *Ḫntj* Saleh (1977) fig. 41, pl. 11; *Hnkw/Jj...f* (II) Kanawati *Gebrawi* I (2005) pl. 55; *Nj-ꜥnḫ-Ppjj* Varille (1938) pl. 5-6.
622 Edel–Wenig (1974) pl. 16, Moussa–Altenmüller (1977) pl. 12.
623 Wild (1966) pl. 111.
624 Moussa–Altenmüller (1977) pl. 12.
625 Duell I (1938) pl. 43, Kanawati (2010) pl. 79.
626 Brewer–Friedman (1989) 38.
627 Marsh vegetation would impede the progress of a dragnet.
628 Van Elsbergen (1997) 48.

629 Respectively, Blackman–Apted *Meir* V (1953) pl. 30, Kanawati I (1980) fig. 12, Kanawati II (1981) fig. 22
630 Compare Blackman–Apted *Meir* V (1953) tomb of *Ppj-ꜥnḫ/□nj-km* pl. 30 with tomb of *Nj-ꜥnḫ-Ppjj-km* pl.13.
631 Kanawati–Abder-Raziq (2003) pl. 54.

who gut the fish are shown squatting on the ground or seated on a raised mat or stone as they cut open a fish on a low table. Images of the flattened fish around the men working with knives were the artists' solution to the problem of conveying the information that the gutted fish were then laid out to dry.[632] The process would have been carried out as soon as possible after the fish had been landed to prevent contamination of the flesh by micro-organisms. This is conveyed by showing the process taking place in the marshes close to the bank of the river or pond. The images are located close to fishing scenes, either on the same register or immediately above or below, with the figure gutting the fish portrayed seated beneath a stand of papyrus.

10.5.1 Historical development of images and themes

Dynasty 4

Fishermen gutting, cleaning and drying fish are depicted in the following tombs:

Nfr-mꜣꜥt	Medum	IV.E
Rꜥ-ḥtp	Medum	IV.E
Nb.j-m-ꜣḫtj	Giza	IV.L

In these early tombs the marsh scenes are not grouped together but appear on registers interwoven with images of agriculture, desert animals and butchery. They do, however present the theme of fish gutting as an important activity that has to be carried out as soon as possible after the catch. In *Rꜥ-ḥtp*'s tomb the marshland location is established by a clump of papyrus behind the figure cleaning a fish.[633] On the south wall of *Nb.j-m-ꜣḫtj*, where there is a predominance of marsh scenes, two men are shown working as a team to process mullet; one man cleans a fish while the other is cutting a fish open.[634]

Mid Dynasty 5

Fishermen gutting, cleaning and drying fish are depicted in the following tombs:

GIZA
Jj-mrjj
Nj-kꜣw-Ḥwt-ḥr
Rꜥ-ḫꜥ.f-ꜥnḫ

SAQQARA
Wr-jr.n-Ptḥ
Nj-ꜥnḫ-H̱nmw and *H̱nmw-ḥtp*
Rꜥ-m-kꜣ.j

ABUSIR
Ptḥ-špss

In the tombs of *Jj-mrjj*, *Nj-ꜥnḫ-H̱nmw* and *H̱nmw-ḥtp*, and *Rꜥ-ḫꜥ.f-ꜥnḫ* the gutting and drying scenes are closely associated with other marshland activities, in particular with fishing with nets and traps. The importance attached to the drying process is emphasised in the marshland scene of *Jj-mrjj*; two men wearing the kilt of an overseer present fish to the tomb owner standing before the panel of marshland registers. Beneath, a man sits with his back to a papyrus clump gutting fish that presumably have come straight from the dragnet depicted in front of him.

The scenes in the tomb of *Nj-ꜥnḫ-H̱nmw* and *H̱nmw-ḥtp* provide new features. The three men squatting to gut the fish also extract roe while a fourth slightly smaller figure tends a large container of fish with the inscription, 'Cooking the fish' (*fsj(t) mḫjt*).[635] The vertical inscriptions either side of this panel of registers read, on the right hand side, 'Viewing the marshland, the marsh thickets and the marsh water, the catching of fish and the netting of birds. More beautiful (to see) than all things' and on the left, 'Viewing the marshland, the marsh thickets and the marsh water, the catching of fish and the netting of birds. More beautiful (to see) than all things' (*mꜣꜣ šꜣw zšw pḥw ḥꜣm mḫjt sḫt ꜣpd nfr (mꜣꜣ) r jḫt nb*).

Second half of Dynasty 5

Fishermen gutting, cleaning and drying fish are depicted in the following tombs:

GIZA
Kꜣpj

SAQQARA
ꜣḫtj-ḥtp (Louvre)
Jj-nfrt
Jrj-n-kꜣ-Ptḥ
Pḥn-wj-kꜣ.j
Ptḥ-ḥtp/Jj-n-ꜥnḫ
Ptḥ-ḥtp/Ṯfj
Nj-kꜣw-ḥr
Nb-j-kꜣw-ḥr/Jdw
Nfr-jrt-n.f
Rꜥ-m-kꜣ.j
Ḥtp-ḥr-ꜣḫtj
Ḥnwt
Spd-ḥtp
Ṯjj

DESHASHA
Jntj

SHEIK SAID
Wr-jr-n,j

ZAWYET EL-AMWAT
Ḫw-ns

Scenes in late Dynasty 5 tombs continue to portray the features established in the earlier part of the dynasty. Seated or squatting workers gut fish laid out on a low table and extract roe. In most instances, the fish are mullet. The gutted fish and strings of roe are laid out to dry. In the tombs of *ꜣḫtj-ḥtp* (Louvre) and *Pḥn-wj-kꜣ.j* this activity is portrayed next to the scenes of fishing. In the tombs of *Jntj* and *Rꜥ-m-kꜣ.j* porters bring a load of fish to be gutted.

Early to Mid Dynasty 6

Fishermen gutting, cleaning and drying fish are depicted in the following tombs:

Mrrj	Saqqara	VI.E
Mrrw-kꜣ	Saqqara	VI.E
Kꜣ-gm-n.j	Saqqara	VI.E
Jbj	Deir el-Gebrawi	VI.M
Ḫntj	el-Khokha	VI.M

In comparison to the number of fish gutting scenes from the tombs of the second half of Dynasty 5 and the prominence given to them, a much smaller proportion of

632 This arrangement allowed the artist to depict the characteristic view of gutted fish laid out to dry on a flat surface. See Schäfer (1986) 190-193 and Robins (1994) 1-13 for discussion of this principle of Egyptian art.
633 Petrie (1892) pl. 18.
634 LD II 12 [b].
635 Moussa–Altenmüller (1977) fig. 12.

Dynasty 6 tombs contain the image. Furthermore, the theme appears to be given less significance. In *Mrrw-k3*'s scene, for example, the activity takes place on a boat where tending to the tomb owner's brother is the main focus.[636] This gutting scene takes place on two papyrus skiffs, one of which floats above basket weirs in which a mass of fish have been caught. A fisherman leans over the side of the skiff to extract the fish, while another is occupied cutting off the head of a fish. A third fisherman lowers a beheaded fish into a bag held by a fourth man. Gutted fish laid out to dry are depicted on sub-registers. In the provincial tomb of *Jbj* two brief images of gutting and drying fish flank a more extended scene of fighting boatmen.[637]

The theme of gutting and drying fish does not appear after mid Dynasty 6.

10.6 The destination of the catch

From the list used in this study 40 tombs contain depictions of bearers transporting fish. These vignettes are associated either with fishing or with gutting fish. Various methods of carrying the fish are shown: one to three fish hung from a horizontal pole or oar supported on the shoulders of bearers, a number of fish in bags or boxes sometimes slung from a shoulder yoke, a single fish carried in the hand.[638] The transported fish appear not to have been gutted, although this is not easy to judge when only the head of the fish is portrayed poking up out of a bag or box[639]

The destination of the fish being transported is not always clear. Specimens shown as whole fish, not gutted and dried, could not have been transported very far. In a few scenes, however, the destination of the catch is indicated. In *Nb.j-m-3ḫtj*, *Rᶜ-m-k3.j*, and *Jntj* the fish are being taken to the location where they are gutted.[640] In other scenes the fish are taken to be presented to the figure of the tomb owner, apparently before they have been gutted and dried.[641] In this case either the tomb owner is in the marshland with the fishermen or quite close to the place of fishing. Owing to the nature of the Nile valley the estate's headquarters and residence of the tomb owner would have been in the vicinity of waters being fished. While the presentation to the tomb owner always consists of a bearer holding either a single fish or one fish in each hand, the scenes of transporting fish to be gutted portray the fish in quantity conveying the impression of a more commercial level of handling the commodity. The scene involving the tomb owner, on the other hand, portrays a formality, the respectful presentation of a fine specimen for inspection.

Occasional scenes suggest further uses of a certain amount of the catch. These include fish hung to dry in the hut of fowlers,[642] fishermen cooking fish in a large pot,[643] fishermen registering their catch with scribes[644] and the marketing of dried fish.[645] As in the case of the birds trapped in the marshes, these scenes reveal that some of the catch was used to feed and pay the workers. The registration of the catch shows that there was a set payment administered by the officials of the *jz.t* of the *pr-ḏ.t.*[646] It must be presumed that the catch was the property of the tomb owner or his *pr-ḏt*, and was used to reward dependants and workers, or as payment in kind for craftsmen such as the builders and artists of his tomb, and to barter for commodities that his estate did not produce, as well as to provide meals for his household. Perhaps it is for these mundane uses that much of the fish was gutted and dried. This suggests that there was no taboo on fish, only that it was an everyday commodity without any high ritual value.[647]

10.7 Work force in fishing scenes

The depictions of the men engaged in the actual work of fishing, gutting and carrying loads of fish are never named and rarely given a label.[648] There are, however, images of men who appear to be overseers and are sometimes so labelled. These men are depicted wearing the formal overseer's kilt, leaning on a staff or checking a fish specimen or waving an arm at a toiling fisherman. In the tombs of *Snfrw-jn-jšt.f, Wr-jr-n.j, Ppjj-ᶜnḫ-ḥrj-jb* and *Ppjj-ᶜnḫ/Ḥnj-km,* they are given a title, *jmj-r3 wḥᶜ.*[649] In *Nj-ᶜnḫ-Ḫnmw* and *Ḫnmw-ḥtp*'s tomb the stout figure in charge of the teams of fishermen hauling in the dragnet is entitled *jmj-r3 wḥᶜt* (overseer of the fishing team).[650] In the tombs of *Ṯjj* and *K3-gm-n.j* the labels, *jmj-r3 wḥᶜ ḏ.*t[651] and *jmj.w-r3 wḥᶜt pr-ḏt* link figures of overseers with the estate.[652] The appearance of these supervising officials makes it clear that the fishermen work for the tomb owner's *pr-ḏt*.

The titles of other minor officials provide evidence that relates the fishing operations with the estate. In the scene of the distribution of fish to fishermen in the tomb of *Mḥw*, officials in charge of the activity are labelled *ḫrp jzw.t* and *jmj-r3 jzw.t.* In *K3-gm-nj* the images of officials with their titles and utterances, beneath the fishing scene, offer further evidence of the estate's administration of

636 Duell I (1938) pl. 42, 43, Kanawati (2010) pl. 79.
637 Kanawati *Gebrawi* II (2007) pl. 67.
638 See TABLE 13: FISH AND FISHING – METHODS AND MANAGEMENT.
639 Gutted fish did not have their heads cut off.
640 Respectively, LD II 12 (b), Hayes (1953) fig. 54, Petrie (1898) pl. 5, Kanawati–McFarlane (1993) pl. 9, 33.
641 See TABLE 13: FISH AND FISHING – METHODS AND MANAGEMENT.

642 Wild II (1953) pl. 122.
643 Moussa–Altenmüller (1977) fig. 12.
644 von Bissing I (1905) pl. 20, Altenmüller (1998) pl. 37. Van Elsbergen interprets these scenes as the fishermen being paid for their work with a proportion of the catch.
645 *Tp-m-ᶜnḫ* Smith (1942) 515-16, fig. 5.
646 See Chapter 13 for a brief discussion of the *pr-ḏt*.
647 Handoussa (1988) 105-10.
648 *wḥᶜ* – see Jones I (2001) 1486, p. 404 and *znᶜḥw n pr-ḏ.t* (an angler in *K3-gm-nj* described as 'fish-catcher of the funerary estate'); see also Jones II (2001) 3326, p. 908.
649 Respectively, Borchardt (1964) CG 1772, Davies *Sheikh Saïd* (1901) pl. 12, Blackman *Meir* IV (1924) pl. 30, Blackman–Apted *Meir* V (1953) pl. 30.
650 Moussa–Altenmüller (1977) fig. 12.
651 *Ṯjj* Wild (1953) pls. 118-19.
652 Respectively, Wild (1953) pls. 118-19, von Bissing I (1905) pl. 19.

fishing. In the two subregisters of squatting men, the six at the rear are called *ẖrp jz.t*, while the four figures in the front *ʿre jmj-r3 jzw.t*, and the two scribes, who record the the catch, are titled *sš jzw.t.*[653]

Other indications of the tomb owner's proprietorial interest in the depicted fishing activities occur in a number of vertical inscriptions that separate the figure of the tomb owner from the panel of registers. Most of these statements do not refer solely to fishing but as fishing activities occupy part or all of one of the registers in the panel it may be assumed that they are included.[654] The statements referring specifically to fishing are included in the scenes of *Nj-ʿnẖ-H̱nmw* and *H̱nmw-ḥtp*,[655] *Mḥw*,[656] *Mrw/Ttj-snb*,[657] *Ppjj-ʿnẖ-ḥrj-jb*:[658] and *Nj-ʿnẖ-Ppjj.*[659]

10.8 Summary

- Over 26 fish species are portrayed, either in nets, traps, hooked on a line, gutted or being transported.[660]

- Mullet appear in over 80 per cent of gutting scenes.

- Other frequently depicted species are *Tilapia, Synodontis schall*, *Citharinus*, *Tetrodon fahaqa* and *Lates niloticus.*

- Although it would have been possible to catch fish all year round, the scenes of extracting roe and the sheer quantity of fish being caught suggest that the depicted activities were seasonal, perhaps timed to the months (May to November) when mullets produce roe and gather in great shoals to make for the open sea where they spawn.

- The most frequently portrayed method of fishing is by dragnet (seine), which required teams of men, a boat and a sturdy net to which sinkers and floats were attached. Although the essence of the dragnet scene does not change over time, there is a development of detail and in the portrayal of the physical effort that the activity demanded.

- Operating the dragnet entailed coordination and direction. This was supplied by an overseer, usually depicted standing between the two teams of men on the river bank.

- The depicted methods of catching fish were designed to catch fish in quantity, but were not necessarily the most efficient way of catching certain species that like the Nile perch could grow too large for a net or weir, or inhabited the depth of the river or lake or, on the other hand, preferred shallow water. Consequently, fishermen are also depicted using the smaller weirs in shallow water, hand held nets and lines with hooks.

- The large weir and basket traps are only depicted in the tombs of the highest Memphite officials, and only three of these portray all four methods of fishing. The use of hand-held nets appears in a greater variety of tombs, although still limited to the highest officials. On the other hand, fishing with hook and line occurs in a variety of tombs.

- The organised handling of the catch is further illustrated in the scenes of gutting and drying fish and stripping and drying the roe. These images are either located very close to the fishing scenes or include the portrayal of a sack of fish being brought to the men processing the fish.

- The catch is clearly the property of the estate. It is pictured being registered and transported either for gutting and drying, to be presented to the tomb owner or to officials of the *pr-ḏt.* The catch was used to feed dependants of the estate, as payment to workers and as local currency.

- The titles of minor officials and their presence in scenes of fishing and the distribution of fish show that the activity of fishing and the catch were controlled by estate authorities.

653 von Bissing I (1905) pls. 18-19.

654 See *Sẖm-k3.j* Simpson (1980) fig. 4, pl. 5 d-e; *3ẖtj-ḥtp* Ziegler (1993) p. 145; *3ẖtj-ḥtp* Davies *Ptahhetep I* (1900) pl. 13-14; *Jj-nfrt* Schürmann (1983) fig. 9a-b; *K3-ḥj.f* Junker VI (1943) fig. 42.

655 'Viewing the marshland, the marsh thickets and the marsh water, the catching of fish, and the netting of birds. More beautiful (to see) than all things': Moussa–Altenmüller (1977) fig. 12.

656 'Viewing the very great delivery of fish, the work of the marshes and the fishing tally of the *ṯz.t* of the *pr-ḏ.t*': Altenmüller (1998) pl. 30.

657 'Viewing all the work of the marshes with the catch of birds and fish': Lloyd (1990) pl. 8.

658 '[Viewing] the work of the marshes and the very great catch of birds and fish': Blackman *Meir* IV (1924) pl. 8.

659 'Viewing the work of the marshes and the very great catch of birds and fish on the part of ...': Blackman–Apted *Meir* V (1953) pl. 13.

660 The number of different species portrayed in Old Kingdom tombs is inexact because a number of depictions of fish cannot be identified with certainty: Van Elsbergen (1997) 133-307.

Figure 34: Methods of catching fish: dragnet, weir and basket net, tomb of Mereruka, Saqqara. Kanawati (2010) pl. 79.

Figure 35: Catching fish with a line and hooks, tomb of Mereruka, Saqqara. Kanawati (2010) pl. 79.

Figure 36: Fishing with weirs, gutting and drying the catch, tomb of Mererukai, Saqqara. Kanawati (2010) pl. 79.

Figure 37: Gutting and drying a catch of fish, tomb of Mereruka, Saqqara. Kanawati (2010) pl. 79.

Figure 38: Transporting the catch, tomb of Mereruka, Saqqara. Kanawati (2010) pl. 68.

CHAPTER 11 WILDFOWL AND POULTRY

11.1 Focus of enquiry: birds as an economic resource for a great estate

TABLE 14: BIRDS – TRAPPING, POULTRY YARDS, PARADES

The data:
- The iconography of trapping and handling wildfowl by estate workers
- The iconography of fowl yards
- Inscriptions accompanying the iconography.

Areas of research:
- Bird species most frequently depicted in the marshlands
- The supply of these birds
  - Trapping technology
  - Seasonal nature of trapping
  - Transport of catch
- Maintenance of flocks of poultry
  - Fowl yards
  - Feeding the poultry.

Great rivers offer a variety of resources, fish, birds and vegetation. In the environment of the Nile valley, where arable land and pasturage are limited, the wetlands bordering the river, the distributaries of its Delta, subsidiary channels and oxbow lakes were particularly valuable. Like fish, wild fowl provide a cheap supply of high quality protein. Consequently scenes of trapping birds and poultry yards are investigated as evidence of the resources of tomb owners.

Birds appear in a variety of scenes decorating the walls of Old Kingdom tombs: as presentations by offering bearers and figures representing domains, in a number of marsh themes, and in poultry yards where they are shown maintained and fattened for the table. Not all these scenes present birds as an actual resource. Images of plucked and trussed fowl in offering lists or brought to the tomb owner by bearers essentially have the ritual significance of providing sustenance for the ka of the deceased. Nor does the representation of the tomb owner standing on a skiff in the marshland and aiming a throwing stick at birds provide a record of his resources. This scene, together with that of the tomb owner spearing fish, may be interpreted as a symbol referring to the unity of the two lands[661] or a metaphor for the tomb owner as a successful provider for his family and household, as well as a representation of a reality.[662] Regardless of the interpretation favoured, the birds depicted in these scenes, whether in the hand of the tomb owner or roosting in or fluttering over a papyrus thicket, do not represent the official's accumulated assets.

The following scene genres portray birds as a resource:[663]
- The motif of wildfowl trapped in a clapnet shows birds being caught alive in large numbers in the marshlands. The clapnet is operated by teams of men under the control of a signaller who directs them with silent signals. (*Figure 39*)
- The motif of transferring the trapped birds to cages to be transported by bearers.
- The representation of the poultry yard, usually with a rectangular pond in the centre, depicted teeming with birds fed grain by attendants and frequently being force-fed. (*Figure 40, 41*)
- A procession of birds, sometimes accompanying a parade of cattle and desert animals presented to the tomb owner. The birds may be shown being controlled by an attendant who wields a long stick.

These scenes representing the organised, large-scale trapping and maintenance of birds, provide images of birds that may be considered as a resource. The themes emphasise the quantities of birds acquired and flocks maintained on a permanent basis by the deceased. For the most part, the depicted birds appear to belong to species (ducks and geese) named in offering lists. The species caught in the clapnet often cannot be identified with any certainty, but the images mainly appear to be water birds that vary somewhat in size and are generally accepted as types of duck and geese, the species most commonly depicted in the poultry yard. Only a few birds may be depicted in a parade but, as in the case of the other animals being presented to the tomb owner, they represent a much larger quantity and may be associated with a caption that provides a number in the thousands.[664]

11.1.1 Bird species most frequently depicted

Old Kingdom bird species (Egyptian names)

wḏꜥ	demoiselle crane	Wb 1, 407.14
ḏꜣ.t	grey crane	Wb 5, 516
ꜥw	crane	Wb 1, 170.3
gꜣ	[a bird]	Wb 5, 149
bḏꜣ	[a duck]	Wb 1, 488
ḥp	[a duck]	Wb 3, 69

661 Brewer–Friedman (1989) 79.
662 van Walsem (2006) 7-79.

663 Scenes of netting songbirds in trees are not treated as depictions of 'resources' because there are no scenes showing how or by whom the birds are to be used.
664 The number usually given is *ḥꜣ*, which must either be understood as 'a very large number of...' or, if a more 'funerary' meaning is applied, as 'an eternal supply of...'.

mnw.t	pigeon	Wb 2, 79.
smn	Nile goose	Wb 4, 136
ṯrp	white-fronted goose	Wb 5, 387
bsbs	[a duck]	Wb 1, 477
mȝš		Godron, *BIFAO* 56, 1957, 19 f.
ḥḏ	[a goose or duck]	Wb 3, 210

In scenes of processions of birds and occasionally in poultry yard scenes, the birds may be identified by a caption. The classifying principle determining the employment of these labels, however, does not appear to be based on physical features of the birds. Labels appear to be some kind of descriptor rather than naming species of avifauna, as the same word may be applied to birds of different shape and size, while birds of the same size and shape may be given a variety of labels. These descriptors presumably refer to principles of categorizing birds in use in the Old Kingdom, but the data is insufficient to establish these principles. A few labels, however, do seem to refer to a single species. For example, *wḏʿ* is only applied to birds with the features of demoiselles cranes, although *ḏȝ.t, gȝ* and *ʿw* are used to identify other cranes. The term *mnw.t* is applied only to birds that appear to be either pigeons or doves. The naming of ducks and geese is more complicated, partly because the outline shapes of the two are very similar, partly because tomb artists sometimes did not distinguish between the sizes of the two breeds of birds and partly because each is given a number of different terms. Geese may be labelled *r, rȝ, ḥḏ, smn* and *ṯrp*. Ducks may be labelled *bsbs, bḏȝ, mȝš, ḥp, ḥḏ, s, sw, sr, st, sȝ.t* and *ṯrp*. It is not always possible to judge whether a bird labelled *r, rȝ* or *s, st, sȝ.t* is a species of duck or goose.[665] Consequently, scholars have sometimes disagreed over the identification of birds and their Egyptian labels. For example, *bsbs* is translated by the Worterbuch as 'goose or duck', by Edel and Boessneck as 'white-headed duck', and by Schäfer as *Erismatura leucocephala*.[666] For the purpose of this study, it is sufficient to note the frequency with which birds that must be either species of geese or ducks, or cranes or doves are portrayed in clapnet, fowl yard and parade scenes.

11.1.2 Bird species most frequently depicted in 'resource' scenes

Ducks

Pintail (*Anas acuta*)

The Pintail breeds across the northern hemisphere and flocks of the species winter in the Nile valley and Delta, the Faiyum and the major Egyptian oases. The species accepts and readily breeds in captivity. The Pintail can be identified in outline by the shape of its tail feathers and consequently appears as the type of duck most frequently represented in Old Kingdom art. It is mainly migratory, breeds in northern zones and flocks migrate up and down the Nile valley in autumn and spring.[667]

Common Teal (*Anas crecca)*

This bird is likely to have been another species of duck trapped in quantity in the marshlands of the Old Kingdom, as it breeds across the middle latitudes of the west Palearctic and adapts to any kind of shallow, peaceful wetland. It is one of the smallest dabbling ducks, highly gregarious and strongly migratory. Flocks from Russia winter in the Nile valley.[668]

Goose (*Anser* sp.)

While large numbers of geese are frequently depicted in clapnet and poultry yard scenes, lack of detail and colour in their depiction usually makes it impossible to specify the type of goose portrayed. In accompanying captions Egyptian words such as *ṯrp, ḥḏ, ḥtʿȝ* and *r* refer to individual birds or groups, but the absence of pictorial detail makes it difficult to identify distinguishing features. The careful delineation and colouring of the geese in the chapel of *Jtt* at Medum [the Egyptian Goose (*Alopochen aegyptiaca*), Bean Goose (*Alopochen aegyptiaca*), Greylag Goose (*Anser anser)*, White-fronted Goose (*Anser albifrons*), Lesser White-fronted Goose *(Anser erythropus),* Barnacle Goose *(Branta leucopsis),* Brent Goose *(Branta bernicla),* Red-breasted Goose (*Branta rubicollis*) and the Spur-winged Goose *(Plectropterus gambensis)]* are exceptional.[669] As in the case of ducks, it is usually not possible to identify the species of geese depicted in clapnet and fowl yard scenes owing to a lack of sufficient defining morphological details. According to Houlihan, only the White-fronted Goose of the Medum series can again be identified in funerary art.[670] Altogether it has been estimated that at least nine species of goose were known in ancient Egypt,[671] but it has often proved impossible to identify individual types in representations.

Most species of geese breed in northerly regions and are winter visitors to Egypt. Geese are not the most nutritious of fowl as most species are only 39% meat, but Egyptians made good use of their fat for cooking, preservation of meat and in recipes for medicines.[672]

Egyptian Goose (*Alopochen aegyptiaca*)

This bird is not a true goose but a gooselike duck and consequently is not easily identified in outline in funerary art.[673] It is a resident of Egypt breeding in the

665 Vandier *Manuel* V (1969) p. 400-401.
666 Vandier *Manuel* V (1969) p. 401-405.

667 Cramp (1977) Vol. 1, 521-524.
668 Cramp (1977) Vol. 1, 494-497.
669 Davies, Nina M. (1936) pl. 1.
670 The white-fronted goose can be identified in *Kȝ.j-m-ʿnḫ* (G4561) and *Mḥw*: Houlihan (1986) 59.
671 These include: the Egyptian goose, bean goose, greylag goose, white-fronted goose, lesser white-fronted goose, barnacle goose, brent goose, red-breasted goose, and the spur-winged goose. See: Ikram (1995) *Choice Cuts*, p. 27.
672 Ikram (1995) 27.
673 The first positive identification of *Alopochen aegyptiaca* is in the tomb of *Mḥw*: Houlihan (1986) 64.

Upper Nile Valley and wintering further south along the Nile. While the species frequents marshlands, it will also nest in trees. Its breeding season is March and April.[674]

Cranes

Common Crane (*Grus grus*)

This bird breeds in the upper middle latitudes of the west Palearctic and is migratory in northern Europe. Flocks from Sweden, Finland and Russia fly south along the Nile valley to winter in the Sudan and Ethiopia. As some winter in the Nile Valley it is both a winter resident (October to March) and a migratory visitor on its way to or returning from regions further south in the spring and autumn.[675]

Demoiselle crane (*Anthropoides virgo*)

This species breeds in central Asia but is migratory and a winter visitor to the Nile Valley.

Doves

Turtle dove *(Streptopelia turtur)* and

Laughing Dove *(Streptopelia senegalensis)*

While it is not always possible to distinguish between this bird and species of pigeons in funerary art, Houlihan concludes that Old Kingdom images were most likely to be one of two species of doves, either the Turtle Dove or the Laughing Dove *(Streptopelia senegalensis)* as any identifying features indicate these two species.[676] Wild Rock Pigeons *(Columba livia)* are native to Europe, North Africa, and southwestern Asia (the palearctic region) and are likely to have been a popular species for the table as they have been domesticated for some thousands of years, but may have been considered too commonplace to be included in funerary scenes. Both dove species are residents breeding in Upper and Lower Egypt and the oases in springtime.

11.2 Trapping waterbirds: the clapnet scene (*Figure 39*)

Clapnet scenes not only show birds being captured in large numbers, they provide evidence that netting birds in the marshland was a well organised activity intended to augment the tomb owner's poultry supply. The method of trapping is by a team of men who haul on a rope to spring the trap of a rigidly framed hexagonal net. A team of men prepare to pull a rope attached to the clapnet. On a silent command from a signaller who usually stands behind a clump of marshland vegetation, the team of men haul on the rope. A variety of signals are depicted: the signaller may hold a strip of cloth with his arms outstretched at his sides, probably the command to close the clapnet, or he may raise an arm. The team's action is synchronised and the men may be shown falling or having fallen backwards on to the ground in their effort to pull the trap closed as suddenly as possible. The setting up of the clapnet,[677] and its mechanism are technically complex.[678] The clapnet is laid over an oval shaped stretch of water that may be bounded by an edging of wetland vegetation, presumably to represent an oxbow lake in the marshes. The trapped birds appear to swim calmly within the clapnet's framework, providing the appearance that they have been given time to settle before the trap is sprung, so that the greatest possible number of birds would be caught. A heron or bittern, sometimes both, is frequently pictured beside the clapnet. Through the ages, fowlers have used tame or tethered herons and bitterns as decoys to lure flocks of water birds into their traps. Old Kingdom scenes record the successful entrapment of large numbers of birds, in particular geese, ducks and cranes, often with a tame or tethered decoy bird standing beside the clapnet (*Figure 39*).

There are over 90 extant Old Kingdom clapnet scenes on record, although only those in 69 tombs have sufficient provenance to be used in this study.[679]

11.2.1 Historical development of images and themes

Clapnet scenes are depicted in the following numbers of tombs:

IV.E	4	VI.E	13
IV.L	2	VI.M	4
V.E	1	VI.L	6
V.M	9	VIII	3
V.L	29		

Dynasty 4

Clapnet scenes are depicted in the following tombs:

Rꜥ-ḥtp	Medum	IVE
Nfr-mꜣꜥt	Medum	IVE
Jtt	Medum	IVE
Nfr...w	Medum	IVE
Mrs-ꜥnḫ III	Giza	IV.L
Nb.j-m-ꜣḫtj	Giza	IV.L

The early Dynasty 4 scenes contain the basic features of the theme. One to four men are depicted either bending their legs or remaining upright as they pull the clapnet closed, but there is no signaller. The birds, difficult to identify as only their outline survives, are probably geese and pintails.[680] The birds swimming within the confines of the clapnet are mainly depicted in overlapping groups of two or three, an artistic device that suggests large numbers.[681] There are no images of captured birds being transferred to cages. Instead, retainers and family members are shown carrying pintails from the net, holding the birds by the neck.

[674] Cramp (1977) Vol. 1 447-450.

[675] Cramp (1977) Vol. 2 618-620.

[676] Houlihan (1986) 105.

[677] See top register in the tomb of *Tjj*: Wild II (1953) pl. 87.

[678] For a detailed explanation of the mechanism of the clapnet see Vandier *Manuel* V 320-34.

[679] See TABLE 14: BIRDS – TRAPPING, POULTRY YARDS, PARADES.

[680] Their tail feathers identify some of the images of birds in the hands of men near the clapnet as pintail ducks: Harpur (2001) fig. 93, p. 100; Junker XII (1955) fig. 2 (lower).

[681] Schäfer (1986) 188.

Certain features that in time become a standard accompaniment of the theme, however, appear in the scene in the tomb of *Mrs-ꜥnḫ* III, perhaps due to her exalted status. A signaller is portrayed standing with his arms outstretched from his sides in front of the team pulling the clapnet closed. A piece of cloth held in his hands is stretched tautly across his body. The clapnet is clearly placed on a pond or oxbow lake, as the water has a marked boundary and an edging of lotus and papyrus. The captured birds, some depicted individually and others in overlapping groups of two or three, are mainly geese and pintails. A worker lifts a bird from the net, presumably to hand it to a figure who is placing live birds in a cage. The associated caption is brief but refers to the tomb owner and the activity: *mꜣꜣ sḫt ꜣpd(.w) m mḥt* ('Viewing the trapping of birds in the marshes.'). A further statement above the signaller reads, *rdj sḫt* ('Working the trap.')

There is a further development in the tomb of *Nb.j-m-ꜣḫtj* where the theme is divided between two scenes. On the lower register, a team of men pull the clapnet closed on command from the signaller, while two registers above, the clapnet is being emptied of its catch. The birds are carried to a group of squatting scribes, who register the details of the catch.

Early to Mid Dynasty 5

Clapnet scenes are depicted in the following tombs:

Sḫm-kꜣ-Rꜥ	Giza	V.E
Jj-mrjj	Giza	V.M
Wr-jr.n-Ptḥ	Saqqara	V.M
Ptḥ-špss	Abusir	V.M
Nj-ꜥnḫ-Ḫnmw and *Ḫnmw-ḥtp*	Saqqara	V.M
Nfr and *Kꜣ-ḥꜣ.j*	Saqqara	V.M
Rꜥ-m-kꜣ.j	Saqqara	V.M
Rꜥ-ḥtp	el-Hammamiya	V.M
Rꜥ-ḫꜥ.f-ꜥnḫ	Giza	V.M
Ḫwfw-ḥꜣ.f II	Giza	V.M

Clapnet scenes in tombs of the first half of Dynasty 5 continue to include the features introduced towards the end of the previous dynasty. Like the depiction in *Nb.j-m-ꜣḫtj*,[682] *Sḫm-kꜣ-Rꜥ* has two clapnet scenes representing two stages of trapping.[683] In the first scene, four men pull the trap closed and in the second, a man is shown taking a bird out of the net, an image that first appeared in *Mrs-ꜥnḫ* III.[684] In five scenes the men on the rope are depicted leaning forward as they pull the net closed, a stance first shown in *Nb.j-m-ꜣḫtj*. Marsh vegetation is represented by lotus flowers and papyrus umbels.

In mid Dynasty 5 new features appear. In the scene of *Nj-ꜥnḫ-Ḫnmw* and *Ḫnmw-ḥtp* the men pulling the rope are hidden behind a stand of papyrus and fall backwards on to the ground in the effort to close the net suddenly. The signaller gives the command to haul by raising one arm.[685]

One scene from this time period indicates the link between the trapping of the birds and their destination. On the register beneath the clapnet scene in the tomb of *Nj-ꜥnḫ-Ḫnmw* and *Ḫnmw-ḥtp* workmen are shown presenting quantities of live birds to the two tomb owners. The inscription to the right side reads: *mꜣꜣ šꜣw zšw pḥw ḥꜥm mḥjt sḫt ꜣpd(w) nfr r jḫt nb(t)*, while that on the left reads: *mꜣꜣ šꜣw zšw pḥw ḥꜥm mḥjt sḫt ꜣpd(w) nfr mꜣꜣ r jḫt nb(t)* ('Viewing the marshland, the thickets of the marshes and the waters of the marshes, the catching of the fish and the trapping of birds, more beautiful to see than anything.)

Later Dynasty 5

Clapnet scenes are depicted in the following tombs:

GIZA	
Jj-nfrt (PM 299)	*Ḫww-wr*
Jtj-sn	*Snḏm-jb/Jntj*
Jttj	*Snḏm-jb/Mḥj*
Nj-mꜣꜥt-Rꜥ	*Sḫm-kꜣ.j*
Nfr (1)	*Sšm-nfr* IV
SAQQARA	
ꜣḫtj-ḥtp (Louvre)	*Ptḥ-ḥtp/Jj-n-ꜥnḫ*
ꜣḫtj-ḥtp (D64)	*Nfr-jrt-n.f*
Jj-nfrt (PM 616)	*Ḥtp-ḥr-ꜣḫtj*
Jrj-n-kꜣ-Ptḥ	*Ḫnwt*
Jrw-kꜣ-Ptḥ	*Spd-ḥtp*
Pḥn-wj-kꜣ.j	*Sḫm-ꜥnḫ-Ptḥ* (PM 454)
Ptḥ-ḥtp I	*Kꜣ.j-m-nfrt* (D23)
Ptḥ-ḥtp (II) (*Ṯfj*)	*Ṯjj*
DESHASHA	
Jntj	
SHEIKH SAID	
Wr-jr.n.j	
ZAWYET EL-AMWAT	
Ḫw-ns	

Where the clapnet scenes have survived intact in tombs dated to the second half of Dynasty 5, they normally contain most of the features that had been introduced by the first half of the dynasty. The teams of haulers may be depicted leaning forward to pull on the rope, or falling backwards on to the ground. Apart from two instances, signallers use their outstretched arms and a piece of cloth to give the command to pull. In the scene in the tomb of *Ṯjj*, however, the tomb owner himself is the signaller, perhaps suggesting the importance of the action or just that he enjoyed bird trapping.[686]

The species of trapped birds are difficult to identify but appear to represent pintails ducks and geese. The nets are shown full of birds, often swimming peacefully in overlapped formations, although the poorly preserved scene in the tomb of *Snḏm-jb/Mḥj* contains images of birds fluttering within the net.[687]

682 LD II 12.
683 LD II 41a.
684 Dunham–Simpson (1974) fig. 4.
685 Moussa–Altenmüller (1977) fig. 12.
686 Wild II (1953) pl.120.
687 LD *Erg*. pl. 14.

All the stages involved in catching birds by this means and then transporting them do not appear in any one tomb, but the process of trapping the birds and transferring them to the deceased's fowl yard can be traced across a number of tombs. The first stage appears in the tomb of *Ṯjj* with eight men setting up the trap. The second and third registers show *Ṯjj* as the signaller giving the order to pull to the upper team of haulers, who, on the lower register have fallen on their backs. The birds in the clapnet, who were swimming peacefully in the upper net, take alarm and try to fly away when, in the lower scene, the net has closed on them. Scenes in *Ptḥ-ḥtp/Ṯfj* also develop the actions of trapping birds in a clear sequence.[688] Two clapnets, one above the other, show the two main stages in catching the birds. In the lower image, the team, seated on the ground, prepare to pull the rope to shut the trap, as the decoy heron pictured beside the net has fulfilled its function of attracting a flock of birds. The upper scene shows the team lying back on the ground, having pulled the rope. In vignettes in the tomb of *Pḥn-wj-k3.j* men extract birds from the net and hand them to others who put the birds in carrying cages.[689] Further vignettes show porters carrying the birds either in the cages hung from a shoulder yoke or resting on the head of the porter. The sequence of ideas is completed by the depiction of presenting birds to the deceased on both registers, the lower of which contains scenes of fighting boatmen with a cage of birds on the third boat.

Early Dynasty 6

Clapnet scenes are depicted in the following tombs:

GIZA
- *Ḫnmw-ntj*

SAQQARA
- *Jnw-Mnw*
- *ʿnḫ -m-ʿ-Ḥr;*
- *Wḏ3-ḥ3-Ttj/Ššj/Nfr-sšm-Ptḥ*
- *Mḥw*
- *Mṯṯj*
- *Ḥzj*
- *Sʿnḫ-wj-Ptḥ*
- *Sšm-nfr/Jwfj*
- *K3-gmnj*

DESHASHA
- *Jttj-Šdw*

ZAWYET EL-AMWAT
- *Nj-ʿnḫ-Ppjj*

NAGA ED-DER
- *Tw3w*

Clapnet scenes in the great Saqqara tombs of early Dynasty 6 continue to present the elaboration of detail that was becoming popular in later Dynasty 5. Double scenes depict sequential stages of clapnet trapping.[690] The first scene shows the team of standing haulers taking a grip on the rope ready to pull on the signal. The second scene shows the result of the action. In the tomb of *ʿnḫ -m-ʿ-Ḥr*, for example, the men have fallen on their back in the effort to pull the net closed, the birds are trapped and being taken from the trap to be presented to the tomb owner. An overseer of fowlers supervises the process.[691]

In the double scenes of *Sʿnḫ-wj-Ptḥ* the hauling team wear headdresses of lotus flowers, which to the modern eye might suggest camouflage if it were not for the fact that the teams of fighting boatmen often wear the same. The panels that have survived from the tomb of *K3-gmnj* not only present the two sequential clapnet scenes, but the register immediately above depicts the deceased's poultry yards. Above this register are the remains of force-feeding birds. These three registers provide an overview of *K3-gmnj*'s poultry resources, showing how the numbers were maintained, how the birds were kept and how they were prepared for the table.[692] The tomb of *Mḥw* contains the largest number of clapnet scenes.[693] On the south wall of Room One, clapnets are depicted five times. The top three clapnets have been pulled closed and the hauler teams have fallen backwards on to the ground. The clapnet below, occupying a register as high as the three higher clapnet scenes together, presents an earlier stage of the action. The standing haulers lean forward taking a good grip on the rope awaiting the order to pull. Beneath the net are two decoy herons and two fluttering pintails. The lowest clapnet contains four cranes, two of which are demoiselle cranes, with three more birds above the net. Two herons, a crane and a pintail(?) stand beneath. This clapnet is differently constructed, perhaps to trap the larger birds without injuring them. The depiction of *Mḥw*'s bird resources continues on the register below the cranes. Offering bearers bring before the deceased live birds held in their hands and in cages slung from shoulder yokes. Another, more conventional clapnet scene is depicted on the north wall of Room Two.

The three clapnet images in *Sʿnḫ-wj-Ptḥ* show the stages of the trapping process. On the upper register the birds are swimming without alarm within the clapnet, the signaller has his arms stretched out wide holding the cloth, and the standing team of haulers begin to pull the rope to close the net. Immediately beneath this scene, in the second clapnet image, the haulers have fallen back on to the ground and the birds flutter in alarm at being trapped within the closed net. The third image shows the team still holding the net closed but the signaller is now extracting a bird from the net.[694]

Generally, the birds caught in the nets in Dynasty 6 tombs appear to be pintails or geese, as do those that fly off, but the birds depicted in the vegetation around the lake are more clearly differentiated. One or two decoy birds, usually a heron or bittern, appear in eight of the scenes. All the clapnets, with the exception of the net containing cranes in the tomb of *Mḥw*, are depicted crowded with birds in small overlapping, layered groups, suggestive of larger numbers. The variation in the size of the birds indicates more than one species but their outline is that of ducks or geese.

688 Paget–Pirie (1898) pl. 32.
689 LD II 46.
690 Tombs with more than one clapnet scene: *ʿnḫ-m-ʿ-Ḥr, Wḏ3-ḥ3-Ttj/Ššj/Nfr-sšm-Ptḥ, Mḥw, Mṯṯj, Sʿnḫ-wj-Ptḥ, K3-gmnj.*
691 Badawy (1978) fig.33, Kanawati–Hassan (1997) pl.42.
692 Harpur–Scremin (2006) Context Drawing 14.
693 Altenmüller (1998) pls. 6, 7, 31(b).
694 Kanawati–Abder-Raziq (1998) pl. 69, 75.

Mid Dynasty 6 to Dynasty 8

Clapnet scenes are depicted in the following tombs:

Jbj	Deir el-Gebrawi	VI.M
Jdw I	Dendera	VI.M
Jḥjj	el-Khokha	VI.M
Ppj-ʿnḫ-ḥrj-jb	Meir	VI.M
Ppj-ʿnḫ/Ḥnj-km	Meir	VI.L
Nj-ʿnḫ-Ppjj	Meir	VI.L
Ḫnj/Špsj-pw-Mnw	el-Hawawish	VI.L
K3(.j)-ḥp/Ṯtj-jkr	el-Hawawish	VI.L
Ṯtj-ʿ3	el-Hawawish	VI.L
Ḏʿw	Deir el-Gebrawi	VI.L
Ḫw-ns	Qubbet el-Hawa	VIII
Mrjj-ʿ3	el-Hagarsa	VIII
Gḥs3/Nbjj	el-Hawawish	VIII

The clapnet scenes of mid to late Dynasty 6 in the Meir tombs[695] and those of the el-Hawawish nomarchs, *Ḫnj/Špsj-pw-Mnw* and *K3(.j)-ḥp/Ṯtj-jkr*,[696] continue to provide much of the detail of the early Dynasty 6 Memphite scenes. The teams of haulers fall back on to the ground in their effort to pull the net closed, and layered images of ducks and geese swim placidly within the confines of the nets, which are packed full of birds lured by decoy herons and bitterns depicted beside the nets.

Few clapnet scenes in the tombs of other provincial administrators of this period have survived with much detail. However, the scene is the tomb of *Ḏʿw* at Deir el-Gebrawi was sufficiently well preserved when Davies recorded it for him to comment that the scene lacked action and detail.[697] The four haulers are represented standing and holding the rope attached to the net. Within the clapnet about nine individual birds are depicted either swimming or fluttering with open wings. One curious detail is the image of two birds with bound wings in front of the net. Davies thought they may have been decoys but their uneasy actions would not have lured a flock of wary birds flying over them. While herons and bitterns are sometimes shown as tethered lures, this would be the only image of birds with bound wings used as a decoy. It is more likely that the birds had been caught and were bound so that they could not fly away while the trappers were busy with the clapnet.

The inclusion of the associated scenes of birds placed in cages and transported from the marshland, either slung from a shoulder yoke or on a skiff, makes it clear that the birds were not trapped merely for immediate consumption by the teams of workers and household members in the marshlands. Despite the occasional vignette positioned near the clapnet scene of workers plucking and cooking birds, the sequence of action is made clear in many of the great tombs. The captured birds are taken live from the net, placed in cages and transported, either to be presented live to the deceased or directly to his fowl yards.

11.3 Poultry yard scenes

11.3.1 Historical development of images and themes

Mid Dynasty 5 to mid/late Dynasty 6

Fowl yard scenes are depicted in the following numbers of tombs:

V.M	1	VI.M	1
V.L	2	VI.L	1
VI.E	9		

Mid Dynasty 5 to mid/late Dynasty 6

Fowl yard scenes are depicted in the following tombs: (The inclusion of force feeding of birds is indicated by an asterisk.)

*Ptḥ-špss**	Abusir	V.M
*Spd-ḥtp**	Saqqara	V.L
*Ṯjj**	Saqqara	V.L
*Wḏ3-ḥ3-Ttj/Ššj /Nfr-sšm-Ptḥ**	Saqqara	VI.E
*Mrjj-Ttj/Mrj**	Saqqara	VI.E
*Mrrj**	Saqqara	VI.E
*Mrrw-k3**	Saqqara	VI.E
*Mṯtj**	Saqqara	VI.E
*Nj-k3w-Jzzj**	Saqqara	VI.E
Ḥzj	Saqqara	VI.E
*K3.j-jrr**	Saqqara	VI.E
*K3-gmnj**	Saqqara	VI.E
Jbj	Gebrawi	VI.M
*Ppj-ʿnḫ/Ḥnj-k m**	Meir	VI.L

Species depicted in fowl yards: pintail ducks, geese, cranes, pigeons. Ten of these tombs portray force feeding of birds.

In the fowl yards scenes, ducks and geese (it is frequently not possible to differentiate between the two species in these scenes) are depicted freely pecking grain from the ground, preening, flapping their wings or peacefully swimming in the rectangular pond situated in the middle of the yard, all actions conveying the impression of contented flocks of birds that are well housed and cared for. Nine of the scenes depict the birds in a well-constructed fowl yard. Part of the yard may have been under cover as these scenes show vertical poles with a U-shaped hook at the top that would have supported cross-beams.[698] Workers bring sacks of grain to the yard or sprinkle grain around the yard. Cranes, which appear in nine of the scenes, are given the same treatment as the flocks of ducks and geese.

These fowl-yard scenes are mainly confined to a short period of time. The earliest extant scenes date to the

[695] *Ppj-ʿnḫ/Ḥnj-km* Blackman–Apted *Meir* V (1953) pl. 30, *Nj-ʿnḫ-Ppjj-km* Blackman–Apted *Meir* V (1953) pl. 13.

[696] *Ḫnj/Špsj-pw-Mnw* Kanawati II (1981) fig. 22, *K3(.j)-ḥp/Ṯtj-jkr* Kanawati I (1980) fig. 12.

[697] Davies *Gebrâwi* II (1902) p. 5 and pl. 4.

[698] *Ṯjj* Épron–Daumas I (1939) pl. 7, 8, 27-30, 32; *Wḏ3-ḥ3-Ttj/Ššj/Nfr-sšm-Ptḥ* Capart (1907) pl. 88; *Mrrj* Davies et al (1984) pl. 6; *Mrrw-k3*: Duell I (1938) pl. 53B; Kanawati (2010) pl. 83; *Ḥzj* Kanawati–Abder-Raziq (1999) pl. 56; *K3-gmnj* Harpur–Scremin (2006) Drawing 14; *Jbj* Davies *Gebrâwi* I (1902) pl. 13-16; Kanawati *Gebrawi* II (2007) pl. 72.

second half of Dynasty 5. Furthermore, although this is the period of time with the largest number of tombs containing scenes of 'daily life' and officials' resources, only three late Dynasty 5 tombs present the poultry yard scene. The majority of these scenes occur in the early Dynasty 6 tombs of the highest officials buried in the Teti cemetery, and only two are found in the tombs of the provincial administrators, *Jbj* and *Ppj-ꜥnḫ/Ḥnj-km* who is dated to the second half of the Dynasty.[699]

Force-feeding of birds by teams of workers is a feature of the poultry yard theme. In Old Kingdom funerary scenes four types of farmed animals are shown being hand or force- or hand-fed: cattle, desert dwelling antelope, striped hyenas and birds in poultry yards. While the reality of these methods of feeding the first three kinds of animals may be doubted,[700] the accuracy of the representation of force feeding farm birds need not be challenged as it is a practice carried out today in a number of countries. In the scenes of *Ṯjj*, *Wḏꜣ-ḥꜣ-Ttj/Ššj//Nfr-sšm-Ptḥ*, *Mrjj-Ttj/Mrj* and *Mrrw-kꜣ* pigeon and cranes as well as geese and ducks are shown being force fed,[701] while the scenes of *Ṯjj*, *Mrjj-Ttj/Mrj* and *Mrrw-kꜣ* in particular suggest that the deliberate fattening of birds was a specialised and organised undertaking by teams of workers who prepared and baked the balls of dough to be crammed down the bird's throat by other workers.

11.4 Procession of birds

Birds represented in these scenes are ducks, geese, cranes, pigeons and smaller birds whose species cannot be identified.

11.4.1 Historical development of images and themes

Date	No of scenes	Date	No of scenes
IV.E	1	V.L	24
IV.L	4	VI.E	3
V.M	7		

Dynasty 4
Bird procession scenes are depicted in the following tombs:

Nfr-mꜣꜥt (LG 57)	Giza	IV.E
ꜣḫj	Giza	IV.L
Mrs-ꜥnḫ III	Giza	IV.L
Nb.j-m-ꜣḫtj	Giza	IV.L
Ḫntj-kꜣw.s	Giza	IV.L

The birds depicted in procession in Dynasty 4 are ducks, geese and cranes. In the scene in *Mrs-ꜥnḫ* III the birds, leading the parade of animals, occupy the same wall panel as a clapnet scene and are positioned immediately above a register of skiffs returning from the marshes, one of which has a cage of birds. This (lowest) register is completed by a scene of sheep treading seed into newly ploughed land. This feature, together with the inscription above the first skiff, suggests that the depictions represent a season following the subsidence of the annual inundation: *prt m šꜣ ḥnꜥ jn mḥtjw.s jn.sn ḫt nbt nfrt jnnt n špst m kꜣr sḫt* 'Coming out of the marsh with lotus blossoms by her marsh-men, that they may bring to her every good thing, which is brought to a noblewoman as the work of the field.'[702]

Early to Mid Dynasty 5
Bird procession scenes are depicted in the following tombs:

Jj-mrjj	Giza	V.M
Ptḥ-špss	Abusir	V.M
Nj-ꜥnḫ-Ḫnmw and *Ḫnmw-ḥtp*	Saqqara	V.M
Nfr and *Kꜣ-ḥꜣ.j*	Saqqara	V.M
Rꜥ-ḫꜥ.f-ꜥnḫ	Giza	V.M

The birds depicted in processions in these tombs are geese (*Ptḥ-špss*, *Nj-ꜥnḫ-Ḫnmw* and *Ḫnmw-ḥtp*, *Nfr* and *Kꜣ-ḥꜣ.j*, and *Rꜥ-ḫꜥ.f-ꜥnḫ*), ducks (*Ptḥ-špss* and *Nj-ꜥnḫ-Ḫnmw* and *Ḫnmw-ḥtp*), cranes (*Jj-mrjj* and *Ptḥ-špss*) and pigeons (*Nj-ꜥnḫ-Ḫnmw* and *Ḫnmw-ḥtp*).[703]

In these scenes, the birds parade before the tomb owner in twos and threes. The three cranes before *Jj-mrjj* and his father are specified as *ḏꜣt, ꜥw* and *wḏꜥ*, with the number *ḫꜣ* ('thousand') attached to each. Weeks suggests that the third bird is a demoiselle crane (*grus virgo*) although he translates *ḫꜣ wḏꜥ* as 'a thousand grey cranes'.[704] In the tomb of *Nj-ꜥnḫ-Ḫnmw* and *Ḫnmw-ḥtp*, a procession of three birds completes each of the four panels containing registers of parades of cattle and desert animals on the two passageways to the offering chamber. The ducks are labelled *sr* or *s* or *st* or *ṯrp*, and the geese, *ṯrp* or *rꜣ* and each has the number *ḫꜣ*. As in the case of the cranes in the tomb of *Jj-mrjj*, the precise import of the labels is unclear.

The procession of geese before *Nfr* and his wife is associated with the presentation of a scroll and scribes making a record. Moussa and Altenmüller identify the (unlabelled) birds on the upper register as *anser anser* and those beneath as *anser albifrons* and note that this south wall of the chapel is largely devoted to the presentation of offerings to the deceased.[705] Similarly, the parade of birds on the south wall of his tomb is closely associated with the presentation of offerings to the figures

699 Vandier *Manuel* V (1969), 398 suggests that the concentration of these scenes in the Memphite tombs of high officials is due to their possession of great estates in the Delta, where the birds would have been captured.

700 Evans (2006) 130-31.

701 *Ṯjj* Épron–Daumas I (1939) pl. 6, 19; *Wḏꜣ-ḥꜣ-Ttj/Ššj//Nfr-sšm-Ptḥ* Wreszinski (1936) pl. 81; *Mrjj-Ttj/Mrj* Kanawati–Abder-Raziq (2003) pl. 48; *Mrrw-kꜣ* Duell I (1938) pl. 52; Kanawati (2010) pl. 83.

702 Dunham–Simpson (1974) fig. 4.

703 *Ptḥ-špss* Verner (1986) pl. 40; *Nj-ꜥnḫ-Ḫnmw* / *Ḫnmw-ḥtp* Moussa–Altenmüller (1977) pl. 82, 83, 84 and 85; *Nfr* and *Kꜣ-ḥꜣ.j* Moussa–Altenmüller (1971) pl. 24(a); *Rꜥ-ḫꜥ.f-ꜥnḫ* LD II 10 [b]; *Jj-mrjj* LD II 50 [b]; Weeks (1994) fig. 31.

704 Weeks (1994) 38 discusses the problem of translating the terms used to describe the birds.

705 Moussa–Altenmüller (1971) pl. 24a, p. 29.

of *Rꜥ-ḫꜥ.f-ꜥnḫ* and his wife seated before an offering table.[706]

Like the scenes of Dynasty 4, the inscriptions do not give any indication that birds are maintained in poultry yards or in any great number. *ḥꜣ*, rather than a specific number, is attached to the images of the birds suggesting, as it does in offering table scenes, an 'eternal or unlimited supply of …'. However, the representation of the bird procession itself suggests that the birds have been accustomed to human contact.

Later Dynasty 5

Bird procession scenes are depicted in the following tombs:

GIZA

Jttj	*Sḫm-ꜥnḫ-Ptḥ*
Nj-mꜣꜥt-Rꜥ	*Kꜣ.j- m- ꜥnḫ*
Nfr	*Kꜣ.j-m-nfrt*
Nfr (1)	*Kꜣ.j-dwꜣ*
Rꜥ-wr II	*Kꜣpj*
Sꜥnḫ-n-Ptḥ	*Kꜣ-ḥj.f*
Snḏm-jb/Jntj	*Kd-nfr*
Snḏm-jb/Mḥj	

SAQQARA

ꜣḫtj-ḥtp (E 17)	*Ḥtp-kꜣ.j/Tp-kꜣ.j*
Ptḥ-ḥtp/ (II)	*Spd-ḥtp*
Ṯfj	*Kꜣ.j-m-nfrt*
Mꜣ-nfr	*Kꜣ.j-m-rḥw*
Rꜥ-špss	*Ṯjj*

The birds in procession in tombs dated to the second half of Dynasty 5 are depicted in greater numbers than in earlier tombs and include images suggesting a greater range of species. These additional birds, usually depicted at the end of the row of birds, are yet more difficult to identify, even though their size and shape vary. They are usually smaller than the cranes, ducks and geese leading the parade of birds, and while some images have the outline of turtle-doves or pigeons, others may be chicks.[707]

A typical location for the procession of birds is immediately beneath or following the parade of cattle and desert animals, as occurs in the tombs of *Ptḥ-ḥtp* (II) *Ṯfj*), *Mꜣ-nfr*, *Nfr* (1), *Rꜥ-wr* II, *Rꜥ-špss*, *Snḏm-jb/Jntj*, *Sḫm-ꜥnḫ-Ptḥ*, *Kꜣ.j-m-nfrt* (Giza), *Kꜣpj*, *Kꜣ-ḥj.f* and *Ṯjj*.[708]

Dynasty 6

Bird procession scenes are depicted in the following tombs:

Nj-ꜥnḫ-Ḫnmw	Giza	VI
Mrjj-Ttj/Mrj	Saqqara	VI.E
Mr.f-nb.f	Saqqara	VI.E
Mrrj	Saqqara	VI.E
Ḥm-rꜣ/Jzj (I)	Gebrawi	VI.E
Ppjj-ꜥnḫ/Ḥnj-km	Meir	VI.L

Considering the proliferation of poultry yard scenes in early Dynasty 6, there are surprisingly few bird procession scenes in the tombs of this date, and only two (*Mrjj-Ttj/Mrj* and *Mrrj*) from the Teti Cemetery. Moreover, the concept of linking the bird procession with the parade of other animals is lost. In *Nj-ꜥnḫ/Ḫnmw* and *Ḫnmw-ḥtp* the bird parade is part of the offering table scene,[709] while in *Mrrj* and *Ppjj-ꜥnḫ/Ḥnj-km* it is associated with the poultry yard and force-feeding themes.[710] In *Ḥm-rꜣ/Jzj* (I) registers of birds are associated with the scene of the tomb owner fowling in the marshes.[711]

11.5 Summary

- Birds are portrayed as a staple resource in three scene genres which are used in this study: birds trapped in a clapnet; the poultry yard; procession of birds.
- Although the birds in these scenes may be identified by a caption, the principle by which these labels are applied is not based on species as birds of apparently the same morphology and size may be given a variety of labels.
- The most frequently depicted birds are ducks (pintail and common teal), geese, Egyptian geese, cranes (common and demoiselle), doves and pigeons.
- The use of a clapnet to trap quantities of birds without killing them must have been developed before Dynasty 4, as the device appears in the earliest Old Kingdom tomb scenes dating to the reign of Snefru.
- Most of the elements that became regular features of the theme were introduced between late Dynasty 4 and mid Dynasty 5. These include the signaler holding a strip of cloth, inclusion of two scenes to suggest the sequence of actions, the nets filled with images of overlapping birds, a figure taking birds from the net, live birds being transported in cages, an overseer of fowling, pond vegetation and the presence of decoy birds.
- From late Dynasty 5 on, the number of haulers tends to increase from three or four figures to five or six figures. *Nj-ꜥnḫ-Ppjj-km* of Meir has the greatest number (seven).[712] This trend, however, is not uniform as *Ḥnj/Špsj-pw-Mnw* and *Kꜣ(.j)-ḥp/Ṯtj-jkr* of el-Hawawish, for example, only depict three haulers.[713]

[706] LD II 10.

[707] For example, *Ṯjj* Wild III (1966) pl. 169, *Ptḥ-ḥtp* (II) *Ṯfj* Paget–Pirie (1898) 30 pl. 31 and *Mꜣ-nfr* LD II 69.

[708] *Ptḥ-ḥtp* (II) *Ṯfj*) Paget–Pirie (1898) pl. 31, *Mꜣ-nfr* LD II 70, *Nfr*(1) Junker IV (1940) fig. 14, *Rꜥ-wr* II Junker III (1938) fig. 48, *Rꜥ-špss* LD II 61b, *Snḏm-jb/Jntj* LD II 74b, Brovarski (2001) fig. 50-51, *Sḫm-ꜥnḫ-Ptḥ* Badawy (1976) fig. 20, *Kꜣ.j-m-nfrt* LD II 91c, Badawy (1976) fig. 29; *Kꜣpj* Roth (1995) pl. 167; *Kꜣ-ḥj.f* Junker VI (1943) fig. 40, *Ṯjj* Wild III (1966) pl. 114.

[709] Moussa–Altenmüller (1977) pl. 82-85.

[710] *Mrrj* Davies et al (1984) pl. 6 [A], *Ppjj-ꜥnḫ/Ḥnj-km* Blackman–Apted *Meir* V (1953) pl. 22 [2].

[711] Kanawati *Gebrawi* I (2005) pl. 50.

[712] Blackman–Apted *Meir* V (1953) pl. 13.

[713] Respectively, Kanawati II (1981) fig. 22, Kanawati I (1980) fig. 12.

- In the earliest scenes only a limited number of birds are depicted in the clapnet but from late Dynasty 5 on, the clapnet is often shown teeming with birds.

- The inclusion of the associated scenes of birds placed in cages and transported from the marshland makes it clear that the birds were not trapped for immediate consumption. The sequence of actions is that the captured birds are taken live from the net, placed in cages and transported either to be presented live to the deceased or taken to his fowl yards.

- Thirteen tombs, dated from mid Dynasty 5 to late Dynasty 6, present scenes of fowl yards, which have a partial cover and a central pond. In ten of these scenes birds are being force-fed.

- Birds, usually ducks, geese and cranes in Dynasty 4, but with the addition of turtle doves or pigeons in Dynasty 5, are depicted in parade scenes. In Dynasty 6 birds are rarely included with the parade of animals.

Figure 39: TRAPPING WATERFOWL WITH A CLAPNET, TOMB OF ANKHMAHOR, SAQQARA.
Kanawati–Hassan (1997) pl. 42.

Figure 40: A FOWL YARD, TOMB OF MERERUKA, SAQQARA.
Kanawati–Abder-Raziq (2004) pl. 48

Figure 41: BIRDS BEING FORCE-FED, TOMB OF MERERUKA, SAQQARA.
Kanawati (2010) pl. 83(c)

CHAPTER 12 PAPYRUS

12.1 Focus of enquiry: value of the papyrus plant to an official's estate

TABLE 15: PAPYRUS GATHERING SCENES
TABLE 16: PAPYRUS SKIFFS – CONSTRUCTION AND USE

The data:

- The iconography and associated inscriptions of the themes representing the exploitation of papyrus
- Other data: material finds

Areas of research:

- Properties of Cyperus papyrus *L.*, *Cyperaceae*
- Representation of the themes of the exploitation of papyrus by estate workers:
  - papyrus gathering
  - construction and use of papyrus craft
  - manufacture of cordage
  - fodder for cattle.
- Other uses.

Cyperus papyrus *L., Cyperaceae*

The papyrus plant is a strong-growing sedge capable of reaching over five metres in height; it grows best in fresh water swamps. A typical thicket of the plant consists of a tangled mass of rhizomes below ground supporting tall fibrous stalks each topped in maturity by an umbellate inflorescence. Papyrus is a perennial whose stalks mature within a year, after which they age, ultimately collapsing to become entangled with the network of rhizomes. This process creates a platform that floats on the marsh waters. The constant presence of aging, bent and falling stalks among the young upright growth produces densely tangled thickets, very different to the tidy images of rows of vertical stalks seen in so many Old Kingdom tombs.[714] In pharaonic times the Delta provided vast areas of papyrus thickets, while more limited stands were available along the Nile valley and subsidiary channels and around oxbow lakes of the water basins lying laterally to the river.[715]

12.1.1 Uses of the papyrus plant

The value of the papyrus plant for Egyptians extended from the mundane and utilitarian to the aesthetic and spiritual. It was not merely manufactured into writing paper but provided transport, fuel, nutrition for cattle and perhaps for people in times of famine, and was the raw material for essential items such as clothing, cordage, sails and baskets, as well as for comforts like matting, seating for stools and chairs and sandals to protect the feet. The form of this invaluable plant contributed artistic inspiration for ornate pilasters and columns that decorated mighty monuments and features to adorn household furniture.

Papyrus grew abundantly in Egypt in ancient times and all parts of the plant were exploited. Rhizomes could be used for fuel and are edible. The triangular shaped stalks were put to a myriad of uses. They were, for example, lashed together to makes rafts and skiffs. The tough outer layer of the stalk was worked into many useful items, while writing paper was manufactured from the inner pithy core. Papyrus has a further use that made a vital contribution to the Old Kingdom economy, providing domestic herds with a nutritious fodder that in dry seasons is higher in protein than grass. Accessibility, however, could not have been easy. Marshland vegetation, which may be abundant, is a difficult environment for narrow-hoofed grazing animals, particularly those that need to lie down to ruminate, while their continuous exposure to wet conditions increases the danger of footrot.[716] Nevertheless, richly nourishing papyrus was essential to the feeding of herds of cattle and other domestic animals.[717] For a considerable part of each year papyrus must have been the main source of fodder in the Nile valley.[718] Even in the Delta the predominant flora were papyrus, bulrushes and reeds. In the hot, dry summer, when pasturage was eaten out, cattle herds were driven down to the marshlands to feed on the stands of papyrus. When the inundation came, they were driven back to higher ground although natural pasturage would not have been available for another two months or so. Collection and storage of papyrus fodder for this period would have been essential.

The acquisition of papyrus as a resource may not have had the high status or value that Egyptian society attributed to herds of livestock and field crops, but it was an invaluable, cheaply procured asset in plentiful supply that met a great variety of working and everyday needs. It is not surprising that the pictograph of the plant was a symbol for Lower Egypt (𓇇), much of which was covered by papyrus thickets. Many of these uses are represented in officials' tombs in scenes and vignettes associated with marshland, although they also appear in other themes. Timber river boats, for example, are depicted with sails and rigging which could have been manufactured from papyrus, Furniture with what could be papyrus seating appears in offering table scenes, and lists and accounts written on scrolls may be depicted in any

[714] Ryan (1988) 132-3.
[715] Ryan (1988) p. 133, left plate.
[716] Ryan (1988) 133-140.
[717] Muthuri–Kintamario (1989) 23-30.
[718] Saad–Sami (1967) 467-503.

scene with scribes at work: lists of items in presentation scenes, the tomb owner inspecting his resources, recording of census details. The manufacture of skiffs, cordage and containers in which to transport fish, birds, young mammals and foodstuffs would have contributed to the owner's wealth as a means of acquiring and handling resources, but whether items such as mats carried and used by workers should be so regarded is doubtful. Matting for workmen was a useful product of the papyrus plant, yet it does not meet the definition of 'officials' resources', that is, a product or agency adding materially to the assets of the elite. Whether officials claimed ownership or exclusive use of stretches of papyrus stands is not clear, but both the acquisition of papyrus and the construction of skiffs and cordage that facilitated the extraction and transport of other marshland resources may certainly be regarded as being of value to the estate.

An important use of papyrus was the manufacture of scrolls for recording. Virtually every extant scene representing the resources of an estate includes record making by scribes or the presentation of a record to the tomb owner. The efficient management of an estate clearly depended on the maintenance of records of the production, usage and handling of resources. Recording of storage, distribution and consumption would have been essential as much of the estate's vital resources depended on an annual cycle of production. However, scenes do not include the manufacture of the necessary papyrus scrolls.

12.1.2 The handling of papyrus depicted in Old Kingdom tombs

Papyrus is shown being handled in various ways, as follows:

- collection of papyrus stalks (*Figure 42*)
- construction and use of papyrus skiffs and rafts (*Figure 42, 43 and 44*)
- as fodder for domestic animals
- manufacture and use of cordage
- use of containers
- manufacture of mats and use of mats and backrests for workmen and overseers
- as writing material
- seating of chairs and stools for officials (possible).

Only the first four of the above items are dealt with here. It is not possible to judge from the iconography whether any depicted container is made from papyrus and while there are a considerable number of scenes of the manufacture of matting,[719] mats are only depicted as a comfort for workers. The remaining items do not meet the definition of papyrus as an asset (a source of wealth).

[719] The Oxford Expedition to Egypt Database (Scene: 1.25) lists eighteen scenes of mat-making, dated from Dynasty 4 to early Dynasty 6.

12.2 Gathering papyrus in the marshes

TABLE 15: PAPYRUS GATHERING SCENES (*Figure 44*)

12.2.1 Historical development of images and themes

The distribution of the gathering papyrus theme in the tombs used in the present study is as follows:

Late Dynasty 4	1 tomb
Early to mid Dynasty 5	4 tombs
Late Dynasty 5	19 tombs
Early Dynasty 6	1 tomb
Mid Dynasty 6 to Dynasty 8	4 tombs

Scenes of collecting papyrus are depicted in the following tombs:

Dynasty 4 to mid Dynasty 5

Nb.j-m-3ḫtj	Giza	IV.L
Pr-sn	Giza	V.E
Jj-mrjj	Giza	V.M
Nfr and *K3-ḥ3.j*	Saqqara	V.M
Ptḥ-špss	Abusir	V.M

Late Dynasty 5

GIZA

Sndm-jb/Jntj	*K3pj*
Sḫm-k3j	

SAQQARA

3ḫtj-ḥtp (Louvre)	*Nb-j-k3w-ḥr/Jdw*
3ḫtj-ḥtp (D 64)	*Nbt*
Jntj	*Ḥtp-ḥr-3ḫtj*
Jrj-n-k3-Ptḥ	*Ḫnwt*
Jḥjj (r/u *Jdwt/Sšsšt*)	*K3.j-m-nfrt*
Ptḥ-ḥtp (II) (*Ṯfj*)	*Ṯjj*

ABUSIR

Tp-m-ʿnḫ

SHEIKH SAID

Wr-jr.n.j

ZAWYET EL-MEITIN

Ḫw-ns

EL-HAWAWISH

Ḥzjj-Mnw

Dynasty 6

Mḥw	Saqqara	VI.E
Jbj	Gebrawi	VI.M
Ḫw.n-wḫ	Quseir el-Amarna	VI.M
Ḳ3r	el-Hawawish	VI.M
Ppj-ʿnḫ/Ḥnj-km	Meir	VI.L

The scene of workers in the marshes gathering and carrying heavy loads of papyrus on their backs extends in time from late Dynasty 4 (*Nb.j-m-3ḫtj*) to the second half of Dynasty 6 (*Ppjj-ʿnḫ/Ḥnj-km* of Meir),[720] with the majority of scenes appearing in tombs of the second half of Dynasty 5. The depiction of this theme does not change materially over time, although marked artistic developments begin to emerge in mid Dynasty 5.

[720] Refer to TABLE 15: PAPYRUS GATHERING SCENES.

Two stages of the papyrus gathering activity are portrayed: plucking the stalks and transporting the bundles. An intermediary stage of tying bundles of stalks together makes a rare appearance.[721] Pulling up a tall stalk of the plant from its roots must have been hard work, but the vignettes show just one labourer doing so, working unaided and without the use of any tool. The lifting and carrying of the heavy loads is the most frequently illustrated activity. Scenes show bearers carrying papyrus stalks in great bundles that are nearly as tall and bulky as themselves. In the early scenes of *Nb.j-m-ꜣḫtj* and *Jj-mrj* the bearers are depicted merely bending forward a little under the weight of the bundle of papyrus on their back.[722] Details of images show greater artistic expression beginning with the scene in *Pr-sn*'s tomb,[723] where each of the six bearers is given a different stance, two of them holding the papyrus in front of their bodies with their arms wrapped round their load. In *Wr-jr.n.j*'s scene three of the bearers strain to lift their loads and a fourth is depicted kneeling, possibly struggling to stand upright under the weight on his back.[724] By early in the second half of Dynasty 5 these scenes have become very expressive of the demands and difficulties of carrying a bundle of papyrus; they present bearers in a variety of postures, bent forward with the upper part of their body horizontal, or an arm reaching behind them to steady the bundle on their back, sometimes clutching the load in both arms. The vignettes make it clear that this task required physical strength and endurance.[725]

The destination of the papyrus carriers provides evidence of the uses to which the stalks were put. Of the 29 scenes of papyrus gathering and carrying used in this study, six clearly show the bearers taking their load to a scene of papyrus skiff construction, while a further three scenes occur on registers immediately beneath or above a skiff building scene. Eight scenes (and a further two possible scenes) show the papyrus carriers facing a figure of the tomb owner.[726] Only two scenes show the bearers taking their load to rope makers and one of these (tomb of *Jntj*) portrays the activities of papyrus carrying, rope making and skiff construction in immediate succession to each other.[727]

Associated captions, which are rare in these scenes, add little to the interpretation of these scenes. The seated figure of *ꜣḫtj-ḥtp* views two registers of papyrus carriers. Above the top line of men is written, *jnt ḫt nb<t> nfr.t jrrt m mḥt jn šdtjw ꜣpdw nb* 'Bringing all kinds of good things made in the papyrus beds by the papyrus gatherers and all kinds of fowl'. Two of the men hold a bird in their free hand. The caption of the lower register reads, *prt m ḏt jnt kꜣt sḫt m mḥt nfr mꜣꜣ r ḫt nb<t>* 'Going forth from the marshland, and bringing works of the fields from the papyrus beds; fair to see above all things'.[728]

12.3 Papyrus skiffs and rafts

Both the construction and use of papyrus skiffs and rafts are represented in marshland scenes. Their transport capacity was, of course, restricted; tomb scenes do not show skiffs carrying more than four or five persons and an occasional large animal. But these simple craft had many advantages over timber boats. They were easy, cheap and quick to construct, while their shallow draft, lighter weight and easy manoeuvrability made them ideal craft for navigating the narrow, twisting waterways of the wetlands. As the Nile subsided, the shallow waters of the marshes were often so thick with mud that that they could only be traversed by very light craft. Skiffs are shown being punted as well as paddled, indicating that one of their advantages was their buoyancy in particularly shallow water.

Two main types of papyrus craft are depicted in the Old Kingdom. The most frequently seen is a simply designed skiff with a hull curving up from the water at either end. It has slightly raised gunwales and, usually, a prow that is elevated but more flattened out than the stern.[729] The hull of the craft consists of flexible papyrus stalks lying lengthwise that appear to have been moulded into the basic shape of a skiff and lashed together with cord at intervals that decrease towards prow and stern. The triangular structure of the papyrus stalks would have allowed them to be fitted snugly against each other, helping to make the craft watertight and give it buoyancy. The gunwales were sometimes reinforced with cordage. As it is not always clear whether a depicted skiff is a true papyrus craft or a papyriform timber model, the principle adopted in this study is to judge skiffs to be made of timber if they are depicted with a carved finial, being crafted with woodworking tools, being rowed, which requires oarlocks on the gunwales, rather than paddled or punted, transporting the deceased to the necropolis or having other funerary purposes, or with any superstructure such as a shelter or rigging.[730] The skiffs are flat-bottomed punts, so they should be depicted with

721 *Jḫjj* (r/u *Sšsšt/Jdwt*) Macramallah (1935) pl. 6-7 and Kanawati–Abder-Raziq (2003) pl. 54, *Ptḥ-ḥtp* (II) (*Ṯfj*) Paget–Pirie (1898) pl. 33 [lower] and Davies *Ptahhetep I* (1900) pl. 3, 21, 23.

722 Respectively, LD II 12b and LD *Erg*. pl. 5; Weeks (1994) fig. 33.

723 Petrie–Murray (1952) pl. 8, Harpur (1985) fig. 7.

724 Davies *Sheikh Saïd* (1901) pl. 12.

725 In fourteen scenes the bundles do not show the individual stalks. Vandier *Manuel V* (1969) 461-465 speculates whether these are depictions of stalks wrapped in sacking, as the rounded bottom of these bundles suggests.

726 See TABLE 15: PAPYRUS GATHERING SCENES.

727 Petrie (1898) pl. 5, Kanawati–McFarlane (1993) pl. 33.

728 Davies *Ptahhetep II* (1901) pl. 13-14.

729 While the basic shape of the skiff does not vary, there are modifications, from the prow and stern depicted both curving upward quite steeply to a flatter, more gradually curved prow. For most of the Old Kingdom their bottom is just slightly rockered, but by the second half of Dynasty 6 the hull of the skiff appears to have become much flatter, with only the tip of prow and stern curving upward. These developments may have added to the stability of the craft when the figures on board were engaged in activities such as fishing or jousting. For a detailed analysis of the various shapes of skiffs supporting the figure of the tomb owner, see Woods (2008) 61-63, Appendix 3: Life Span of Old Kingdom Marsh Scenes and Individual Features, Nos. 116 – 120.

730 Vandier *Manuel* V 697-698, Junker II (1934) 66, von Droste zu Hülshoff (1980) 24-26, Landtröm (1970) 34-36, Harpur (1987) 200-201.

standing figures using poles to propel them or paddlers facing forward, not rowers who face the stern.

Depicted papyrus rafts are of an even simpler design as only the prow curves above the water so that there is no contoured stern. According to this design, they are half raft and half skiff. They appear mainly as a one-man craft, usually employed by fishermen, who are angling with line and hooks.[731] Highly manœuvrable, the small raft would have been particularly useful in the waterways between the papyrus 'islands' as its shortened length would have facilitated navigation in the narrow, winding channels, while its lighter weight would have enabled it to float in shallow water and to be picked up and carried from one waterway to another when there was insufficient water depth to float the craft.

Elements of the papyrus boat theme

Papyrus craft appear in a range of marshland scenes:

- skiffs under construction, sometimes with men bringing up loads of papyrus stalks (*Figure 42*).
- the tomb owner on a skiff, either standing to hunt birds, harpoon fish or to pull papyrus, or merely passive (*Figure 43*).
- a number of men on one or two craft pulling up a catch in fishing traps (*Figure 34 and 36*).
- a single figure angling with line and hook; these craft are usually rafts rather than skiffs *(Figure 35)*.
- men on one to three skiffs peacefully taking marshland resources back to their camp or to their master's estate (*Figure 44*).
- teams of four or five men armed with poles and staves standing on their skiffs to joust with each other
- one or two craft conducting a herd of cattle across a stretch of water (*Figure 18*).

12.4 Construction of papyrus skiffs

Craft in constant use may not have had a long life, making it necessary for new boats to be regularly constructed.

12.4.1 Historical development of images and themes

The distribution of scenes of papyrus skiff construction in the tombs used in this study is as follows:

Dynasty 4	2 tombs
Early to mid Dynasty 5	4 tombs
Late Dynasty 5	17 tombs
Early Dynasty 6	1 tomb
Mid Dynasty 6 to Dynasty 8	1 tomb

Scenes of papyrus skiff construction are depicted in the following tombs:

Dynasty 4 to Mid Dynasty 5

Jtt	Medum	IV.E
Nb.j-m-3ḫtj	Giza	IV.L
Ḫnmw-ḥtp (PM 579)	Saqqara	V.E
Ptḥ-špss	Abusir	V.M
Jj-mrjj	Giza	V.M
Nfr and *K3-ḥ3.j*	Saqqara	V.M

Late Dynasty 5

GIZA	
Snḏm-jb/Jntj	*K3pj*
SAQQARA	
3ḫtj-ḥtp (Louvre)	*Nbt*
3ḫtj-ḥtp (D64)	*Ḥtp-ḥr-3ḫt*
Jrj-n-k3-Ptḥ	*Ḫnwt*
Jḥjj (*r/u Jdwt/Sšsšt*)	*K3.j-m-nfrt*
Ptḥ-ḥtp (II) (*Ṯfj*)	*Ṯjj*
Nb.j-k3w-ḥr/Jdw	
DESHASHA	
Jntj	
SHEIKH SAID	
Wr-jr.n.j	
ZAWYET EL-MEITIN	
Ḫw-ns	
EL-HAWAWISH	
Ḥzjj-Mnw	

Dynasty 6

K3-gmnj	Saqqara	VI.E
K3.j-ḥp/Ṯtj-jḳr	el-Hawawish	VI.L

While papyrus skiffs would have been in use long before the Old Kingdom, the earliest portrayal of skiff construction dates to Dynasty 4.[732] Most depictions, however, are found in tombs of the second half of Dynasty 5; this theme appears in 20 of the Dynasty 5 monuments used in this study.[733] The early stages of skiff construction are never shown. Instead, three to five men are portrayed working on a partially built skiff with its basic flat-bottomed hull curving upward towards prow and stern, which are already in place. Tools must have been needed for some of the work, such as cutting the papyrus stalks to size, but these are not shown. Instead the workmen are invariably depicted engaged in lashing the papyrus stalks together, pulling taut and tying ropes that are wound around the hull at regular intervals. The shaping of the curved ends of the hull appears to have been achieved by the use of timber props to hold up the prow and stern. More than one skiff is depicted being constructed in a number of scenes,[734] suggesting that the craft could be quickly and cheaply replaced.

12.5 The papyrus skiff in use: tomb owner on the skiff

There are over 130 extant scenes of the tomb owner in the marshland standing or seated on a skiff, but only those clearly showing the craft are used in this study.

[731] Moussa/Altenmüller (1977) pl. 31, fig. 12.

[732] *Nb.j-m-3ḫtj* LD II 12 [b]; Hassan IV (1943) fig. 77, pl. 37.

[733] See TABLE 16: PAPYRUS SKIFFS – CONSTRUCTION AND USE.

[734] *W*: Petrie (1898) pl. 5; Kanawati–McFarlane (1993) pl. 33; *Wr-jr.n.j* Davies *Sheikh Saïd* (1901) pl. 12; *Ptḥ-ḥtp* (II) *Ṯfj* Paget–Pirie (1898) pl. 32 and Davies *Ptahhetep I (*1900) pls. 21, 25 [upper], 26 [upper]; *Ḥzjj-Mnw* Kanawati IV (1983) fig. 17; *Ḫw-ns* LD II 106 [a], Varille (1938) fig. 4; *Snḏm-jb/Jntj* LD *Erg.* pl. 20, Brovarski (2001) fig. 38-9; *K3pj* Roth (1995) pl. 157-8; *Ṯjj* Wild II (1953) pl. 75 [B].

These include the tomb owner himself fishing or fowling, the 'pleasure cruise' and the tomb owner 'pulling papyrus'.

12.5.1 Historical development of images and themes

The distribution of scenes of the tomb owner on a papyrus skiff in the tombs used in this study is as follows:

Late Dynasty 4:	1 tomb
Early to mid Dynasty 5:	5 tombs
Late Dynasty 5:	22 tombs
Early Dynasty 6:	20 tombs
Mid Dynasty 6 to Dynasty 8:	13 tombs

(*Figure 43*)

Scenes clearly depicting the skiff in the following tombs:

Dynasty 4 to Mid Dynasty 5

Mrs-ꜥnḫ III	Giza	IV.L
Snb	Giza	V.E
K3(.j)-ḫnt	el-Hammamiya	V.E
K3(.j)-ḫnt (*Jwfj*)	el-Hammamiya	V.E
Jj-mrjj	Giza	V.M
Nj-ꜥnḫ-Ḫnmw and *Ḫnmw-ḥtp*	Saqqara	V.M

Late Dynasty 5

GIZA

3ḫtj-mrw-nswt	*Ḫww-wr*
J3sn	*Snḏm-jb/Mḥj*
Jj-nfrt	*Sšm-nfr* IV
Nj-m3ꜥt-Rꜥ	*K3.j- m-ꜥnḫ*

SAQQARA

3ḫtj-ḥtp (PM 633)	*Nbt*
Jj-nfrt/Š3.n.f	*Nfr-jrt-n.f*
Jrj-n-k3-Ptḥ	*Rꜥ-špss*
Jrw-k3-Ptḥ	*Ḥtp-ḥr-3ḫtj*
Jḥjj (r/u *Sšsšt/Jdwt*)	*K3.j-m-nfrt*
Jtj-sn	*Ṯjj*
Mṯtj	

ZAWYET EL-MEITIN
Ḫw-ns

Early Dynasty 6

GIZA
Mrjj-Rꜥ-mrjj- Ptḥ-ꜥnḫ/Nḫbw
Mrjj-Rꜥ-nfr/Ḳ3r

SAQQARA

Jnw-Mnw	*Mḥw*
ꜥnḫ.n.s-Mr.y-Rꜥ II	*Nj-k3w-Jzzj*
Wr-nw	*Rmnj/Mrwj*
Mr.f-nb.f	*Ḥzj*
Mrrj	*Sꜥnḫ-wj-Ptḥ*
Mrrw-k3	*K3-gmnj*

DAHSHUR
Jn-Snfrw-jšt.f

DESHASHA
*Jttj-Šdw*Deir el-Gebrawi

Hnḳw/Jj…f (II)	*Ḥm-r3/Jzj* (*I*)
Hnḳw/Ïttj (I)	

QUSEIR EL-AMARNA
Ḫw.n-wḫ

ZAWYET EL-AMWAT
Nj-ꜥnḫ-Ppjj

Mid Dynasty 6

MEIR
Ppj-ꜥnḫ-ḥrj-jb

DEIR EL-GEBRAWI
Jbj

EL-HAWAWISH
B3wj

NAGA-ED-DER
Mr.w/Jj.w

Late Dynasty 6 to Dynasty 8

Ppjj-ꜥnḫ/Ḥnj-km	Meir	VI.L
Ḥzjj-Mnw/Zzj	el-Hawawish	VI.L
Ḫw-ns	Qubbet el-Hawa	VI.L
Ḥnj/Špsj-pw-Mnw	el-Hawawish	VI.L
S3bnj I	Qubbet el-Hawa	VI.L
S3bnj II	Qubbet el-Hawa	VI.L
K3(.j)-ḥp/Ṯtj-jkr	el-Hawawish	VI.L
ꜥnḫ.t(j).f(j)	Moalla	VIII
Ḥwj/Ṯti-jkr	Gohaina	VIII

In most of these scenes the craft is clearly constructed of papyrus, but where no cordage binding the hull is depicted, the skiff may be wooden, as Harpur suggests.[735] This cannot be verified.[736] However, as thirteen of the fifteen scenes where the skiff lacks any sign of lashing are either fishing and fowling or papyrus pulling scenes (themes invariably set in the marshes), it is highly likely that the represented craft were papyrus skiffs, the craft best adapted to the wetland.

As might be expected, the skiff carrying the tomb owner is depicted in its most elaborate form. There is usually decking on which the tomb owner stands and the gunwales may be trimmed with cordage.[737] This craft undergoes a number of design modifications through the Old Kingdom, from having prow and stern curving upward to a flattened prow.[738] In early Dynasty 6 at Saqqara, the tomb owner's skiff becomes more streamlined with the hull elongated and a lower prow and stern closer to the surface of the water. This modification, pronounced in the tombs of *Mrrw-k3* and *K3-gmnj*,[739] is reversed later in Dynasty 6, when changes in the shape of skiffs go in the opposite direction with the side of the hull

---

735 Harpur (1977) p. 200. This interpretation is accepted by Woods (2008) 61-63.

736 Scenes where the tomb owner's skiff has no lashing: *Mrs-ꜥnḫ* III Dunham–Simpson (1974) fig. 4; *K3(.j)-ḫnt* (*Jwfj*) El-Khouli–Kanawati (1990) pl. 42 [b, B], 50-1; *Nj-ꜥnḫ-Ḫnmw* / *Ḫnmw-ḥtp* Moussa–Altenmüller (1977) pls. 4, 5, 74, figs. 5, 6; *3ḫtj-mrw-nswt* Smith (1946) fig. 194; *3ḫtj-ḥtp* Petrie–Murray (1952) pl.6 [lower]); *J3sn* Simpson (1980) fig. 30; *Nj-m3ꜥt-Rꜥ* Roth (1995) pl. 185; *Ḫww-wr* LD II 43 [a]; *K3.j- m- ꜥnḫ* Junker IV (1940) pl. 2, 11; *K3.j-m-nfrt* Simpson (1992) figs. 4-5, pls. 2 [a], 3, 4, 6 [a], pl. A; *Snḏm-jb/Mḥj* LD *Erg*. 11 [upper], 12 [upper] and Brovarski (2001) fig. 100-101, 102-3; *Sšm-nfr* IV Junker XI (1953) pl. 60, figs. 1, 16d.

737 These details are extensively analysed in Woods (2008) 61-63 and Appendix 3.

738 Woods (2008) 61-63.

739 Duell II (1938) pls. 127-129; Harpur–Scremin (2006) 7-28, 353-364, 491 [3].

appearing to deepen. Finally, in the southern provinces both the prow and stern turn upward at a virtual ninety-degree angle to the body of the craft, perhaps in response to the different nature of the Nile in these provinces, where the river and islands have rocky foreshores that would be hazardous to craft with elongated prow and stern.

12.6 The papyrus skiff in use: papyrus skiffs and rafts used by labourers[740] (*Figure 44*)

12.6.1 Historical development of images and themes

The distribution of scenes of papyrus skiffs and rafts used by workers in the list of tombs used in this study is as follows:

Mid to late Dynasty 4:	2 tombs
Early to mid Dynasty 5:	8 tombs
Late Dynasty 5:	36 tombs
Early Dynasty 6:	22 tombs
Mid Dynasty 6 to Dynasty 8:	10 tombs

Labourers and fishermen appear on skiffs in a variety of themes:

- Scenes of the tomb owner enjoying the pleasures of the marshland contain vignettes of men harpooning a hippopotamus or fishing from a much smaller skiff.
- In fishing scenes men fish from a skiff or papyrus raft with a hand-held net or control a dragnet.
- Hunters or porters are portrayed standing on skiffs laden with marshland products such as live birds contained in boxes, bundles of papyrus, baskets of fruit, lotus flowers and occasionally a large mammal which may be a calf or even a fully grown steer.
- Boatmen on groups of skiffs are depicted belabouring each other with staves and poles.
- One or two skiffs always accompany the herd of cattle swimming across a stretch of water.

These working craft are not all of the same shape or size; nor do they mirror the shape of the skiff carrying the owner of the tomb, when they appear in the same scene. The skiff may be a raft, that is, a papyrus craft with a prow but lacking a shaped stern.[741] This raft appears in tombs from mid Dynasty 5 to mid Dynasty 6 but was probably in use before and after the Old Kingdom. The skiffs depicted with both prow and stern shaped and raised above the water surface, vary in size from simple two-man craft to much larger transports. As the largest skiffs could carry a team of four or more men, together with produce that might include a large animal, they must have been very stable. Although this feature is never depicted, the larger working skiffs, like those carrying the tomb owner, would have needed flat, wide bottoms that would allow quite large animals to stand in them as well as men standing and even attacking each other with poles. Nevertheless, the incentive to be the first boat or team of boats home must have been very strong,[742] as there would always have been a chance of losing the skiff's cargo overboard in the heat of battle. This possibility may account for the reduced loads of products carried by the boats involved in the fighting.[743]

By mid Dynasty 6, labourers and fishermen portrayed on skiffs in provincial scenes are less likely to be depicted going about their business on extended registers devoted to marshland occupations. Instead, perhaps because the provincial tombs have less wall space than their earlier Memphite counterparts, scenes of boatmen jousting or returning with skiffs filled with marshland products or conducting a herd are sometimes reduced to vignettes on narrow registers grouped around the depiction of the tomb owner on his skiff, while depictions of craft returning with produce disappear in the tombs of late Dynasty 6 to Dynasty 8. This tendency may, of course, be due to the fact that marshland activities had less prominence in the middle and southern Upper Egyptian provinces. One use of the papyrus skiff, however, to conduct the fording cattle, continued to be portrayed into Dynasty 8.[744] Leading a herd of cattle out of the marshland was an essential annual task. There are at least 58 scenes of this activity,[745] which appear in tombs from Dynasty 3 to Dynasty 8 at Memphis and in Upper Egypt,[746] although the theme is given its greatest prominence in the tombs of Dynasty 5 and early Dynasty 6. Not all scenes show the herd swimming through water accompanied by one or two skiffs; some portray an earlier or later stage with a walking herder driving or leading the herd either into the water or back on to dry land. However, the implication is that the herds have to be conducted across a stretch of water, for which skiffs were needed.[747]

12.7 Rope and cord making

Analysis of surviving ancient Egyptian cordage has identified a range of plant and animal products used to manufacture rope of all types and diameters.[748] However, Old Kingdom scenes of rope making all show the use of the papyrus plant and the same manufacturing technique, the twisting of individual fibre yarns following by a

740 See TABLE 16: PAPYRUS SKIFFS – CONSTRUCTION AND USE.

741 See TABLE 16: PAPYRUS SKIFFS – CONSTRUCTION AND USE, in which the papyrus rafts are marked by "R".

742 Bolshakov (1993) 29-39.

743 The only boats with fighting crews to be depicted carrying loads equal to those seen on boats returning peacefully are in the tomb chapels of *3ḫtj-ḥtp*: Ziegler (1993) 80-1, 128, 131 and *Sḫm-k3*: Simpson (1980) fig. 4. In some cases the fighting boats carry no products and perhaps are protecting those with a cargo. See *Mrs-ʿnḫ* III Dunham–Simpson (1974) fig. 4, *Ḥzj* Kanawati–Abder-Raziq (1999) pl. 52 and *Nj-k3w-Jzzj* Kanawati–Abder-Raziq (2000) pl. 47.

744 See, for example, the tombs of *Mrjj-3ʿ* Kanawati (1995) pl. 37, 39-40); *W3hj* Kanawati (1995) pl. 20, 23; *B3wj* Kanawati VII (1987) fig. 18.

745 The Oxford Expedition to Egypt Database, Scene: 1.7, Cattle fording a canal or stream.

746 This theme is dealt with in this chapter: 12.8 Papyrus as cattle fodder, p. 115.

747 Two tombs actually portray the two stages in conducting the herd: *Rʿ-špss* LD II 60 and *Ḥzj* Kanawati–Abder-Raziq (1999) pl. 55.

748 Studies cite dom palm, halfa, flax and leather: Ryan (1987); Lucas–Harris (1999) 160-1; Singer et al (1954) 451-55; Teeter (1987) 71-77, pl. 7-9.

process of counter-twisting a number of yarns together to create a rope that will maintain its shape.

12.7.1 Historical development of scenes

The distribution of scenes of workers making rope in the tombs used in this study is as follows:

Mid Dynasty 5:	1 tomb
Late Dynasty 5:	8 tombs

Scenes of workers making rope are depicted in the following tombs:

Dynasty 5

Jj-mrjj	Giza	VM
ꜣḫtj-ḥtp (Louvre)	Saqqara	VL
Jntj	Deshasha	VL
Ptḥ-ḥtp (II)/*Ṯfj*	Saqqara	VL
Nfr (1)	Giza	VL
Sꜣ-jb	Giza	VL
Kꜣ.j-m-nfrt	Saqqara	VL
Kꜣ-ḥj.f	Giza	VL
Ṯjj	Saqqara	VL

Rope-making scenes, which are rare and date to the period mid to late Dynasty 5, are all Memphite except for one scene, that of *Jntj* at Deshasha.[749] The more extensive treatment of the theme shows the successive stages in the rope-making process. In the first stage two kneeling men face each other to stack lengths of fibre ready for manufacture. There are two steps in actually making the rope. Two men, one standing and the other seated on the ground, face each other and hold the ends of the fibre to twist it tightly in one direction. The final step involves two men, again facing each other, twisting a number of yarns of fibre together in the opposite direction.

According to Teeter, associated captions use technical terms to describe the stages of rope-making.[750] For example, the first stage is described in the scene of *Kꜣ.j-m-nfrt* as 'preparing the fibres' (*sšš smꜥ*).[751] The phrase 'twisting the fibres' (*nꜥt smꜥ*)[752] is used to describe the actual rope-making in the tombs of *ꜣḫtj-ḥtp, Kꜣ.j-m-nfrt* and *Kꜣ-ḥj.f*,[753] while the variation *nꜥt šsw n sp* in the scene of *Ptḥ-ḥtp* (II) (*Ṯfj*)[754] may be translated as 'twisting the ropes for boat-building' in reference to the scene of skiff construction on the register immediately below.

Cordage, even of many types and diameters, may seem a modest resource, but it played a vital role in the economy of an official's estate. Strong, flexible ropes were needed for a variety of purposes as metal wire or cable was not available. In Old Kingdom iconography cordage is shown in use for bird and fish traps, lashing for papyrus skiffs and rigging for boats powered by sails, to hang containers and jars, control domestic animals, and for many forms of haulage, but in all Old Kingdom scenes of rope-making the raw material consists of the outer husk of papyrus stalks.

12.8 Papyrus as cattle fodder

Portrayals of cattle swimming or wading through a stretch of water do not provide direct evidence that papyrus was used as cattle fodder, but in Dynasty 5, the scene of a herd of cattle wading or swimming through water accompanied by herders on skiffs becomes a frequent inclusion in the funerary repertoire of important officials.[755] It is argued in Chapter 4 'Domestic Cattle' that these scenes should be interpreted as bringing the herd out of the marshes to higher ground at the onset of the annual inundation and that the reason for taking the cattle into the wetland was to allow them to feed on its vegetation, notably papyrus,[756] during the hot summer months when estate pastures had dried off. This is portrayed as an arduous and dangerous activity, which would not have been undertaken unless the need for this source of cattle fodder was imperative. If this interpretation is correct, the papyrus here had a double utility providing both the means of transport for herders and nourishment for the herd.

We can only surmise how the herd was fed in the weeks between the return to higher ground and the appearance of new pasture following the subsiding of the flood water. Piles of papyrus stalks on skiffs returning from the marshes and rectangular bundles carried by herders leading animals in the parade of cattle suggest that quantities of papyrus were taken up to the higher ground to feed the cattle, but it is unlikely that the vast quantities of fodder needed by a large herd could have been transported. Perhaps herds were never of the number cited in tombs.

12.9 Summary

- The papyrus plant is a strong growing perennial that produces densely tangled thickets in marshy conditions.
- All parts of the papyrus plant could be used and served a great variety of working and everyday needs: as fuel, material for skiffs and rafts, to make cord and writing paper.
- Papyrus provided nutritious fodder for cattle herds, which were driven to the lower marshlands for the hot drier months. Piles of papyrus stalks on skiffs may represent fodder taken to feed the cattle when the

[749] Petrie (1898) pl. 5, Kanawati–McFarlane (1993) pl. 33.
[750] Teeter (1987) 75-76.
[751] Simpson (1992) fig. 6, pls. 3 [a, b], 5 [c], pl. A.
[752] Dunham (1935) 300-309.
[753] Respectively, Ziegler (1993) pp. 128, 131; Simpson (1992) fig. 6, pl. 3 [a, b], 5 [c], pl. A; Junker VI (1943) fig. 43.
[754] Paget–Pirie (1898) pl.32; Davies *Ptahhetep I* (1900) pl. 21, 25 [upper].

[755] The earliest surviving scene occurs in the Dynasty III tomb of *Ḥzjj-Rꜥ*: Quibell (1913) pl. vii, and the next in the late Dynasty 4 tomb of *Nb.j-m-ꜣḫtj*: LD II, 12b, where part of the scene has been lost due to the construction of a doorway. See TABLE 4: CATTLE CROSSING WATER.
[756] A vignette of an oxen eating papyrus appears in the tomb of *Sꜣ-jb*: Roth (1995) pl. 181.

herds returned from the wetlands to higher ground at the onset of the inundation.

- Although papyrus was not in itself a resource of high status, it contributed to the tomb owner's wealth as a means of acquiring and handling resources and manufacturing items useful for the work of the estate.

- The portrayed theme of workers gathering papyrus is attested from late Dynasty 4 to the second half of Dynasty 6, with the majority of scenes occurring in tombs of the second half of the Fifth Dynasty. It appears in both Memphite and provincial tombs.

- The theme of gathering papyrus is presented in two stages: plucking stalks and carrying the bundles of papyrus on men's backs. The very large stalks pictured being carried by workers were intended for the construction of skiffs and rafts.

- Marshland scenes include images of the construction and use of papyrus skiffs and rafts. These crafts, cheap and easy to construct with a shallow draught, were ideal for navigating the marshland waterways. They had a myriad of uses: for fishing, conveying marshland products and conducting herds of cattle through the waterways. Scenes of skiff construction are attested from early Dynasty 4 to late Dynasty 6 and appear in both Memphite and provincial tombs.

- Papyrus skiffs were used by the tomb owner and labourers. Scenes of papyrus skiffs in use extend from late Dynasty 4 to late Dynasty 6. The more elaborate skiff carrying the tomb owner undergoes design modifications through the Old Kingdom.

- Working craft carrying labourers and herders are much simpler, without decking. They vary in size and shape from a one-man raft to much larger craft carrying a team of men and produce.

- Scenes of jousting boatmen suggest that these larger craft were stable, with a wide flat bottom. Ample scenes of men returning with laden skiffs or conducting a herd across water begin to be reduced to minor vignettes in mid Dynasty 6.

- Old Kingdom scenes depict rope and cord making from papyrus and are clustered into the period of mid to late Dynasty 5. In the absence of metal wire or cable, cordage played a vital role in the estate's economy.

Figure 42: COLLECTING PAPYRUS FOR THE CONSTRUCTION OF A SKIFF, TOMB OF NEFER AND KAHAY, SAQQARA.
Line drawing courtesy of The Asutralian Centre for Egyptology.

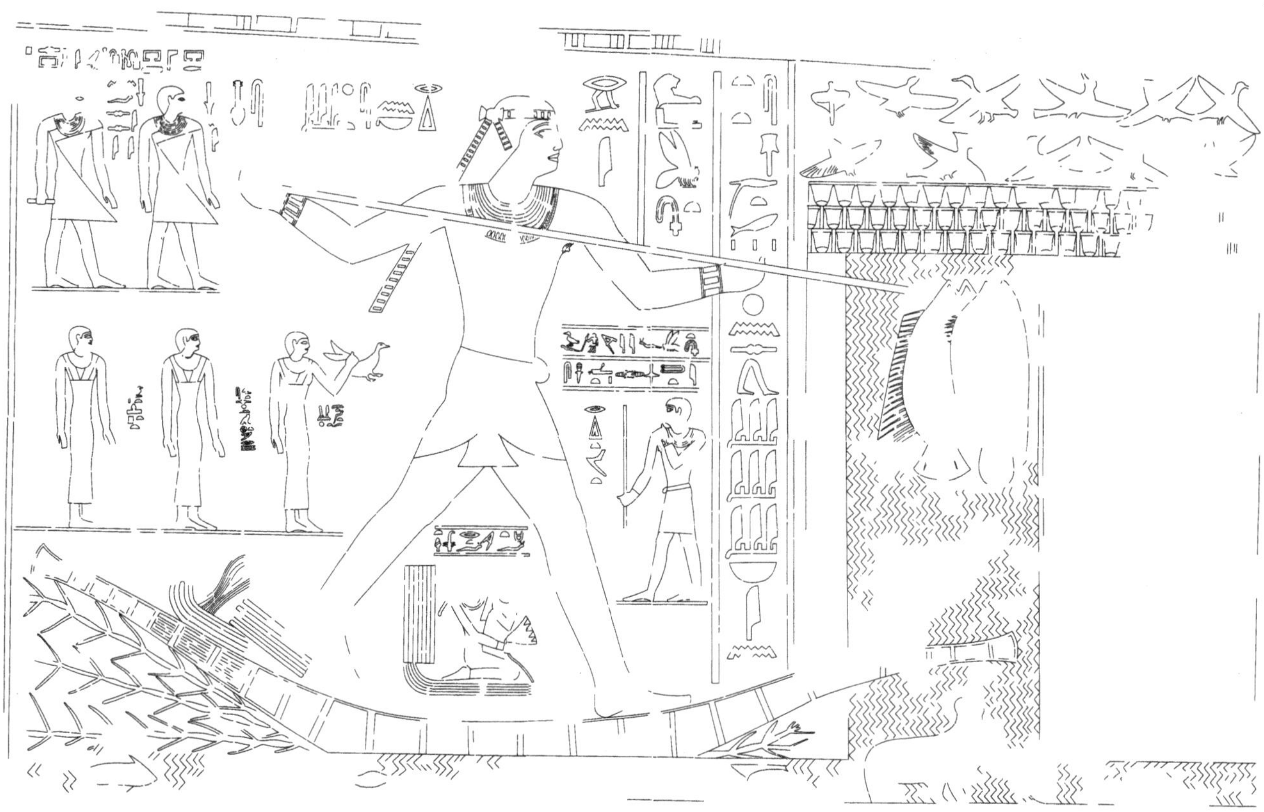

Figure 43: THE TOMB OWNER SPEARING FISH FROM A PAPYRUS RAFT, TOMB OF KHENI, EL-HAWAWISH.
Kanawati (1981) fig. 18.

Figure 44: WORKERS RETURNING WITH SKIFFS LADEN WITH MARSH PRODUCTS, TOMB OF NIKAUISESI, SAQQARA.
Kanawati–Abder-Raziq (2000) pl. 47.

Figure 45: THE TOMB OWNER VIEWING THE WORK ON HIS ESTATE, TOMB OF PEPIANKH THE MIDDLE, MEIR. Kanawati *Meir* (2011) pl. 84.

SECTION FIVE

THE RURAL ESTATE

CHAPTER 13 MANAGEMENT OF THE ESTATE

13.1 Focus of enquiry: managing a large estate

TABLE 17: TOMB OWNER VIEWING THE WORK AND PRODUCE OF HIS ESTATE

Data:

- The columns of inscription separating the figure of the tomb owner from the registers of estate scenes
- The portrayal of supervisors and their responsibilities.

Areas of research:

- The tomb owner's relationship to the scenes of the rural estate portrayed in his tomb
- The organization of the estate management

13.2 The relationship of the tomb owner to the estate as portrayed in his tomb

The pictorial and inscriptional data in tombs provide evidence that the tomb owners were beneficiaries of the resources of rural estates. However, the precise relationship of tomb owners to the rural estates portrayed in their tombs is unclear. In tomb inscriptions officials make general claims regarding the resources of the portrayed estate in terms that fail to establish the level of personal ownership that might be involved, whether an estate was held in usufruct from the crown, held in return for funerary and priestly duties or a stipend attached to an appointed position in the royal administration.[757] A further area of ignorance is whether an official was given the management of a royal estate from which it was designated that he should receive benefits. A small number of tomb inscriptions tell us:

- officials could inherit property.
- enjoy the benefits of landed properties.
- in some circumstances hand these benefits on to their heirs.[758]
- the king might bestow land and settlements (*njwwt*) on an official.[759]

Yet it is unclear whether tomb owners were in all cases the administrators of the estates that provided their benefits. For example, do depictions of personified domains represent individually owned or managed estates or do they imply a complex redistributive system from which officials benefitted? On the other hand, should Moreno Garcia's interpretation of domain figures as "représentent tant l'extension territoriale du pouvoir idéal d'un fonctionnaire (la Haute et la Basse Égypte, les *ḥwwt* et les *nwwt),* que la reconnaissance de son rôle bienfaisant dans le maintien de l'ordre, de la prospérité et de la stabilité" be accepted? [760]

Statements in autobiographies and extended inscriptions in tomb chapels, some of which accompany wall scenes, are no more enlightening. Of particular relevance is the question of the *pr-ḏt*. From Dynasty 4 to early Dynasty 6 tomb owners frequently describe themselves as 'viewing' the work 'of the *pr-ḏt*' or 'of his *pr-ḏt*'. '*pr-ḏt*' is usually translated as 'funerary estate' or after Perepelkin as 'estate'.[761]

Ḳ3r of Edfu

Ḳ3r was appointed by Merenre to Edfu as administrator (*ḥrj-tp ꜥ3 n sp3t*) of Upper Egypt Nome 3, an office requiring him to reside in his nome, where, according to his autobiography, the resources of his *pr-ḏt* were available to him. It is clear that he considered these resources to be his personal wealth as he stresses his ethical behaviour by stating that he used the resources of his *pr-ḏt* to aid others:[762]

(a) *jj.n(.j) n dmj(.j) m ḥ3t ḥrj-tp nb n Šmꜥw mj-ḳd.f*
'I came to my town ahead of any overlord of the entire South.'

(b) *jw ḫ3.n(.j) šmꜥw m pr(.j) n ḏt n ḥḳr gm.n(.j) (m) sp3t tn*
'I measured out the Upper Egyptian grain of my *pr ḏt* for the hungry man whom I found in this province.'

(c) *jr rmṯ nb gm.n(.j) m sp3t tn ṯ3bt nt kjj r.f jnk ḏb3 s(j) n nb.s m pr(.j) n ḏt*
'As for any man against whom I found in this province a debt of grain to another one, it was I who repaid it to its owner from my *pr n ḏt.*'

(d) *jnk ḳrs rmṯ nb n sp3t tn jwtj s3.f m ḥbz m jšt(.j) n ḏt*
'It was I who buried every man of this province who had no son with cloth from possessions of (my) *ḏt.*'

757 In the infrequent instances where the nature of the relationship between tomb owner and a parcel of landed property is made clear, such as those described in the tombs of *Ṯntj* Goedicke (1970) pp. 122f, pl. 13 and *Nj-k3-ꜥnḫ* Goedicke (1970) pp. 131ff, pl. 14, the land in question, a royal grant, was held in return for priestly and funerary services.

758 This may have been the implication in *Ṯntj*'s arrangements for his wife and the 'brother of the estate': Weeks, Noel (1983) 6-8.

759 As in the case of *K3.j-m-nfrt* Goedicke (1970) 44ff, pl. 5.

760 Moreno Garcia (1999) 117.

761 Perepelkin (1986).

762 Urk I, 251-5.

Comment:
(a)
Unlike statements associated with scenes in tomb chapels, the possessive adjective (*.j*), which in any case is understood, does not appear necessarily to indicate ownership of landed property. Rather, together with the use of *dmj* (not *njwt*) it seems to refer to the town in which *Ḳꜣr* settled and made his headquarters.

(b), (c) and (d)
Ḳꜣr's references to providing from his *pr(.j) n ḏt* and *jšt(.j) n ḏt* for people in trouble indicate that he had an entitlement to an income from one or more estates. However, his legal relationship to such estate property is not clear. It is uncertain whether his *pr-ḏt* was held in return for royal or priestly duties.

Ḳꜣr's titles included:
(from his architrave) *smr wꜥtj, ḥrj-tp [ꜥꜣ] n spꜣt, smr wꜥtj, ḫrj-ḥbt, jmj-rꜣ ḫnt(w)-š pr-ꜥꜣ, jmj-jb n nswt.*
sš nswt, sꜣb ꜥḏ-mr, nj nst ḫntjt, ḥrj-tp nswt pr-ꜥꜣ mꜣꜥ, smr wꜥtj, ḥrj-sštꜣ n mdt nb(t) štꜣt jwt m r-ꜥꜣ n ꜣbw, jmj-rꜣ wp(w)t nb(t) nt nswt.

Finally, according to his false door inscription he was promoted to *ḥꜣtj-ꜥ, jmj-rꜣ Šmꜥw, ḥrj-sštꜣ n pr dwꜣt.*

None of these titles suggests a tie with any local temple from which he might have received an income. A temple to Horus may have existed at this time at Edfu. Although none of its remains have been found, there was a tradition that the Third Dynasty vizier, Imhotep, designed the first stone temple at Edfu, which was dedicated to Horus, Hathor of Dendera, and their son, Herumatawy. It is likely that as the senior official resident at Edfu *Ḳꜣr* either received a stipend in kind from the crown[763] or, as he had been sent to Edfu by the crown, more likely that that he held estate/s in usufruct from the crown and were part of the system of remuneration known as the *pr-ḏt*.

Jbj of Deir el-Gebrawi

Jbj provides evidence that landed property could be owned and inherited:

jr.n(.j) nn m njwwt nt ḏt(.j) m wꜥb m ḥtp dj nswt rdj.n n(.j) ḥm n nb(.j) [r] jrt n(.j) ... nn ḫr mrt nt ḏt(.j) mḥ m kꜣw m ꜥnḫt m ꜥꜣw m jr[t.n.j m ḫpš.j ḏs.j] wp(w)-ḥr ḫt jt(.j) sk w(.j) m ḥkꜣ ḥwt nt pr-šnꜥ nt ꜣst stꜣt 203 rdj[t.n n.j] ḥm.n nb(.j) r snḫt(.j)[764]

'I have made this from the towns of my private estate in a pure manner, and from the royal offerings which the majesty of my lord has given to me, in order to do for me ... this stocked with *mrt* servants of my private estate and filled with oxen, goats and donkeys as things which I acquired by my own means, apart from the property of my father, when I was chief of the estate of the storehouse/labour establishment, and of the 203 arouras of land which the majesty of my lord has given to me in order to strengthen/enrich me.'

Comment:
Kanawati translates *ḏt(.j)* as 'my private estate' and Moreno Garcia's inserts *jr[t.n.j m ḫpš.j ḏs.j]* as 'which I acquired by my own means'. While Kanawati questions whether this insertion matches the available space,[765] *Jbj* is making a distinction between the property inherited from his father and 203 arouras bestowed by the king, which may have been part of his *pr-ḏt* and was probably bestowed on him in connection with his position as administrator of the province {*r snḫt(.j)*}. *Jbj*'s earlier statements (up to the words 'apart from **the property of my father**,') seem to apply to property conferred on him by the crown.

The above inscriptions are evidence that there was more than one way of holding rural land. *Ḳꜣr*'s claims of providing help for the unfortunate from his *pr-ḏt* make it clear that officials regarded the resources of their *pr-ḏt* as personal property.

Vertical captions relating to scenes

Columns of inscriptions frequently separate the figure of the tomb owner from a panel of registers portraying rural activities. In these columns the tomb owner is said to be 'viewing' (*mꜣꜣ*) either:

- the presentations/offerings (*nḏ.t-ḥr/prt-ḫrw*) brought by/from the settlements (*njwwt*) of the *pr-ḏt*

or

- the activities and work carried out on the settlements (*njwwt*)[766] of the *pr-ḏt.*

A small number of tomb owners very closely associated with the crown refer to 'viewing' offerings from the Residence but only in the case of *Nbt* (Munro, (1993) pl. 13) does the related panel of registers contain scenes of a rural estate.[767]

In a majority of panels the tomb owner is stated to be 'viewing' (*mꜣꜣ*) certain rural activities depicted on the registers or offerings either as a parade of livestock or items carried by offering bearers. The terms describing these scenes fall into three categories:

- terms descriptive of the offerings or of the activities being viewed
- terms descriptive of the source of the offerings or activities being viewed
- terms denoting the geographical location from which the offerings derive or denoting the location of the activities being viewed.

[763] As probable son of *Jzj*, a previous governor of Edfu, *Ḳꜣr* may have inherited privately held property but there is no evidence of this.
[764] Moreno García (1997) 29.

[765] Kanawati *Gebrawi* II (2007) 53, note 309.
[766] When the term *ḥwwt* (estate) is included, the tomb owner is normally described as 'viewing' presentations or offerings. The only instance when *ḥwwt* refers to 'viewing' activities and work is in the tomb of *Jḥjj* (r/u *Sšsšt/Jdwt*). See TABLE 17: TOMB OWNER VIEWING THE WORK AND PRODUCE OF HIS ESTATE.
[767] *Mr-jb.j* LD II 22a,b; *Nbt* Munro (1993) pl. 13, 14; *Mrrw-kꜣ* Duell I (1938) pl. 96-99 and Duell II (1938) pl. 113-116; *Kꜣ-gmnj* von Bissing II (1911) pl. 37-39, 41.

Terms denoting offerings or activities being viewed

nḏ.t-ḥr (Wb 2.373.1-10) (gift(s)
pr-ḫrw (Wb 1.529.7-530.5)
prt-ḫrw (Wb 1.528.11) (invocation offering)
pḫr(t) (Wb 1.548.11-12) (reversionary offerings)

When these terms are used to describe the scenes that the tomb owner is depicted viewing, the themes on the corresponding panel of registers usually consist of offerings: animal parades, offering bearers including figures of domains, the recording by scribes and the presentation of accounts to the tomb owner. These terms rarely relate to registers with scenes of estate work, although there are a few exceptions:

- *Sḫm-kꜣ-Rꜥ,* who uses *jnw* ('gifts') instead of *nḏ.t-ḥr*, is depicted viewing one register of offerings and two registers of marshland activities with the statement 'Viewing the gifts of his chiefs and herdsmen, his fowlers and fishermen, which are brought from the settlements of (his *pr-ḏt*) for ...' *(mꜣ(ꜣ) jnw n ḥkꜣw(.f) njwwt.f wḥꜥw.f jnnt m njwwt pr(-ḏt(.f) n).*[768]
- On the south wall of his chapel *Ptḥ-ḥtp* (II) (*Ṯfj*) uses the term *nḏ.t-ḥr* although the corresponding registers contain scenes of young men training or playing games, hunters and bearers returning from the hunt and cattle husbandry, as well as three registers of the animal parade.[769]
- *Sšm-nfr* IV has two registers of harvest activities coupled with the statement 'Viewing the presentations which are brought ... from Lower and Upper Egypt by' (*(m)ꜣꜣ nḏ.t-ḥr jnnt ... Tꜣ-mḥw Šmꜥw jn ..*).[770]

While the three terms, *nḏ.t-ḥr, prt-ḫrw* and *pḫr(t)* continue to appear in other contexts, from mid Dynasty 6 these terms denoting offerings from the *pr-ḏt* system are rarely used. Only *nḏ.t-ḥr* appears, and only on one occasion in a 'viewing' inscription in the tomb of the Meir administrator, *Ppj-ꜥnḫ/Ḥnj-km.*[771] In the provinces of Upper Egypt in Dynasty 6, column inscriptions beginning *mꜣꜣ* much more frequently accompany panels of estate activities and omit altogether the terms referring to offerings. The absence of the terms appears to be associated with the corresponding absence of the term *pr-ḏt.*

Terms denoting the source of the offerings or activities being viewed

The terms most frequently used are:
njwwt (njw.t): Wb 2, 210.6-212.4 (community, settlement, town, city)
ḥwwt (ḥw.t): Wb 3, 1.4-3.5 (larger) house; estate (administrative unit); temple ("mansion")
pr-ḏ.t (Wb 5, 510.5) (estate; funerary estate)

njwwt / njwt

According to TABLE 17: TOMB OWNER VIEWING THE WORK AND PRODUCE OF HIS ESTATE, the term *njwwt* is in frequent use in the column of inscriptions from early Dynasty 5 to early Dynasty 6:

Dynasty 4	in two tombs
Dynasty 5.E-M	in eight tombs
Dynasty 5.L	in twenty-six tombs
Dynasty 6.E	in ten tombs
Dynasty 6.M – Dynasty 8	in five tombs

After early Dynasty 6 the term only appears in: *Bꜣwj* (BA 48) of el-Hawawish, *Ppj-ꜥnḫ-ḥrj-jb* and *Ppj-ꜥnḫ/Ḥnj-km,* the powerful administrators of Meir.

The preponderance of use of the term *njwwt* or *njwwt.f* is confined to the tomb owner viewing parades of animals and offerings. Of the 51 vertical inscriptions containing the term *njwwt,* 37 refer to panels that only contain depictions of animal parades, offerings and offering bearers and scribes making records. When the tomb owner views a panel of registers depicting a mixture of offerings and rural estate activities, the term *njwwt* may be used as the source of 'presentations'. There are only two exceptions where the term *njwwt* is used to denote the source of a panel of depictions exclusively of estate activities.[772]

Although the term *njwt* is regularly translated as 'town, Dorf, Stadt', Parlebas insists that this translation provides too definite a meaning, as the ideogram Gardiner 03 is regularly used as a determinative of localities of varying size from small communities inhabiting mounds above flood level to the royal palace and the capital.[773] Used to describe the source of offerings for the deceased, the plural term *njwwt* is better translated as 'communities'. The reference must be to localities that are rural as they supply an assortment of live animals and agricultural products and thus would be communities with small populations. These communities are frequently described as 'of the *pr-ḏt*' (*nt pr-ḏt*), so it has to be assumed that they were committed to supplying provisions for the tomb owner. Whether they may be considered as part of his property depends on the meaning attached to the term 'of the *pr-ḏt*' (*nt pr-ḏt*).

ḥwwt / ḥwt

The term *ḥwwt* is also cited as a source of offerings, although only by the highest officials and members of the royal family. In the list of vertical inscriptions used in this study the term does not appear until towards the end of Dynasty 5, is not seen after early Dynasty 6, and is not used in provincial tombs.

768 Hassan IV (1943) fig. 58; LD II, 42.
769 Paget–Pirie (1898) pl. 31, 33.
770 The remaining registers have been lost: Junker XI (1953) fig. 74b.
771 Blackman–Apted *Meir* V (1953) pl. 29.
772 *Jj-mrjj* Weeks (1994) fig. 40 and *Jtj-sn* Hassan V (1944) fig. 122-23.
773 This comment refers to the many tomb reports in which this translation is given: Parlebas (1983) 199.

Tombs with vertical inscriptions containing the term '*ḥwwt*'

Late Dynasty 5

Jḥjj (r/u *Sšsšt/Jdwt*)	*Rʿ-špss*
Ptḥ-ḥtp (LS 31)	*Ḫnwt*
Ptḥ-ḥtp (II) (*Ṯfj*)	*Sndm-jb/Mḥj*
Nbt	

Early Dynasty 6[774]

Ptḥ-špss II	*Mrrw-k3*
Mrjj-Ttj/Mrj	*Mḥw*
Mrw/ttj-snb	*S3bw/Jbbj*
Mrrj	*K3-gmnj*

According to Moreno Garcia the term *ḥwt* does not refer to open land but to a rural unit of the crown with economic, administrative and defensive functions. The headquarters of the *ḥwwt* were consequently buildings and, linked with the *prw-šnʿ* which were centres for the storage and management of farm products, they played an important role in the rural milieu. Moreno Garcia points out that they were authorised to provide food from their storage to the expedition led by *Ḥr-ḫw.f*. Moreno Garcia's conclusion is that the *ḥwt* was never alienated from the crown to be bestowed as private property, although it might be attached to the funerary service of an individual. Its essential role was to support the crown's control over a region. It also functioned as a residence for the district administrator, a centre for the organization of local public works and for the storage and provision of goods and livestock needed in the crown's service.[775]

Moreno Garcia also notes that the *ḥwt* as an administrative unit was established in the late Fifth Dynasty as successor to the earlier and much larger *ḥwt ʿ3t*. This is supported by the data from the vertical inscriptions. The earliest official to introduce the term *ḥwwt* in his vertical inscriptions is *Rʿ-špss*, whose tomb dates to the reign of Djedefre:

m33 ḥsb ḥk3w ḥwwt njwwt.f T3-Mḥw Šmʿw[776]
'Viewing the account of the chiefs of his estates and towns of Lower and Upper Egypt'

However, Moreno Garcia also states that the *ḥwt* was never included in a *pr-ḏt*.[777] Yet the phrase *ḥwwt.f njwwt.f n(t) pr-ḏt* is occasionally used. *Ptḥ-ḥtp* (II) (*Ṯfj*), for example, seems to connect *ḥwwt* with the *pr-ḏt* on four occasions:

m33 jw3 jnw ḏḥwtt m ḥwwt ḥwwt-k3 nt Mḥw Šmʿw pr-n-ḏt
'Viewing the cattle, the contribution of the feast of Thoth from the estates and temples of Upper and Lower Egypt, of the *pr-n-ḏt*'[778]

m33 nḏ.t-ḥr jnw n ḥwwt njwwt Mḥw Šmʿw pr-ḏt n ...
'Viewing the offering brought by the estates and towns of Lower and Upper Egypt of the *pr-ḏt* for'[779]

m33 sjptj n jw3 n mḏwt nt ḥwwt ḥwwt-k3 nt pr-n-ḏt
'Viewing the counting of the cattle of the stalls of the estates and temples of the *pr-ḏt*.'[780]

m33 nḏ.t-ḥr 3pdw jnw n ḥwwt ḥwwt-k3 nt Mḥw Šmʿw pr-n-ḏt
'Viewing the presentation (of) fowl that is brought by the estates and temples of Upper and Lower Egypt, of the *pr-ḏt*.'[781]

The association of the two terms, *ḥwwt* and *pr-ḏt*, also appears in the tombs of *Ptḥ-ḥtp* (PM 653 LS 31), *Nbt* (PM 624), *Ḫnwt* (PM 623), *S3bw/Jbbj* (PM 460 E 1,2 H 3) and *K3-gmnj* (PM 521).

The apparent contradiction between Moreno Garcia's conclusions and the evidence of the column inscriptions may be resolved by the hypothesis that while the *ḥwt* was a crown estate it also provided a stipend, which the official appointed to administer it considered as his *pr-ḏt*.

The *pr-ḏt*

Interpretation of the term *pr-ḏt* is a complex issue that has been considered at length by a number of scholars.[782] There is general agreement that it relates to a system whereby officials were granted access to resources, but the exact nature of the system is not clarified by investigation of the columns of text that separate the figure of the tomb owner from the panel of registers that he 'views'.

According to TABLE 17: TOMB OWNER VIEWING THE WORK AND PRODUCE OF HIS ESTATE, the term *pr-ḏt* disappears from the columns of inscriptions, probably early in the reign of Pepy I. The provincial officials of mid to late Dynasty 6 do not refer to the source of what they 'view'. However, two provincial officials of Dynasty 8, *W3hj* and *Mrjj-ʿ3* of el-Hagarsa, refer to livestock of their own property (*ḏs.f*). The implication may be that the *pr-ḏt* system was no longer applied in the provinces.

In general, the use of the discussed terms in the Fifth and Early Sixth Dynasty Memphite tombs and their disappearance from inscriptions that begin '*m33*' in the second half of the Sixth Dynasty in the provinces suggest that a different set of conditions governing the tomb owner's relations with the resources of his estates came to prevail in the provinces from mid Dynasty 6 on. These new conditions may have been connected with an abandonment of the *pr-ḏt* system or a recognition

774 The term *ḥwwt* also appears in the tombs of *Nj-k3w-Jzzj* Kanawati–Abder-Raziq (2000) pl. 58, 59; *Ḥzj* r/u *Sšm-nfr* Kanawati–Abder-Raziq (1999) pl. 62; *Nḏt-m-pt*, mother of *Mrrw-k3/Mrj* Kanawati–Hassan (1996) pl. 41, 42 but in horizontal captions above lines of offering bearers.

775 Moreno Garcia (1999) passim.

776 LD 2 62.

777 Moreno García (1999) 217.

778 Paget–Pirie (1898) pl. 31-33.

779 Paget–Pirie (1898) pl. 31-33.

780 Paget–Pirie (1898) pl. 31-33.

781 Paget–Pirie (1898) pl. 31-33.

782 An in-depth investigation of the term is beyond the scope of the present study.

that it did not suit the circumstances of provincial administrators.

13.3 Management of estates

The range of titles of figures depicted carrying out various estate activities suggests that each resource was handled by a hierarchy of workers with specialised skills.

Cattle management: associated titles		*Jones no.*[783]
jmj-r3 mḏt	(overseer of the cattle stall)	573, 574
jmj-r3 jt jḥ	(overseer of grain and cattle)	318
jmj-r3 jḥw	(overseer of cattle)[784]	286
jmj-r3 ꜥb nb	(overseer of all cattle)	338
jmj-r3 ṯzwt	(overseer of herds)	992
mnjw	(herdsman)	1589
nr	(custodian of the cattle herd)	1812

A hierarchy of cattlemen is depicted in many tombs. Figures portrayed as overseers in scenes of husbandry and leading animals in a procession are depicted as older men who are using their experience to supervise the work performed by farm hands. If portrayed supervising the birth of a calf, they are shown standing in front of the parturient cow to check on the animal's demeanour, as in the scene in the tomb of *J3sn*[785] where a figure wearing the fringed kilt of a supervisor holds out a hand to the animal as it bellows in pain.[786] Another supervisory role, which is depicted in the parade of livestock, is of a figure following the ox using a staff to direct or control the animal. Again, the implication is that the supervisor's experience is needed to deal with the situation. Figures portrayed supervising these activities are not always given a title but their level of responsibility is indicated by their dress or stance or a caption giving their speech. In *Jj-mrjj*, for example, on the highest register of a panel in Room 2, two overseers are depicted although given no title.[787] The supervisor standing to the left watches two workers pounding a mat; on the far right a seated figure is offered food or drink from a shallow bowl. On the bottom register of this panel a figure holding the staff of an overseer supervises the birth of a calf. Supervisors may be depicted using a staff to deal with a situation that requires attention. In the scene in the vestibule of *Nj-ꜥnḫ-Ḫnmw/Ḫnmw-ḥtp* two figures wearing the fringed kilt of an overseer each use a staff to control cattle being led towards the tomb owners. The caption has been lost.[788] This concept of an overseer standing at the rear of an animal to control it is reproduced on the two lowest registers in *Rꜥ-Špss*[789] and again in *Sšm-nfr* III.[790] In *Ḥtp-ḥr-3ḫtj* a supervisor, defined by the style of his kilt, uses both hands to guide the oxen from the rear. This supervisor is clearly shown to be an older man, balding and with a thickening figure.[791]

Ordinary herders depicted managing cattle or leading animals in the parade of livestock are usually anonymous figures without titles. Nevertheless they are portrayed having specialist knowledge and expertise. They are shown delivering calves, milking cows, controlling fighting bulls and leading the herd across a stretch of water to higher ground. Older men are depicted handling calves, perhaps to judge which bull calves were to be stall fed as *jw3.w*. Due care of cattle by herders is emphasised in an unusual vignette in the tomb of *3ḫtj-ḥtp* where a herder lassoes the rear legs of a *ng3w* ox that is showing an aggressive interest in a very young calf.[792] A particular class of cattlemen, identified as 'marsh dwellers' by Brovarski,[793] are sometimes depicted with a beard and hair grown longer than usual. These men may have been true residents of the marshland. One such bearded figure appears in a vignette in the tomb of *M3-nfr* where he appears to be coaxing a dappled ox to lie down, presumably as a prelude to hand feeding the animal.[794] Most herders, however, are depicted clean shaven with short hair wearing a brief kilt which as no apron and is often tucked into the belt.

Pictorial evidence suggests that both the herders and their overseers were specialist workers who lived with their herd. In the tomb of *Jj-mrjj* the topmost register of the panel of registers portraying aspects of cattle husbandry suggests this association with vignettes of herders making mats, baking bread and roasting a fowl, while a senior herder sits on the ground reclining against a wicker. Above the roasting fowl is a subregister depicting a group of containers and a looped mat.[795] A scene of cattle fording a stretch of water in the tomb of *Nfr* and *K3-ḥ3.j* portrays herders with mats looped over their shoulder, again suggesting that the herders lived with their charges when the cattle were in the marshland as these mats were presumably used as bedding[796]

Field crops: workforce

Associated titles		*Jones no.*
jmj-r3 šnwt	(overseer of the granary)	916
nḫt-ḫrw n šnwt nt ḫnw	(crier of the granary)	1827
sš šnwt	(scribe of the granary)	3203
ḫrj-tp šnwt	(granary attendant)	2892
djwt	(grain winnowers/sievers)	3728
jrj md3(w)t	(letter carrier/scribal assistant)	1167
jrj ḫt	(custodian of property)	1200
ḫrp ḫ3(jw)	(controller of measurements)	2670

The workforce depicted in the theme of grain production consists of four groups: labourers preparing the ground and planting the crop, harvesters, workers dealing with the harvest and those storing the crop.[797]

783 Jones (2001).
784 This title is sometimes made more specific as *jmj-r3 jḥw n ṯntt* (overseer of the Thenet cattle) or *jmj-r3 jḥw Dḥwty* (overseer of the cattle of the Ibis-nome).
785 Simpson (1980) fig. 30.
786 Evans (2006) Appendix.
787 The eastern section of the south wall of Room 2: Weeks (1994) fig. 34.
788 Moussa–Altenmüller (1977) fig. 17.
789 LD *Erg*. 41.
790 Junker III (1938) fig. 8b.
791 Mohr (1943) fig. 35.
792 Davies *Ptahhetep II* (1901) pl. 22.
793 Brovarski (2001) 44 note 204.
794 LD II 66.
795 Weeks (1994) fig. 34.
796 Moussa–Altenmüller, (1971) pl. 5.
797 Cultivation of the growing crop is never portrayed.

Farm labourers

Labourers are portrayed carrying out every stage of the operation from planting crops to storage of the harvest. Women are often portrayed winnowing the crop. More so than in the case of cattle herders, the figures of the farm workers who till the soil, reap and transport the harvest, are ciphers lacking any descriptive label or name. Captions either designate the work being undertaken or give the speech of labourers as they urge each other to greater effort.

Workers in scenes of ploughing and planting seed are depicted engaged in an organized, defined series of actions. One worker handles the plough while a second figure drives forward the team of oxen. The figure sowing the seed may hold out a handful of grain to encourage the flock of sheep to follow him. The sheep are controlled by a team of herders, armed with a whip and a staff. These workers are not given occupational titles beyond 'team/s of the *pr-ḏt*' (*jz.wt pr-ḏt*). In general, the field hands are anonymous ciphers, figures representing this particular phase in the agricultural sequence.

This remains generally true of harvesting scenes despite the inclusion of vignettes and captions showing that the workers were regarded as more than mere ciphers. Harvesters may be depicted taking a drink,[798] or being entertained by a flute player or piper.[799] Inscriptions suggest that harvesters were required to guarantee their efficiency.[800] Occasionally a distinction is made between the teams of harvesters. Captions in the provincial tombs of *Wr-jr.n.j* of Sheikh Said and *Ḫw-ns* of Zawyet el-Maiyetin specify harvesting by teams described as *jz.wt nt pr-ḏt, jz wr* and *ḥmw nswt*. These labels have been discussed by Vandier, Perepelkin and Siebels. Vandier suggests that the reapers called 'servants of the king' either were allotted to tomb owners or that reapers and the land they worked on were crown property administered by a landowner. [801] Perepelkin argues that the tomb owners controlled and therefore could make use of the crown's workforce in the districts they administered.[802] Siebel's view is that the three labels (*jz pr-ḏt, jz wr, ḥmw nswt*) denote that the harvest yield was to be divided three ways, for the tomb owner's mortuary cult, his household and the taxes to be paid to the crown.[803] The more frequent description of harvesters, however, is to describe them as teams of the *pr-ḏt*. The inclusion of this term *pr-ḏt* in harvest scenes in the tombs of *Wr-jr.n.j, Nfr* (1), *Rᶜ-m-k3.j, Ḫww-wr, Ḫw-ns, Spd-ḥtp, Sḫm-k3j* and *K3.j-m-nfrt*,[804] coincides with the use of the term in the vertical inscriptions. In particular, *Wr-jr.n.j, Nfr* (1), *Ḫw-ns* and *Sḫm-k3j* describe themselves 'viewing' the harvesting carried out in the *pr-ḏt*.[805] Women workers, shown winnowing and sieving, are sometimes given the label *djwt* ('fivers/field workers').[806]

Overseers of farm workers

Scenes of harvesting flax and grains may be portrayed with the figure of an overseer standing to one side but the role of these figures supervising the work of the harvesters has to be assumed from their stance – usually depicted leaning on a staff – and their fringed kilt, as they are rarely given a title.[807] Overseers are usually anonymous. The only significant exceptions to this anonymity are the minor officials in the grain storage scenes where the care needed in both storing grain and issuing it from the granaries is stressed. The minor officials in these scenes have a variety of roles and titles. While the labourers who carry the baskets of grain to the granaries, pour their contents into the granary or remove a measured quantity are anonymous, the responsible officials are frequently titled. They may be labelled *sš* or *sš šnwt* ('scribe of the granary'),[808] *nḫt-ḫrw šnwt* ('crier of the granary'), *jrj md3t* ('letter carrier'/'scribal/ assistant') or *jrj ḫt* ('custodian of property') or *ḫrp ḥ3(jw)* ('controller of measurements'). The most important figure in this theme is the steward (*jmj-r3 pr*), who is sometimes provided with an assistant titled *rḫt*.[809] The *jmj-r3 pr* clearly had the final responsibility for the correct accounting of the harvest as he is depicted presenting the accounts to the tomb owner.[810]

***jmj-r3 pr* ('steward') and *ḏ3ḏ3t nt pr-ḏt* ('assessors of the *pr-ḏt*')**[811]

The portrayal of a range of minor officials and workers engaged in tasks connected to the theme of storing grain in granaries presents a scene of scrupulous supervision and recording of the grain harvest. This concept is further represented, for example, in the tomb of *Ṯjj* by the depiction of a group of officials, described as *ḏ3ḏ3t nt pr-ḏt*, approaching the threshing scene, emphasising the need to record and preserve the grain harvest. Loss due to theft or careless handling would have been most likely after the grain had been threshed and winnowed.

The hierarchical nature of the chain of responsibility is exemplified in the presence of overseers at important stages in harvesting the crops and culminates in the

[798] Moussa–Junge, (1975) fig. 4, Mohr (1943) fig. 47, Davies (1891) pl. 16.

[799] Mohr (1943) fig. 48; Sharawi–Harpur (1988); van de Walle (1978) pl. 11; Junker *Gîza* VI (1943) fig. 43 facing p. 136; Wild III [2] (1966) pls 136 [A], 152-53; Kanawati–Abder-Raziq (1999) pl. 52; Junker VI (1943) fig. 43.

[800] See Chapter 8, Field Crops – Cereals: 8.5 Inscriptional data, p. 76.

[801] Vandier VI (1952) 111.

[802] Perepelkin (1986) 165-166.

[803] Siebels (2000) 181 and note 144.

[804] Respectively, Davies *Sheikh Saïd* (1901) pl. 16; Junker VI (1943) fig. 14, 17; Hayes I (1953) fig. 57; LD II 43 [a]; LD II 106 [b], 107; Harpur (1986) figs. 1-7; Simpson (1980) fig. 4; Simpson (1992) fig. 15, pls. F. G.

[805] See TABLE 17: TOMB OWNER VIEWING THE WORK AND PRODUCTS OF THE ESTATE.

[806] *djwt* 'field labourer': *Wb.* 5, p. 421,1. *K3-ḥj.f* Junker VI (1943) fig. 47.

[807] *Wr-jr.n-Ptḥ* James (1961) pl. 21, *Pḥn-wj-k3.j* LD II 47, *Nfr* (1) Junker IV (1940) fig. 14, *Nfr-sšm-Ptḥ/Sḫntjw* Moussa–Junge (1975) fig. 4b, *Sšm-nfr* IV Junker XI (1953) fig. 74a.

[808] For this and following titles see above: Field crops: workforce.

[809] Siebels (2000) 410 and Note 40 translates *rḫt* as 'he who keeps the numbers'.

[810] *Mrs-ᶜnḫ* III Dunham–Simpson (1974) fig. 4, *Nj-ᶜnḫ-Ḫnmw/Ḫnmw-ḥtp* Moussa–Altenmüller (1977) fig. 24.

[811] Épron I (1939) pl. 150; Steindorff (1913) pl. 122.

portrayal of the grain storage scene. It is further emphasised by association with a theme usually called 'the rendering of accounts', in which a number of men are brought before scribes. The men, often depicted as elderly, balding and with sagging stomach muscles, are being presented to scribes. On a register immediately above a scene of filling granaries in the tomb of *Nj-ꜥnḫ-H̱nmw/H̱nmw-ḥtp* four elderly men titled 'chiefs of the towns of Lower Egypt' present themselves to an *jmj-rꜣ pr*. Above the scene is written *sš m jt tnt ḫꜣ bdt tnt šnt jt* 'Reporting (on) the grain. A payment of 1000 (sacks) of wheat and 100 (sacks) of barley.'

The marshlands

Scenes set in the marshlands also demonstrate a hierarchy of responsibility. Figures of overseers appear in the following themes:

- Fishing. An overseer is sometimes portrayed controlling the two teams of men who haul in the dragnet. See Table 13: Marshland Resources: Fish.
- In two rare scenes the catch of fish is shown being checked and registered before officials described as *ḏꜣḏꜣt nt pr-ḏt*.[812]
- Trapping wildfowl. The figure of an overseer is depicted controlling the operation of the clapnet. He is frequently shown giving the signal to close the net. See Table 14 Marshland Resources: Birds: Trapping, Poultry-yards, Parades'.

Rendering of accounts

This theme portrays elderly village chiefs, sometimes titled *ḥkꜣ* or *ḥkꜣ njwwt* or *ḥkꜣ ḥwt*, brought in a line before scribes of the estate. The purpose of their appearance before the scribes is to face an examination on the accounts and records of the resource for which they are responsible. These scenes are set in the context of grain storage and the presentation of livestock so it may be assumed that all resources had to be accounted for. This responsibility seems to have become more onerous over time. The scenes of mid Dynasty 5 present the village chiefs in a standing posture but bowing in deference and subservience, as in the portrayal in the tomb of *Nj-ꜥnḫ-H̱nmw/H̱nmw-ḥtp*.[813] By the end of the dynasty, however, the mood of the scene occasionally becomes more severe with the chiefs depicted grovelling, sometimes being hauled forward by guards. In the tomb of *Sndm-jb/Mḥj* the village chiefs are portrayed being led before the scribes by armed guards holding them by the scruff of the neck, forcing them to bend parallel with the ground.[814] By early Dynasty 6 physical punishment is depicted with miscreant chiefs tied to a post receiving a beating or fettered, awaiting punishment. In the scene in the tomb of *Mrrw-kꜣ/Mrj* three chiefs are in a kneeling position, three more are being forced into this position by armed guards and one, held down by force to a post, is being beaten by another man.[815] In the slightly later tomb of *H̱ntj-kꜣ* five estate chiefs are brought before the tomb owner, two of whom are forced into a crawling position. Behind them two other men, presumably also chiefs, are pictured seated on the ground and tied to a post to be beaten by two other men wielding rods.[816] It is difficult to account for this accretion of the portrayal of violence. Perhaps there were years of poor harvest with the consequent temptation to steal grain from the estate, or these high officials wished to demonstrate their authority and power.

The role of the tomb owner

Beyond the reception of offerings and accounts, the tomb owner is never portrayed taking an active role in management of his estate.[817] In the column of inscription his activity is described as '*mꜣꜣ*', usually translated as 'viewing'; perhaps a more apt translation would be 'inspecting'. In his autobiography *Mrjj-Rꜥ-mrjj-Ptḥ-ꜥnḫ/Nḫbw* refers to managing his brother's property when he worked for his brother and looking after his brother's *pr-ḏt*:

> 'When he (*Nḫbw*'s brother) was appointed royal master builder, I (*Nḫbw*) was in charge of his town for him and I did everything excellently for him. When he was appointed as sole companion and royal master builder in the two houses, I took charge of all his property for him, as there was more in his house than that of any noble. When he was promoted to overseer of works, I followed up on all his commands to his satisfaction in these matters. I looked after matters for him in his funerary estate for a period of more than twenty years.'[818]

It may be assumed that important officials like *Nḫbw*'s brother were fully occupied with their administrative duties. Thus they appointed a subordinate, most probably a trusted family member, to manage their estates and themselves only paid occasional visits of inspection.

13.4 Summary

- In the column inscriptions three terms denote the offerings or activities viewed by the tomb owner: *nḏ.t-ḥr* (gift(s); *pr(t)-ḫrw* (invocation offering); *pḫr(t)* (reversionary offerings). These are animal parades, offering bearers, figures of domains, scribes recording and the presentation of accounts to the tomb owner.

- In the column inscriptions three terms denote the *source* of the offerings or activities being viewed: *njwwt* (*nw.t n.t*) (community, settlement, town, city); *ḥwwt* (*ḥw.t*) (larger) house; estate (administrative unit); *pr-ḏt* (estate; funerary estate). The term *njwwt* or *njwwt.f* is mainly confined to the tomb owner viewing parades of animals and offerings and in this context applies to rural communities.

[812] *Kꜣ-gmnj* von Bissing I (1905), pl. 17, 18, 19, 21 and Harpur (2006) 15, 16; *Mḥw* Altenmüller (1998) pl. 34, 36, 37.
[813] Moussa–Altenmüller (1977) fig. 24.
[814] Brovarski (2001) 150-51.
[815] Kanawati (2010) pl. 77, 78.
[816] James–Apted (1953) pl. 9. James speculated that the men are answering 'charges of tax evasion and other crimes', ibid. 21.
[817] There is one exception; in his Saqqara tomb *Ṯjj* is depicted as the signaller in a clapnet scene: Wild II (1953) pl. 87.
[818] Strudwick (2005) 267: Cairo Text JE 44608, lines 5-6.

- In the column inscriptions the term *ḥwt* does not appear until towards the end of Dynasty 5, is not seen after early Dynasty 6 and is not used in provincial tombs.

- Although Moreno Garcia has argued that the *ḥwt* was never included in a *pr-ḏt*, the column inscriptions separating the figure of the tomb owner from a panel of estate activities suggests that some officials did include the *ḥwt* in their *pr-ḏt*. The term *pr-ḏt* relates to a system whereby officials were granted access to resources, although it is unclear whether tomb owners were the owners or administrators of the estates pictured in their tombs. The inscriptional material combined with the iconographic provides evidence that they received a 'stipend' through the system of the *pr-ḏt*. The term disappears from the columns of inscriptions, probably early in the reign of Pepy I so it may no longer have applied in the provinces.

- Hierarchies of workers and overseers responsible for each resource carried out the day-to-day management of the pictured rural estates.

- The estate 'chiefs' (*ḥḳ3* or *ḥḳ3 njwwt* or *ḥḳ3 ḥwt*) were ultimately answerable for the full presentation of these resources to the estate. Scribes recorded these accounts, which were then presented to a superior estate official. These may have been the accounts finally presented to the tomb owner. Punishment of miscreant chiefs is included in scenes of the presentation of accounts in three early Dynasty 6 tombs.

SECTION SIX

INTERPRETING THE DATA

RATIONALE

It was proposed in Chapter One that the decorative programmes of Old Kingdom tomb chapels were intended to be multi-functional. A belief in the efficacy of 'magic' pervaded the Egyptian world rather as our belief in 'science' does today.[819] Unlike our acceptance of the limited potency of science, however, Egyptian society regarded magic as a creative and transforming force that was effective in the after-life as well as on earth. The belief in the existence of the ka after the death of the body combined with the belief in the power of magic to provide a doctrine concerning the regeneration of the ka after death, specifically that the appropriate rites and the tomb could provide the ka with eternal existence. Tombs of officials were machines of magic in which the deceased individual would be transformed into an 'akh'.[820] The evidence that this belief system was responsible for the Egyptian tomb and all its elements is indirect but it provides the hypothesis that best satisfies the available data. By providing protection for the body and thus the ka of the deceased, statuary that the ka could inhabit, ritual inscriptions that empowered[821] and a scenic programme that could be 'lived' by the ka, the tomb was both a bridge between death and the afterlife and a means of creating the afterlife for the deceased tomb owner.

The process of regenerating the ka of the deceased required the active cooperation of the living. The ceremony of the 'opening of the mouth' and other rites were imperative to restore the ka to 'life'.[822] The decorative programme of the tomb chapel encouraged visitors to participate in these all-important rites by informing them about the career, achievements and status of the deceased. Wall scenes and inscriptions were an important means of communication with earthly visitors on whom the regeneration of the ka depended. Visitors were entreated to support the cult of the ka by speaking the offering formula, which included the name of the deceased, thus keeping the tomb owner 'alive'. Interest and pleasure in viewing the decorative programme of the tomb chapel attracted visitors, and the presentation of scenes recalled for the living, or informed them of a life lived well,[823] according to the doctrine and moral tenets of Maat, all of which would motivate visitors to carry out the cultic rites that the ka required.

[819] Ritner (2001) 321-26, 333-36.

[820] Greven (1954), Hornung (1992) 175-184.

[821] When hieroglyphs were inscribed, they were *mdw-nṯr*, 'the words of (the) god' with the power both to supply the ka of the deceased and to do harm. Lacau (1914), Wimmer (1998) 347, te Velde (1985–1986) 63-72.

[822] Roth (1992) 113-147, Roth (1993) 57-79.

[823] te Velde (1982) 143, Baines (1994) 75.

This belief in the tomb and the ka is reflected in Old Kingdom tombs but it does not explain why the 'daily life' element of the decorative programme was confined to the private aspects of the life of the deceased. Why, for example, did the highest officials in the land provide extended scenes with detailed images of mundane activities on their rural estates but relegate the crowning achievements of their careers to a list of titles, which was the custom until officials began to refer to their careers in their autobiographies late in Dynasty 5? Even when they included reference to stages in their service to the crown, they do not provide corresponding pictorial images. In their lifetime officials had served the crown: administering the royal treasury or a province, supervising the collection of taxes or organising great public works; they acted as agents of the crown carrying out the king's obligation to maintain Maat.[824] Scenes of 'daily life' and work on rural estates, however, presented the tomb owner inhabiting a world where he was master and every other figure a dependant. Thus the depicted overseers, scribes and labourers were carrying out the tomb owner's obligation to Maat. In this milieu the tomb owner's contribution to the maintenance of Maat was personal and direct. He was responsible through his agents for the successful exploitation of all the natural resources of the different environments that comprised the estate. It was with this aspect of the tomb owner's life work that many visitors would be familiar and thus able to judge and appreciate how he was represented.

The decorative programme incorporating the theme of a well managed, productive estate presenting the tomb owner as its beneficiary supported the vision both of an earthly life well lived and a world the ka could inhabit. As the programme had the further intention of impressing visitors to the tomb in order to encourage their support of the cult of the deceased, every inscribed word and every image presenting the life work of the deceased as an effective but ethical administrator and member of his class had to be a truthful, if selective portrayal. For example, in his tomb at Meir *Ppjj-ꜥnḫ-ḥrj-jb* is at pains to inform the visitor of his ethical life. Although he does not provide any explanatory detail of the problems he might have faced, in his long autobiography he insists,

> 'I was never placed under guard. I was never imprisoned.
> With regard to everything the witnesses said in the presence of the officials,

[824] Teeter (1997), Teeter (2001) 319-21.

> I always came away from the matter with success,
> the matter having been thrown back on those who spoke (against me),
> since I had been cleared in the presence of the officials.
> For they had maliciously spoken against me.'[825]

Visitors to the tomb would read that the deceased was now an excellent and potent 'akh' with the power to punish desecrators of the tomb but also to support and reward the living. To be reminded of the good life led by the deceased and his beneficence, visitors needed to read and see what they knew to be true. A graffito on the entrance to the late Dynasty 6 tomb of *Nj-ꜥnḫ-Ppjj* of Saqqara makes this point. On the right jamb of the doorway *Nj-ꜥnḫ-Ppjj* warns robbers that he was now an excellent and powerful 'akh',

> 'With regard to any man who shall do anything evil to this tomb of mine of the necropolis or who shall take a stone from its appointed place,
> I shall contest with him in the noble and excellent court of the Great God, the lord of the necropolis.
> I shall seize his neck like a bird's, and the fear of me will be instilled in him, so that the living on earth shall be fearful of excellent akhs.
> I am an excellent akh …'

However a visitor was ready to desecrate this tomb. A graffito in front of the figure of the tomb owner reads,

> 'You have locked me up, you have beaten my father; I am now happy, as how can you escape my hand? My father is satisfied.'[826]

It seems that there was one visitor to *Nj-ꜥnḫ-Ppjj*'s tomb who did not think the deceased had always behaved ethically and was prepared to record his opinion.

Based on the proposition that scenes in tombs of officials reflect 'daily life' as it was experienced on individual estates, the primary aim of this study has been to document and investigate the funerary iconography and inscriptional data presenting the resources of rural estates and their management. The material has been classified according to depicted species of livestock and plants and the environments supporting them. As the period of study spans the entire Old Kingdom, the data presented in previous chapters has been organized on a chronological basis on the assumption that changing trends may have been reflected in the data. Interpretation of the data follows this principle.

[825] Sethe (1933) *Urk.* I, 221-4. Translation: Strudwick (2005) 370, No. 270.

[826] Hassan *Saqqara* II (1975) fig. 3; Translation: Strudwick (2005) 228-9, No. 143.

CHAPTER 14 RESOURCES OF FOUR ENVIRONMENTS

14.1 The four environments exploited by estates

The resources portrayed in Old Kingdom tombs were the product of four different natural environments:

- The desert fringe was the habitat of ungulates and many smaller creatures adapted to a comparatively arid environment. Domestic goats appear to be depicted in the ecotone between communities of the true desert fringe and those of the pastoral/arable regions. The browsing/grazing capacity of goats would have enabled the domestic species to survive in either environment.
- The true pastoral environment would only have supported cattle on a seasonal basis. Consequently cattle herds had to be led annually to the lower-lying land where, if necessary, they could survive on papyrus and sedges. Sheep and donkeys would have survived on a combination of the ecotone and pasturage, although grazing animals, sheep and donkeys can also exist on poorer quality pasturage than domestic cattle.
- The well inundated regions of the valley floor enabled field crops to be cultivated on an annual basis.
- The marshland offered fish and fowl, sources of cheaply acquired foods but rich in high quality protein, as well as the multi-purpose papyrus which provided summer fodder for cattle herds.

When the annual inundation was within normal range, successful exploitation of these environments supplied Egyptian society with adequate nutritional sustenance. This included provision for a large workforce engaged seasonally on constructions and an expanding non-productive ruling class that enjoyed a diet which included luxuries and an appropriate lifestyle. Large estates were needed to provide a food surplus sufficient to feed these sections of the population, as subsistence peasant farming would not have produced an adequate surplus. Furthermore, productive soil was severely limited and the vagaries of inundation levels had to be met and managed each year. Thus efficient management of available resources was vital both for a rural estate and the country as a whole. This vital task was in the hands of the elite officials who, as well as administering the country, managed the great estates, whether they were crown or temple property or owned individually. Whether there existed individual peasant plots and smaller family properties in the Old Kingdom is not known.

The ruling class of officials were the greatest beneficiaries of the activities by which the estates exploited the four environments that comprised the valley. Particularly from mid Dynasty 5 on, these elite administrators reflect this situation in the decoration of their tombs. Consequently the decorative programmes of tomb chapels do not reflect the pattern of the economy as a whole but in part are devised to illustrate the successful management, wealth, economic benefits and status enjoyed by the tomb owner. For example, fattened desert ungulates, prime beef cattle and well-fed hyenas, which are represented particularly in the tombs of the highest officials, would not normally have been a source of nourishment available to ordinary people. Depictions of butchery of cattle and desert ungulates are without exception presented as ritual events of significance for the tomb owner. Ordinary estate workers portrayed in scenes are ciphers to illustrate work procedures. They are not depicted slaughtering prime cattle and the scenes portraying them cooking fowl and fish, baking bread and brewing beer, skinning goats and stewing goat-meat are implicitly for the benefit of the tomb owner and his family. There must have existed great disparity between the living conditions of the lower social classes and those of the ruling class, which would account for the contradictions between the pictorial funerary data reflecting the lifestyle of an elite class and the material data from faunal remains such as found at Kom el-Hisn, where a wider social spectrum is represented.[827] Consequently the funerary art in private tombs is a selection of the realities of 'daily life', rather than a balanced reflection of successes and disasters and life styles of both rich and poor.

The principle used here to interpret the data is based on the proposition that scenes found in funerary contexts represent matters of vital importance for tomb owners as well as their individual tastes and interests. This entitles the researcher to infer the level of significance that tomb owners attached to a motif from the frequency of its occurrence, its location within the tomb and from the amount of space and detail devoted to it. While the same themes and motifs appear in tomb after tomb, scenes are never an identical copy from one tomb to another. Consequently the essential elements in the endless variation in iconographic and inscriptional presentation have to be identified so that the artistic treatment of different resources may be classified and compared. This process makes it possible to specify how different resources were used, the degree of value attached to each and to identify any signs indicative of changing trends.

The animals represented as resources in Old Kingdom tombs are, with a single exception (striped hyena), species dependent on pasture of varying quality and/or

[827] Redding *Kom el-Hisn* (2002).

foliage.[828] Cattle, goats, donkeys and sheep are portrayed as domesticated herds and flocks; the desert ungulates depicted in parades were probably wild animals accustomed to some degree of human contact, perhaps caught very young during a hunt, reared and fattened in an enclosure. The variations in the treatment of the themes depicting these species and modifications over time are the basis of many of the following interpretations.

14.2 Resources of the desert environment

Scholars have identified a variety of elements in the relationship of Egyptians with the wild animals that inhabited the Nile valley in pharaonic times and earlier. The focus here, however, is whether tomb owners regarded desert animals as a resource. Inscriptional material from autobiographies, statements by tomb owners and royal decrees does not offer much data by which to assess the importance attached to desert animals. Consequently, the value attributed to these species has to rely on deductions from wall scenes and their associated captions, the degree of prominence given to the presentation of individual animals before the tomb owner and the presence or otherwise of scenes showing husbandry and care of desert animals. This includes the significance of desert hunting scenes. To judge whether or not desert animals were regarded as an economic resource or were included in the tomb's decorative programme for some other reason, it is necessary to assess why the animals were hunted when all estates kept herds of domesticated livestock. Were desert antelope captured in considerable numbers during hunts and taken back to the estate to be bred and maintained as a herd?

The desert hunt scene

Scenes from the early Dynasty 4 tomb chapels at Medum and Saqqara suggest that the purpose of the hunt was to drive desert ungulates into fenced enclosures constructed where the hunting took place.[829] Thus these stockades must have been located in the desert, as the terrain depicted in the chapel of *Jtt* suggests.[830] Dogs are not portrayed killing antelope but nipping at their heels, rather as cattle dogs do today. These early scenes do not make the purpose of this activity clear; they show little violent slaughter of desert ungulates and do not indicate any other fate for the entrapped animals. However, comparing them with the royal hunting scene of Sahure[831] and scenes from the second half of Dynasty 5 and later, suggests a reason for driving animals into a stockade located in the desert. The Sahure scene portrays the king outside of the enclosure shooting the animals trapped within, indicating that the purpose of enclosure was to herd the animals where the king could be portrayed slaughtering them in numbers. From mid to late Dynasty 5 to early Dynasty 6 hunting scenes show the animals within the stockade being savagely attacked by hunters and packs of hunting dogs, with the tomb owner always depicted outside of the scene viewing the action. Despite the inclusion in the scenes of animals engaged in natural activities such as copulation, smaller animals that are not the object of the hunt, and lions attacking ungulates, the purpose of the scene is to illustrate the second stage of the hunt, the slaughter of the larger entrapped creatures. As in the case of the Sahure scene, the primary aim of the slaughter cannot have been for meat to be taken back to the estate as the action takes place in the desert, presumably at a distance from the estate.[832] Furthermore, packs of hunting dogs are pictured tearing at the animals. Although tomb owners are never portrayed taking part in the action it may be assumed that the primary aim of the hunt and slaughter was to provide sport for the tomb owner or to present a concept/belief symbolically or both.[833] Perhaps for reasons of respect or protocol, tomb owners were not prepared to appear to usurp the royal function by being depicted wielding a bow and arrow.

The young of desert animals

From mid Dynasty 5 on, two themes concerning antelope fawns are portrayed in hunting scenes: hidden behind a clump of vegetation and being carried by returning hunters in baskets hanging from shoulder yokes. A third and fourth theme also appear in later Dynasty 5 of the young of antelope and striped hyenas being carried in the arms or on the shoulders of bearers. These images appear in scenes of bearers leaving the hunt, returning to the estate and presentations to the tomb owner, sometimes as images in the parade of animals. Combining the images and scenes sequentially suggests a further reason for the desert hunt: to capture, bring back to the estate and hand rear the young of desert ungulates. Mature ungulates are not portrayed being brought back. Indeed, this could only have been achieved in stout cages as they would have been wild animals untrained to human direction.[834] In the most complete scene of live animals being brought back from the hunt, that of *Ptḥ-ḥtp* (II) (*Ṯfj*)[835], ungulate calves are carried in baskets or in hunters' arms; cages, which have to be hauled by porters, are reserved for a lion and a leopard.

---

828 The exception is the hyena, which also appears in parades and in scenes where it is being fattened for human consumption.

829 See Chapter 3 – Desert Animals: 3.5 The desert hunt, p. 22-24.

830 Petrie (1892) pl. 27, Harpur (2001) fig. 88.

831 Berlin 21783, Borchardt (1913) pl. 17.

832 The terrain depicted is not true desert of undulating sand dunes but uneven scrubland with sparse, drought-resistant flora that may have occurred around Lake Moeris and Delta fringes.

833 Not until the mid Dynasty 6 tombs of *Jbj* of Deir el-Gebrawi and *Mrw* of Naga ed-Der is a hunter depicted armed with a bow and arrow. This may suggest that for most of the Old Kingdom the weapon was a prerogative of the crown.

834 The Oxford Expedition to Egypt: Scene Details Database cites in the category "Desert and desert-related activities: 2.2 Hunters returning from desert with game" a number of tombs with images of mature desert ungulates being led by bearers. It is unlikely that these are images of untamed animals taken from the desert because they are depicted as amenable to handling by humans. It is more probable that they represent animals that have been accustomed to human handling and thus illustrate the end-product of the desert hunt, fawns that have been reared by estate workers and are being presented to the tomb owner. See, for example, Kanawati–Abder-Raziq (2004) pl. 46.

835 Paget–Pirie (1898) pl. 32-33.

Thus scenes of desert hunting indicate two purposes for the hunt: as sport for king and tomb owner with perhaps an ulterior symbolic significance, and to bring back alive the young of selected desert animals, mainly ungulates, to be hand reared and ultimately sacrificed either as ritual offerings for the tomb owner or to enrich his table.

Did Egyptians domesticate desert ungulates?

The many images of herders carrying the young of desert animals in boxes hanging from a shoulder yoke or larger young in their arms give the impression that some specimens were trained to accept human contact. The constancy with which a variety of desert ungulates and the striped hyena are portrayed in procession before the tomb owner further suggests that these species were to some extent accustomed to human contact. Whether this treatment was combined with a programme of true domestication is, however, questionable. Unlike the domesticated cattle pictured walking placidly behind their herder, the desert ungulates are quite often depicted shying and refusing, giving a clear impression of being reluctant to proceed in the parade. In procession, the unease of the desert animals is illustrated in images of ungulates balking with both front legs stretching forward and the head lowered indicating fear and refusal to advance.[836] In many instances the herder who leads them has to turn back to control his animal by grasping its horns and muzzle.[837] Sometimes the herder with his legs planted wide apart and his back arched or his entire arm hooked around the animal's horns shows the strain of controlling the animal, while a second herder at the rear leans forward pushing or guiding the animal with both hands on the animal's rump.[838] Many desert ungulates on parade require two herders to control them. On the other hand, the men leading cattle appear undisturbed; they face forward without needing to turn back to keep check on their charge. A single herder may lead a group of two or three cattle. Desert ungulates on parade are depicted requiring more attention, although there are a limited number of scenes of recumbent ungulates being hand fed by a herder. This motif, however, may not have been intended to be 'read' literally; it may simply contain the information that some desert specimens were deliberately fattened for slaughter.[839]

Any attempt to domesticate desert ungulates implies the maintenance of herds, which would have required a breeding programme, but there is no evidence of this. A contrast between the portrayal of domestic cattle and desert ungulates shows cattle frequently portrayed in scenes of husbandry. Herders aid with birth, milk the cows, handle young calves and lead herds across a stretch of water. Cattle husbandry scenes include herders attempting to control bulls mating cows and calves tethered to hoops driven into the ground. Such themes of herders giving skilled and close attention to their charges are not depicted with desert animals. The birth of a desert animal is never aided by a herder. Milking is never portrayed, and images of copulation and birth are confined to scenes of the desert hunt and animals confined in a stockade.[840] A variety of animals may be depicted in these scenes; ten different species are portrayed copulating in the tomb of *Nj-mꜣꜥt-rꜥ*, some of which are small animals that were not sufficiently prized to be included in the parade of animals.[841] There does not appear to be any intention to depict selective breeding. However, although the evidence does not indicate that estates carried out programmes to domesticate any of the desert ungulate species, the capture of their young and the inclusion of adults in the parade of livestock suggest that they did 'tame' the animals, that is, accustom individuals to handling by humans.

Did tomb owners maintain herds of desert ungulates?

The propositions above, that the purposes of desert hunts were to slaughter mature animals and capture ungulate young alive, and that there is no evidence of breeding and domestication programmes, do not support a contention that herds of desert animals were maintained on estates. Scenes and inscriptions in many tombs suggest that estates held considerable herds of domesticated cattle, goats, donkeys and sheep, and in the case of cattle moved herds from one pasturage to another, which would have been difficult in the case of semi-wild animals. Despite the large figures cited for herds of gazelle and oryx in two tombs at el-Hawawish,[842] it is probable that desert animals were kept in small numbers because of the lack of year round pasturage and the difficulty of moving large herds to keep them supplied with pasturage. Maintaining a herd of any size would have been a costly undertaking as the animals would have needed fodder brought to them for part of the year (height of summer and inundation). Natural pasturage in any enclosed area stocked by grazing animals would have soon been eaten out after the inundation subsided. Indeed the viability of keeping herds of the numbers cited even for part of the year has to be doubted, although it is possible that in the middle provinces of Upper Egypt there were tracts of land still available for exploitation.[843]

Are the numbers given for captive herds valid?

Very few tombs provide numbers for herds of desert animals and those given are remarkably large. *Rꜥ-špss* II provides the number '1000' for each of the animals in his parade of livestock, including four desert ungulates, and birds depicted in the parade scene.[844] The repetition of such 'round numbers' makes them suspect. The only

[836] *Nfr-mꜣꜥt* LD II 17.

[837] Evans (2006) 72-73 identifies this action as a method of controlling nervous animals.

[838] *Nbt* Munro (1993) pl. 17.

[839] Evans (2006) 147ff.

[840] *Nj-ꜥnḫ-H̱nmw/H̱nmw-ḥtp* Moussa–Altenmüller (1977) pl. 40, *Ṯjj:* Wild (1953) pl. 127, *Mrrw-kꜣ/Mrj* Duell I (1938) pl. 25, *Mrjj-ttj/Mrj* Kanawati–Abder-Raziq (2004) pl. 46.

[841] Roth (1995) fig. 189. This scene depicts caracals, hedgehogs, weasels and foxes in the act of copulation, as well as at least four different species of ungulates.

[842] See TABLE 5: LIVESTOCK NUMBERS.

[843] The autobiography of *Hnḳw/Jj…f* (II), redated by Kanawati to early Dynasty 6, provides evidence that land was available in this nome. Davies *Gebrâwi* II (1902) pl. 24 and Kanawati *Gebrawi* I (2005) 72, pl. 66.

[844] Junker III (1938) fig. 48.

other numbers for desert animals are found in the el-Hawawish chapels of *Ḫnj/Špsj-pw-Mnw* and *K3(.j)-ḥp/Ṯtj-jkr* of the second half of Dynasty 6. *Ḫnj/Špsj-pw-Mnw* attaches the number '10,000' to both the dorcas gazelle and the oryx depicted in the parade of animals;[845] *K3(.j)-ḥp/Ṯtj-jkr* has the numbers '1023' against a dorcas gazelle and '1000' against the oryx in his animal procession.[846]

As these animals would have had to be kept in a fenced stockade at all times, it would have been necessary to provide them with hand cut fodder for part of each year. Domesticated cattle could be managed as a herd allowed to range over pasturage and be moved from one location to another, providing herders protected them day and night. These practices could not have been applied to the semi-wild desert ungulates, which suggests that desert animals could not have been held in large captive herds.

These comparisons between the depiction of domestic cattle and desert ungulates suggest that the desert ungulates were accustomed to human contact rather than domesticated, and perhaps even this level only applied to a small number of animals. The hunting forays probably took place in winter/spring as, whether or not portrayed in desert enclosures, pairs of animals are depicted mating, females are giving birth and young ungulates hide in clumps of grass and bush. There is no evidence of controlled breeding programmes. The desert animals were carried off as very young animals in containers slung from a yoke across a hunter's shoulders,[847] their fate to be reared in captivity and finally slaughtered. Perhaps some individuals were kept to for breeding. It is certain that a ritual value was assigned to desert ungulates but it is doubtful whether they were regarded as a resource by which tomb owners measured their wealth.

14.3 Resources of the pastoral environment

14.3.1 Cattle

The analysis of cattle depictions and associated inscriptions in tombs from Dynasty 4 to Dynasty 8 has illustrated the high status assigned to herds, particularly in the second half of the Old Kingdom. At least from mid Dynasty 5 on, a large herd was a primary indicator of a tomb owner's wealth. This aspect is less clear in the first half of the Old Kingdom as data from the cruciform chapels of early Dynasty 4 and the L-shaped chapels of mid Dynasty 4 to early Dynasty 5 do not single out cattle as officials' most highly prized resource. The prominence given to cattle in tomb iconographic programmes begins to emerge in mid Dynasty 5 with portrayals of livestock parades giving pride of place to the species. Also from mid Dynasty 5 on, cattle are portrayed in a variety of themes: husbandry, hand feeding, fording a stretch of water and finally in scenes of pairs of bulls fighting each other. Furthermore, cattle are almost always depicted as the subject of ritual butchery, pulling ploughs and even occasionally threshing the harvested cereals. In many tombs cattle are depicted with varied horn arrangement and colour markings. No other breed of livestock receives so diverse a range of depictions. In the provincial tombs of Dynasty 6 emphasis on the size of herds (cited herd numbers) is added, further making it clear that they were the prized livestock. Their place in animal parades and the variety of themes in which they appear make cattle the most consistently represented of all livestock in Old Kingdom tombs.

Cattle in the parade of animals

In the tombs of Dynasty 4 to early Dynasty 5, with the exception of the scene in the tomb of Queen *Mrs-ꜥnḫ* III, the portrayal of the parade of cattle amounts to a depiction of one or two animals. From mid Dynasty 5 on, cattle parades begin to include more animals and to develop the trend, first seen in *Mrs-ꜥnḫ* III, of presenting animals with distinctive individual features such as deformed horns or polled. In the second half of Dynasty 5 a line of three or more animals, some of which may be desert ungulates, is one of the themes most consistently represented in tomb chapels, appearing in 42 of the late Dynasty 5 tombs used in this study. The most frequent depiction of paraded cattle is of heavy bodied animals labelled *jw3* led by herders holding a short rope that appears to be attached in some manner to the animal's jaw.

This presentation theme may be associated with the concept of offerings for the ka, but its enlargement and development in the second half of Dynasty 5 reflects the changing emphasis in the pictorial programme of tombs, best identified by comparing the programmes of L-shaped chapels of Dynasty 4 and early Dynasty 5 with those of late Dynasty 5, when the representation of more worldly and material themes, mainly of the production and administration of resources on rural estates, becomes more dominant.[848]

Herd numbers

Herd numbers are cited in only a small number of Dynasty 5 tombs.[849] In Dynasty 6 they become a frequent

[845] Kanawati II (1981) fig. 26. The number '10,000' is also applied to the cattle in the parade.

[846] Kanawati I (1980) fig. 14. The total number of 3226 is given for the oxen of *K3(.j)-ḥp/Ṯtj-jkr*.

[847] See TABLE 2: MANAGEMENT OF DESERT ANIMALS.

[848] A comparison of the pictorial programmes of *Ḫwfw-ḫꜥ.f* I (PM 188-90), *Mr-jb.j, Sš3t-ḥtp/Ḥtj, Nswt-nfr, Wḫm-k3.j* with those of *K3.j-m-nfrt* (PM 467-8), *Nfr-jrt-n.f* and *Nj-k3w-Ḥr* shows a marked change of themes. The iconography is preserved on all four walls of each of the above chapels and shows a new range of motifs. For the earlier four chapels the pictorial programme is largely occupied with themes of family members and offerings. In the later three chapels scenes of activities on the estate are introduced. In *Nfr-jrt-n.f* and *Nj-k3w-Ḥr* much of the space of two walls (north and east walls) is devoted to registers displaying themes connected with work on the rural estate. Reflected in almost all late Dynasty 5 tomb chapels, this marked change appears to be more than a response to additional wall space.

[849] *Rꜥ-ḫꜥ.f-ꜥnḫ* cites a herd of 1055 (LD II 9), *Nfr-b3w-Ptḥ* adds 1000 to the captions above cattle on each of three registers (Weeks [1994] fig. 24, pl. 11a; LD II, 57c) and *Rꜥ-wr* II adds 1000 to the three captions above cattle on the fourth register (Junker III [1938] fig. 48). The bald numbers of 1000 given by *Nfr-b3w-Ptḥ* and *Rꜥ-wr* II

feature in captions accompanying lines of cattle depicted in the tombs of Upper Egyptian administrators. Enormous numbers, beyond the carrying capacity of an estate or even a province, are occasionally cited.[850] The very largest of these numbers refer to the cattle census and probably relate to a group of provinces.[851] However, even the smaller numbers of tens of thousands, perhaps cited for a single province, are difficult to accept. With the disappearance of reliable seasonal precipitation, the possibility of providing year-round pasturage and adequate fodder for cattle herds numbered in tens of thousands is questionable. With the need to keep an adequate amount of the most fertile and watered land under crop, any significant increase in herd numbers on most estates could only have been produced by exploiting new lands, as described by *Hnḳw/Jj...f* (II) of Deir el-Gebrawi.[852] Although the actual numbers are unacceptable, it is not necessary to agree with Moreno Garcia who argues that 'les quantités fantastiques de produits ou de bestiaux censés appartenir aux défunts' undermine the historical value of pictorial programs of tombs.[853] Numbers have always been used two ways, to provide an exact accounting and to give a general idea of relative quantity.[854] It is probable that figures cited in tombs were of the latter kind as so many are clearly suspect.[855] It has been calculated, based on the work of Butzer, that the total fertile land of the Nile valley only amounted to 16,100 square kilometers in the Old Kingdom.[856] While the middle nomes of Upper Egypt contained the broadest stretch of fertile land apart from the Delta, only a proportion of this land would have been assigned to pasturage, and would have only provided seasonal fodder. Consequently the numbers cited by *Ppj-ꜥnḫ-ḥrj-jb* totalling over three quarters of a million head of cattle appear to be much inflated, although they refer to the official census and, probably, to *Ppj-ꜥnḫ-ḥrj-jb*'s position as *jmj-rꜣ Šmꜥ m spꜣwt ḥrjwt-jb* ('overseer of Upper Egypt of the central provinces')[857] In fact, all the cited herd numbers must have been inflated in view of the seasonal nature of natural pasturage. The need either to hand feed or move herds according to an annual cycle casts doubt on these vast numbers. Furthermore, the figures vary enormously from those cited by one generation of tomb owners to the next. For example, at el-Hawawish the total numbers cited in *Kꜣ(.j)-ḥp/Ṯtj-jkr*'s tomb are 3226 cattle and 2023 desert ungulates,[858] while, *Ḫnj/Špsj-pw-Mnw*, who was either his immediate predecessor or successor, cites a total of 30 000 head of cattle and 20 000 desert ungulates.[859] That 20 000 wild ungulates could have been held in enclosures and adequately fed when seasonal pasturage died out in the height of summer is not credible. An acceptable reading of the herd numbers is that they represent an emphatic statement that the crown's policy of increasing herds in Upper Egypt had been implemented with outstanding success.

The provincial emphasis on increased herds does not apply solely to cattle. *Hnḳw/Jj...f* (II) and *Jbj* of Meir[860] and *Ppjj-ꜥnḫ-ḥrj-jb*[861] and *Ppjj-ꜥnḫ/Ḥnj-km*[862] of Meir refer to large and small cattle either in autobiographical statements or captions. *Kꜣ(.j)-ḥp/Ṯtj-jkr* of el-Hawawish refers to numbers of gazelle and oryx, and *Ḳrrj* of el-Hawawish proudly announces the size of the herd of donkeys he has achieved.[863] The emphasis on increasing the production of provincial rural estates in Dynasty 6, primarily of cattle, was extended to other livestock, although there is no evidence of a corresponding trend or policy to expand the production of field crops.

Cattle varieties

In general, the structural changes sometimes observed in zooarchaeological remains,[864] as well as the depiction of cattle with differently shaped horns,[865] have been attributed either to breeding programmes or to sub-species. Yet there is no conclusive evidence that there were different breeds of cattle in the Old Kingdom. It is possible that attempts were made to breed polled cattle, which were highly valued. However, the complex genetic knowledge needed to carry out a successful programme of breeding polled cattle would not have been available, although it is possible that some estates had a degree of success by mating polled males and females. The interest in breeding polled cattle may have waned a little in the second half of Dynasty 6 when the major preoccupation became the production of increased herd numbers and probably the size of the animals. Old Kingdom data is too limited and uncertain to accept the claim that a breed of shorthorns was introduced at this time.

Terms appearing in associated captions and used to identify individual animals refer either to gender, function or a characteristic feature. Gender labels are applied to animals with appropriate physical attributes; yet almost all cattle depicted in parades of animals appear to be oxen with only the preputial sheath of the male visible. The terms *jwꜣ* and *ngꜣw* denote function, with *jwꜣ*

are suspect and may be a reflection of the *ḫꜣ* ('thousand') applied to the items of the short offering list associated with the offering table.

850 Refer TABLE 5: LIVESTOCK NUMBERS.

851 In the case of *Ppj-ꜥnḫ-ḥrj-jb* of Meir the enormous numbers cited for the census refer to the 'middle provinces' (*jrt jrw n kꜣw ꜥwt n spꜣwt ḥrjw-jb*): Blackman *Meir* IV (1924) pl. 16.

852 In his autobiography *Hnḳw/Jj...f* refers to filling the river banks of his nome with cattle and its pastures with small cattle (*jw gr mḥ.n(.j) wdbw.s m mnmnt mḫrw.s m ꜥwt*): Davies *Gebrâwi* II (1902) pl. 24 and Kanawati *Gebrawi* I (2005) 72, pl. 66.

853 Moreno García (1999) 63.

854 This citing of vast numbers to indicate relative quantity was not confined to herd numbers. In his decorated burial chamber, *Kꜣ.j- m-ꜥnḫ* recorded over four million timber boats of various sizes and designs. Kanawati (2001) pl. 36.

855 This applies to both the enormous numbers cited in Dynasty 6 provincial tombs and the bald numbers of 1000 given by *Nfr-bꜣw-Ptḥ* and *Rꜥ-wr* II.

856 Following Butzer (1976). p 83 and Allen (1997) 145, note 13.

857 Blackman *Meir* IV (1924) 1-3.

858 Kanawati I (1980) 26, fig. 14.

859 Kanawati II (1981) 39-40, fig. 26.

860 *Hnḳw/Jj...f* (II) Kanawati *Gebrawi* I (2005) fig. 56, 66-67; *Jbj* Kanawati *Gebrawi* II (2007) fig. 73, 75b.

861 Blackman *Meir* IV (1924) fig. 16.

862 Blackman–Apted *Meir* V (1953) pl. 32, 41.

863 Kanawati VI (1986) 48, fig. 20b.

864 Boessneck (1953).

865 Andersson (1912) 145-164.

being an ox selected for slaughter. A *ng3w* was most probably an ox destined to be a beast of burden pulling the heaviest loads, although images of working animals are rarely labelled. The principles on which cattle are labelled apply to the primary purpose of the animal, not to type or breed.

Management of cattle herds

Few images represent details of husbandry in the early Old Kingdom tombs; certainly there are no sustained scenes of cattle management until the tomb of *Jj-mrjj* in mid Dynasty 5. Images and scenes portraying aspects of cattle management appear much more frequently in the second half of Dynasty 5 and the reign of Teti. In this time period, copulation, birth and care of calves, milking and the fattening of oxen are the most frequently depicted features. Two husbandry activities that first appear in mid to later Dynasty 5 are fording cattle and fighting bulls, with the latter given particular prominence in Dynasty 6. Fording scenes represent herds driven seasonally to the marshland and back and bull fighting scenes represent the aim of identifying the strongest bulls, probably for reproductive purposes. Vignettes and captions suggest that estates carried out a policy of controlled breeding, one aspect of which was mating cows with the largest and most powerful bulls.

The daily management of the herd generally receives less attention in the period from the beginning of Dynasty 6 to Dynasty 8, particularly in provincial tombs. If scenes of the herd crossing water and pairs of bulls fighting are excluded, other husbandry features only appear as brief vignettes in 29 of the tombs of this long period of time. In the chapel of *Jbj,* for example, a bull is depicted mounting a cow and a second cow is being milked but both vignettes are on registers in which a preceding animal is being presented, perhaps to the tomb owner. The text above the ox on the lower register reads *jrt jrw mḏwt* ('making the inventory of the cattle stalls'), which suggests a cattle count rather than husbandry.[866]

Annual movement of cattle

Cattle are grazers that depend on a pastoral environment. The climate of Egypt, the topography of the Nile valley and the annual inundation could not provide year-round grasslands as pasturage. While estates in or bordering on the Delta would have been able to move their herds to the lower, better watered areas during the height of summer, all estates further south would have had to rely on the papyrus stands of the marshland fringing the river, streams and surviving oxbow lakes.

The term used in inscriptions to describe the herds as leaving the 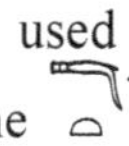, usually translated as *T3-mḥw* ('Lower Egypt/the Delta') is best translated as *mḥjt* ('marshland'). It would not have been feasible to move many large herds north and south, all at the same time of year, as it is unlikely there would have been available fodder for the journey. Yet the increasing frequency of scenes of cattle crossing a stretch of water in the Fifth Dynasty in the tombs of high officials suggests that this activity was an important, annual feature of the management of cattle herds.[867] The portrayal of the herd crossing a stretch of water did not necessarily entail a journey south from the marshes of the Delta. Inscriptional evidence suggests that movement was from lower-lying marshland to higher pastures, rather from Lower to Upper Egypt. The frequency of portrayal is maintained in Dynasty 6, although at Meir and Deir el-Gebrawi its depiction seems to lose the significance of the activity. This reduction of the scene, however, is not the universal trend in Upper Egypt as the traditional representation is maintained in the el-Hawawish tombs of Upper Egypt Nome 9.

Apart from the later Dynasty 6 departure from the traditional representation noted above, the scenes of swimming and wading cattle confirm and elaborate on the impression drawn from other scenes of husbandry that good management of the cattle herds was of especial importance to officials. They portray the watchfulness of herders taking their cattle across a crocodile infested channel and the successful conclusion of the journey. The scenes represent the care bestowed on the herds by experienced and skilled herders in a dangerous episode that occurred annually in the management of the herd.

Fighting bulls

Old Kingdom images of fighting bulls are only found in the tombs of Upper Egypt. Interpretations of the bullfighting motif range from entertainment for the tomb owner,[868] to a natural clash between bulls during the mating season[869] or a method of choosing the strongest animals for a programme of selective breeding.[870] Herders brandishing sticks are interpreted as encouraging the bulls to fight. While Kanawati prefers to interpret the theme in a purely secular light as a contest organised for the enjoyment of the tomb owner,[871] Galan's conclusion, drawn from bull fighting scenes of the Old Kingdom through to the New Kingdom, is that these scenes represent a deliberately arranged encounter which has a "marked religious-funerary" association and symbolism.[872] Galan's opinion that the image of the bullfight was 'a metaphor for challenges that the deceased had to overcome to confirm the right to his

866 Kanawati *Gebrawi* II (2007) pls. 54, 73.

867 See TABLE 4: CATTLE CROSSING WATER.

868 Kanawati (1991) 51-58.

869 Vandier (1969) 58-59.

870 Klebs (1922) 88. Blackman suggests that the two bulls in the tomb of Wekh-Hetep were part of the Hathor ceremony fighting over a cow representing the goddess. But the interpretation of a Middle Kingdom scene may not apply to the Old Kingdom. See Blackman *Meir* II (1915) 24-25.

871 Kanawati (1991) 57. Kanawati's interpretation of the motif as a representation of a secular event corresponds more closely than that of Galan with Old Kingdom scenes and captions but his conclusion that the bull fights were organised as entertainment is less certain. For example Old Kingdom iconography and inscriptions equally support the interpretation of bulls encouraged to fight as part of a breeding programme.

872 To support this view Galan accepts that scenes of the tomb owner fishing and fowling and goats browsing among trees are to be interpreted as having an essentially symbolic meaning: Galan (1994) 92.

status as leader in the afterlife'[873] derives from a reliance on Middle Kingdom and later texts and scenes rather than from Old Kingdom data,[874] as does the earlier argument of Lloyd that the bullfight symbolises the struggle between Horus and Seth, that is, between Order and Chaos.[875] Galan refers to the inclusion of the bullfighting image in the tomb and its association with other scenes, to which he attributes a symbolic significance, as support for his thesis of the metaphorical nature of the image.[876]

Galan's reasons for the inclusion of themes in tomb decoration and the significance of the other scenes he cites are open to question. Old Kingdom bullfighting scenes contain no reference or allusion to a symbolic or metaphorical reason for their inclusion in the tomb repertoire.[877] In nine of the thirteen Old Kingdom tombs that include the motif the tomb owner appears watching the combat; on eight occasions he is depicted in a relaxed stance leaning on his staff. Kanawati notes that chapel wall scenes depict both 'the useful' and 'the entertaining'.[878] The captions associated with the Old Kingdom images of combat give the speech of the men in charge, implying enjoyment of the contest and particular interest in the outcome,[879] but the close proximity of scenes of husbandry or the parade of cattle suggests that the activity had a purpose beyond entertainment. Images of cattle copulating and cows giving birth attended by herders indicate that estates ran breeding programmes.[880] The frequency with which provincial administrators provided herd numbers in Dynasty 6 further suggest that increasing the herd was an important policy. Finding the strongest bull to father the calves would have been a consideration in order both to identify the most active and fertile of the bulls and to produce the strongest animals. That cows giving birth are depicted in distress and the actual birth of the calf is always attended by a herder suggest that cows were producing particularly large offspring, having been mated with the most powerful bulls.[881] That the Old Kingdom bullfighting scene is a metaphor with symbolic significance has to rely on data derived from many hundreds of years later than the Old Kingdom. Evidence from Old Kingdom tombs suggests that contemporaries included the motif mainly to emphasise and elaborate on their success in building up large, healthy herds. All the data relating to bullfights come from Upper Egypt, where all but one of the scenes are located in the Sixth Dynasty tombs of the provincial administrators, who emphasise their successful breeding programmes by citing the size of herds.

The themes of cattle fording and bulls fighting represent significant activities in the management of cattle herds. The increased portrayal of the seasonal movement of the herd across waterways occurs in the second half of the Fifth Dynasty and the motif of pairs of fighting bulls becomes more popular from mid to late Dynasty 6. Both activities involving an element of danger or violence would have been notable occurrences. There is a further development at Meir and Deir el-Gebrawi, where the scene of cattle crossing water appears to lose the significance of the activity and at Meir is only included in one tomb, that of *Ppj-ʿnḫ/Ḥnj-km*.[882] Such localised trends are difficult to interpret but the increasing absence of cattle-fording scenes at Meir and Deir el-Gebrawi may have been connected with the opening up of unexploited land in two provinces where the Nile valley expands to provide a greater extent of fertile land.

The role of cattle in the economy of the estate and the state

Moreno Garcia has raised the question of the true role of cattle in the economy of the Old Kingdom state.[883] He asks whether the animals pictured and referred to in tombs are privately owned or a reward from the crown or part of an establishment granted by the crown to an official who would manage the estate. Moreno Garcia also considers the economic reasons that might explain the crown's interest in encouraging the increase of herds and using cattle in the exploitation of provincial resources in Dynasty 6. His findings are that the crown's interest was to develop resources in provinces with land available for exploitation, as in the middle nomes of Upper Egypt, and to provide for expeditions and workers of the crown. The autobiographical statements of provincial administrators such as *Ḳ3r* of Edfu, *Hnkw/Jj…f* (II) and *Jbj* of Deir el-Gebrawi and the enormous census numbers provided by the Meir administrators support the conclusion as it relates to the Sixth Dynasty. Yet these conclusions only appear to be part of the story when these questions and findings of Moreno Garcia are applied to the decorative programmes of the Memphite tombs of Dynasty 5 and early Dynasty 6. The detailed scenes of cattle husbandry, in particular of the annual journey of the herds to the marshland and back, and those with vignettes of herders living with their charges, suggest that cattle were an essential part of the balanced, mixed farming economy of a large estate. Scenes of cattle being transported by boat suggest they were also valuable as taxation in kind transported to the capital so that the workers employed on royal constructions could be fed

[873] Galan (1994) 96. van Walsem (2006) 303-305, who deals with this issue, also questions the logic of Galan's interpretation that the theme symbolises a vital afterlife belief, when it is only represented in a minority of provincial tombs.

[874] Galan (1994) 92-3 does cite the Pyramid Texts but the references do not relate to bulls in combat with each other.

[875] Lloyd (1978) 613-15.

[876] Galan (1994) 91-2.

[877] van Walsem (2006) 304-305.

[878] Kanawati (1991) 54.

[879] These captions have been collected by Galan (1994) 84 – 86.

[880] In the registers beneath *Ḥnj/Špsj-pw-Mnw*'s bull fighting scenes there are depictions of a bull mating with a cow and a cow giving birth: Kanawati II (1981) fig. 20.

[881] A suggestion by Linda Evans.

[882] The herd follows two skiffs, with a calf tied to the stern of the second craft. The three men on the skiff seem to be ignoring the following cattle although a caption over their heads records the herder in the prow warning of a lurking crocodile. The two men in the first skiff are bringing a variety of offerings received by *Ḥnn-jt* (*Ppj-ʿnḫ/Ḥnj-km*'s son), while men on two skiffs behind the herd are harpooning hippopotami: *Ppj-ʿnḫ/Ḥnj-km* Blackman–Apted (1953) pl. 30.

[883] Moreno García *RdE* 50 (1999) 241-42.

with high quality nourishment. Thus cattle were both a valuable resource enjoyed by the tomb owner and, as the findings of the Giza Mapping Project make clear, also a vital state resource. The rate of this 'taxation in kind' was no doubt based on census returns but the level of this charge and the extent to which the crown depended on it cannot be assessed.
The analysis of the presentation of cattle in scenes of parades and husbandry demonstrates that cattle were an important element both in the funerary ritual of tomb owners and in the strategy to exploit the various types of land of the Nile valley. All ruminants can turn pasture into commodities such as meat and milk, which are valuable to humanity, but in the context of the Nile valley domesticated cattle were the most productive species that could be driven in herds from upland pastures in summer when natural grasslands dried out, across fields of stubble or left fallow and safely induced to swim across stretches of water to marshlands where they could survive on grasses and sedges rich in amino-acids. The ability to drive the cattle down to the marshlands each summer meant that herds could be much larger than if retained year round in upland pastures, when and where they would have had to be culled or would have required supplementary fodder once the seasonal pasture dried out.

14.3.2 'Small Cattle': Sheep

From early Dynasty 4 to Dynasty 8 sheep are only portrayed in a parade of animals in six of the tombs used in the study.[884] Apart from two scenes of sheep on the threshing floor, all other scenes of sheep portray them as a flock driven over ploughed land. Whereas cattle, goats and desert animals are represented as having intrinsic value, as in the case of donkeys sheep are shown as animals whose worth lies essentially in the work they perform for human beings.

With the exception of three Memphite and two provincial scenes, sheep disappear altogether from the funerary record in the Sixth Dynasty. The Memphite scenes date to early Dynasty 6, while the two provincial scenes, dated to late Dynasty 6, only represent sheep in a parade of animals.[885]

14.3.3 'Small Cattle': Goats

Prior to the second half of the Fifth Dynasty goats are only securely attested in three tombs[886]. In two of these tombs, those of *Nfr* and *K3-ḥ3.j* and *Nj-ʿnḫ-H̱nmw* and *H̱nmw-ḥtp*, they are portrayed as a small herd on scrubland reaching up into trees to eat the foliage.[887] The goats are accompanied by a herder, who uses a long pole hooked at one end to pull down high branches for his charges, a depiction in both chapels.

In the Memphite tombs of the second half of Dynasty 5 and early Dynasty 6 scenes of browsing goats are perhaps a little more prolific although still comparatively rare.[888] The portrayal of husbandry of the herd remains the dominant motif in these scenes but the portrayed degree of care is of a limited nature, In particular it concentrates on the provision of adequate fodder for the herd from terrain that appears to provide little pasturage. Only the chapel of *Ḥtp-ḥr-3ḫtj* contains a scene of the herd grazing on herbage; on the upper sub-register four goats nibble at a variety of low growing plants.[889] In all the other tombs of this period, the herd is portrayed browsing on trees and bushes, either rearing up to nibble from the lower branches or depicted within the tree canopy or on sub-registers level with the canopy. Herders protect the flock and knock down branches to make foliage accessible. A herder is sometimes depicted in the vignette of a female giving birth but does not attempt to help the female in the manner portrayed for parturient cows, although in two scenes (*Nj-ʿnḫ-H̱nmw* and *H̱nmw-ḥtp* and *3ḫtj-ḥtp*)[890] the herder raises a stick as though to protect the new born kid from a waiting predator. In six scenes, a herder is portrayed skinning the carcass of a goat that hangs from a tree, perhaps indicating that slaughtering a goat was a perquisite of goat herders.

Goats may appear to browse on the canopy of trees that have been chopped down[891] but it is doubtful that the trees were being felled in order to give the herd of goats access to the higher foliage. It is most unlikely that entire trees, scarce in Egypt, would have been sacrificed for such a temporary purpose especially when goats can climb trees and balance on branches to browse on the foliage. It is much more likely that the herd was taken to where trees were being felled for other purposes so that they could browse on the foliage once the tree had fallen, a thrifty use of a timber by-product.[892] This may in part explain the statements of *Hnḳw/Jj…f* of Deir el Gebrawi who claims that as ruler of U.E. 12 he developed the exploitation of unused land:

> *jw grg.n(.j) j3t.s nb m rmt mnmnt ʿwt m bw m3ʿ n ḏd(.j) grg jm*
>
> 'I settled all its mounds with men, large cattle and small cattle, truly, and I do not tell lies therein'

and

> *jw gr mḥ.n(.j) wḏbw.s m mnmnt mḫrw.s m ʿwt jw gr ss3.n(.j) wnšw nw ḏw ḏrjwt nt pt m ḫ3w n ʿwt j mr(.j) 3ḫ s ntj jm.s*
>
> 'Moreover, I filled its banks with cattle, and its pastures with small cattle. Furthermore, I satisfied the

884 TABLE 6: PORTRAYAL OF FLOCK OF SHEEP.
885 TABLE 6: PORTRAYAL OF FLOCK OF SHEEP.
886 There may have been a herd of goats depicted in the tomb of *Nb.j-m-3ḫtj* but very little of the scene survives: Hassan IV (1943) fig. 75). The third tomb is that of *Sḫm-k3-Rʿ* LD *Erg.* 36, Hassan IV (1943) fig. 57, Harpur *JEA* 71 (1985) fig. 10.
887 *Nfr/K3-ḥ3.j* Moussa–Altenmüller (1971) pl. 1, 4, 18, 19 and *Nj-ʿnḫ-H̱nmw / H̱nmw-ḥtp* Moussa–Altenmüller (1977) pl. 20-1, fig. 8.
888 In the period of the second half of Dynasty 5 to early Dynasty 6 herds of goats appear in six Memphite and four provincial tombs. SEE TABLE 7: MANAGEMENT OF DOMESTIC GOATS.
889 Mohr (1943) fig. 21.
890 *3ḫtj-ḥtp* Ziegler (1993) drawing 137-9.
891 *Nb-k3w-ḥr/Jdw* Hassan *Saqqara* I (1975) fig. 12.
892 See Klebs (1914) 24, Schäfer (1986) 253-255, Evans (2006) 217-22 and Vandier *Manuel* V (1952-1978) 86-92 for discussion of the relationship between tree felling scenes and depictions of browsing goats.

jackals of the mountain and the kites of the sky with the hides of the small cattle, for I wished the man in it to benefit.'[893]

Goats would have been ideal animals to support the clearing of new land for farming and settlement.

From mid Dynasty 6 on and into Dynasty 8 the thematic emphasis changes. Herds of goats are included with cattle and desert animals in the parade of animals before the tomb owner, which may imply that goats were now regarded as a more important resource in Upper Egypt. In fact, in the provinces of Meir and Deir el-Gebrawi herds of goats may have taken the place of desert ungulates as the animals that could survive in the more arid terrain of the estate and as browsers did not compete with cattle for grazing land.

14.3.4 Donkeys

Donkeys appear in two major motifs: working for man and, less frequently, in the parade of livestock. Most portrayals of donkey are associated with the grain harvest. In this sequence donkeys appear in two themes, both of which represent the animals at work, either as a means of transporting the produce of the estate or threshing the grain harvest.[894] Unlike portrayals of cattle, and to a lesser extent goats, donkeys are never depicted in scenes indicating husbandry; they are not shown mating or giving birth and foals only appear walking beside their mother. Donkeys are never portrayed grazing freely.[895]

The harvest themes are seasonal, yet donkeys would have been needed throughout the year. On a large estate there would have been many uses for donkeys, working singly or in a pack. At certain times of the year donkeys would have been needed to transport large quantities of fodder that must have been required by the herds of cattle and captured desert animals, as well as constant quantities of grain for the poultry yard. Pack animals would surely have been needed on hunting expeditions for wild animals, although this is only portrayed once, in the chapel of *Nj-ꜥnḫ-Ḫnmw* and *Ḫnmw-ḥtp*, in a vignette of a crate of baby antelopes on the back of a donkey.[896]

The value of a herd of donkeys

Donkey power and transport would have been essential on a large estate. Yet the size of herds claimed by tomb owners seems out of proportion to the needs of an individual estate. The dwarf *Snb* added the figures '12 017 males' and '10 200 females' above the depiction of three donkeys.[897] On the other hand, the figure given by *Wr-jr.n.j* of Sheik Said is only 100 while that of *Rꜥ-ḫꜥ.f-ꜥnḫ* is 760.[898] If the very large numbers are accepted literally, two questions arise. Why was such a large herd of donkeys needed on an estate? Could an estate credibly feed thousands of pack animals? Certainly large herds were required for the transport of commodities beyond individual estates.[899] There is ample evidence of Old Kingdom activity to and from oases, mines and quarries as well as expeditions to neighbouring lands.[900] In early Dynasty 5 King Sahure recorded that he brought back as booty from his Libyan campaign 223,400 donkeys.[901] While the actual figures may be questioned, it is likely that they represent the information that estates maintained particularly large donkey herds in order to provision expeditions to neighbouring lands and to quarries and mines. Harkhuf, who made four expeditions to lands south of Egypt, records the return from his third journey with 300 donkeys.[902] If he received his commission in Memphis and set out on his journey from the capital, he would not have needed a donkey herd until he left the river to strike across country. This would also apply to all other expedition leaders, who would have collected their pack animals at one of the departure points from the Nile. Snefru and Khufu, for example, sent teams through the Eastern Desert to the Wadi Hammamat and to the Sinai[903]. The names of Snefru, Khufu Sahure, Neuserre, Menkauhor, Djedkare, Pepy I and Pepy II are inscribed at Wadi Maghara in the south west of the Sinai peninsula[904]. The men who led expeditions in the Old Kingdom would have taken large herds of pack animals to transport food and water and bring back the heavy loads produced in mines and quarries or bales of tribute and trade from surrounding tribes. Donkeys were thus important to the economy in many ways, although in officials' tombs, apart from rare exceptions referred to above, their value is only reflected in agricultural scenes until Dynasty 6, when clear evidence of breeding programmes occurs. At el-Hawawish a *sḥḏ ḥm-nṯr* and *ḥrj-tp nswt pr ꜥ3* named *Qrrj* recorded that he had acquired a number of female donkeys from which he bred a herd of 200.[905] In the second half of the dynasty herds of donkeys begin to appear on registers of animals being presented for the census to the tomb owner. *Ppj-ꜥnḫ/Ḥnj-km* of Meir and *Ḏꜥw* of Deir el-Gebrawi include donkey herds among their parade of animals but do not give numbers of their

893 Davies *Gebrâwi* II (1902) pl. 9, 15 and Kanawati *Gebrawi* 1 (2005) pl. 66-67.

894 See TABLE 8: REPRESENTATIONS OF DONKEYS. In three scenes donkeys are pictured carrying the tomb owner. See also Moussa–Altenmüller (1977) pl. 42, 43; LD II 43 [a]; Hassan *Gîza* V (1944) fig. 104; Martin (1979) pl. 33 [90].

895 In order to get sufficient nourishment, donkeys either have to graze for six to seven hours a day but if required to work, they have to be supplied with additional fodder. Evans (2006) 196-197; Aganga–Letso–Oaganga (2000) "Feeding donkeys" *LRRD* **12**(2)

896 Moussa–Altenmüller (1977) fig. 13.

897 Junker V (1941) fig. 19.

898 *Rꜥ-ḫꜥ.f-ꜥnḫ* LD II 9.

899 No camels appear in Old Kingdom scenes or are referred to in inscriptions, so it may be concluded that they were not used in Egypt in this period, although this surmise has been challenged by Ripinsky (1985) 134–41. Camels had probably been domesticated in the Arabian peninsula between 3000 and 2500 BC: Compagnoni, B. and M. Tosi, (1978) "The camel: Its distribution and state of domestication in the Middle East during the third millennium B.C. in light of the finds from Shahr-i Sokhta" in Meadow–Zeder (1978) pp. 119–128

900 Frank Förster (2007), Bloxam–Storemyr (2002).

901 Borchardt (1913) pl. 1.

902 Sethe (1903) I 120–31, Lichtheim (1973) 23–7.

903 Kaper–Willems (2002) 79–94.

904 Sethe (1903) 7-8, 32, 53-54, 55, 56, 91-92, 112-13.

905 Kanawati VI (1986) 48, fig. 20b.

herds as they do for other domestic animals.[906] In Dynasty 8 *Mrjj-ꜥꜣ* of el-Hagarsa also portrays donkeys among the parade of domestic animals.[907]

14.4 Resources of the inundated regions of the Nile valley

It is possible that there was a multi-level agrarian economy in operation in the Old Kingdom consisting of large parcels of land (referred to as *njwwt* and *ḥwwt* in inscriptions) with herds of cattle, producing considerable surpluses of grain and flax and a subsistence economy practised by peasants on much smaller parcels of land. There may also have been a middle rank of farmers represented in the Middle Kingdom by the property of Hekanakhte.[908] In the Old Kingdom, however, the available evidence applies only to the great properties. These estates supported the tomb owner, his immediate family and dependants, as well as the overseers and farmhands who worked on the estate, and would have paid taxes to the crown.

14.4.1 Cereals

Both Murray and Samuels maintain that barley and emmer wheat are the only two grain crops proved to have been in use in pharaonic times and that these two crops provided the bread and beer which were the dietary staples of the entire population.[909] In coming to these conclusions Murray and Samuels use evidence drawn from the entire period, from pre- to post-pharaonic times. The iconographic data relating to estate activities in the Old Kingdom, together with associated captions, reinforce their conclusion. This data consists of images showing grain crops to be the most economically important of all the estate's resources. A graphic composition, designed to depict each stage of the process of harvesting and storage, represents teams of field hands at work, with the storage of the precious crop checked at every step by overseers with scribes and criers keeping account.

The pictorial material, however, presents a number of problems of interpretation:

- It is difficult to distinguish which grains were most frequently grown, although captions referring to barley (*jt*) and emmer wheat (*bdt*) support Murray's suggestion that barley was more important in the Old Kingdom.[910] Yet the available quantitative data are largely from the Predynastic period. The evidence from the Nagada Predynastic site shows a 2:1 ratio in favour of barley over emmer wheat, perhaps because barley has a greater resistance to varied degrees of ground moisture and to saline conditions.[911] In the past, material finds from the Pre-dynastic and Dynastic periods was not carefully collected and classified. In addiion, there are problems of translation. The term *swt*, for example, has been variously translated as wheat and barley.[912]
- As with the theme of planting seed,[913] scenes illustrating the final stages of preparing the grain for storage (winnowing and sweeping) are difficult to interpret with precision, perhaps because this part of the process is inadequately understood. This has led to previous misreading of the artistic record. Murray and Samuel insist that both barley and emmer wheat grains were stored still hulled, that is, in spikelet form, and that the additional processing was not applied until the grain was removed from the granary to be ground for baking.[914]
- Neither images of crops nor associated captions provide clear information regarding extent of land under crop or harvest yields.

Historical development of the iconography of field crop production

In the tombs of early Dynasty 4, the minimal representation of the motif of agriculture contrasts with the more expansive and detailed treatment given to depictions of marshland resources and desert hunting. This may suggest that the final transfer of Egyptian society's reliance on animal husbandry, fishing, fowling and hunting to a critical dependence on cereals is reflected in Old Kingdom iconography. On the other hand, the impression given by the iconography that agriculture held less interest for the early Dynasty 4 high officials than did the hunting in the marshland and desert may be a reflection of the personal interests of an aristocracy still wedded to elite pursuits rather than to the responsibilities of a bureaucracy.

From mid Dynasty 5 on, the proportion of wall space devoted to agriculture is enlarged so that in many tomb chapels it equals or exceeds the space occupied by marshland resources. This supports the contention that grain provided the resource that was essential for the survival of the estate and the material well being of the tomb owner, family and dependants.

The organization of field crop production

The cultivation of field crops on these estates is portrayed in scenes grouped thematically around the concepts of planting, harvesting and storage. At least one scene representing field crop production is included in all but three tombs used in this study. More so than for any other resource, from mid Dynasty 5 onwards the iconography of grain crop production is organised in a sequence of

906 *Ppjj-ꜥnḫ/Ḥnj-km* Blackman–Apted *Meir* V (1953) pl. 32, *Ḏꜥw* and *Ḏꜥw/Šmꜣj* Davies *Gebrâwi* II (1902) pl.6.

907 *Mrjj-ꜥꜣ* Kanawati (1995) pl. 37.

908 James (1962). The average size of a farm in the New Kingdom was only one and a quarter hectares (five arouras): O'Connor (1983) 227. With low yielding strains of crops this would scarcely have supported a family of eight from one harvest to the next. Family farms, if they existed, may have been larger in the Old Kingdom.

909 Murray (2000) 505-536, Samuel (2000) 537-576.

910 Murray (2000) 512.

911 Wetterstrom (1993) 165-226.

912 Germer (1985) 14:210, Darby et al (1977) 490-91.

913 It is not always possible to work out the relationship between the plough, sower and the flock of sheep: Siebels (2000) 96 and note 151.

914 Samuel (2000) 538, 545; Murray (2000) 527.

detailed scenes. The method of ploughing, the number of workers needed to control the plough, the work of the sower, and herders controlling the flock of sheep all portray the process of soil preparation and sowing. The elaboration of detail is even greater in harvesting scenes: the different techniques needed to harvest flax and cereals, the bundling of the various crops, transporting sheaves of barley and wheat for threshing and winnowing, and the final, carefully supervised storage of the grain.[915] If the piling up of sheaves and of threshed grain are counted as separate motifs within the harvesting theme, at least ten consecutive stages are depicted. No tomb includes every stage of this sequence, but when the scenes from different tombs are combined, every step in the sequence of grain production is presented in detail.[916] No other resource is given such detailed treatment. The motif of soil preparation and planting covers the longest time-span, from early Dynasty 4 to Dynasty 8,[917] while harvesting themes appear infrequently until mid Dynasty 5, from which time they are as regularly included in tomb repertoires as the ploughing theme.[918] The theme of storing the grain harvest in granaries begins to appear in mid Dynasty 5[919] but is never so frequently depicted, although it continues to the end of the Old Kingdom.

While work crews are typically called the crew/s of the estate (*jz/wt pr-ḏt*) a number of provincial inscriptions describing the harvesting of wheat and barley make a distinction between crews of reapers by giving them different labels (*jz wr, ḥmw nswt*). Various reasons have been suggested for the difference in labels. The reapers called 'servants of the king' may have been allotted to tomb owners by the crown or the land they worked on was crown property administered by a land owner. Again, tomb owners may have controlled and therefore could make use of the crown's workforce in the regions they administered. On the other hand, the three labels (*jz pr-ḏt, jz wr, ḥmw nswt*) may denote that the harvest yield was to be divided three ways, for the tomb owner's mortuary cult, his household and the taxes to be paid to the crown.

Changes in the representation of techniques of arable farming

The two modifying elements in the scenes of ploughing and sowing are the depiction of men using hoes and of a flock of sheep driven over the soil. As there are only six scenes of hand-held hoes, all of which are Memphite and span the period mid Dynasty 5 to early Dynasty 6, there is insufficient evidence to account for the appearance, disappearance and precise purpose of this method of soil preparation. As Murray notes, the order in which the preparatory activities of tilling, sowing and trampling by sheep is unclear from the iconography but may have depended on the condition of the soil.[920] The other major modification is the disappearance of the flock of sheep early in Dynasty 6. As sheep disappear from Memphite as well as provincial tombs at this time, their absence may not reflect specifically provincial circumstances.[921] Dispensing with a flock of sheep to trample the seed into the ground may signify a new, improved planting technique or the use of soils holding less moisture, in which case using sheep would have been disadvantageous. Precipitation, probably never very significant, was decreasing,[922] but changes to the level of the inundation and primitive basin irrigation may have produced reduced herbage, while the development of the cattle industry in Dynasty 6 would have increased the demand for grazing land for cattle. Sheep had insignificant value for elite officials. They are rarely included in the parade of livestock and wool was not used for cloth making. In Dynasty 6 many officials may have dispensed with their flocks.[923] Other features of planting and harvesting field crops, with the exception of men using hoes and driving flocks of sheep, continue to be portrayed through to Dynasty 8.

The only change to the theme of harvesting is the disappearance of the herd of donkeys returning to the field for bales of grain. The final Old Kingdom scenes to include this feature occur in the Memphite tombs of *Mrrw-k3*, *Mḥw* and *Mṯtj*[924]. Its ensuing absence may be due to lack of wall space, as it is absent in the smaller Memphite tomb of *Ḥzj* and in all Dynasty 6 tombs of Upper Egypt, which are typically rock-cut with less wall space suitable for decoration. There appears to have been no lack of donkeys in the provinces as they are depicted hauling bales of grain in thirteen Upper Egyptian tombs.[925]

Egypt's staples

Huge dumps of pottery fragments, largely consisting of bread mould and beer jar debris offer strong evidence that the products of the grain crops provided Egypt's staple

915 The only omission is the tending to the growing crop. There are no scenes of weeding or irrigation.

916 No tomb contains a scene of weeding or cultivating the growing crop.

917 It is probable that the subject was included in every decorated Old Kingdom tomb.

918 A harvesting scene occurs in the Medum tomb of *Nfr...w*: Petrie (1892) pl. xxviii [6], but this tomb (No. 22) may be one of a later group of tombs. See Petrie (1892) 18-19.

919 The motif may be portrayed in the tomb of *Mrs-ʿnḫ* III Simpson–Dunham (1974) pl. 12(b).

920 Murray (2000). The inclusion of both methods of soil preparation, ploughing and hoeing, may indicate that soil requiring more work was being exploited and that hoeing was an adjunct to the preparation of the soil.

921 A suggestion by Harpur (1987) 163.

922 Hassan (2005) 2-3.

923 The omissions may reflect developments in farming according to the thesis of Richard Redding, that as a community increases its dependence on growing cereals, so its production of sheep will decline while at the same time cattle and goats will become more important economically. His premise is based on four conditions: the particular dietary requirements of the three species, their physiology, the secondary products of these animals and use as beasts of traction and burden, and labour needs. More cattle will be needed to support activities associated with increased cereal production, while providing more milk and superior dung for cooking. See Redding (1984) pp. 161-70 and Redding (1991) pp. 20-30.

924 Respectively, Duell II (1938) pls. 168-70; Altenmüller (1998) pl. 22 [b]-25; Kaplony (c1976) 22-4, No. 3.

925 *Qrrj* of El Hawawish: Kanawati VI (1986) pl. 8b, fig. 20b, p. 48, stated that he 'acquired a (herd) of breeding she-asses, and they produced 200 asses' (*jw jr.n (.j) ḥmwt bnnt jr rdj.n.sn ʿ3 200*).

food and drink.[926] However, the paramount value of the grain crop to the tomb owners who administered the great estates is demonstrated by the portrayal of the care taken in planting cereals and handling the harvest, while the organization of the storage of the crop reveals the effort made to achieve accountability in its preservation. The grain harvest had to be rationed out so that it would feed the estate's dependants for an entire year.[927] Unlike any loss of wildfowl and fish, losses of grain could not be made up until the following harvest. The cereal crop not only provided the staple for humans, it also must have provided fodder for herds in the form of post harvest stubble and cakes of bread, and feed for farmed poultry. Furthermore, it was a commodity that could be taxed in kind, although the extent of such taxation cannot be assessed.

14.4.2 Flax

The field crops depicted in wall scenes are flax (*mḥꜥ*), barley (*jt*)[928] and emmer wheat (*bdt*). Ploughing and seed sowing scenes rarely indicate which crop is being planted.[929] Harvesting scenes, however, clearly distinguish flax from cereals.[930] The harvesting of flax is first attested in early to mid Dynasty 5 and continues to appear until late in Dynasty 6. It is not confined to Memphite tombs and is portrayed in tombs of the Upper Egyptian middle provinces and as far south as Qubbet el-Hawa.[931] Scenes of the flax harvest only appear in 31 tombs,[932] a much smaller number than for the grain harvest.[933] The reason for the reduced number of flax scenes may have been a matter of available arable land. On all estates the first priority would have been to grow sufficient grain to ensure the community's survival until the following harvest. Many estates may not have had surplus arable land that could be used for flax. Furthermore, the marked decrease in number of tombs with flax harvest scenes in Dynasty 6 suggests a reduction of arable land available for growing flax. All the Dynasty 6 Memphite scenes are in tombs of viziers, while four of the five provincial scenes are in the tombs of the greatest provincial administrators. This may indicate increased pressure on many estates to produce sufficient grain and that only the highest officials had estates large enough to provide surplus arable land for a flax crop.

Although only the actual flax harvest is portrayed in Old Kingdom tombs,[934] there is a considerable body of Old Kingdom material and inscriptional evidence that linens of various qualities were produced, some of which were of very fine. Linen had a variety of uses: for the household, as mummy bandage, sails for boats and for rope making, while the flax seed was crushed to produce linseed oil. As a medium of exchange linen was used as payment for services.[935]. Abundant flax harvests provided officials with a resource with which to purchase goods and labour. The state's interest in linen is reflected in the number of titles relating to linen held by important officials and the responsibility of the collection and redistribution of linen as funerary equipment for officials.

From mid Dynasty 5 on the amount of wall space in tombs devoted to the production of grain indicates that elite tomb owners were representing the economic importance of these resources. While flax was a crop with many uses, the grain harvest was indispensable for the maintenance of society, which is also reflected in the degree of detail with which each crop is represented.

14.5 Marshland resources

14.5.1 Papyrus

Importance of papyrus as an economic resource for the great estates

Although no inscription makes it explicit, it may be accepted that scenes of gathering of papyrus, the skiffs and the items they transport represent the property of the tomb owner's estate. It is highly unlikely that the tomb owner would have used large sections of chapel walls to portray the activities of labourers working on their own behalf or that he would have pictured himself viewing scenes that did not relate to his own assets. However, no scene of collecting papyrus or using it as a raw material contains a caption specifically relating to the interests of the tomb owner. Papyrus scenes located on chapel walls are either part of a panel or a register devoted to a range of marshland activities or to estate work in general. Where the tomb owner is depicted viewing these activities, the vertical inscription identifying his interest is either a general statement such as (*m33/ḫnz*) *pḥww zšw jn z3t nswt nt ẖt.f rn.s Sšsšt* ('[Viewing/traversing] the marshlands and ponds by the king's daughter of his body, her great name, Seshseshet')[936] or, if more specific, refers to a major rather than an auxiliary activity involving the collection and usage of papyrus, such as *m3[3 sḫt] 3pdw m*

926 The massive dumps excavated by the Giza Plateau Mapping Project in 2005, and the bakeries and vats excavated in 1991 and 2007 must have been in use to feed the thousands of workers brought each year to Giza.

927 This procedure is best illustrated in the Hekanakhte Papers, Letter II: James (1962).

928 In *Snḏm-jb/Jntj* barley from Upper and Lower Egypt is distinguished as *ḫ3 Šmꜥ* ('a thousand of Upper Egyptian barley') and *ḫ3 mḥj* ('a thousand of Lower Egyptian barley'). Wb. 4, p. 476, 8–477, 7; LD *Erg*. pl. 21; Brovarski (2001) 68 and note 465.

929 Exceptions: *Rꜥ-ḥtp* (*sk3 bdt* 'Ploughing emmer wheat') Petrie *Medum*, pl. 12; *MḥW* Altenmüller (1998), fig. 8, p. 143.

930 The flax plant is pulled out of the soil, while cereals are cut with sickles.

931 Old Kingdom scenes of the grain harvest have a longer life span as their portrayal continues at el-Hagarsa to Dynasty 8. Furthermore, grain harvest scenes appear more frequently, with a provincial life span from mid Dynasty 5 to Dynasty 8.

932 van Walsem is referring only to the Memphite scenes included in the Leiden Mastabase. *IBAES VI*, 292.

933 The Oxford Expedition to Egypt: Scene-details Database, which includes provincial tombs, records 48 flax harvest scenes and 73 grain harvest scenes.

934 Scenes in the Abusir tomb of *Ftk-t3* represent the sequence of soaking, softening and drying the fibre and the ensuing processes of spinning and weaving, which otherwise do not appear until the Middle Kingdom: LD *Erg*. 40 [c]; Bárta (2001) figs. 3.11, 3.18.

935 See Roth (1994) 236-9 for a list of tomb owners who record that they paid for work on their tomb in part with linen.

936 Kanawati–Abder-Raziq (2003) pl. 17 and 54.

mḫt ('Viewing the trapping of birds from the marshlands.').[937]

Papyrus supported both the physical work and the administration of the estate. It provided transport across the waterways so that the other resources of the marshlands such as fish, birds and feed for cattle herds could be exploited and underpinned this work by providing the raw material for containers and rope for cartage, as well as writing material needed for the recording and maintenance of accounts. The variety of types of skiffs depicted, together with their diverse uses, highlight the extent to which an estate depended on the marshland and its waterways and how vital these were to the good running of estates. As the product of a natural ecosystem, papyrus was an invaluable raw material which in turn made possible the maintenance of large herds, the addition of high quality protein (fish and fowl) to the diet of workers and families dependant on the estate and, to a lesser degree, afforded estate workers and perhaps the tomb owner with a means of exchange on local markets.

14.5.2 Bird life

As a 'bridge' between Eurasia and Africa, the Nile valley is one of the most important corridors for birds of the Palearctic migrating seasonally from their northern breeding grounds. Even though today much of the Delta has been drained for agriculture, lakes in the Nile Delta still attract large flocks, as do the Fayoum and other major oases and the Nile valley. Since the drying up of the Sahara and with vast expanses of desert to the east and west of the river, these locations have provided habitats for birds that winter in Egypt and life-saving stopovers for flocks wintering further south in Africa.

While many bird species are depicted in a variety of marshland scenes, a comparatively small number of species, mainly water birds, are depicted trapped in clapnets and transported to poultry yards in Old Kingdom scenes. These were species that could be caught in large numbers, maintained as flocks in poultry yards and fattened for the table: geese and ducks, the precise species of which it is usually impossible to identify, cranes and turtle doves. Many of these species are migratory and flocks would have been attracted to the wetlands of the Delta, the Nile river and valley marshlands and large oases, either to winter there or as a stopover on their flight north or south.[938] Some species such as the greylag and the white fronted goose and the pintails and teals, breed in northern lands but migrate in large flocks to winter further south and domesticate easily. As well as being good eating, geese and ducks are a rich source of fat for which Egyptians had many uses. Although it is only from late Dynasty 5 to mid Dynasty 6 that these birds are depicted in the tombs of the highest officials in poultry yards and being force-fed, they appear in bird processions throughout the Old Kingdom, from Dynasty 4 to Dynasty 6. With their long stick-like legs and long beaks, cranes (*Grus grus* and *Anthropoides virgo*) can be more readily identified in wall scenes. They, too, are migratory visitors to Egyptian marshlands that readily accept domestication.

The supply of birds

With waterbirds being the main species that Egyptians domesticated, the wetlands of the Delta, oases and the marshy fringes of the oxbow lakes and channels along the Nile valley provided the locations for large-scale trapping. The clapnet, designed for use on a stretch of water, was an efficient method for trapping birds that could be lured to land on pools and streams. Scenes present teams of men operating clapnets designed to catch birds in large quantities. Each team has a leader, who silently gives the signal to the team to haul on the rope that will close the net when the birds have settled. He, and sometimes a second figure, extract the birds from the net, transferring them to cages to be presented to the deceased and taken to his poultry yard. The scenes present a picture of orderly, systematic activity with the human participants organised according to an accepted division of labour. The importance of this activity is highlighted in two ways. Tombs with two, sometimes three depictions of the clapnet and team of haulers present the two main stages in operating the net successfully, while in a number of scenes an overseer of birds supervises the operation.

The trapping of birds was a seasonal practice, taking place in spring and autumn when great flocks of migrating fowl flew along the Nile valley. Birds could only have been lured into the clapnet in large numbers when large flocks flew overhead, migrating north in spring and south in autumn to winter. Clapnet scenes show flocks of birds above the net as well as papyrus and lotus in blossom.

Maintenance of flocks

Twelve of the poultry yard scenes are associated with images of attendants force-feeding birds. The owners of most of these tombs were among the very highest officials in the land.[939] Their fowl yards appear to have been highly organised undertakings operating on a considerable scale. The structures and layout of the poultry yards, the use of workers to feed the flocks and maintain the property suggest that this was an enterprise for the greatest estates, which had the capability to employ poultry yard workers throughout the year. These estates would also have to be sufficiently productive to have surplus quantity of grain to feed the flocks. Poultry yard scenes present depictions of workers pouring sacks of grain on to the ground and birds swimming contentedly in rectangular ponds, whose water level

937 *Jj-mrjj:* LD *Erg*. pl. 6; Weeks (1994) fig. 40, pl. 25 (a, b).

938 It is possible to distinguish the Pintail from the shape of its tail feathers.

939 Six of these officials, *Ptḥ-špss, Mrrw-k3, K3-gmnj, Ḥzj, Mrrj, Mrjj-Ttj/Mrj,* were viziers; *Ṯjj* was *jmj-r3 sš ʿ nswt, jmj-r3 k3t nbt nt nswt; Wḏ3-ḥ3-Ttj/Ššj/Nfr-sšm-Ptḥ* was *jrj pʿt, ḥ3tj-ʿ* and *Nj-k3w-Jzzj* was *jmj-r3 k3t nbt nt nswt*. The two provincial administrators were powerful officials. As well as being close to the royal family, *Jbj* was *jrj pʿt, ḥ3tj-ʿ, jmj-r3 Šmʿw, jmj-r3 prwj-ḥḏ, jmj-r3 šnwtj,* and *Ppj-ʿnḫ/Ḥnj-km* was *t3tj z3b ṯ3tj* and *jmj-r3 Šmʿw*.

would have had to be regularly replenished. Not only did the poultry yard supply fowl for the table throughout the year, it also provided specimens whose flesh had been tenderised by captivity and diet. Birds caught in the wild can be eaten immediately but their flesh is tough unless the carcass is allowed to hang for some days allowing enzymic and bacterial action to tenderise the meat and improve its flavour.

The considerable reduction in scenes of the poultry yard and the procession of birds in the second half of Dynasty 6 may indicate the inability of even the greatest estates to maintain large flocks of birds. The year-round maintenance of flocks of fattened poultry would have been a luxury as well as a resource.

14.5.3 Fish and fishing

Scenes from Old Kingdom tombs provide evidence that many fish species, from the huge Nile perch to shoals of mullet and to various species of catfish was harvested in abundance from the waters of the Nile valley.[940] The species of most fish depictions can be identified from the shape of body and fins. Over 26 species are portrayed, either in nets, traps, hooked on a line, gutted or being transported.[941] Mullet appears to have been the most popular fish. It figures in more tombs than any other species and in over 80 per cent of gutting scenes.[942] Other frequently depicted species are *Tilapia, Synodontis schall*, *Citharinus*, *Tetrodon fahaqa* and *Lates niloticus*. The majority of depicted species (15 out of 26) appear in at least fifteen tombs of the list used in this study.

Fishing by dragnet is the most consistently portrayed technique, as well as being the method shown to catch fish in largest quantities. It also required the greatest degree of organization. A boat was needed to take one side of the net out from land and two teams, sometimes directed by an overseer of fishermen, pulled in the net heavy with the catch. The scenes of dragnet fishing do not show any critical change over time. The numbers of men and fish depicted increase until they reach a maximum in the scenes of high officials in the Teti cemetery in early Dynasty 6. The representation of other fishing techniques such as basket traps is largely confined to the period mid Dynasty 5 to early Dynasty 6, while images of single fishermen holding a line with hooks continues through Dynasty 6.

The theme of gutting, cleaning and drying fish first occurs in the tombs of early Dynasty 4.[943] In the tombs of Dynasty 5 it is closely associated with the actual landing of the catch either by being placed alongside the fishermen hauling in a dragnet or by the image of porters bringing fish to be gutted. The theme loses its prominence in the tombs of the great officials who served Teti and almost disappears from the tombs of the powerful provincial magnates. For example, it is absent from the tombs of the administrators of Meir and el-Hawawish and at Deir el-Gebrawi only appears as a minor theme on a register where the focus is fighting boatmen.

Depictions of bearers transporting fish occur in 34 of the list of tombs used in this study. That the scenes of fishing represent the organised work of the estate, rather than the activities of individuals, is further emphasised by these vignettes of the transport of fish from the banks of the river and streams. Very large fish are either depicted being carried attached to a pole or a paddle held on the shoulders of two workmen[944] or by hand. Smaller fish are transported in bags or baskets or carried by hand.[945] Portrayals of the recording by scribes and distribution of fish by the overseers of fishermen, the distribution of fish by the team leaders to the workers and the final accounting in the presence of the *ḏ3ḏ3.t n pr-ḏt* make it clear that the fishing scenes represent the work of the estate.[946]

Fish was an important resource for the estate. It fed the tomb owner, his family and dependants as well as being used to pay workers and feed them and their families. Dried fish may have been an element in the local 'currency'. However, there is no evidence that fish from an estate was part of the estate's taxation in kind to the crown, presumably because even gutted and dried, it would not have lasted long enough to be transported in a usable condition. Consequently, fish was not a commodity that directly linked the estate with the economy of the state as did cattle, grain and flax.

14.6 Assessment

The detailed images of crop production, particularly of cereals, demonstrate very clearly that the protection and preservation of the harvest was of vital importance to an estate. The frequency of scenes of cattle parades and husbandry display most clearly the status value attached to cattle, while the value attached to hunting of desert animals and their inclusion in animal parades appears to have been primarily ritualistic. The additional portrayals of the flax harvest, gathering and usage of papyrus, the catching of fowl and fish present a picture of a well managed and well balanced estate economy.

940 At least nineteen different species are depicted in the tomb of *Ṯjj*, seventeen in the tomb of *Mrrw-k3* and eleven in the tomb of *K3-gm-nj*.

941 The number of different species portrayed in Old Kingdom tombs is inexact because a number of depictions of fish cannot be identified with certainty. See Van Elsbergen (1997) 133-307.

942 Van Elsbergen (1997) Ch. 3.

943 *Nfr-m3ʿt* Petrie (1892) pl. 18, *Rʿ-ḥtp* Petrie (1892) pl. 12.

944 In the case of *Rʿ-ḥtp* of Medum the pole is a paddle: Petrie (1892) pl. 12. *Nb.j-m-3ḫtj* exhibits three fish suspended from the same pole, as do *Rʿ-ḥʿ.f-ʿnḫ* and *Jntj* of Deshasha: respectively, LD II 12[b], LD II pl. 9 and Petrie (1898) pl. 5. In the scene of *3ḫtj-ḥtp* (Louvre) a bag holding a quantity of fish is slung from the pole: Ziegler (1993) 132.

945 In the scene of *3ḫtj-ḥtp* (Louvre) a bag holding a quantity of fish is slung from the pole: Ziegler (1993) 132. *K3-gmnj*'s scene demonstrates a variety of methods of transporting fish. Porters carry fish hanging from poles borne by one and two men, in woven baskets, suspended from rope and held in the hand: Harpur (2006) 16.

946 Perepelkin (1986) 173-74, van Elsbergen (1997) 102.

CHAPTER 15 A CHANGING ENVIRONMENT?

15.1 Change over time

The second objective of this study has been to identify changes over time in the pictorial and inscriptional record. This is in accord with the proposition that consistent and substantial changes in the portrayal of an important theme are likely to reflect economic developments affecting social conditions and attitudes, perhaps even crown policies. Such modifications may reflect significant economic or climatic trends and thus contribute to evidence of a socio/economic/political dynamic. The attempt to associate historical change with inferences drawn from funerary iconology, however, requires a chronology for the data dependent on an accepted dating of tombs. As this is in some instances a debateable issue, the monuments and thus the data they contain have been organised into broad periods of time that allow a broad chronological picture to emerge.

This section began with a brief description of the developed philosophy of the ka as it is encountered in the tombs of Dynasties 5 and 6. To what extent these ideas had been worked out by the early Fourth Dynasty is difficult to judge. Many of the features essential to this doctrine appear in the early tombs, which were provided with burial chambers and false doors, furnished with statuary placed within the serdab,[947] inscribed with ritual inscriptions and career defining titles and decorated with scenes of rural life. However, a dynamic that can be perceived but not easily explained was at work. The principal place of burial for high officials shifted from Medum and Saqqara[948] to Giza in mid Dynasty 4, and then to Saqqara in Dynasty 5. These shifts are associated with changing 'preoccupations' in tomb iconography, as the hunting, fishing and fowling themes occupying significant space in the cruciform chapels of Medum and Saqqara give way to ritual and family dominated iconography of the early L-shaped chapels. A further cluster of changes appears in mid Dynasty 5. Although architectural and iconographic modifications may be first seen at Giza,[949] these changes are more closely associated with the Saqqara necropolis. Additional rooms expand the area of the chapel providing increased wall capacity and a shift in iconographic emphasis to scenes in particular portraying mundane 'daily life' on rural estates. The decorative programme now incorporated the theme of a well-managed, productive estate and presented the tomb owner as its beneficiary, thus supporting the vision of a life well lived and a world the ka could inhabit.

This level of analysis avoids the search for cryptic symbolisms. It is based on the principle of allowing the programme of decoration to tell us what the broader preoccupations of tomb owners were at a given period of time and what aspects of his life the tomb owner wished to communicate to the living and to be remembered for. This approach raises questions concerning the dynamic at work in Old Kingdom society, namely, economic, social, political and ideological changes that might have affected and modified the preoccupations reflected in funerary iconology. Funerary and afterlife beliefs may have remained essentially the same throughout the Old Kingdom, yet the needs, expectations and viewpoint of the class that produced this art would not have been static and would have responded to significant social change.

Many modifications in funerary arrangements are perceptible:

- The location of chosen necropolis shifted from Medum and Saqqara to Giza in the Fourth Dynasty, and then from Giza to Saqqara in Dynasty 5, to Saqqara South and to the provinces of Upper Egypt in Dynasty 6.
- Tomb architecture was modified: from cruciform to L-shaped chapels, to chapels with extended east/west walls, to multi-room chapels, chapels constructed below ground level and the rock-cut tombs of the provinces of Upper Egypt.
- Changes in tomb elements such as the false door, construction of the burial chamber and the shaft leading to it were introduced.
- There was a significant increase in the number of officials with the status required to construct decorated tombs in the Memphite necropleis in the second half of the Fifth Dynasty, followed by the proliferation of tombs of provincial administrators in Dynasty 6.
- The apparent changes in the preoccupations dominating the iconography of tomb chapels went through four major phases: from desert and marshland hunting in early Dynasty 4 to family and ritual centred themes from mid Dynasty 4 to early Dynasty 5, to the incorporation of themes representing rural estates and 'daily life' from mid Dynasty 5 on, and in the provincial tombs of Dynasty 6 to a preoccupation with the size and increase in numbers of herds of livestock. The preoccupations of the tomb builders of Saqqara South are difficult to assess because, according to Jequier, only burial chambers were constructed. The above ground constructions are undecorated and do

947 Petrie (1892) 11-22.

948 The cruciform Saqqara tombs: *Mṯn* (Plan PM 49) and *3ḫtj-ḥtp* (Plan PM 47).

949 One of the earliest chapels to present these changes is that of *Jj-mrjj* in Cemetery G 6000. Weeks (1994) 31-57, 71-2, fig. 4, 6, 25-46,

not appear to have been laid out as funerary chapels.[950]

15.2 Modifications in the scenes of work on rural estates

(1) *Resources of the desert and desert fringe environment*

- Scenes of hunting wild animals of the desert are prominent in the tombs of early Dynasty 4 but are missing from the record of tombs from mid Dynasty 4 to mid Dynasty 5. When the theme appears again in the tombs of the mid Fifth Dynasty it exhibits a marked increase in the portrayal of violence that further intensifies in early Dynasty 6. This may represent a different method of hunting, different priorities in staging desert hunts, or most likely different stages of the hunt. In the early scenes single hunting dogs are pictured nipping at the heels of fleeing ungulates,[951] whereas hunting scenes portray dogs attacking the same species by the neck in the later tombs and packs of dogs tearing the animals apart.[952] Other violent elements are introduced into the later tombs. For example, lions are depicted attacking ungulates[953] usually with their jaws clamped on to the animal's muzzle.[954]
- Other motifs are introduced in mid Dynasty 5. Ungulate calves are portrayed hiding in tufts of grass and scrub, large and small desert animals are depicted copulating and suckling young, and hunting episodes are sometimes pictured taking place within palisade enclosures containing desert animals that do not normally inhabit the same type of desert country, such as the mountain goat (ibex) and the plains dwelling oryx and gazelle.
- A new theme that emphasises the value of ungulate calves is associated with the hunt in the second half of the Fifth Dynasty. Hunters as bearers are portrayed returning from the hunt carrying live ungulate young either in containers or in their arms and also carrying ungulate young in the parade of livestock before the tomb owner.[955]

(2) *Resources of the pastoral environment*

Images of cattle husbandry, such as copulation, birth of calves and milking of cows, first appear in mid Dynasty 5 but become much less frequent and are given much less attention in the provincial chapels from mid Dynasty 6 on. This trend contrasts with the theme of bulls fighting each other. Although confined to the provinces of Upper Egypt, the latter theme, first attested in later Dynasty 5, appears in most of the tombs of the higher provincial administrators and the attention and and theerhaps the most significant change in the pastoral theme is the increased emphasis on numbers and the increase in the size of herds that is expressed in the provincial tombs of Dynasty 6. While this element applies primarily to cattle, other livestock are included and the amount of wall space devoted to it increases. At the same time, there appears to have been a reduced interest in the theme of cattle fording a stretch of water in certain Upper Egyptian provinces, notably at Meir and Deir el-Gebrawi.
As flocks of sheep must have grazed on pasturage, although never depicted doing so, their virtual disappearance from the pictorial record can be included in changes to the portrayal of resources. Apart from rare inclusions in the parade of livestock, flocks of sheep are mainly depicted in ploughing scenes but disappear from this theme in early Dynasty 6.

Goats are browsers that can survive on foliage as well as grazing on sparse vegetation. Consequently, a marked increase in their depiction may signify a modification in priorities assigned to livestock resources. In the period from mid Dynasty 5 to early Dynasty 6 portrayals of herds of goats are included in less than ten per cent of the tombs used in this study. In the succeeding period, from mid Dynasty 6 to Dynasty 8 they appear in forty per cent of tombs, most of which are provincial. In addition, goats are included in the parade of livestock and the 'cattle' count in the tombs of some of the highest provincial officials in Dynasty 6.

(3) *Resources of the agrarian environment*

As noted above, sheep disappear from scenes of ploughing in early Dynasty 5, although working with the plough continues to be portrayed. While ploughing vignettes appear in the tombs of early Dynasty 4, detailed scenes portraying the sequence of activities involved in the harvesting of grain, from reaping to storage in granaries, rarely appear until mid Dynasty 5. From this date on, however, 50 tombs contain depictions of at least three stages in this sequence.

Two phases in the sequence, the pack of donkeys being driven to the field and the winnowing/sweeping action are not seen after early Dynasty 6, possibly due to the reduced wall space of the rock-cut provincial tombs.

Scenes of storage of grain have a similar chronological distribution to those of dealing with the harvested crop, although in all periods of time they appear less frequently.

(4) *Resources of the marshland environment*

There are only a few changes in scenes of marshland resources that may have significance. Scenes of multiple fishing techniques using weirs and basket traps are first depicted in mid Dynasty 5 in the tomb of *Nj-ʿnḫ/H̱nmw* and *H̱nmw-ḥtp* and thereafter appear very occasionally until early Dynasty 6.

950 Jéquier (1929) passim.
951 Petrie (1892) pl. 17, 27; LD II 6.
952 Paget–Pirie (1898) pl. 32, Roth (1995) fig. 189, Duell I (1938) pl. 24-5, Kanawati (2006) pl. 47.
953 Junker XI (1953) fig. 63.
954 Paget–Pirie (1898) pl. 32, Duell I (1938) pl. 25, Kanawati–Abder-Raziq (2004) pl. 46.
955 TABLE 2: MANAGEMENT OF DESERT ANIMALS.

Tombs with two or more clapnet scenes are generally confined to Dynasty 6, although two Dynasty 4 tombs portray double scenes.[956]

15.3 Presentation of accounts by chiefs of the estate

Scenes of chiefs rendering accounts to scribes of the estate appear from mid Dynasty 4 on. They may be depicted led forward and forced to bow low by an official holding a stick, even being dragged forward or made to crawl. Scenes of actual punishment by beating, however, do not appear until the second half of Dynasty 5.[957]

15.4 Vertical inscriptions

These inscriptions, beginning with the term *m33* ('viewing'), normally separate the figure of the tomb owner from a panel of registers on which are displayed scenes of parades of livestock and offering bearers and work on rural estates. On rare occasions the inscriptions are horizontal but their position makes it clear that they refer to the tomb owner 'viewing' the associated scenes.

- The term *pr-ḏt* appears in early Dynasty 4. It is frequently included in Dynasty 5 inscriptions but becomes rare early in Dynasty 6, when it only appears in the tombs of *Jttj-Šdw, Mḥw, S3bw/Jbbj* and *K3-gmnj*. Apart from the inscription in the tomb of *Jttj-Šdw,* occurrences are confined to Memphite tombs.
- In Dynasty 8 the term *pr-ḏt* gives way to *m jšt.f ḏs.f* ('his own property') in the inscription of *W3hj* and to *m jšt.f ḏs.f m wn-m3ʿ* ('from his own property in reality') for *Mrjj-ʿ3*. Both tombs are at el-Hagarsa.
- Another term with a limited life span in these inscriptions is *ḥwwt* ('estates'), which is included in the vertical inscriptions from late Dynasty 5 to early Dynasty 6.

15.5 Other theme changes

To the above changes may be added the appearance and disappearance of lesser themes, which only occur in a limited number of tombs and are confined to a limited chronological time period:

- Scenes of preparing soil by farm labourers working with hoes held in the hand appear in six Memphite tombs from mid Dynasty 5 (*Jj-mrjj*)[958] to early Dynasty 6 (*Mr.f-nb.f*).[959]
- The theme of hand feeding desert ungulates appears in a small number of Memphite tombs of the very highest officials who date to the end of Dynasty 5 and the beginning of Dynasty 6. Of the six tomb owners, five (*Ptḥ-ḥtp, Rʿ-špss, K3-gmnj, Mrrj* and *Mrrw-k3*)[960] were viziers.
- The theme of force feeding the striped hyena is included in five Memphite tombs, also of the very highest officials (*Ṯjj, K3-gmnj, Mrrj, Mrrw-k3* and *Jnw-Mnw*) who date to the second half of Dynasty 5 and the beginning of Dynasty 6.

Themes with so limited a number of depictions do not offer adequate data to support hypotheses relating to trends in funerary iconography, although the coincidence of tombs with depictions of desert ungulates and hyenas, also desert dwellers, being hand or force fed does suggest a willingness by the highest officials of this period to present images of a richly rewarded life. The first tomb to exhibit the last of these themes, that of *Ṯjj*, was undoubtedly decorated by a craftsman who was a true creative artist. It is probable that these images were copied by the viziers of early Dynasty 6 as the concept behind the representations developed and expanded the vision of the opulent lifestyle considered appropriate to their power and social status.

15.6 Significance of changes

The major shifts in the representation of estate resources can be identified but it is much more difficult to understand the significance of these changes. Some, such as the disappearance of sheep from the pictorial record or the greater frequency of the representation of herds of browsing goats, may have been brought about by changing climatic conditions such as the disappearance of precipitation or even a progressive reduction in the annual inundation. Continually diminished inundations would have reduced the amount of land available for pasturage and thus would have produced responses on affected estates. There is some evidence that Dynasty 6 experienced such a change. According to Bárta,[961] the evidence for climatic change can be perceived as early as the reign of Teti. His data is scientific: evidence of species of beetles found in two separate contexts, the burial chamber of the vizier *Ḳ3r*, who dates to the time of Teti, and a shaft (Shaft E) in the tomb complex of *Jntj*, dated to early Pepy II. These beetle species (Family Tenebrionidae: *Prionotheca coronata coronata* and *Sclerum orientale* and Family Carabidae: *Poecilus* (sbg. *Ancholeus*) *pharao*) are associated with the semi-arid deserts of North Africa. Specimens of the species *Poecilus* (sbg. *Ancholeus*) *pharao*, only found in saline habitats, were identified glued in the mummification bowl of *Jntj*. The discovery of these and other beetle species associated with arid, saline environments at Abusir has led Bárta to conclude that the mummification of *Ḳ3r* and *Jntj* was carried out in a salt rich context, which suggests that aridification was affecting the environment as early as the beginning of Dynasty 6.

956 Two clapnet scenes appear in the tombs of *Nfr-m3ʿt:* Petrie (1892) pl. 18 and Harpur (2001) fig. 75 and *Nb.j-m-3ḫtj* LD II 12a, 14b.

957 The earliest scenes of the punishment of estate chiefs appear in the tombs of *Tp-m-ʿnḫ* II Borchardt (1937) fig. 52 [CG 1541], Smith (1942) fig. 5; *3ḫtj-ḥtp* (Louvre) Ziegler (1993) 135-5 Drawing; and *Ḥzjj-Mnw* Kanawati IV (1983) fig. 12-13.

958 LD II 51, Weeks (1994) fig. 39.

959 Mysliwiec (2004) pl. 21, 66.

960 Scenes of hand feeding desert ungulates appearing in the tombs of the above: *Ptḥ-ḥtp* LD II 102(b), *Rʿ-špss* LD II 62, *K3-gmnj/Mnj* von Bissing I (1905) pl. 11-13, *Mrrj* Davies et al (1984) pl. 6(A), *Mrrw-k3/Mrj* Duell II (1938) pl. 152.

961 Bárta (2008).

Bárta associates the data from the tombs of *K3r* and *Jntj* with evidence of the height of Nile floods from the Palermo Stone and the discovery of layers of sand over Early Dynastic structures around Abusir Lake, which may indicate a reduction in the area covered by lake water. The data from the Abusir tombs is perhaps the strongest evidence for aridification as other evidence for a widespread climate change cannot be correlated with the Egyptian Sixth Dynasty with sufficient chronological precision.[962] Data from the array of geoarchaeological methods used, drill cores and examination of lake sediments for example and of course carbon-14 dating, do not provide evidence within time limits that can match those of an Old Kingdom Dynasty, which are also uncertain.[963]

While pictorial evidence showing an increase in the portrayal of goats and a decrease in that of sheep in Dynasty 6 may support Bárta's hypothesis, the markedly increased interest in developing the cattle industry in the provinces of Upper Egypt, which is very clearly portrayed in the tombs of the provincial administrators, suggests a more complex dynamic. Any significant growth in cattle numbers would have inevitably demanded a corresponding expansion of pasturage. The herd figures supplied by the provincial administrators may be unreliable but they point to a statistical trend of increasing cattle numbers. Evidence from the autobiographical statements of *K3r* of Edfu and *Jbj* of Deir el-Gebrawi suggests that their concentration on building up herd numbers was a response to crown policy. Both men were appointed to their respective provinces by Merenre.[964] The statement by *Hnkw/Jj…f* of Deir el-Gebrawi that he developed under-used land does not refer to the king. However, *Hnkw/Jj…f* has recently been re-dated by Kanawati to the first half of the reign of Pepy I, making him a predecessor of *Jbj*.[965] Such a statement in a high official's autobiography dated to the first half of the Sixth Dynasty implies that the work of development would find royal approval.

Although changes and changing trends that may be inferred from scenes portraying the resources of rural estates are perhaps too narrowly focused to provide solid evidence, they offer significant hints relating to socio-economic developments and can provide valuable supporting or corrective data for hypotheses developed from different sources.

962 Hassan (1981b) 1142-5, Hassan (1986b) 483-501.
963 Kitchen (1991) 201-208.

964 *K3r* Urk. 1, 254:1-3; *Jbj* Davies *Gebrâwi* II (1902) pl. 23.
965 Kanawati *Gebrâwi* I (2005) 11-20, 63.

CHAPTER 16 CONCLUSIONS

The aims of this study are twofold: to identify, document and analyse the resources that supported the tomb owner and enabled him to devote himself to the service of the crown and to identify changes over time in the presentation of these resources.

The high officials of Old Kingdom Egypt comprised both the nobility and the upper level of the state's bureaucracy. Consequently the wealth of this class is not to be found in acquisition by force of arms or trade but in inheritance and rewards for administrative service. As servants of the crown, these officials would not have farmed their own land, and probably did not even personally manage the properties for which they were ultimately responsible.[966] Establishing the source(s) of their wealth, however, depends on data that rarely offers direct and precise information as to the nature and origins of the resources that enabled them to devote their life's work to the royal administration and to provide for the considerable entourage that many officials depict in their tombs. Yet maintenance of a system of successful and productive estates was essential to the efficient administration of the country. The crown could not rely on a land-hungry, impoverished, dissatisfied bureaucracy to maintain order, dispense justice, execute royal policies and collect taxes.

This study of the resources of the great servants of the crown is based on the proposition that funerary data providing scenes of 'daily life' in Old Kingdom private tombs reflect the actual work of a rural estate. The data, drawn from the iconographic and inscriptional material of 190 tombs, have been organised into sequential periods of time. The scenes have also been grouped according to theme. As no two scenes within a theme are identical, it has been necessary to identify the essential elements portrayed in the scenes as a basis for analysis and interpretation. These initial steps have been followed by identifying substantial variations in each scene in order to highlight and analyse points of contrast and comparison.

This approach, which involves accepting funerary data 'at face value' has provided a comprehensive picture of the resources of a great estate that were available to high officials. It has also provided the means of assessing the relative value officials assigned to these resources.

The picture that emerges from the study is that the consistent exploitation of the four environments of the Nile valley and Delta provided estates with a balanced and largely self-sufficient rural economy. Scenes of rural estates concentrate largely on the management of livestock, crops and exploitation of the marshland. They rarely include buildings or man-made structures and only workshop scenes present images of craftsmen working with materials (stone, metals and semi-precious stones) which are not the product of the estate. Cattle afforded the tomb owner with invested wealth that he could measure. There was a supply of cheaply acquired high protein foods, as well as the staple of cereals, for the estate's labourers and their families, and cheap and luxury foodstuffs for other dependants. At the same time, the products of livestock, crops and marshlands provided some of the raw materials for the tools needed to exploit the various environments and other material items of daily life. All this work was carried out by a hierarchically ordered workforce of labourers and overseers with specialized skills and with a significant division of labour. For example, the herders who lived year round with prized cattle would not have been available to harvest the annual crops. The labourers who harvested the cereals and flax were required to work deftly, efficiently and with speed to get in the crop at exactly the right stage of growth. Frequently depicted supervising this work, overseers are also shown supplying scribes with information and reporting to superiors.

The attention in terms of wall space and detailed images given to cattle and grain crops makes it clear that these were the most highly valued resources. This is signified by the pride of place given to cattle in the parade of livestock and the cited size of herds particularly in the Sixth Dynasty, the sequence of detailed images of harvesting the grain crop and the care portrayed in the scenes of storing the grain. While the tomb owner's depiction of the elements of cattle husbandry and the grand parade of animals suggest that cattle herds rather than grain crops were his prized asset, no depictions of other resources match the continuity of detailed scenes of planting, harvesting and storing of the grain crop. Such a comparison between the portrayal of cattle herds and of cereal crops suggests that high officials placed different kinds of values on the two resources. Large, healthy herds of cattle were a measure of the tomb owner's wealth and status. The premier position of cattle in many scenes of the parade of livestock and the consistent portrayal of the slaughter of cattle, together with the position of this image associating it very closely with the tomb owner at the offering table, indicate the ritual importance assigned to the species. At the same time, as the administrator responsible for an efficiently managed estate, the tomb owner also needed to make clear that the production of cereals was successfully achieved. By the time of the Old Kingdom cultivated cereals were society's staple food and the basic element in the diet of all who depended on the estate. Consequently, to demonstrate his effective

[966] See Chapter 1, footnote 7.

administration and capacity to provide for his dependants, the tomb owner chose to portray the detail of individual stages of cereal production.

A comparison of depicted themes from one tomb to another highlights the difference in level of resources available to officials of varying rank. This is especially noticeable for officials whose tombs date to the end of the Fifth and beginning of the Sixth Dynasties. Only a small number of the very highest officials include scenes of well constructed and well supplied fowlyards, which would have been costly to maintain, or of the force feeding of hyenas. The purpose of depicting the latter theme would have been to demonstrate the high standing and appropriate lifestyle of the tomb owner.

Not all the funerary data presents evidence that can be clearly interpreted. The evidence regarding degree of socio/economic stress experienced by officials in Dynasty 6, from the reign of Pepy I onwards, is puzzling, suggesting that the situation confronting the royal administration was much more complex than a consequence of a single major factor such as climate change. Pictorial evidence from the tombs of the Sixth Dynasty administrators encapsulates the contradictions. On the one hand the data shows an increased interest in certain species of domesticated animals, namely goats that would have been able to exploit an increasingly semi-arid environment, and the virtual disappearance of another species, sheep that were less valued except as working animals and so might have been considered expendable. These trends may indicate a reduction of grazing land due to increasing desiccation. On the other hand, the data present a trend of expanding herds of cattle that would have required more lush pasturage. At the same time a picture emerges of increasing preoccupation with grain harvests and the careful management and storage of grain, which would support the hypothesis of a growing need to conserve the crops that provided society's staple food and drink. Against the above and Bárta's evidence of increasing aridification is the late Dynasty 6 or early Dynasty 8 statement by *ʿnḫ-tj.fj*, nomarch of Hieraconpolis, that he found the neighbouring province of Edfu (Upper Egypt 2) 'inundated like a marsh, abandoned by him who belonged to it'.[967]

The present study supports the proposition that the later Old Kingdom experienced long-term aridification as this is suggested by changes to farming practice. It would have increased pressure on food supplies and by reducing the amount of available farmland would have aggravated the land hunger of the upper bureaucracy. It is likely that Memphite-based officials had been facing growing impoverishment since the reign of Teti, perhaps due to the expansion of the bureaucracy which was a feature of the later Fifth Dynasty.[968] By the reign of Pepy I there is evidence that even some of the highest servants of the crown were in difficulties.[969] If to this smouldering situation were added bouts of torrential rainfall, evidence of which has been found in recent years,[970] the consequences could have been catastrophic. There is no evidence of destruction of mudbrick structures beyond the Saqqara necropolis and rainstorms may have been localized. Yet heavy precipitation in the inhabited areas would have caused much distress and, depending on the season, could have ruined the harvest or at least stores of crops. The pressure to control viable farming land would have intensified, creating a situation of acute political and social instability.

While the pictorial and inscriptional data do not provide clear evidence of the nature of the historical factors operating in the Old Kingdom, an integrated analysis of a wider range of data may contribute to the present meagre understanding of the dynamic of the age. For example, the remarkable increase in the number of decorated tombs in the second half of the Fifth Dynasty suggests a corresponding increase in the number of officials who would have considered themselves entitled to appropriate remuneration through the *pr-ḏt* system. This would have led to pressure on crown resources and a consequent reduction in what the crown could bestow on its servants. The marked preoccupation of tomb owners of the second half of the Fifth Dynasty and early Sixth Dynasty with presenting themes that document the resources of their rural estates implies a concern in this life as well as for the hereafter with acquiring an appropriate level of material well being. Consequently the reduction of decorative programmes and size of tomb chapels in the Memphite necropoleis of the later reign of Teti and in the reign of Pepy I may suggest that quite early in Dynasty 6 high officials based on Memphis were facing reduced circumstances.

Although the data used in this study only offer suggestions rather than hard evidence, this method of investigating the wall scenes has also highlighted historical 'trends' that may relate to the socio/economic dynamic of the Old Kingdom, In addition, tomb iconography makes it possible to amass data in quantity, which contributes to research into the working of the economy of the Old Kingdom. The approach, an examination of the 'broad picture' presented by the iconography in wall scenes of funerary chapels has provided a method of analyzing the operation of the great estates of the Old Kingdom as economic rural units. While their precise contribution to the national economy cannot be measured, the pictured estates, comprising *njwwt,* and for a limited period of time *ḥwwt*, were important elements in the economy and administration of the country. They provided a valuable medium through which the crown tapped surplus production and labour.

967 Lichtheim (1973) pp. 85.

968 Naguib Kanawati identifies and examines evidence for official discontent in *Conspiracies in the Egyptian Palace: Unis to Pepi I* (2003).

969 El-Fikey (1980). Among the tombs in the necropolis of Pepy II at Saqqara South Gustave Jéquier found indications that later burials of high officials were significantly impoverished: Jéquier (1929).

970 Kuraszkiewicz (2005) 169-175; Trzciński, Kuraszkiewicz, Welc (2007) 194-208.

The products of the great estates were able to make so significant a contribution to the national economy because Egypt was blessed with a reliable, navigable river. Not only did the Nile link all the main centres of Egypt, it provided the only viable, cost-effective means for the continual movement of heavy goods in very large quantities. This transport facility applied to the produce of the rural estates as well as material for the kings' building programmes. The two features, the surplus food production by the great estates and the capability of river-borne transport, underpinned the civilization of mankind's first unified state.

BIBLIOGRAPHY

A

Aganga–Letso–Oaganga (2000)
Aganga, A. A., Letso, M. and Oaganga, A. O., 'Feeding donkeys', *Livestock Research for Rural Development,* 12 (2) (2000). <www.cipav.org.co/lrrd/lrrd12/2/agan122.htm>.

Allam (1988)
Allam, S., 'Taxe (?) sur le bétail dans l'Egypte ancienne' in *Stato economia lavoro nel Vicino Oriente antico,* Istituto Gramsci toscano. Seminario diorientalistica antica. F.Angeli (Milano, 1988).

Allam (1994)
Allam, S., (ed.) *Grund und Boden in Altägyptischen* (Tübingen, 1994).

Allen (2005)
Allen, J. P., *The Ancient Egyptian Pyramid Texts,* Society of Biblical Literature, Atlanta (2005).

Allen (2006)
Allen, J. P., 'Some aspects of the non-royal afterlife in the Old Kingdom' in Bárta M., (ed.) *The Old Kingdom Art and Archaeology* (2006).

Allen (1997)
Allen, R. C., 'Agriculture and the Origins of the State in Ancient Egypt', *Explorations in Economic History* 34, 135–154 (Kent, Ohio, 1997).

Altenmüller (1998)
Altenmüller, H., *Die Wanddarstellungen im Grab des Mehu in Saqqara,* Archäologische Veröffentlichungen 42 (Mainz-am-Rhein, 1998).

Andersson (1912)
Andersson, E., *La denomination égyptienne des boeufs sans cornes,* Sphinx 16, (Upsala, 1912).

Arkell–Ucko (1965)
Arkell, A. J., and Ucko, P. J., Review of 'Predynastic Development in the Nile Valley', *Current Anthropology,* Vol. 6, No. 2 (1965) 145-166.

Arnold–Bourriau (1993)
Arnold, D. and Bourriau, J. (eds.), *An introduction to Ancient Egyptian Pottery* (Mainz am Rhein, 1993).

B

Balter (2007)
Balter, M., 'Seeking Agriculture's Ancient Roots', *Science* 29 June 2007, Vol. 316. No. 5833, 1830-1835.

Badawy (1976)
Badawy, A., *The Tombs of Iteti, Sekhemankh-Ptah, and Kaemnofert at Giza,* University of California Publications: Occasional Papers No. 9: (Berkeley, Los Angeles and London, 1976).

Badawy (1978)
Badawy, A., *The Tomb of Nyhetep-Ptah at Giza and the Tomb of Ankhmahor at Saqqara,* University of California Publications: Occasional Papers No. 11: (Berkeley, Los Angeles and London, 1978).

Baer (1962)
Baer, K., 'The Low Price of Land in Ancient Egypt', *Journal of the American Research Center in Egypt,* Vol. 1 (1962).

Baer (1963)
Baer, K., 'An Eleventh Dynasty Farmer's Letter to his Family', *Journal of the American Oriental Society,* Vol. 83 (1963).

Baines (1994)
Baines, J., 'On the Status and Purposes of Ancient Egyptian Art', *Cambridge Archaeological Journal* 4 (1994) 67-94.

Baines (1988)
Baines, J., 'Literacy, social organization and the archaeological record: the case ofearly Egypt' in Gledhill, J., Bender, B., and Larsen, M. T., (eds.), *State and Society: The Emergence and Development of Social Hierarchy and Political Centralization* (London 1988) 192-214.

Baines (1989)
Baines, J., 'Ancient Egyptian concepts and use of the past: 3rd to 2nd millennium B.C. evidence' in R. Layton (ed.), *Who Needs the Past?* (London, 1989) 131-149.

Baines (1990)
Baines, J., 'Restricted knowledge, hierarchy and decorum: modem perceptions and ancient institutions', *Journal of the American Research Center in Egypt* 27 (1990) 1-23.

Baines (1991)
Baines, J., 'Society and religious practice' in Shafer, B. E., (ed.), *Religion in Ancient Egypt: Gods Myths and Personal Piety* (London, 1991) 123-200.

Barsanti (1900)
Barsanti, A., 'Fouilles autour de la Pyramide l'Ounas': I: Le Mastaba de Samnofir', *Annales du Service Antiquités de l'Égypte* 1 (1900) 149-190.

Barta (1963)
Barta, W., *Die altägyptische Opferliste: von der Frühzeit bis zur Griechischrömischen Epoche* (Berlin, 1963).

Barta (2001)
Barta, M., *Abusir V, The Cemeteries of Abusir South I, The Excavations of the Czech Institute of Egyptology,* Roman Mísek (Prague, 2001).

Barta (2006)
Barta M. (ed.), *The Old Kingdom Art and Archaeology: Proceedings of the Conference held in Prague, May 31-June 4 2004* (Prague 2006).

Bárta- Coppens- Krejčí (2008)
Bárta, M., Coppens, F., Krejčí, J. (eds.), *Abusir and Saqqara in the Year 2005. Proceedings of the Conference Held in Prague (June 27–July 5, 2005)*. Czech Institute of Egyptology, Prague 2008.

Barta (2008)
Barta, M. and Bezdek, A., 'Beetles and the decline of the Old Kingdom', in Bárta, M. and Vymazalova, H. (eds.), *Chronology of the Late Fourth and the Third Millenniums B.C.E.* Proceedings of the Conference held in Prague, June 11–14, 2007 (Prague 2008).

Beckerath (1984)
Beckerath, J. von, *Handbuch der ägyptischen Königsnamen, Series: Münchner ägyptologische Studien: Heft 20,* Deutscher Kunstverlag (München 1984).

Beckerath (1997)
Beckerath, J. von, *Chronologie des pharaonischen Ägypten: Die Zeitbestimmung der ägyptischen Geschichte von der Vorzeit bis 332 v. Chr.*, MÄS 46 (Mainz, 1997).

Begelsbacher-Fischer–Hottinger (1985)
Begelsbacher-Fischer, B. L. and Hottinger, A., *Ägypten* (Zurich, 1985).

Beja-Pereira (2004)
Beja-Pereira, A. et al, 'African origins of the domestic donkey', *Science* (2004) 304:1781.

Bianchi (1995)
Bianchi, R. S., 'Ancient Egyptian reliefs, statuary, and monumental paintings', in J.M. Sasson (ed.), *Civilizations of the Ancient Near East,* IV, (New York, 1995) 2533-2554.

Bianchi (1997)
Bianchi, R. S., 'An elite image', in E. Goring. N. Reeves, J. Ruffle (ed.), *Chief Of Seers. Egyptian Studies in Memory of Cyril Aldred,* (London, 1997) 34-48.

Bietak (1996)
Bietak, M. (ed.), *Haus und Palast im Alten Ägypten,* Verlag der Österreichischen Akademie der Wissenschaften, (Wien, 1996).

Binford (1962)
Binford, L. R., 'Archaeology as Anthropology', *American Antiquity* 28 (1962) 217–225.

Binford (1972)
Binford, L. K., *An Archaeological Perspective* (New York 1972).

Binford–Binford (1968)
Binford, L R. and Binford S. R. (eds), *New Perspectives in Archaeology* (New York, 1968) [papers from a symposium entitled The Social Organization of Prehistoric Communities, held at the 64th Annual Meeting of the American Anthropological Association at Denver in November 1965].

Blackman Meir II (1915).
Blackman, A. M., *The Rock Tombs of Meir, Volume II: The Tomb-chapel of Senbi's son Ukh-Hotp (B, no. 2)* (London 1915).

Blackman Meir IV (1924)
Blackman, A. M., *The Rock Tombs of Meir, Volume IV: The Tomb-chapel of Pepionkh (D, No. 2),* Egypt Exploration Society (London, 1924).

Blackman–Apted Meir V (1953)
Blackman, A. M. and Apted, M. R., *The Rock Tombs of Meir, Volume V: The Tomb-chapels A, No. 1 (Ni-ankh-Pepi), A, No. 2 (Pepionkh), A, No. 4, D, No. 1, E, Nos. 1-4,* Egypt Exploration Society (London, 1953).

Bloxam (2006)
Bloxam, E., 'Miners and mistresses', *Journal of Social Archaeology*, Vol. 6, No. 2, (2006) 277-303.

Bloxam–Storemyr (2002)
Bloxam, E. and Storemyr, P., 'Old Kingdom Basalt Quarrying Activities at Widan el-Faras, Northern Faiyum Desert', *Journal of Egyptian Archaeology*, Vol. 88 (2002) 23-36.

BMSAES
British Museum Studies in Ancient Egypt and Sudan.

Boessneck (1953)
Boessneck, J., *Die Haustiere in Altägypten* (Munich, 1953).

Boessneck (1988)
Boessneck, J., *Die Tierwelt des alten Ägypten: untersucht anhand kulturgeschichtlicher und zoologischer Quellen* (München, c1988).

Bolshakov (1993)
Bolshakov, A. O., 'The Scene of the Boatmen Jousting in Old Kingdom Tomb Representations', *Bulletin de la Société d'Égyptologie*, Genève, 17 (1993).

Borchardt (1910-13)
Borchardt, L., *Das Grabdenkmal des Königs Sahure* 2 vols. (Leipzig, 1910-13).

Borchardt (1964)
Borchardt, L., *Catalogue général des antiquités égyptiennes du Musée du Caire. Denkmäler des Alten Reiches im Museum von Kairo, Nr. 1542-1808, Teil II, Text und Tafeln zu Nr. 1542-1808*, Service des Antiquités de l'Égypte (1964).

Boulenger (1907)
Boulenger, G.A., *Zoology of Egypt: The Fishes of the Nile* (London 1907).

Brass (2003)
Brass, M., 'Tracing the Origins of the Ancient Egyptian Cattle Cult' in Eyma, A. K. and Bennett, C. J., (eds.) *A Delta-man in Yebu* (2003).

Brewer et al (1994)
Brewer, D. J., Redford, D. B. and Redford, S., *Domestic Plants and Animals: The Egyptian Origins* (Warminster, 1994).

Brewer/Friedman (1989)
Brewer, D. J. and Friedman, R. F., *Fish and Fishing in Ancient Egypt* (Warminster, 1989).

Brovarski, (2001)
Brovarski, E., *The Senedjemib Complex, Part I: The Mastabas of Senedjemib Inti (G 2370), Khnumenti (G 2374) and Senedjemib Mehi (G 2378),* Giza Mastabas Volume 7, Art of the Ancient World, Museum of Fine Arts (Boston 2001).

Brunner-Traut (1977)
Brunner-Traut, E., *Die altägyptische Grabkammer Seschemnofers III. aus Gîsa,* Eine Stiftung des Geheimen

Hofrats Dr. h. c. Ernst von Sieglin an die Tübinger Universität (Mainz, 1977).

Butzer (1976)
Butzer, K. W., *Early Hydraulic Civilization in Egypt: A Study in Cultural Ecology* (Chicago and London, 1976).

C

Cagle (2001)
Cagle, A. J., *The Spatial Structure of Kom el-Hisn: An Old Kingdom Town in the Western Nile Delta, Egypt,* PhD Thesis (University of Washington, 2001).

Cagle (2002)
Cagle, A., 'Differential Consumption of Pig vs. Sheep/ Goats at the Old Kingdom Site of Kom el-Hisn', Paper presented at the 53rd Annual Meeting of the American Research Center in Egypt (Baltimore, 2002).

Capart (1907)
Capart, J., *Une Rue de Tombeaux à Saqqarah.* 2 vols. (Brussels 1907).

Carneiro (1970)
Carneiro, R. L., 'A theory of the origin of the state', *Science* 169 (1970) 733-8.

Caton-Thompson–Gardner, (1934)
Caton-Thompson, G. and Gardner, E.W., *The Desert Fayum*, Royal Anthropological Institute (London 1934).

Chenal-Vélardé, (1997)
Chenal-Vélardé, I., (1997) 'Le boeuf domestique en Afrique du Nord', in *Animals and People in the Holocene of North Africa*, A. Gautier (ed.) Archaeozoologica IX/1.2 (1997).

Cherpion, (1989)
Cherpion, N., *Mastabas et hypogées d'Ancien Empire: le problème de la datation Connaissance de l'Égypte Ancienne,* Études 2 (Bruxelles 1989).

Childe, (1936)
Childe, V. G., *Man makes himself* (London 1936).

Clutton-Brock–Grigson (1984)
Clutton-Brock, J. and Grigson, C. (eds.), *Animals and Archaeology: 3. Early Herders and their Flocks,* vol. 3, British Archaeological Reports, International Series (Oxford 1984).

Cohen–Armelagos, (1984)
Cohen, M. N. and Armelagos, G. J. (eds.), *Paleopathology and the Origins of Agriculture* (Orlando 1984).

Compagnoni–Tosi (1978)
Compagnoni, B. and Tosi, M., 'The camel: Its distribution and state of domesticationin the Middle East during the third millennium B.C. in light of the finds from Shahr-I Sokhta' 119–128, in *Approaches to Faunal Analysis in the Middle East*, edited by R.H. Meadow and M.A. Zeder. Peabody Museum Bulletin no 2, Peabody Museum of Archaeology and Ethnology (New Haven, 1978).

Coulson–Leonard (1981)
Coulson, W. D. E. and Leonard, A. Jr., *Cities of the Delta, vol. 1, Naukratis* (Malibu, 1981).

Crabtree–Ryan (1991)
Crabtree, P. J. and Ryan, K. (eds.), *Animal Use and Culture Change, Supplement to Volume 8,* University Museum of Archaeology and Anthropology (Philadelphia, 1991).

Cramp (1977)
Cramp, S., *Handbook of the Birds of Europe, the Middle East and North Africa: The Birds of the Western Palearctic,* 9 volumes (Oxford, 1977).

D

Darby et al (1977)
Darby, W. J., Ghalioungui, P., Grivetti, L., *Food: the gift of Osiris,* Academic Press, (London, New York, 1977).

Davies (1891)
Davies, N. de G., *The Rock Tombs of Sheikh Saïd,* Archaeological Survey of Egypt, Tenth Memoir (London, 1901).

Davies Ptahhetep I (1900)
Davies, N. de G., *The Mastaba of Ptahhetep and Akhethetep at Saqqareh, Volume I: The Chapel of Ptahhetep and the Hieroglyphs*, Egypt Exploration Fund (London, 1900).

Davies Ptahhetep II (1901)
Davies, N. de G., *The Mastaba of Ptahhetep and Akhethetep at Saqqareh, Part II: The Mastaba, The Sculptures of Akhethetep,* Egypt Exploration Fund (London, 1901).

Davies Sheikh Saïd (1901)
Davies, N. de G., *The Rock Tombs of Sheikh Saïd,* Egypt Exploration Fund (London, 1901).

Davies Gebrâwi I (1902)
Davies, N. de G., *The Rock Tombs of Deir el-Gebrâwi, Volume I: Tomb of Aba and Smaller Tombs of the Southern Group*, Egypt Exploration Fund (London, 1902).

Davies Gebrâwi II (1902)
Davies, N. de G., *The Rock Tombs of Deir el-Gebrâwi, Volume II: Tomb of Zau and Tombs of the Northern Group*, Egypt Exploration Fund (London, 1902).

Davies–Gardiner (1920)
Davies, N. de G. with a chapter by Gardiner, A. H., *The tomb of Antefoker, Vizier of Sesostris I, and of his wife, Senet* (No. 60) (London, 1920).

Davies et al (1984)
Davies, W. V., El-Khouli, A., Lloyd, A. B., and Spencer, A. J., *Saqqâra Tombs Volume I, The Mastabas of Mereri and Wernu*, Egypt Exploration Society (London,1984).

Davies, Nina M. (1936)
Davies, N. M., *Ancient Egyptian Paintings*, University of Chicago Press (1936).

Decker–Herb (1994)
Decker, W. and Herb, M., *Bildatlas zum Sport im alten Ägypten: Corpus der bildlichen Quellen zu Leibesübungen, Spiel, Jagd, Tanz und verwandten Themen,* 2 volumes (Leiden, 1994).

de Morgan, Cat. Des monuments I (1894)
de Morgan, J., *Catalogue des monuments et inscriptions de l'Égypte Antique. Première série. Haute Égypte. I. De la frontière de Nubie á Kom Ombos,* Service des Antiquités de l'Égypte (Vienna, 1894).

de Morgan (1903)
de Morgan, J., *Fouilles à Dahchour en 1894-1895*, Volume II, Adolphe Holzhausen (Vienna, 1903).

de Rachewiltz (1960)
de Rachewiltz, B., *The Rock Tomb of Irw-k3-Pth*, Documenta et Monumenta Orientis Antiqui, Volumen Nonum (Leiden, 1960).

Derchain (1976)
Derchain, P., 'Symbols and Metaphors in Literature and Representations of Private Life', *RAIN* No. 15 (Aug., 1976) Royal Anthropological Institute of Great Britain and Ireland.

Dodson–Ikram (2008)
Dodson, A. and Ikram, S., *The tomb in Ancient Egypt: royal and private sepulchers from the early dynastic period to the Romans* (London, c 2008).

Duell I (1938)
Duell, P., *The Mastaba of Mereruka,* University of Chicago Oriental Institute publications 31, University of Chicago Press (Chicago, 1938).

Duell II (1938)
Duell, P., *The Mastaba of Mereruka,* University of Chicago Oriental Institute publications 39, University of Chicago Press (Chicago, 1938).

Dunham–Simpson (1974)
Dunham, D. and Simpson, W. K., *The Mastaba of Queen Mersyankh III. G 7530-7540,* Giza Mastabas, Volume I. Department of Egyptian and Ancient Near Eastern Art, Museum of Fine Arts, Boston (Boston, 1974).

Dunham (1935)
Dunham, D., 'A "Palimpsest" on an Egyptian Mastaba Wall', *American Journal of Archaeology*, vol. 39 (1935) 300-309.

Dyson (1955),
Dyson, R. H., 'Archaeology and the Domestication of animals in the Old World', *American Anthropologist* (new Series), vol. 5 (1955).

E

Edel (1944)
Edel, E., 'Untersuchungen zur Phraseologie der ägyptischen Inschriften des Alten Reiches', *Mitteilungen des Deutschen Archäologischen Instituts, Abteilung Kairo* 13 (1944) 1-90.

Edel (1953)
Edel, E., 'Inschriften des Alten Reiches: III. Die Stele des *Mḥw-3ḫtj*', *Mitteilungen des Instituts für Orientforschung* I, Deutsche Akademie der Wissenschaften zu Berlin, Institut für Orientforschung (Berlin, 1953) 327-33.

Edel (1958)
Edel, E., 'Inschriften des Alten Reiches: IX, Inschriften des *Mṯti*', *Zeitschrift für Ägyptische Sprache und Altertumskunde* 83 (1958) 11-17.

Edel–Wenig (1974)
Edel, E. and Wenig, S., *Die Jahreszeitenreliefs aus dem Sonnenheiligtum des Königs Ne-user-Ra,* Staatliche Museen zu Berlin, Mitteilungen aus der ägyptischen Sammlung 7 (Berlin, 1974).

El-Fikey (1980)
El-Fikey, Said, *The tomb of the vizier Re'-wer at Saqqara,* Egyptology Today, 4 (Warminster 1980).

El-Khouli–Kanawati (1988)
El-Khouli, A. and Kanawati, N., *Excavations at Saqqara: North-West of Teti's Pyramid,* Volume II (Sydney, 1988).

El-Khouli–Kanawati (1990)
El-Khouli, A. and Kanawati, N., *Quseir el-Amarna: The Tombs of Pepy-ankh and Khewen-wekh,* The Australian Centre for Egyptology: Reports 1 (Sydney, 1989).

El-Khouli–Kanawati (1990)
El-Khouli, A. and Kanawati, N., *The Old Kingdom Tombs of El-Hammamiya,* The Australian Centre for Egyptology: Reports 2 (Warminster, 1990).

El-Masry (2004)
El-Masry, Y., 'Two Old Kingdom Rock-Tombs at Gohaina', *The Bulletin of the Australian Centre for Egyptology* 15 (2004) 89-106.

Épron–Daumas I (1939)
Épron, L. and Daumas, L., *Le Tombeau de Ti, Fasc. I, Les approches de la chapelle.* Mémoires publiés par les Membres de l'Institut Français d'Archéologie Orientale du Caire, Tome LXV (Cairo, 1939).

Epstein (1971)
Epstein, H, *The Origin of Domestic Animals in Africa,* I-II (New York, 1971).

Erman (1971)
Erman, A., 1894 (Reprint 1971), *Life in ancient Egypt.* Translated by H. M. Tirard, and B. Blom (New York, 1969).

Evans (2006)
Evans, L., *The Representation of Animal Behaviour in Old Kingdom Tomb Scenes from Giza and Saqqara,* unpublished PhD thesis (Macquarie University Sydney, 2006).

Evans (2010)
Evans, L., *Animal Behaviour in Egyptian Art: Representations of the Natural World in Memphite Tomb Scenes,* Australian Centre for Egyptology, Studies 9 (Oxford 2010).

Evans–Peacock (1981)
Evans, L.T. and Peacock, W.J. (eds.), *Wheat science today and tomorrow*, Cambridge University Press (Cambridge, 1981).

Eyma–Bennett (eds.) (2003)
Eyma, A. K. and Bennett, C. J. (eds.), *A Delta-man in Yebu* (EEF, 2003)

Eyre (1987)
Eyre, C. J., 'Work and the Organisation of Work in the Old Kingdom' in Powell, M. A., (ed.) *Labor in the Ancient Near East*, American Oriental Society (1987) 5-47, 167-221.

Eyre (1994)
Eyre, C., 'Feudal Tenure and Absentee Landlords' in Allam, S. (ed.) *Grund und Boden in Altägyptischen* (Tübingen, 1994) 107-133.

Eyre *JEA* 80 (1994)
Eyre, C., 'The Water Regime for Orchards and Plantations in Pharaonic Egypt', *Journal of Egyptian Archaeology* 80 (1994) 57-80.

Eyre (1995)
Eyre, C., 'The Agricultural Cycle, Farming, and Water Management in the Ancient Near East' in Sasson, J. M. (ed.), *Civilizations of the Ancient Near East*, Volume I (1995) 175-189.

Eyre (1999)
Eyre, C., 'The Village Economy in Pharaonic Egypt', in Bowman, A. K. and Rogan, E., (eds.) *Agriculture in Egypt from Pharaonic to Modern Times* (1999) 33-60.

Eyre (2008)
Eyre, C., *Abstracts of Papers*, University of the Aegean, (Rhodes, 22-29 May 2008), 88-90.

Faltings (1991)
Faltings, D., 'Die Bierbrauerei im AR', *Zeitschrift für Ägyptische Sprache und Altertumskunde* (Leipzig/Berlin, 1991) 104-116.

F

Feucht (1992)
Feucht, E., 'Fishing and Fowling with the Spear and Throw-stick Reconsidered', in U. Luft, *The Intellectual Heritage of Egypt Studies presented to László Kákosy by friends and colleagues on Occasion of his 60th Birthday,* Studia Aegyptiaca 14 (Budapest, 1992) 157-169.

Firth–Gunn (1926)
Firth, C. M. and Gunn, B., *Teti Pyramid Cemeteries,* Volumes 1-II, Service des Antiquités de l'Égypte: Excavations at Saqqara, Institut Français d'Archéologie Orientale (Cairo, 1926).

Fischer (1959)
Fischer, H. G., 'A scribe of the army in a Saqqara mastaba of the early Fifth Dynasty', *Journal of Near Eastern Studies* 18 (1959) 233-72.

Fitzenreiter–Herb (2006)
Fitzenreiter, M. and Herb, M. (eds.), *Dekorierte Grabanlagen im Alten Reich: Methodik und Interpretation* (London, 2006).

Förster (2007)
Förster, F., 'The Abu Ballas Trail: a Pharaonic Donkey-caravan Route in the Libyan Desert (SW-Egypt)' in Bubenzer, O., Bolten, A. and Darius, F. (eds.), *Atlas of Cultural and Environmental Change in Arid Africa,* Africa Praehistorica 21, Heinrich-Barth-Institut (Köln, 2007) 130–3.

Förster (2007)
Förster, F., 'With donkeys, jars and water bags into the Libyan Desert: the Abu Ballas Trail in the late Old Kingdom/ First Intermediate Period' in *BMSAES* 7 (2007) 1–39, <http://www.thebritishmuseum.ac.uk/research/publications/bmsaes/issue_7/foerster.aspx>

Frazer (1903)
Frazer, G., 'The Early Tombs of Tehneh', *Annales du Service des Antiquités de l'Égypte* 3 (1903) 122ff.

G

Galan (1994)
Galan, J. M., 'Bullfight Scenes in Ancient Egyptian Tombs', *Journal of Egyptian Archaeology* 80 (1994).

Gamer-Wallert (1970)
Gamer-Wallert, I., *Fische und Fischkulte im Alten Ägypten* (Wiesbaden, 1970).

Gautier (1997)
Gautier, A. (ed.), *Animals and People in the Holocene of North Africa*, Archaeozoologica IX/1.2. (1997).

Geller (1992a)
Geller, J. R., 'From prehistory to history: beer in Egypt' in Friedman, R. and Adams, B. (eds), *The Followers of Horus*, Egyptian Studies Association Publication 2, Oxbow Books (Oxford, 1992) 19-26.

Geller (1992b)
Geller, J. R., *Predynastic Beer Production at Hierakonpolis, Upper Egypt: Archaeological Evidence and Anthropological Implications*, PhD thesis (Washington University 1992).

Germer (1985)
Germer, R., *Flora des pharaonischen Agypten*, Sonderschrift, Deutsches Archaologisches Institut Abteilung Kairo (Mainz, 1985) 14:210.

Geus–Thill (1985)
Geus, F. and Thill, F., (eds), *Mélanges offerts à Jean Vercoutter*, Éditions Recherche sur les Civilisations (Paris, 1985).

Ghoneim, (1977)
Ghoneim, W., *Die ökonomische Bedeutung des Rindes im alten Ägypten*, Habelts Dissertationsdrucke, Heft 3, Reihe Ägyptologie 3 (Bonn, 1977).

Goedicke (1956)
Goedicke, H., 'Zu *ỉmj-r3 šmꜥ* und *tp-šmꜥ* im Alten Reich', *Mitteilungen des Instituts für Orientforschung* 4 (1956), 1-10.

Goedicke (1958)
Goedicke, H., 'Zwei Inschriften aus dem Grab des *Mttj* aus Sakkara', *Zeitschrift für Ägyptische Sprache und Altertumskunde* 83 (1958) 18-27.

Goedicke (1967)
Goedicke, H., *Königliche Dokumente aus dem alten Reich* (Wiesbaden, 1967).

Goedicke (1970)
Goedicke, H., *Die privaten Rechtsinschriften aus dem Alten Reich* (Wien, 1970).

Goyon– Josset (1988)
Goyon, J-Cl. and Josset, P., *Un corps pour l'éternité: autopsie d'une momie* (Paris, 1988).

Greven (1954)
Greven, L., *Der Ka in Theologie und Königskult der Ägypter des Alten Reiches*, ÄF (Gluuckstadt-Hamburg-New York, 1954).

Grinder–Krausman–Hoffmann (2006)
Grinder, M. I., Krausman, P. R. and Hoffmann, R. S., '*Equus asinus, Mammalian Species' No. 794* (2006).

Grzimek's Encyclopedia (1990)
Grzimek, B., *Grzimek's Encyclopedia of Mammals*. 5 volumes (New York, 1990).

H

Haaland (1993)
Haaland, R., 'Aqualithic sites of the Middle Nile', *Azania* 28 (1993)

Hafez (1969)
Hafez, E. S. E., *The Behaviour of Domestic Animals* (London, 1969).

Hall–Lambert (1922)
Hall, H. R. and Lambert, E. J., *Hieroglyphic Texts from Egyptian Stelae, &c., in the British Museum*, Volume VI (London, 1922).

Handoussa (1988)
Handoussa, T., 'Fish offering in the Old Kingdom' in *Mitteilungen des Deutschen Archäologischen Instituts, Abteilung Kairo* 44 (1988) 105-10.

Hannig (2003)
Hannig, R., *Ägyptisches Wörterbuch I: Altes Reich und Erste Zwischenzeit*, Kulturgeschichte der Antiken Welt 98, Hannig-Lexika 4, (Mainz/Rhein, 2003).

Harlan (1981)
Harlan, J. R., 'The early history of wheat: Earliest traces to the sack of Rome', in Evans, T. and Peacock, W. J. (eds.), *Wheat science today and tomorrow*, Cambridge University Press (Cambridge, 1981).

Harpur (1985)
Harpur, Y. M., 'The Identity and Positions of Relief Fragments in Museums and Private Collections: Miscellaneous Reliefs from Saqqâra and Giza', *Journal of Egyptian Archaeology* 71 (1985) 27-42.

Harpur (1986)
Harpur, Y. M., 'The identity and positions of relief fragments from museums and private collections: The identity and positions of five reliefs from Saqqara', *Mitteilungen des Deutschen Archäologischen Instituts, Abteilung Kairo* 42 (1986) 59-66.

Harpur (1987)
Harpur, Y. M., *Decoration in Egyptian tombs of the Old Kingdom: studies in orientation and scene content,* Series: Studies in Egyptology, distributed by Methuen, Routledge and Kegan Paul (London, 1987).

Harpur (2001)
Harpur, Y. M., *The Tombs of Nefermaat and Rahotep at Maidum: Discovery, Destruction and Reconstruction, Egyptian Tombs of the Old Kingdom,* Volume One, Oxford Expedition to Egypt (Oxford, 2001).

Harpur–Scremin (2006)
Harpur, Y. M. and Scremin, P., *The Chapel of Kagemni: Scene Details* (Oxford, 2006).

Harris–Hillman (1989)
Harris, D. R. and Hillman, G. C., *Foraging and Farming: The Evolution of Plant Exploitation* (London, 1989).

Hassan F. A. (1981)
Hassan, F. A., 'Historical Nile floods and their implications for climatic change', *Science* 212 (1981) 1142-5.

Hassan F. A. (1986b)
Hassan, F. A., 'Holocene lakes and Prehistoric settlements of the Western Faiyum', *Journal of Archaeological Science* 13 (1986) 483-501.

Hassan F. A. (1988)
Hassan, F. A., 'The Predynastic of Egypt', *Journal of World History* 2 (1988).

Hassan F. A. (1997)
Hassan, F. A., 'The Dynamics of a Riverine Civilization: A Geoarchaeological Perspective on the Nile Valley, Egypt', in *World Archaeology, Vol. 29, No. 1, Riverine Archaeology*, pp. 51-74, (London, June 1997).

Hassan F. A. (2005)
Hassan, F. A., 'A River Runs Through Egypt: Nile Floods and Civilization', *Geotimes* (2005).

Hassan I (1932)
Hassan, S., *Excavations at Gîza 1929-1930. Volume I,* The Faculty of Arts of the Egyptian University (Oxford, 1932).

Hassan II (1936)
Hassan, S., *Excavations at Gîza 1930-1931, Volume II.* The Faculty of Arts of the Egyptian University (Cairo, 1936).

Hassan III (1941)
Hassan, S., *Excavations at Gîza 1931-1932, Volume III.* The Faculty of Arts, Fouad I University (Cairo, 1941).

Hassan IV (1943)
Hassan, S., *Excavations at Gîza 1932-1933, Volume IV.* Service des Antiquités de l'Égypte (Cairo, 1943).

Hassan V (1944)
Hassan, S., *Excavations at Gîza 1933-1934, Volume V.* Service des Antiquités de l'Égypte (Cairo, 1944).

Hassan VI (1948)
Hassan, S., *Excavations at Gîza 1934-1935, Volume VI [2].* The Faculty of Arts, Fouad I University, Service des Antiquités de l'Egypte (Cairo, 1948).

Hassan VII (1953)
Hassan, S., *Excavations at Gîza 1935-1936, Volume VII.* Antiquities Department of Egypt (Cairo, 1953).

Hassan VIII (1953)
Hassan, S., *The Great Sphinx and its Secrets, Volume VIII, Historical Studies in the Light of Recent Excavations* (Cairo, 1953).

Hassan IX (1960)
Hassan, S., *Excavations at Gîza 1936-1938, Volume IX,* Antiquities Department of Egypt (Cairo, 1960).

Hassan X (1960)
Hassan, S., *The Great Pyramid of Khufu and its Mortuary Chapel, Volume X, with Names and Titles of Volumes I-X of the Excavations at Gîza 1938-1939.* Antiquities Department of Egypt (Cairo, 1960).

Hassan Saqqara I (1975)
Hassan, S., *Excavations at Saqqara, 1937-1938, Volume I: Mastaba of Neb-Kaw-Her* (Cairo, 1975).

Hassan Saqqara II (1975)
Hassan, S., *Excavations at Saqqara, 1937-1938, Volume II: Mastabas of Ny-aankh-Pepy and Others* (Cairo, 1975).

Hassan, Saqqara III (1975)
Hassan, S., *Excavations at Saqqara, 1937-1938, Volume III: Mastabas of Princess Hemet-Ra and Others* (Cairo, 1975).

Hawass (1996)
Hawass, Z., 'The Workmen's Community at Giza', in Bietak, M. (ed.) *Haus und Palast im Alten Ägypten,* Verlag der Österreichischen Akademie der Wissenschaften (Wien, 1996) 53-67.

Hayes (1953)
Hayes, W. C., *The Scepter of Egypt: Part I: From the Earliest Times to the End of the Middle Kingdom* (New York, 1953).

Helck (1975)
Helck, W., *Wirtschaftsgeschichte des alten Ägypten im 3. und 2. Jahrtausend vor Chr.* (Leiden, 1975).

Helck–Otto (1975)
Helck, W. and Otto E., (eds.) *Lexikon der Ägyptologie* (1975).

Hodder (1989)
Hodder, I., 'Writing archaeology: Site reports in context', *Antiquity* 63 (1989) 268-274.

Hodder (1993)
Hodder, I., 'Social cognition' in *Cambridge Archaeological Journal* 3 (1993) 253-257.

Hodder (2003)
Hodder, I., *Reading the Past. Current approaches to interpretation in archaeology* (Cambridge, 2003).

Hope–McFarlane (2007)
Hope, C. and McFarlane, A., *Akhmim in the Old Kingdom Part II: Pottery, Decoration Technique and Colour Conventions,* The Australian Centre for Egyptology: Studies 7 (Sydney, 2007).

Hornung (1992)
Hornung, E., *Idea into image: Essays on Ancient Egyption Thought,* translated by E. Bredeck (New York, 1992).

Houlihan (1986)
Houlihan, P. F., The Birds of Ancient Egypt (Warminster, 1986).

Houlihan (1987)
Houlihan, P. F., "Some Remarks on Deer (Cervidae) in Ancient Egypt" in *Journal of Egyptian Archaeology* 73 (1987).

I

Ikram (1995)
Ikram, S., *Choice cuts: meat production in ancient Egypt,* Series: Orientalia Lovaniensia analecta 69 (Leuven, 1995).

Ikram (2001)
Ikram, S., 'The iconography of the hyena in ancient Egyptian art' in *Mitteilungen des Deutschen Archäologischen Instituts, Abteilung Kairo* 57, 127-140 (2001).

J

Jacquet-Gordon (1962)
Jacquet-Gordon, H., *Les Noms des Domaines Funéraires sous l'Ancien Empire Égyptien* (Cairo, 1962).

James–Apted (1953)
James, T. G. H., *The mastaba of Khentika, called Ikhekhi,* Egyptian Exploration Society (London, 1953).

James, (1961)
James, T. G. H., *Hieroglyphic Texts from Egyptian Stelae &c. in the British Museum,* Volume I [2] (London, 1961).

James (1962)
James, T. G. H., *The Hekanakhte papers and other early Middle Kingdom documents,* Metropolitan Museum of Art, Egyptian Expedition (1962).

Jéquier (1929)
Jéquier, G., *Tombeaux de particuliers contemporains de Pepi II,* Impr. de l'Institut français d'archéologie orientale (Le Caire, 1929).

Jones (2001)
Jones, D., *An Index of Egyptian Titles, Epithets and Phrases of the Old Kingdom,* 2 volumes (Oxford, 2001).

Junker I (1929)
Junker, H., *Gîza, Band I, Die Mastabas der IV. Dynastie auf dem Westfriedhof,* Akademie der Wissenschaften in Wien (Wien and Leipzig, 1929).

Junker II (1934)
Junker, H., *Gîza, Band II, Die Mastabas der IV. Dynastie auf dem Westfriedhof,* Akademie der Wissenschaften in Wien (Wien und Leipzig, 1934).

Junker III (1938)
Junker, H., *Gîza, Band III, Die Mastabas der vorgeschrittenen V. Dynastie auf dem Westfriedhof,* Akademie der Wissenschaften in Wien (Wien and Leipzig, 1938).

Junker IV (1940)
Junker, H., *Gîza, Band IV, Die Mastaba des Kai-em-anch,* Akademie der Wissenschaften in Wien (Wien und Leipzig, 1940).

Junker V (1941)
Junker, H., *Gîza, Band V, Die Mastaba des Seneb und die umliegenden Gräber,* Akademie der Wissenschaften in Wien (Wien und Leipzig, 1941).

Junker VI (1943)
Junker, H., *Gîza, Band VI, Die Mastabas der Nefer, Kedfi, Kahjef und die westlichanschliessenden Gräbanlagen,* Akademie der Wissenschaften in Wien (Wien und Leipzig, 1943).

Junker VII (1944)
Junker, H., *Gîza, Band VII, Der Östabschnitt des Westfriedhofs, Erster Teil,* Akademie der Wissenschaften in Wien (Wien und Leipzig, 1944).

Junker (1947)
Junker, H., *Gîza, Band VIII, Der Östabschnitt des Westfriedhofs, Zweiter Teil*, Akademie der Wissenschaften in Wien (Wien, 1947).

Junker IX (1950)
Junker, H., *Gîza, Band IX, Das Mittelfeld des Westfriedhofs*, Österreichische Akademie der Wissenschaften (Wien, 1950).

Junker X (1951)
Junker, H., *Gîza, Band X, Der Friedhof südlich der Cheopspyramide, Westteil*, Österreichische Akademie der Wissenschaften (Wien, 1951).

Junker XI (1953)
Junker, H., *Gîza, Band XI, Der Friedhof südlich der Cheopspyramide, Östteil*, Österreichische Akademie der Wissenschaften (Wien, 1953).

Junker XII (1955)
Junker, H., *Gîza, Band XII, Schlussband mit Zusammenfassungen und Gesamt-Verzeichnissen von Band I-XII*, Österreichische Akademie der Wissenschaften (Wien, 1955).

K

Kanawati (1977)
Kanawati, N., *The Egyptian Administration in the Old Kingdom: Evidence on its Economic Decline* (Warminster, 1977).

Kanawati Reforms (1980)
Kanawati, N., *Governmental Reforms in Old Kingdom Egypt* (Warminster 1980).

Kanawati I (1980)
Kanawati, N., *The Rock Tombs of El-Hawawish: The Cemetery of Akhmim, Volume I,* The Macquarie Ancient History Association (Sydney 1980).

Kanawati II (1981)
Kanawati, N., *The Rock Tombs of El-Hawawish: The Cemetery of Akhmim, Volume II,* The Macquarie Ancient History Association (Sydney 1981).

Kanawati III (1982)
Kanawati, N., *The Rock Tombs of El-Hawawish: The Cemetery of Akhmim, Volume III,* The Macquarie Ancient History Association (Sydney, 1982).

Kanawati IV (1983)
Kanawati, N., *The Rock Tombs of El-Hawawish: The Cemetery of Akhmim, Volume IV,* The Macquarie Ancient History Association (Sydney, 1983).

Kanawati V (1985)
Kanawati, N., *The Rock Tombs of El-Hawawish: The Cemetery of Akhmim, Volume V,* The Macquarie Ancient History Association (Sydney, 1985).

Kanawati VI (1986)
Kanawati, N., *The Rock Tombs of El-Hawawish: The Cemetery of Akhmim, Volume VI* (Sydney, 1986).

Kanawati VII (1987)
Kanawati, N., *The Rock Tombs of El Hawawish: The Cemetery of Akhmim, Volume VII,* The Ancient History Documentary Research Centre (Sydney, 1987).

Kanawati VIII (1988)
Kanawati, N., *The Rock Tombs of El-Hawawish: The Cemetery of Akhmim, Volume VIII* (Sydney, 1988).

Kanawati IX (1989)
Kanawati, N., *The Rock Tombs of El-Hawawish: The Cemetery of Akhmim, Volume IX* (Sydney, 1989).

Kanawati (1991)
Kanawati, N., 'Bullfighting in Ancient Egypt', *The Bulletin of the Australian Centre for Egyptology,* Vol. 2 (1991) 51-58.

Kanawati (1993)
Kanawati, N., *The Tombs of El-Hagarsa: Volume I,* The Australian Centre for Egyptology: Reports 4 (Sydney, 1993)

Kanawati (1995)
Kanawati, N., *The Tombs of El-Hagarsa. Volume III,* The Australian Centre for Egyptology: Reports 7 (Sydney, 1993).

Kanawati Giza I (2001)
Kanawati, N., *Tombs at Giza: Volume 1: Kaiemankh (G4561) and Seshemnefer I (G4940) Volume I,* The Australian Centre for Egyptology: Reports 16 (Warminster, 2001).

Kanawati Giza II (2002)
Kanawati, N., Tombs at Giza. Volume II. Seshathetep/Heti (G 5150), Nesutnefer (G 4970) and Seshemnefer II (G 5080), Australian Centre for Egyptology: Reports 18 (Warminster, 2002).

Kanawati (2003)
Kanawati, N., *Conspiracies in the Egyptian Palace: Unis to Pepi I,* Routledge (NewYork, 2003).

Kanawati Gebrawi I (2005)
Kanawati, N., *Deir el-Gebrawi: Volume I: The Northern Cliff,* The Australian Centre for Egyptology: Reports 23 (Oxford, 2005).

Kanawati (2006)
Kanawati, N., *The Teti Cemetery at Saqqara, Volume VIII: The Tomb of Inumin,* The Australian Centre for Egyptology: Reports 24 (Oxford, 2006).

Kanawati Gebrawi II (2007)
Kanawati, N., *Deir el-Gebrawi, Volume II: The Southern Cliff,* The Australian Centre for Egyptology: Reports 25 (Sydney, 2007).

Kanawati (2009)
Kanawati, N., *The Teti Cemetery at Saqqara, Volume IX: The Tomb of Remni,* The Australian Centre for Egyptology: Reports 28 (Oxford, 2009).

Kanawati (2010)
Kanawati, N., Woods, A., Shafik, S. and Alexakis, E., *Mereruka and his Family Part III:1: The Tomb of Mereruka* (Oxford 2010).

Kanawati (2011)
Kanawati, N., Woods, A., Shafik, S. and Alexakis, E., *Mereruka and his Family Part III:2: The Tomb of Mereruka* (Oxford 2011).

Kanawati Meir I (2011)
Kanawati, N., *The Cemetery of Meir, Volume I: The Tomb of Pepyankh the Middle*, The Australian Centre for Egyptology: Reports 31 (Oxford 2011).

Kanawati–Abder-Raziq (1998)
Kanawati, N. and Abder-Raziq, M. *The Teti Cemetery at Saqqara, Volume III: The Tombs of Neferseshemre and Seanhkuiptah,* The Australian Centre for Egyptology: Reports 11 (Warminster, 1998).

Kanawati–Abder-Raziq (1999)
Kanawati N. and Abder-Raziq, M., *The Teti Pyramid Cemetery at Saqqara, Volume V: The Tomb of Hesi,* The Australian Centre for Egyptology: Reports 13 Wiltshire, 1999).

Kanawati–Abder-Raziq (2000)
Kanawati N. and Abder-Raziq, M., *The Teti Pyramid Cemetery at Saqqara, Volume VI: The Tomb of Nikauisesi,* The Australian Centre for Egyptology: Reports 14 (Warminster, 2000).

Kanawati–Abder-Raziq (2003)
Kanawati, N. and Abder-Raziq, M., *The Unis Cemetery at Saqqara, Volume II: The Tombs of Iynefert and Ihy (reused by Idut),* The Australian Centre for Egyptology: Reports 19, (Oxford, 2003).

Kanawati–Abder-Raziq (2004)
Kanawati, N. and Abder-Raziq, M., *Mereruka and his Family, Part 1: Meryteti,* The Australian Centre for Egyptology: Reports 21 (Oxford, 2004).

Kanawati–Abder-Raziq (2008)
Kanawati, N. and Abder-Raziq, M., *Mereruka and his Family, Part 2: The Tomb of Waatet-khet-hor,* The Australian Centre for Egyptology: Reports 26 (Oxford, 2008).

Kanawati–Hassan (1996)
Kanawati, N. and Hassan, A., *The Teti Cemetery at Saqqara, Volume I: The Tombs of Nedjet-em-pet, Kaaper and Others,* The Australian Centre for Egyptology: Reports 8 (Warminster, 1996).

Kanawati–Hassan (1997)
Kanawati, N. and Hassan, A., *The Teti Cemetery at Saqqara, Volume II: The Tomb of Ankhmahor,* The Australian Centre for Egyptology: Reports 9 (Warminster, 1997).

Kanawati–McFarlane (1992)
Kanawati, N. and McFarlane, A., *Akhmim in the Old Kingdom, Part I: Chronology and Administration,* The Australian Centre for Egyptology: Studies 2 (Sydney, 1992).

Kanawati–McFarlane (1993)
Kanawati, N. and McFarlane, A., *Deshasha. The Tombs of Inti, Shedu and Others*, The Australian Centre for Egyptology: Reports 5 (Sydney, 1993).

Kaper–Willems (2002)
Kaper, O.E. and Willems, H., 'Policing the Desert: Old Kingdom Activity around the Dakhleh Oasis', in Friedman, R. (ed.), *Egypt and Nubia: Gifts of the Desert*, The British Museum Press (London, 2002).

Kaplony (c1976)
Kaplony, P., *Studien zum Grab des Methethi,* [Riggisberg] Abegg-Stiftung (Bern, c1976).

Kayser (1954)
Kayser, H., *Die Mastaba des Uhemka. Ein Grab in der Wüste.* Reprint of the Zeitschrift des Museums zu Hildesheim, Band 15 (Hannover, 1954).

Keeling–Gonyou (2001)
Keeling, L. J. and Gonyou, H. W., *Social Behaviour in Farm Animals* (Wallingford, 2001).

Kees (1932)
Kees, H., 'Beiträge zur altägyptischen Provinzialverwaltung und der Geschichte des Feudalismus,' in *Nachrichten von der Gesellschaft der Wissenschaften zu Göttingen* (1932) 85-119 and 579-598.

Khaldi (1984)
Khaldi, T., (ed.) *Land tenure and social transformation in the Middle East,* American University of Beirut (Beirut, 1984).

Kitchen (1991)
Kitchen, K. A., 'The Chronology of Ancient Egypt', *World Archaeology*, Vol. 23, No. 2, Chronologies (Oct., 1991) 201-208.

Klebs (1914)
Klebs, L., 'Die Tiefendimension in der Zeichnung des Alten Reiches', *Zeitschrift für Ägyptische Sprache und Altertumskunde* 32 (1914).

Klebs (1915)
Klebs, L., *Die Reliefs des alten Reiches (2980-2475 v. Chr.) Material zur ägyptischen Kulturgeschichte* (Heidelberg 1915).

Klebs (1922)
Klebs, L., *Die Reliefs und Malereien des Mittleren Reiches (VII-XVII. Dynastie ca. 247-1580 v. Chr.) Material zur ägyptischen Kulturgeschichte* (Heidelberg, 1922).

Kloth (2002)
Kloth, N., *Die (auto-)biographischen Inschriften des ägyptischen Alten Reiches: Untersuchungen zu Phraseologie und Entwicklung*, Studien zur Altägyptischen Kultur 8 (Hamburg 2002).

Kozlowski–Ginter (1993)
Kozlowski, J. K., and Ginter, B., 'Holocene changes in the Fayum: Lake Moeris and the evolution of climate in Northeastern Africa', in (eds), Krzyzaniak, L. et al, *Environmental Change and Human Culture* (Poznan 1993)

Krejčí (2008)
Krejčí, J., *Abusir XI. The Architecture of the Mastaba of Ptahshepses,* Czech Institute of Egyptology (Prague, 2008).

Kruuk (1976)
Kruuk, H., 'Feeding and social behaviour of the striped hyaena (Hyaena vulgaris Demarest)', *East African Wildlife Journal* 14, 91-111 (1976).

Krzyzaniak, (1977)
Krzyzaniak, L., *Early farming cultures on the lower Nile : The predynastic period in Egypt,* Panstwowe Wydawnictwo Naukowe (Warszawa, 1977).

Kuper (2003)
Kuper, R., 'The Abu Ballas Trail: Pharaonic Advances into the Libyan Desert', in Hawass, Z. and Pinch Brock, L. (eds.), *Egyptology at the Dawn of the Twenty–first Century: Proceedings of the Eighth International Congress of Egyptologists, Cairo 2000, Vol. 2: History, Religion* (Cairo, New York, 2003) 372–6.

Kuraszkiewicz (2005)
Kuraszkiewicz, *K.O. Remarks on the Development of The Old Kingdom Necropolis*, Polish Archaeology in the Mediterranean XVII, Reports 2005, Polish Centre of Mediterranean Archaeology, (University of Warsaw 2007).

L

LÄ
Helck, W., Otto, E. and Westendorf, W., (eds.), *Lexikon der Ägyptologie,* 7 vols. (Wiesbaden 1972).

Lacau (1914)
Lacau, P., 'Suppressions et modifications de signes dans les textes funéraire', *Zeitschrift für Ägyptische Sprache und Altertumskunde* 51 (1914).

Lacau–Lauer (1965)
Lacau, P. and Lauer, J.-Ph., *La pyramide à degrés. Tome V. Inscriptions à l'encre sur les vases,* Institut français d'Archéologie orientale (Le Caire, 1965).

Landström (1970)
Landström, B., *Ships of the Pharaohs: 4000 years of Egyptian Shipbuilding* (New York, 1970).

Lehner (1992)
Lehner, M., '*Excavations at Giza 1988-1991: The Location and Importance of the Pyramid Settlement'*, The Oriental Institute News and Notes, No. 135 Fall, University of Chicago (Chicago, 1992).

Lehner (2002).
Lehner, M., 'The pyramid age settlement of the Southern Mount at Giza', *Journal of the American Research Center in Cairo* 39: 27-74 (2002).

Lehner (2004)
Lehner, M., *2003-2004 Annual Report, Giza Plateau Mapping Project,* The Oriental Institute, 64-71.

Lehner (2006)
Lehner, M., *2005-2006 Annual Report, Giza Plateau Mapping Project,* The Oriental Institute, 53-54, 55-57, 60.

Lehner (2007)
Lehner M., *Giza Plateau Mapping Project, 2006-2007 Season, Annual Report,* The Oriental Institute, 55, 59-61, 64, 66, 71-74.

Lehner–Wetterstrom, (2007)
Lehner M. and Weterstrom, W., (eds), *Giza Reports, The Giza Mapping Project,* vol. 1, Ancient Egypt Research Associates Inc. (Brighton, MA, 2006).

LD Erg. (1913)
Lepsius, C. R., *Denkmäler aus Aegypten und Aethiopien. Ergänzungsband,* J. C. Hinrich (Leipzig, 1897-1913)

LD II
Lepsius, C. R., *Denkmäler aus Ägypten und Äthiopien. Zweite Abtheilung. Denkmäler des Alten Reiches.* Band III, Blatt i-lxxxi; Band IV, Blatt lxxxii-cliii. (Berlin 1849-1859)

Lichtheim (1973)
Lichtheim, M., *Ancient Egyptian Literature: A Book of Readings: Volume I: The Old and Middle Kingdoms* (Berkeley 1973).

Lichtheim (1988)
Lichtheim, M., *Ancient Egyptian Autobiographies Chiefly of the Middle Kingdom: a study and an anthology* (Göttingen, 1988).

Lichtheim (1992)
Lichtheim, M., *Maat in Egyptian Autobiographies and Related Studies* (Göttingen, 1992).

Lloyd (1978)
Lloyd, A. B., 'Strabo and the Memphite tauromaly' in de Boer, M. B. and Edridge, T.A. (eds), *Hommages à Maarten J. Vermaseren II* (Leiden, 1978).

Lloyd et al (1990)
Lloyd, A. B., Spencer, A. J., and El-Khouli, A., *Saqqâra Tombs. Volume II. The Mastabas of Meru, Semdenti, Khui and Others,* Archaeological Survey of Egypt, Fortieth Memoir, Egypt Exploration Society (London, 1990).

Lucas–Harris (1999)
Lucas, A. and Harris, J. R., *Ancient Egyptian materials and industries,* Dover Publications (Mineola, N.Y. 1999).

Luft (1992)
Luft, U., *The Intellectual Heritage of Egypt: Studies Kákosy*, Studia Aegyptiaca (Budapest, 1992).

M

McFarlane (2000)
McFarlane, A., *The Tomb of Irukaptah. The Unis Cemetery at Saqqara.* Volume I, The Australian Centre for Egyptology: Reports 15 (Warminster, 2000).

McFarlane (2003)
McFarlane, A., *Mastabas at Saqqara: Kaiemheset, Kaipunesut, Kaiemsenu, Sehetepu and Others,* The Australian Centre for Egyptology: Reports 20 (Oxford, 2003).

McGregor (2000)
McGregor, B., Goat Notes B19: *What Do Goats Really Like To Eat?* (2000) Victorian Institute of Animal Science. www.acga.org.au/goatnotes/C001.php

Macramallah (1935)
Macramallah, R., *Le Mastaba d'Idout, Fouilles à Saqqarah.* Service des Antiquités de l'Égypte, Institut Français d'Archéologie Orientale (Cairo, 1935).

Mahmoud (1991)
Mahmoud, O., *Die wirtschaftliche Bedeutung der Vögel im Alten Reich* (Frankfurt am Main - Bern, 1991).

Manuelian (2003)
Manuelian, P. der, *Slab Stelae of the Giza Necropolis*, Publications of the Pennsylvania–Yale Expedition to Egypt, Number 7 (New Haven and Philadelphia, 2003).

Markowitz–Haynes–Freed (2002)
Markowitz, Y. J., Haynes, J. L., and Freed, R. E. *Egypt in the Age of the Pyramids. Highlights from the Harvard University–Museum of Fine Arts Boston Expedition* (Boston, 2002).

Martin et al (1984)
Martin, D. L., Armelago, G. J., Goodman, A. H. and Van Gerven, D. P., 'The Effects of Socioeconomic Change in Prehistoric Africa: Sudanese Nubia as a Case Study' in Cohen, M. N. and Armelagos, G. J. (eds.). *Paleopathology at the Origins of Agriculture (Orlando*, 1984) 193–214.

Martin (1979)
Martin, G. T. M., *The Tomb of Hetepka and Other Reliefs and Inscriptions from the Sacred Animal Necropolis,* North Saqqâra 1964-1973. Texts from Excavations, Fourth Memoir, Egypt Exploration Society (London, 1979).

Martin-Pardey (1976)
Martin-Pardey, E., *Untersuchungen zur ägyptischen Provinzialverwaltung bis zum Ende des Alten Reiches*, Verlag Gebrüder Gerstenberg (Hildesheim, 1976).

Meadow–Zeder (1978)
Meadow, R. H. and Zeder, M. A., (eds), *Approaches to Faunal Analysis in the Middle East, Peabody Museum Bulletin no 2*, Peabody Museum of Archaeology and Ethnology (New Haven, 1978).

Menu (1982)
Menu, B., *Recherches sur l'histoire juridique, économique et sociale de l'ancienne Egypte* [Privately published] (1982).

Menu (1985)
Menu, B., 'Ventes de maisons sous l'Ancien Empire', in Geus, Francis and Thill, Florence (eds), *Mélanges offerts à Jean Vercoutter*. Éditions Recherche sur les Civilisations (Paris, 1985) 249-262.

Menu (1994)
Menu, B., 'Introduction', in Menu (ed), *Les problèmes institutionnels de l'eau en Égypte ancienne et dans l'Antiquité méditerranéenne, XI-XV*, Institut français d'archéologie orientale (Le Caire, 1994).

Moehlman (1998)
Moehlman, P. D., 'Behavioral patterns and communication in feral asses (Equus africanus)' *Applied Animal Behaviour Science* 60 (2-3) (1998).

Moens–Wetterstrom (1988)
Moens, M.-F. and Wetterstrom, W., "The Agricultural Economy of an Old Kingdom Town in Egypt's West Delta: Insights from the Plant Remains", *Journal of Near Eastern Studies,* 47m No. 3 (1988),

Mogensen (1921)
Mogensen, M., *Le Mastaba égyptien de la Glyptothèque Ny Carlsberg*, Gyldendalske Boghandel Nordisk Forlag (Copenhagen, 1921).

Mohr, (1943)
Mohr, H. T., *The Mastaba of Hetep-her-akhti. Study on an Egyptian Tomb Chapel in the Museum of Antiquities,* Leiden (Leiden 1943).

Montet (1924)
Montet, P., *Les Scènes de la vie privée dans les tombeaux égyptiens de l'ancien empire,* Publications de la Faculté des Lettres de l'Université de Strasbourg, Fasc. 24 (Strasbourg, Paris and Oxford, 1924).

Montet, Kemi, 13 (1954)
Montet, P., 'Les Boeufs Égyptiens', *Kemi* 13 (1954).

Moore (1985)
Moore, A. M. T., ''The Development of Neolithic Societies in the Near East.'' *Advances in World Archaeology* 4 (1985) 1–69.

Moreno Garcia (1996)
Moreno Garcia, J. C., Administration territoriale et organisation de l'espace en Égypte au troisième millénaire avant J.-C.: *grgt* et le *titre ʿ(n)ḏ-mr grgt*', *Zeitschrift für Ägyptische Sprache und Altertumskunde* 123 (1996) 117-138.

Moreno Garcia (1997)
Moreno Garcia, J. C., *Études sur l'administration, le pouvoir et l'idéologie en Égypte, de l'Ancien au Moyen Empire* (Liége, 1997).

Moreno Garcia JEA 84 (1998)
Moreno Garcia, J. C., 'La population mrt: une approche du problème de la servitude dans l'Égypte du IIIe millénaire (I)', *Journal of Egyptian Archaeology* 84 (1998) 71-83.

Moreno Garcia (1999)
Moreno Garcia, J. C., *Hwt et le milieu rural égyptien du IIIe millénaire: Économie, administration et organisation territoriale* (Paris, 1999).

Moreno Garcia ZÄS 126 (1999)
Moreno Garcia, J. C., 'Administration territoriale et organisation de l'espace en Égypte au troisième millénaire avant J.-C. (V): *gz-pr*', *Zeitschrift für Ägyptische Sprache und Altertumskunde* 126 (1999) 116-131.

Moreno Garcia RdE 50 (1999)
Moreno Garcia, J. C., 'J'ai rempli les pâturages de vaches tachetées... Bétail, économie royale et idéologie en Égypte, de l'Ancien au Moyen Empire', *Revue d'égyptologie* 50 (1999) 241-257.

Moreno Garcia (2000)
Moreno Garcia, J. C., 'Acquisition de serfs durant la Première Période Intermédiaire: une étude d'histoire sociale dans l'Égypte du IIIe millénaire', *Revue d'égyptologie* 51 (2000) 123-139.

Moreno Garcia (2001)
Moreno Garcia, J. C., 'L'organisation sociale de l'agriculture dans l'Égypte pharonique pendant l'Ancien Empire', *Journal of the Economic and Social History of the Orient*, Vol. 44, No. 4 (2001) 411-450.

Moreno Garcia (2006)
Moreno Garcia, J. C., 'Les temples provinciaux et leur rôle dans l'agriculture institutionnelle de l'Ancien et du Moyen Empire,' in *Cahier de Recherches de l'Institut de Papyrologie et d'Égyptologie de Lille, Cripel* 25 (Lille, 2006) 125-136,

Moussa–Altenmüller (1971)
Moussa, A. M. and Altenmüller, H., *The Tomb of Nefer and Ka-hay. Old Kingdom Tombs at the Causeway of King Unas at Saqqara*. Archäologische Veröffentlichungen 5, Deutsches Archäologisches Institut, Abteilung Kairo, Philipp von Zabern (Mainz am Rhein, 1971).

Moussa–Altenmüller (1977)
Moussa, A. M. and Altenmüller, H., *Das Grab des Nianchchnum und Chnumhotep. Old Kingdom Tombs at the Causeway of King Unas at Saqqara*. Archäologisches Veröffentlichungen 21, Deutsches Archäologisches Institut, Abteilung Kairo, Philipp von Zabern (Mainz am Rhein, 1977).

Moussa–Junge (1975)
Moussa, A. M. and Junge, F., *Two Tombs of Craftsmen. Old Kingdom Tombs at the Causeway of King Unas at Saqqara*. Archäologische Veröffentlichungen 9, Deutsches Archäologisches Institut, Abteilung Kairo, Philipp von Zabern (Mainz am Rhein, 1975).

Munro (1993)
Munro, P., *Der Unas-Friedhof Nord-West, Topographisch-historische Einleitung, Volume I: Das Doppelgrab der Königinnen Nebet und Khenut*, Philipp von Zabern (Mainz am Rhein, 1993)

Murray (2000)
Murray, M. A., 'Cereal production and processing' in Nicholson, P. T. and Shaw, I. (eds.), *Ancient Egyptian materials and technology*, Cambridge University Press (2000) 505-536.

Murray (2005a)
Murray, M. A., 'Provisions for the pyramid builders: New evidence from the ancient Egyptian site of Giza', *Archaeology International* 2004/2005, 38-42.

Murray (2005b)
Murray, M. A., 'Feeding the Town: New Evidence from the Complex of the Giza Pyramid Builders', *General Anthropology, Bulletin of the General Anthropology Division of the American Anthropological Association*, Vol. 12, Nos 1 and 2 (2005) 1, 6-9.

Murray (2007a)
Murray, M. A., 'Questions of continuity: Fodder and fuel use in Bronze Age Egypt' in Fairbairn, A. and Weiss, E. (eds.), *Ethnobotanist of distant pasts: Archaeological and ethnobotanical studies in honour of Gordon Hillman* (Oxford, 2007).

Murray (2007b)
Murray, M. A., 'Main Street Plant Remains' and 'Gallery III.4 plant remains', in Lehner, M. and Wetterstrom, W. (eds.), *Giza Reports: Giza Plateau Mapping Project. Volume 1. Project History, Survey, Ceramics and Main Street and Gallery III.4 Operations* (Boston, 2007) 163-69; 267-58.

Muthuri–Kintamario (1989)
Muthuri, F. M. and Kinyamario, J. L. 'Nutritive Value of Papyrus (Cyperus papyrus, Cyperaceae), a Tropical Emergent Macrophyte', *Economic Botany* 43 (1), New York Botanical Gardens (Bronx New York, 1989), 23-30.

Myśliwiec (2004)
Myśliwiec, K., *Saqqara I: The Tomb of Merefnebef* (Varsovie, 2004).

Myśliwiec and Kuraszkiewicz (2010)
Myśliwiec, K. and Kuraszkiewicz, K. O., *Saqqara IV, The Funerary Complex of Nyankhnefertem* (Warsaw 2010).

N

Newberry I (1893)
Newberry, P. E., *Beni Hasan, Part I* (London, 1893)

Nicholson–Shaw (2000)
Nicholson, P. T. and Shaw, I., (eds.) *Ancient Egyptian materials and technology*, Cambridge University Press (Cambridge University, 2000).

O

O'Connor (1983)
O'Connor, D. J., 'New Kingdom and Third Intermediate Period, 1552-664' in Trigger, B.G. et al (eds.), *Ancient Egypt: a social history* (Cambridge, New York, 1983).

Osborn–Osbornová (1998)
Osborn, D.J. and Osbornová, J. *The Mammals of Ancient Egypt* (Warminster, 1998).

Oxford Database (2006)
Oxford Expedition To Egypt: Scene-details Database, Linacre College, Oxford, (2006). <http://www.oxfordexpeditiontoegypt.com/>

P

Paget–Pirie (1898)
Paget, R. F. E., Pirie, A. A. and Griffith, F. L., *The Tomb of Ptah-hetep, Volume II* (London, 1898).

PAM
Polish Archaeology in the Mediterranean

Parlebas (1983)
Parlebas, J., 'La notion de niout, '(Localité) dans la pensée Égyptienne antique', *CEPOA* I (1983) 199-207.

Peck (1959)
Peck, C. N., *Some Decorated Tombs of the First Intermediate Period at Naga ed-Dêr*, University Microfilms, Department of Egyptology, Brown University (Ann Arbor, Michigan, 1959).

Peet–Loat (1913)
Peet, T. E. and Loat, W. L. S., *The Cemeteries of Abydos II: 1912-13*, EEF (London, 1913).

Perepelkin (1986)
Perepelkin, J. J., *Privateigentum in der Vorstellung der Ägypter des Alten Reichs*, translated by Renate Müller-Wollermann. [s.n.] (Tübingen, 1986).

Petrie (1890)
Petrie, W. M. F., *Kahun, Gurob and Hawara* (London, 1890).

Petrie (1892)
Petrie, W. M. F., *Medum* (London, 1892).

Petrie (1898)
Petrie, W. M. F., *Deshasheh 1897*, Egypt Exploration Fund (London, 1898).

Petrie (1900)
Petrie, W. M. F., *Dendereh* (London, 1900).

Petrie–Murray (1952)
Petrie, H. F. and Murray, M. A., *Seven Memphite Tomb Chapels,* British School of Egyptian Archaeology, Volume LXV (London, 1952).

Phillips–Coventry (May, 2005)
Phillips, A. and Coventry, J., 'Breeding Polled Cattle' in *Agnote* (May, 2005).

Porter–Moss (1974)
Porter, B. and Moss, R. L. B., *Topographical Bibliography of Ancient Egyptian Hieroglyphic Texts, Reliefs and Paintings*, vol. III (Oxford, 1974).

Porter–Moss (1934)
Porter, B. and Moss, R. L. B., *Topographical Bibliography of Ancient Egyptian Hieroglyphic Texts, Reliefs and Paintings*, vols. V, VI (Oxford, 1934).

Posener-Krieger (1968)
Posener-Krieger, P., *Abu Sir papyri* (London, 1968).

Power (2004)
Power, R., 'Deconstructing, Deciphering and Dating the Donkey in Old Kingdom Wall Paintings and Relief', *The Bulletin of the Australian Centre for Egyptology*, Vol. 15 (2004) 131-151.

Q

Quibell (1909)
Quibell, J. E., *Excavations at Saqqara 1907-1908.* Service des Antiquités de l'Égypte, Institut Français d'Archéologie Orientale (Cairo, 1909).

Quibell (1913)
Quibell, J. E., *Excavations at Saqqara (1911-1912): The Tomb of Hesy.* Service des Antiquités de l'Égypte, Institut Français d'Archéologie Orientale (Cairo, 1913).

Quibell–Hayter (1927)
Quibell, J. E. and Hayter, A. G. K., *Excavations at Saqqara. Teti Pyramid, North Side.* Service des Antiquités de l'Égypte. Institut Français d'Archéologie Orientale (Cairo, 1927).

R

Redding (1984)
Redding, R. W., 'Theoretical Determinants of a Herder's Decisions: Modelling Variation in the Sheep/goat Ratio', in Clutton-Brock, J. and Grigson, C. (eds.) *Animals and Archaeology: 3, Early Herders and their Flocks,* vol. 3, British Archaeological Reports, International Series (Oxford, 1984).

Redding (1992)
Redding, R. W., 'Egyptian Old Kingdom Patterns of Animal Use and the Value of Faunal Data in Modeling Socioeconomic Systems', *Paléorient* (1992) Volume 18, Issue 18-2, 99-107.

Redding (1991)
Redding, R. W., 'The role of the pig in the subsistence economy of Aancient Egypt', in Crabtree, Pamela J and Ryan, Kathleen (eds.), *Animal Use and Culture Change, Supplement to Volume 8*, University Museum of Archaeology and Anthropology (Philadelphia, 1991).

Redding Kom el-Hisn (2002)
Redding, R. W., 'Differential Consumption of Pig vs. Sheep/ Goats at the Old Kingdom Site of Kom el-Hisn', Paper presented at the 53rd Annual Meeting of the American Research Center in Egypt (Baltimore, 2002). <www.acagle.net/ARCE2002/PigPaper2Refs.html>

Redding (2008)
Redding, R. W., 'Main Street Faunal Remains' and 'Gallery III.4 Faunal Remains', in *Giza Reports: Giza Plateau Mapping Project. Volume 1. Project History, Survey, Ceramics and Main Street and Gallery III.4 Operations.* Mark Lehner and Wilma Wetterstrom (eds.) (Boston, 2007) 171-78; 263-69.

Redman (1978)
Redman, C. L., *The rise of civilization: from early farmers to urban society in the ancient Near East* (San Francisco, c1978).

Reed (1969)
Reed, C. A., 'The pattern of animal domestication in the Prehistoric Near East', 361-380. In Ucko, P. and Dimbleby, G.W. (eds), *The Domestication and Exploitation of Plants and Animals* (London, 1969).

Reisner (1942)
Reisner, G. A., *A History of the Giza Necropolis, Volume I.* Harvard University Press (Cambridge Massachusetts, 1942).

Ripinsky (1985)
Ripinsky, M., 'The Camel in Dynastic Egypt', *Journal of Egyptian Archaeology* 71 (1985) 134–41.

Ritner (2001)
Ritner, R. K., 'Magic: an Overview', *Oxford Encyclopaedia of Ancient Egypt II* (2001) 321-26; Ritner, R. K., 'Magic and the Afterlife', *Oxford Encyclopaedia of Ancient Egypt II* (2001) 333-36.

Robins (1994)
Robins, G., *Proportion and Style in Ancient Egyptian Art*, University of Texas Press (Austin, 1994).

Rossel (2008)
Rossel, S., et al. 'Domestication of the donkey: Timing, processes, and indicators'. *Proceedings of the National Academy of Sciences* 105(10) (2008) 3715–3720.

Roth (1992).
Roth, A. M., 'The Peseshkef and the 'Opening of the Mouth' ceremony: a ritual of birth and rebirth', *Journal of Egyptian Archaeology* 78, (1992) 113-147.

Roth (1993)
Roth, A. M., 'Fingers, stars and the "Opening of the Mouth": the nature and function of the ntrwy-blades', *Journal of Egyptian Archaeology* 79 (1993) 57-79.

Roth, (1994)
Roth, A. M., 'The practical economics of tomb-building in the Old Kingdom: a visit to the necropolis in a carrying chair', in Silverman, D. P. (ed.) *For his ka: essays offered in memory of Klaus Baer,* Oriental Institute of the University of Chicago (Chicago, 1994).

Roth (1995)
Roth, A. M., *A Cemetery of Palace Attendants, Including G 2084-2099, G 2230 + 2231, and G 2240.* Giza Mastabas, Volume 6. Department of Ancient Egyptian, Nubian, and Near Eastern Art, Museum of Fine Arts (Boston, 1995).

Roth IBAES VI (2006)
Roth, A. M., 'Multiple Meanings in Carrying Chair Scenes' in Martin Fitzenreiter und Michael Herb (Hrsg.) *Dekorierte Grabanlagen im Alten Reich: Methodik und Interpretation, IBAES VI* (Humboldt-Universität zu Berlin 2006).

Roth (2006)
Roth, A. M., 'Little women: gender and hierarchic proportion in Old Kingdom mastaba chapels' in *The Old Kingdom Art and Archaeology. Proceedings of the Conference held in Prague, May 31–June 4, 2004*, pp. 281–296, edited by M. Barta, Czech Institute of Egyptology (Prague, 2006).

Ryan (1987)
Ryan, D. P. and Hansen, D., *A Study of Ancient Egyptian Cordage from the British Museum,* Occasional paper, British Museum 62 (London, 1987).

Ryan (1988)
Ryan, D. P., 'Papyrus', *Biblical Archaeologist* (September, 1988).

S

Saad–Sami (1967)
Saad, S. I. and Sami, S., 'Studies of pollen and spores content of Nile Delta deposits (Berenbal region)', *Pollen Spores* 9, (1967). 467-503.

Sahrhage (1998)
Sahrhage, D., *Fischfang und Fischkult im alten Ägypten, Kulturgeschichte der Antiken Welt* 70 (Mainz/Rhein, 1998).

Said (1993)
Said, R., *The River Nile. Geology, Hydrology and Utilization* (Oxford, 1993).

Saleh (1977)
Saleh, M., Three Old Kingdom Tombs at Thebes: The Tomb of Unas-Ankh no. 413. II. The Tomb of Khenty no. 405. III. The Tomb of Ihy no. 186, I, trans. M. Bullock (Mainz/Rhein, 1977).

Samuel (2000)
Samuel, D., 'Brewing and Baking' in Nicholson, P.T. and Shaw, I., (eds.) *Ancient Egyptian materials and technology,* Cambridge University Press (Cambridge 2000) 537-576.

Schäfer (1986)
Schäfer, H., *Principles of Egyptian Art.* Trans. and edited by J. Baines (Oxford, 1986).

Schürmann (1983)
Schürmann, W., *Die Reliefs aus dem Grab des Pyramidenvorstehers Ii-nefret. Eine Bilddokumentation des Badischen Landesmuseums Karlsruhe* (Karlsruhe, 1983).

Sethe (1933)
Sethe, K., *Urkunden des alten Reichs*, Heft 1-2 (Leipzig, 1933).

Shafik (2010)
Shafik, S., 'Force-feeding animals in Old Kingdom scenes', *Göttinger Miszellen* 227 (2010),

Sharawi–Harpur (1988)
Sharawi, G. and Harpur, Y., 'The identity and positions of relief fragments from museums and private collections. Reliefs from various tombs at Saqqara', *Journal of Egyptian Archaeology* 74 (1988) 57-67.

Shaw et al (1993)
Shaw, T., Sinclair, P., Andah, B. and Okpoko, A. (eds), *The Archaeology of Africa: Food, Metals and Towns,* Routledge (London, 1993).

Siebels (2000)
Siebels, R.J., *Agriculture in Old Kingdom tomb decoration [manuscript] : an analysis of scenes and inscription,.* Unpublished PhD thesis (Macquarie University, 2000).

Silverman (c1994)
Silverman, D. P., (ed.) *For his ka: essays offered in memory of Klaus Baer*, Oriental Institute of the University of Chicago, (Chicago, c1994).

Simpson (1976)
Simpson, W. K., *The Offering Chapel of Sekhem-ankh-ptah in the Museum of Fine, Arts, Boston.* Department of Egyptian and Ancient Near Eastern Art, Museum of Fine Arts (Boston, 1976).

Simpson *Qar and Idu* (1976)
Simpson, W. K., *The Mastabas of Qar and Idu. G 7101 and 7102,* Giza Mastabas Volume 2. Department of Egyptian and Ancient Near Eastern Art, Museum of Fine Arts (Boston, 1976).

Simpson *Sekhem-ankh-ptah* (1976)
Simpson, W. K., *The Offering Chapel of Sekhem-ankh-ptah in the Museum of Fine Arts, Boston.* Department of Egyptian and Ancient Near Eastern Art, Museum of Fine Arts (Boston, 1976).

Simpson (1978)
Simpson, W. K., *The Mastabas of Kawab, Khafkhufu I and II. G 7110-20, 7130-40, and 7150 and subsidiary mastabas of Street G 7100.* Giza Mastabas, Volume 3. Museum of Fine Arts (Boston, 1978).

Simpson (1992)
Simpson, W. K., *The Offering Chapel of Kayemnofret in the Museum of Fine Arts, Boston.* Department of Egyptian and Ancient Near Eastern Art, Museum of Fine Arts (Boston, 1992).

Simpson (1980)
Simpson, W. K., *Mastabas of the Western Cemetery. Part I Sekhemka, Tjetu I, Iasen, Penmeru, Hagy, Nefertjentet and Herunefer, Djaty, Tjetu II and Nemesti.* Giza Mastabas, Volume 4. Department of Egyptian and Ancient Near Eastern Art. In collaboration with the Pennsylvania-Yale Expedition to Egypt. Museum of Fine Arts (Boston, 1980).

Singer et al (1954)
Singer, C., et al (eds), *A History of Technology* (Oxford, 1954).

Smith (1942)
Smith, W. S., 'The Origin of Some Unidentified Old Kingdom Reliefs', *American Journal of Archaeology*, Vol. 46, No. 4 (Oct. - Dec., 1942) 509-531.

Smith (1946)
Smith, W. S., *A History of Egyptian Sculpture and Painting in the Old Kingdom,* the Museum of Fine Arts, Boston. Harvard University Press and Oxford University Press

(Cambridge, Massachusetts and London, 1946), (New York, re-issued (1978).

Smith (1958)
Smith, W.S., 'The Judge goes Fishing', *Bulletin of the Museum of Fine Arts* 56 [304] (1958) 56-63.

Spalinger (1986)
Spalinger, A., 'Baking during the Reign of Seti I', *Bulletin de l'Institut français d'archéologie orientale*, Vol. 86 (1986).

Steindorff (1913)
Steindorff, G., *Das Grab des Ti. Veröffentlichungen der Ernst von Sieglin Expedition in Ägypten.* Zweiter Band. J. C. Hinrichs'sche Buchhandlung (Leipzig, 1913).

Steward (1949)
Steward, J. H., 'Cultural Causality and Law: A Trial Formulation of the Development of Early Civilizations', *American Anthropologist* 51 (1949) 1–27.

Strandberg (2009)
Strandberg, Å., *The gazelle in ancient Egyptian art: image and meaning* (Uppsala, 2009)

Strudwick (1985)
Strudwick, N., *The Administration of Egypt in the Old Kingdom : the highest titles and their holders* (London, 1985).

Strudwick (2005)
Strudwick, N., *Texts from the Pyramid Age,* Society of Biblical Literature (Georgia 2005).

Swinton (2001)
Swinton, J., *The Dating of the Tombs of Officials of the Old Kingdom of Egypt,* Unpublished Masters of Arts Honours Thesis (Macquarie University, 2001).

Swinton-Evans (2012)
Swinton, J. and Evans, L., 'Force-feeding animals in Old Kingdom scenes: A reply', *Göttinger Miszellen* (2012) (forthcoming).

T

Taylor (2001)
Taylor, J. H., *Death and the Afterlife in Ancient Egypt* (London, 2001).

Taylor (1948)
Taylor, W. W., *A study of archaeology,* Memoir series of the American Anthropological Association no. 69, American Anthropological Association (1948).

Teeter (1987)
Teeter, E., 'Techniques and Terminology of Rope-making in Ancient Egypt', *Journal of Egyptian Archaeology* 73 (1987) 71-77, pls. 7-9.

Teeter (1997)
Teeter, E., *The presentation of Maat. Ritual and Legitimacy in Ancient Egypt,* Studies in Ancient Oriental Civilisation 57 (Chicago, 1997).

Teeter (2001)
Teeter, E., 'Maat', *Oxford Encyclopaedia of Ancient Egypt II* (2001).

te Velde (1982)
te Velde, H., 'Commemoration in Ancient Egypt' *Visible Religion I: Commemorative Figures: Papers Presented to Dr. Th. P. van Baaren (*Leiden 1982).

te Velde (1985 – 1986)
te Velde, H., 'Egyptian Hieroglyphs as Signs, Symbols and Gods' in *Visible Religion* 4-5 (1985 – 1986) 63-72.

Trigger (1983)
Trigger, B. G., et al (eds.) *Ancient Egypt: a social history,* Cambridge University Press (Cambridge New York, 1983).

Trzciński, Kuraszkiewicz, Welc (2007)
Trzciński, J., Kuraszkiewicz, K.O., Welc, F. *Preliminary Report on Geoarchaeological Research in West Saqqara,* Polish Archaeology in the Mediterranean XIX, Reports 2007, Polish Centre of Mediterranean Archaeology, (University of Warsaw 2010).

U

Ucko–Dimbleby (1969)
Ucko, P. and Dimbleby, G. W., (eds.), *The Domestication and Exploitation of Plants and Animals* (London, 1969).

V

Vachala (2004)
Vachala, B., *Abusir VIII. Die Reliefs aus der Ptahschepses-Mastaba in Abusir.* Charles University in Prague, Czech Institute of Egyptology (Prague 2004).

Vachala (2004)
Vachala, B., *Abusir VIII. Die Relieffragmente aus der Mastaba des Ptahshepses in Abusir,* Czech National Centre for Egyptology (Prague, 2004).

van de Walle (1978)
van de Walle, B., *La Chapelle funéraire de Neferirtenef.* Musées Royaux d'Art et d'Histoire (Brussels, 1978).

Vandier (1950)
Vandier, J., *Mo'alla. La Tombe d'Ankhtifi et la tombe de Sebekhotep.* Bibliothèque d'Étude, Tome XVIII. Institut Français d'Archéologie Orientale (Cairo, 1950).

Vandier Manuel IV (1964)
Vandier, J., *Manuel d'Archéologie Égyptienne: Bas-reliefs et Peintures Scènes de la Vie Quotidienne IV* (Paris, 1964).

Vandier Manuel V (1969)
Vandier, J., *Manuel d'Archéologie Égyptienne: Bas-reliefs et Peintures Scènes de la Vie Quotidienne V* (Paris, 1969).

Vandier Manuel VI (1978)
Vandier, J., *Manuel d'Archéologie Égyptienne: Bas-reliefs et Peintures Scènes de la Vie Quotidienne VI* (Paris, 1978).

Van Elsbergen (1997)
Van Elsbergen, M. J., *Fischerei Im Alten Ägypten: Untersuchungen zu den Fischfangdarstellungen in den Grabern der 4. bis 6. Dynastie* (Berlin, 1997).

Van Neer (1989)
Van Neer, W., 'Fishing Along the Prehistoric Nile' in L. Krzyzaniak and M. Kobusiewicz (eds.), *Late Prehistory of the Nile Basin and the Sahara.* Poznan: Poznan Archaeological Museum (1989) 49–56.

van Walsem (1989)
van Walsem, R., 'The Mastaba Project at Leiden University' in: S. Schoske (ed.), *Akhten des vierten Ägyptologen kongresses*, München 1985, II (Hamburg, 1989).

van Walsem (1995)
van Walsem, R., 'Iconographic Programmes in Old Kingdom Elite Tombs' in: Eyre (ed.), *Proceedings Seventh International Congress of Egyptologists* (1995).

van Walsem (1998)
van Walsem, R., 'The Interpretation of Iconographic Programmes in Old Kingdom Elite Tombs of the Memphite Area: Methodological and Theoretical (Re)considerations', in C. J. Eyre (ed.), *Proceedings of the Seventh International Congress of Egyptologists, Cambridge, 3-9 September 1995* (Leiden, 1998) 1205-1213.

van Walsem (2005)
van Walsem, R., *Iconography of Old Kingdom Elite Tombs: Analysis and Interpretation, Theoretical and Methodological Aspects*, Mémoires de la Société d'Études Orientales 35, (Leiden, 2005).

van Walsem (2006)
van Walsem, R., 'Sense and Sensibility. On the Analysis and Interpretation of the Iconography Programmes of Four Old Kingdom Elite Tombs' in Martin Fitzenreiter und Michael Herb (Hrsg.*), Dekorierte Grabanlagen im Alten Reich: Methodik und Interpretation, IBAES VI,* (Humboldt-Universität zu Berlin, 2006).

van Walsem, (2008)
van Walsem, R., *Mastabase* [electronic resource]: the Leiden Mastaba project (Leiden, 2008).

Varille (1938)
Varille, A., *La Tombe de Ni-Ankh-Pepi à Zâouyet el-Mayetîn.* Mémoires de l'Institut Français du Caire, Tome LXX, Institut Français d'Archéologie Orientale (Cairo, 1938).

Vartavan (1987)
Vartavan, de C.T., 'Egyptian barley: a reassessment. Wepwawet', *Discussions in Egyptology* 2, ED [i.e. DE] Publications, (Oxford, 1987) 11-14.

Verner (1986)
Verner, M., *Abusir I. The Mastaba of Ptahshepses. I/1 Reliefs.* The Excavations of the Czechoslovak Institute of Egyptology, Charles University (Prague, 1986).

Verner- Callender (2002)
Verner, M. and Callender, V. G., *Abusir VI. Djedkare's Family Cemetery,* Czech Institute of Egyptology (Prague, 2002).

Vogelsang-Eastwood (1992)
Vogelsang-Eastwood, G., *The production of linen in Pharaonic Egypt*, Stitching Textile Research Centre (Leiden, 1992).

von Beckerath (1984)
von Beckerath, J., *Handbuch der ägyptischen Königsnamen* (München, 1984).

von Bissing I (1905)
von Bissing, F. W., *Die Mastaba das Gem-ni-kai, Band I* (Berlin, 1905).

von Bissing II (1911)
von Bissing, F. W., *Die Mastaba das Gem-n-kai, Band II* (Leipzig, 1911).

von Droste zu Hülshoff (1980)
von Droste zu Hülshoff, V., *Der Igel im alten Ägypten* (Hildesheim, 1980).

Vymazalová-Bárta (2008)
Vymazalová, H. and Bárta, M., (eds.) *Chronology and Archaeology in Ancient Egypt (The Third Millennium B.C.)* Czech Institute of Egyptology (Prague, 2008).

W

Wagner (2006)
Wagner, A. P., *Behavioral Ecology of the Striped Hyena (Hyaena hyaena),* Doctoral Dissertation, Montana State University (2006).

Ward (1984)
Ward, W. A., 'Some Aspects of Private Land Ownership and Inheritance in Ancient Egypt, ca. 2500-1000 B.C.' in Tarif Khaldi (ed.), *Land tenure and social transformation in the Middle East,* American University of Beirut (Beirut, 1984) 63-64.

Weeks (1979)
Weeks, K. R. (ed.), *Egyptology and the Social Sciences: Five Studies.* The American University in Cairo Press (Cairo, 1979).

Weeks (1994)
Weeks, K. R., *Mastabas of Cemetery G 6000. Including G 6010 (Neferbauptah); G 6020 (Iymery); G 6030 (Ity); G 6040 (Shepseskafankh),* Giza Mastabas, Volume 5 (Boston 1994).

Weeks, Noel (1983)
Weeks, N., '"Care" of Officials in the Egyptian Old Kingdom', *Chronique d'Égypte* LVIII, Nos. 115-116 (1983).

Wendorf et al (1970)
Wendorf, F., Said, R. and Schild, R., 'Egyptian Prehistory: Some New Concepts', *Science* Volume 169, Issue 3951 (1970) 1161-1171.

Wendorf–Schild (1994)
Wendorf, F. and Schild, R., 'Are the Early Holocene Cattle in the Eastern Sahara Domestic or Wild?' *Evolutionary Anthropology* (1994) 3(4): 118-128.

Wendorf–Schild (1998)
Wendorf, F. and Schild, R., 'Nabta Playa and Its Role in Northeastern African Prehistory', *Journal of Anthropological Archaeology* 17(2) (1998) 97-123.

Wendorf–Schild (2001)
Wendorf, F., Schild, R. et. al (eds.), *Settlement of the Egyptian Sahara. Volume 1: The Archaeology of Nabta Playa,* (New York, London, 2001).

Wengrow (2001)
Wengrow, D., 'Rethinking "Cattle Cults" in Early Egypt: Towards a Prehistoric Perspective on the Narmer Palette', *Cambridge Archaeological Journal* 11:1 (2001), 91–104.

Wenke–Long–Buck (1988)
Wenke, R. J., Long, J. E. and Buck, P. E., 'Epipaleolithic and Neolithic Subsistence and Settlement in the Fayyum Oasis of Egypt', *Journal of Field Archaeology*, Vol. 15, No. 1 (1988) pp. 29-51.

Wetterstrom (1993)
Wetterstrom, W., 'Foraging and farming in Egypt: the transition from hunting and gathering to horticulture in the Nile Valley' in *The Archaeology of Africa: Food, Metals and Towns* (eds T. Shaw, P. Sinclair, B. Andah and A. Okpoko) (1993) 165-226.

Wicker (1997)
Wicker, U., 'Flax and Egypt', *Discussions in Egyptology* 39 (1997) 95-116.

Wild II (1953)
Wild, H., *Le Tombeau de Ti, Fasc. II [1], La chapelle.* Mémoires publiés par les membres de l'Institute Français d'Archéologie Orientale du Caire (Cairo, 1953).

Wild III (1966)
Wild, H., *Le Tombeau de Ti, Fasc. III [2], La chapelle.* Mémoires publiés par les membres de l'Institut Français d'Archéologie Orientale du Caire (Cairo, 1966).

Wimmer (1998)
Wimmer, S., 'Hieroglyphs – Writing and Literature' in Schulz, R. and Seidel, M. (eds.), *Egypt: The World of the Pharaohs* (Cologne, 1998).

Wittfogel (1957)
Wittfogel, K. A., *Oriental Despotism: A Comparative Study of Total Power*, Yale University Press (New Haven, 1957).

Woods (2006)
Woods, A., 'The Composite Fishing and Fowling Scene in the tomb of Remni at Saqqara', *The Bulletin of the Australian Centre for Egyptology*, Vol. 17 (Sydney 2006).

Woods (2008)
Woods, A., *'A day in the marshes': a study of Old Kingdom marsh scenes in the tombs of the Memphite cemeteries.* Unpublished thesis (Macquarie University, 2008).

Wreszinski, (1923)
Wreszinski, W., *Atlas zur altägyptischen Kulturgeschichte. Teil I.* J. C. (Leipzig, 1923).

Wreszinski, (1936)
Wreszinski, W., *Atlas zur altägyptischen Kulturgeschichte. Teil III.* Graber des Alten Reiches. J. C. Hinrichs Verlag (Leipzig, 1936).

Z

Zeder–Hesse (2000)
Zeder, M. A. and Hesse, B., 'The Initial Domestication of Goats (Capra hircus) in the Zagros Mountains 10000 Years Ago'. *Science* 28(7):2254-2257. <archaeology.about.com/od/domestications/qt/goats.htm>

Ziegler (1990)
Ziegler, C., *Catalogue des stèles, peintures et reliefs égyptiens de l'Ancien Empire et de la Première Période Intermédiaire vers 2686-2040 avant J.-C.* Département des Antiquités Égyptiennes, Musée du Louvre. Éditions de la Réunion des Musées Nationaux (Paris, 1990).

Ziegler (1993)
Ziegler, C., *Le Mastaba d'Akhethetep. Une chapelle funéraire de l'Ancien Empire.* Département des Antiquités Égyptiennes du Musée du Louvre. Réunion des Musées Nationaux (Paris, 1993).

Zohary–Hopf (1993)
Zohary, D. and Hopf, M., *Domestication of plants in the Old World. The origin and spread of cultivated plants in West Asia, Europe and the Nile Valley*, Clarendon Press (Oxford, 1993).

TABLES

TABLE 1: DATA DIGEST

MONUMENT				SOURCES PORTRAYED IN TOMB CHAPEL									
TOMB OWNER	PM No. FIELD No.	LOCATION	PREFERRED DATE	Desert Animals	Cattle	Sheep	Goats	Donkeys	Cereal	Flax	Fish	Birds	Papyrus
DYNASTY IV.E													
Jtt	PM (iv) 93 No. 16	Medum	IV.1M		•				•			•	•
Mṯn	PM 493 LS 6	Saq. NSP	IV.1E	•	•								
Nfr…w	PM (iv) 94 No. 22	Medum	IV.1-2?		•			•	•		•	•	
Nfr-m3ʿt	PM (iv) 92 No. 16	Medum	IV.1E-M	•	•				•		•	•	
Nfr-m3ʿt	PM 183, G7060	Giza-EF	IV.E	•	•							•	
Rʿ-ḥtp	PM (iv) 90 No. 6	Medum	IV.1L	•	•				•		•	•	
DYNASTY IV.M													
Nfr	PM 72, G2110	Giza-WF	IV.4	•	•								
DYNASTY IV.L													
Bw-nfr	PM 256	Giza-CF	IV.6-V.1		•								
Mr-jb.j/K3-pw-nswt	PM 71 LG 24	Giza-WF	IV.6-V.1	•	•								
Mrs-ʿnḫ III	PM 197, G7530/40	Giza-EF	IV.4-6	•	•	•			•			•	•
Nj-k3w-Rʿ	PM 232 LG87	Giza-CF	IV.4-6	•	•								
Nb.j-m-3ḫtj	PM 230 LG 86	Giza-CF	IV.4-6	•	•	•	•	•	•		•	•	•
Ḥmt-Rʿ	PM 243	Giza-CF	IV.5-V.1		•								
Dbḥnj	PM 235 LG 90	Giza-CF	IV.5	•	•						•		
DYNASTY V.E													
3ḫtj-ḥtp	PM 200, G7650	Giza-EF	V.2-4								•		
Wḥm-k3.j	PM 114 D117	Giza-WF	V.1-3		•	•		•					
Pr-sn	PM 48 LG 20-21	Giza-WF	V.1-2	•	•						•		•
Nswt-pw-nṯr	PM 278 Giza-CF	Giza-CF	V.2		•				•				
Nswt-nfr	PM 143, G4970	Giza-WF	V.1-2	•	•								
Rʿ-wr	PM 265	Giza-CF	V.3		•			•	•	•			•
Ḫnmw-ḥtp	PM 578 D49	Saq. ESP	V.2-6		•			•	•	•			•
Snb	PM 101	Giza-WF	V.6L-V.3	•	•	•		•					•
Sḫm-k3-Rʿ	PM 233 LG89	Giza-CF	V.2				•		•		•	•	
Sš3t-ḥtp/Htj	PM 149 LG 36	Giza-WF	V.1-2	•	•								
K3.j-ʿpr	PM 501	Saq. NSP	V.1-2								•		
K3.j-nj-nswt (I)	PM 78, G2155	Giza-WF	V.1-3		•								
K3(.j)-ḫnt (I)	PM (v) 8 A2	el-Hammamiya	V.3-6		•								•
K3(.j)-ḫnt (II)	PM (v) 8 A3	el-Hammamiya	V.1-3		•								
K3-pw-nswt/K3j	PM 135, G4651	Giza-WF	V.2-5	•	•								
Ṯntj	PM 141, LG 47	Giza-WF	V.1-5		•								
DYNASTY V.M													
Jj-mrjj	PM 170, G6020	Giza-WF	V.4-6E	•	•	•		•	•		•	•	•
Jj-nfrt	PM 298	Giza-MC	V.5-6	•	•						•		•

TABLE 1: DATA DIGEST

MONUMENT				SOURCES PORTRAYED IN TOMB CHAPEL									
TOMB OWNER	**PM No. FIELD No.**	**LOCATION**	**PREFERRED DATE**	Desert Animals	Cattle	Sheep	Goats	Donkeys	Cereal	Flax	Fish	Birds	Papyrus
Wr-jr.n-Ptḥ	PM 699	Saqqara	V.M	•	•	•		•	•	•	•	•	
Ptḥ-špss	PM 340 Abusir	Abusir	V.6L	•	•				•	•	•	•	•
Nj-ʿnḫ/Ḫnmw and *Ḫnmw-ḥtp*	PM 641	Saq. UPC	V.6-7	•	•	•	•	•	•	•	•	•	•
Nj-m3ʿt-Rʿ	PM 282, G8900	Giza-CF	V.6-7		•								
Nj-k3w-Ḥwt-ḥr	PM 247	Giza-CF	V.6								•		
Nfr and *K3-ḥ3.j*	PM 639	Saq. UPC	V.6	•	•	•	•		•		•	•	•
Rʿ-wr I	PM 158, G5270	Giza-WF	V.6	•	•								
Rʿ-m-k3.j	PM 487 D3=S903	Saq. NSP	V.6 r/u V.7	•	•			•	•		•	•	
Rʿ-ḥtp	PM n/r C5	el-Hammamiya	V.6?		•				•		•	•	
Rʿ-ḥʿ.f-ʿnḫ	PM 207, G7948=LG75	Giza-EF	V.5-6		•	•	•	•	•			•	•
Ḫwfw-ḫ3.f II	PM 190-1, G7150	Giza-EF	V.6						•		•	•	•
Ḫntj-k3w.s	PM 148	Giza-WF	V.M	•	•								
Sšm-nfr II	PM146, G5080	Giza-WF	V.6	•	•							•	
K3.j-m-nfrt	PM 208 LG 63	Giza-EF	V.6-9	•	•	•		•	•			•	
K3.j-dw3	PM 244	Giza-CF	V.6		•								
Dw3-Mnw	PM n/r L6	el-Hawawish	V.6		•				•				
DYNASTY V.L													
3ḫtj-mrw-nswt	PM 80, G2184	Giza-WF	V.9	•	•		•	•	•		•		•
3ḫtj-ḥtp (Louvre)	PM 634	Saq. UPC	V.6-8		•	•	•	•	•	•	•	•	•
3ḫtj-ḥtp	PM 633 E17	Saq. UPC	V.9	•	•							•	
3ḫtj-ḥtp	PM 599 D64	Saq. WSP	V.8L-9	•	•			•	•		•	•	•
J3zn	PM 82, G2196	Giza-WF	V.8-9	•	•	•		•	•				•
Jj-nfrt/Š3.n.f	PM 616	Saq.-UPC	V.9		•	•			•		•	•	•
Jntj	PM (iv) 121	Deshasha	V.9	•	•						•	•	•
Jrj-n-k3-Ptḥ	PM 644	Saq. UPC	V.6-9		•	•		•	•	•	•	•	•
Jrw-k3-Ptḥ	PM 639	Saq. UPC	V.7-8		•						•	•	•
Jḫjj (r/u *Sšsšt/Jdwt*)	PM 617	Saq. UPC	V.9L r/u VI,1E	•	•			•	•		•	•	•
Jtj-sn	PM 252	Giza-CF	V.6		•							•	•
Jttj	PM 193, G7391	Giza-EF	V.7-8		•						•	•	•
Wr-jr.n.j	PM (iv) 188 No.25	Sheikh Said	V.8E-9E		•	•		•	•	•	•	•	•
Pr-nb	PM 497 S913	Saq. NSP	V.8L-9E		•								
Pḥn-wj-k3.j	PM 491 LS15	Saq. NSP	V.6-8E	•	•			•	•	?	•	•	
Ptḥ-ḥtp I	PM 596 D62	Saq. WSP	V.8M-L	•				•	•	•		•	
Ptḥ-ḥtp	PM 653 LS 31	Saq. AJ-ES	V.8-9	•	•				•				
Ptḥ-ḥtp/Jj-n-ʿnḫ	PM 606	Saq. WSP	V.9-VI.1		•	•			•	•	•		•
Ptḥ-ḥtp (II) (*Ṯfj*)	PM 600 D64	Saq. WSP	V.9M-L	•	•						•	•	•

TABLE 1: DATA DIGEST

MONUMENT				SOURCES PORTRAYED IN TOMB CHAPEL									
TOMB OWNER	PM No. FIELD No.	LOCATION	PREFERRED DATE	Desert Animals	Cattle	Sheep	Goats	Donkeys	Cereal	Flax	Fish	Birds	Papyrus
Ptḥ-ḥtp/Ṯfw	PM 605	Saq. WSP	V.9L								•		
Ftk-tꜣ	PM 351 LS1	Abusir	V.6-9	•	•						•		•
Mꜣ-nfr	PM 575 LS17=H2	Saq. ESP	V.9	•	•							•	
Mttj	PM 646	Saq. UPC	V.9		•			•				•	•
Njwtj	PM 133, G4611	Giza-WF	V.8-9	•	•								
Nj-mꜣꜥt-Rꜥ	PM n/r G2097	Giza-WF	V.9	•	•	•		•	•		•	•	•
Nj-mꜣꜥt-Rꜥ	PM 282 Giza-CF	Giza-CF	V.9		•	•		•	•	•	•	•	•
Nj-ḥtp-Ptḥ	PM 94, G2430=LG25	Giza-WF	V..9	•				•	•				
Nj-kꜣw-ḥr	PM 498 S915	Saq. NSP	V.9					•	•		•		•
Nb-j-kꜣw-ḥr/Jdw	PM 627	Saq. UPC	V.9 r/u VI.4-6		•		•		•	•	•		•
Nbt	PM 624	Saq. UPC	V.9	•	•	•			•		•		•
Nfr	PM 258 LG 99	Giza-CF	V.6-9	•		•						•	
Nfr (1)	PM 137 G4761	Giza=WF	V.9-VI.1	•	•	•		•	•	•	•	•	•
Nfr-jrt-n.f	PM 583 D55	Saq. ESP	V.8	•	•	•		•	•	•	•	•	•
Nfr-bꜣw-Ptḥ	PM 169, G6010=LG15	Giza-WF	V.7-8E	•	•	•		•	•				
Nfr-mḏr-Ḫwfw	PM n/r G2240	Giza-WF	V.9		•				•				•
Nfr-ḫw(wj)	PM n/r G2098	Giza-WF	V.8-9	•	•								
Nfr-sšm-Ptḥ/ Sḫntjw	PM 645	Saq. UPC	V.M-L		•	•		•	•	•			
Nḫt-kꜣ.j	PM 240	Giza-CF	V.6-9	•	•								
Nṯr-wsr	PM 485 D1=S901	Saq. NSP	V.6L-7	•	•								
Rꜥ-wr II	PM 162, G5470=LG32	Giza-WF	V.8-9M	•	•			•	•			•	
Rꜥ-špss	PM 494, LS16-S902	Saq. NSP	V.8	•	•				•	•		•	•
Rdj	PM n/r G2086	Giza-WF	V.6L-8E		•								
Ḥm-Mnw	PM n/r M 43	el-Hawawish	V.9	•	•								•
Ḥzjj-Mnw	PM n/r M22	el-Hawawish	V.8L		•		•		•	•	•		•
Ḥtp-ḥr-ꜣḫtj	PM 593 D60	Saq. WSP	V.6L-8E	•	•	•	•	•	•	•	•		•
Ḥtp-kꜣ.j/Tp-kꜣ.j	PM 447 S3509	Saq. NSP	V.9-VI.1	•	•			•		•			
Ḫww-wr	PM 254 LG95	Giza-CF	V.8-9M		•			•	•		•		•
Ḫw-ns	PM (iv) 134 No.2	Zawyet el-Maiyetin	V.8-9		•	•	•	•	•	•	•		•
Ḫnwt	PM 623	Saq. NSP	V.9		•	•		•	•		•		•
Sꜣ-jb	PM n/r G2092,G2093	Giza-WF	V.9		•			•	•				•
Spd-ḥtp	PM 481 D15	Saq. NSP	V.6L-8		•			•	•		•	•	•
Snḏm-jb/Jntj	PM85, G2370=LG27	Giza-WF	V.8-9		•	•		•	•	•	•	•	•
Snḏm-jb/Mḥj	PM87, G2378=LG26	Giza-WF	V.9L	•	•	•		•	•	•	•	•	•
Srf-kꜣ.j	PM (iv) 188	Sheikh Said	V.8E-9E		•				•		•		
Sḥtpw	PM 222	Giza-GIS	V.9		•								

TABLE 1: DATA DIGEST

MONUMENT				SOURCES PORTRAYED IN TOMB CHAPEL									
TOMB OWNER	PM No. FIELD No.	LOCATION	PREFERRED DATE	Desert Animals	Cattle	Sheep	Goats	Donkeys	Cereal	Flax	Fish	Birds	Papyrus
Sḫm-ʿnḫ-Ptḥ	PM191, G7152	Giza-EF	V.6-9	•	•	•		•	•			•	
Sḫm-ʿnḫ-Ptḥ	PM 454 D41	Saq. NSP	V.L8-9	•	•			•	•	•		•	•
Sḫm-k3j	PM 53, G1029	Giza-WF	V.9?	•	•	•		•	•	•	•	•	•
Sšm-nfr III	PM 153, G5170	Giza-WF	V.8E	•	•							•	
Sšm-nfr IV	PM 223 LG53	Giza-GIS	V.8-9	•	•		•	•	•		•	•	•
Štwj	PM106	Giza-WF	V.9	•	•								
K3.j- m- ʿnḫ	PM 131, G4561	Giza-WF	V.8-9E		•				•		•	•	•
K3.j-m-nfrt	PM 467 D23	Saq.-NSP	V.8-9		•				•	•	•	•	•
K3.j-m-nfrt II	PM 263	Giza-CF	V.6-8		•						•		•
K3.j-m-rḥw	PM 485 D2=S905	Saq. NSP	V.8		•			•	•		•	•	•
K3.j-m-ḥzt	PM 542	Saq. TPC	V.9-VI.1E			•			•		•		•
K3.j-m-ṯnnt	PM 195, G7411	Giza-EF	V.8						•				
K3.j-dw3	PM 244	Giza-CF	V.6-9	•	•							•	
K3pj	PM 69, G2091	Giza-WF	V.6-8	•	•				•	•	•	•	•
K3-pw-Rʿ	PM 455-6 D39	Saq. NSP	V.8-9									•	
K3-ḥj.f	PM 76	Giza-WF	V.L	•	•	•		•	•	•	•	•	•
Kd-nfr	PM 56, G1151	Giza-WF	V.9-VI.1					•	•			•	
Tp-m-ʿnḫ	PM 342-3	Abusir	V.6-9		•				•		•		•
Ṯjj	PM 468 D22	Saq. NSP	V.6-8	•	•	•		•	•	•	•	•	•
DYNASTY VI.E													
Jnw-Mnw	PM n/r	Saq. TPC	VI.1L-2E	•	•						•	•	•
Jn-Snfrw-jšt.f	PM 891 No 2	Dahshur	VI.1?					•	•		•		•
Jttj-Šdw	PM (iv) 122	Deshasha	VI.1		•		•	•	•			•	•
ʿnḫ -m- ʿ-Ḥr	PM 512	Saq. TPC	VI.1L-M		•			•	•		•	•	•
W3tt-ḫt-ḥr/Sšsšt	PM 534	Saq. TPC	VI.1M-L	•	•						•		
Wnjs-ʿnḫ	PM (i) n/r el Khokha 413	el Khokha	VI.6								•		
Wr-nw	PM 519	Saq. TPC	VI.1E-L	•									•
Wḏ3-ḥ3-Ttj/Ššj/ Nfr-sšm-Ptḥ	PM 515	Saq. TPC	VI.1L-2E								•	•	•
B3wj	PM n/r G 126	el Hawawish	VI.1L-2E										•
Mrjj-Rʿ-mrjj-Ptḥ-ʿnḫ /Nḫbw	PM 89, G2381	Giza-WF	VI.2								•		•
Mrjj-Rʿ-nfr/Ḳ3r	PM 184, G7101	Giza-EF	VI.2		•						•		
Mrjj-Ttj/Mrj	PM 536	Saqqara	VI.E	•	•						•	•	
Mrw/Bbj	PM (iv) 189 No.20	Sheikh Said	VI.E	•	•								
Mrw/Ttj-snb	PM 520	Saq. TPC	VI.1L-2		•				•	•	•	•	•
Mr.f-nb.f	PM n.r.	Saq. WSP	VI.1L-2L		•				•		•	•	•

TABLE 1: DATA DIGEST

MONUMENT				SOURCES PORTRAYED IN TOMB CHAPEL									
TOMB OWNER	PM No. FIELD No.	LOCATION	PREFERRED DATE	Desert Animals	Cattle	Sheep	Goats	Donkeys	Cereal	Flax	Fish	Birds	Papyrus
Mrrj	PM 518	Saq. TPC	VI.1M-2E	•	•		•				•	•	•
Mrrw-k3/Mrj	PM 525	Saq. TPC	VI.1	•	•	•	•	•	•	•	•	•	•
Mḥw	PM 619	Saq. UPC	VI.2		•	•		•	•	•	•	•	•
Nj-ʿnḫ-Ppjj	PM (iv) 137 L14	Zawyet el-Maiyitin	VI.2				•	•	•		•	•	•
Nj-ʿnḫ-nswt	PM 694, 755	Saq. WSP	VI.2		•				•	•			
Nj-k3w-Jzzj	PM n/r	Saq. TPC	VI.1M	•	•						•	•	•
Nfr-sšm-Rʿ/Ššj	PM 511	Saqqara	VI.E		•								
Nhwt-dšr	PM n/r G95	el-Hawawish	VI.E	•									
Rmnj/Mrwj	PM n/r	Saq. TPC	VI.1L-2E		•								•
Hnkw/Jj …f (II)	PM (IV) 242 N67	el-Gebrawi	VI.2E-M		•						•		•
Hnkw/Ḫttj (I)	PM (IV) 242 N39	el-Gebrawi	VI.E-L	•	•							•	•
Ḥm-r3/Jzj (I)	PM (IV) 243 N72	el-Gebrawi	VI.1L-2E		•			•	•			•	•
Ḥzj r/u Sšm-nfr	PM n/r	Saq. TPC	VI.1L-2E	•	•	•			•	•	•		•
Ḥzj-Mnw/Zzj/ Dwdjj	PM n/r F1	el-Hawawish	VI.1L										•
Ḫw.n-wḫ	PM (iv) 239, 241	Quseir el-Amarna	VI.2										•
H̱nmw-ntj	PM 87, G2374	Giza-WF	VI.1					•	•				•
Sʿnḫ-wj-Ptḥ	PM n/r	Saq. TPC	VI.1L	•	•							•	•
St-k3.j	PM 160	Giza	VI.E1-2		•								
K3.j-jrr	PM 631	Saqqara	VI.2		•								
K3-gmnj/Mmj	PM 521	Saq. TPC	VI.1E	•	•						•	•	•
Tw3w	PM n/r N 359	Naga ed-Der	VI.2								•	•	
Ṯmrrjj	PM (v) 28 N248	Naga ed Der	VI.1L-2E		•				•		•		
DYNASTY VI.M													
Jbj	PM (iv) 243 No.8	Deir el-Gebrawi	VI.3-4E	•	•		•	•	•	•	•	•	•
Jn-jt.f/Bḫnt	PM (v) 18 BA63	el-Hawawish	VI.4E-M	•	•								
Jḥjj	PM 1: 291 No.186	el-Khokha	VI.4E-M	•	•		•					•	
Jdw	PM 185, G7102	Giza-EF	VI.3-4E										•
B3wj	PM n/r BA48	el-Hawawish	VI.E-M		•			•	•				
Ppj-ʿnḫ-ḥrj-jb	PM (iv) 254 D2	Meir	VI.3-4E	•	•		•	•	•		•	•	•
Mmj	PM n/r B12	el-Hawawish	VI.2L		•				•				
Mrw	PM n/r N 3737	Naga ed-Dêr	VI.3-4E	•									•
Nbt	PM n/r H27	el Hawawish	VI.4E-4M		•								
Ḫntj	PM n/r No.405	el-Khokha	VI.4-5		•		•				•		
Sfḫw/Ḫn-Mnw	PM n.r L21	el-Hawawish	VI.4E		•								
Snfrw-ḥtp	PM 96, G3008	Giza-WF	VI.4					•	•				
Sšm-nfr/Jwfj	PM 614	Saq. UPC	VI.2-4?		•								

TABLE 1: DATA DIGEST

MONUMENT				SOURCES PORTRAYED IN TOMB CHAPEL									
TOMB OWNER	PM No. FIELD No.	LOCATION	PREFERRED DATE	Desert Animals	Cattle	Sheep	Goats	Donkeys	Cereal	Flax	Fish	Birds	Papyrus
K3.j-ḥp/Ṯtj	PM n/r M8	el-Hawawish	VI.4E		•				•				•
Ḳ3r	PM n/r L31	el-Hawawish	VI.2L-4E										•
DYNASTY VI.L													
Ppj-ʿnḫ/Ḥnj-km	PM (iv) 247	Meir	VI.4M-L		•	•	•	•	•	•	•	•	•
Mrjj	PM (iv) 35 C2	el-Hagarsa	VI.4-7	•	•								
Mḫw and *S3bnj*	PM (v) 231 A1	Qubbet el-Hawa	VI.6		•			•	•	•			•
Nj-ʿnḫ-Ppjj-km	PM (iv) 247 A1	Meir	VI.4	•	•						•	•	
Ḥzjj-Mnw/Zzj	PM n/r G42	el-Hawawish	VI.4M	•	•		•	•	•				
Ḫnj/Špzj-pw-Mnw	PM n/r H24	el-Hawawish	VI.4L	•	•		•	•	•		•	•	•
S3bnj II	PM n/r EQH 35e	Qubbet el-Hawa	VI.L										•
K3(.j)-ḥp/Ṯtj-jkr	PM (v) 19 H26	el-Hawawish	VI.4M	•	•		•	•	•		•	•	•
Ṯtj-ʿ3	PM n/r H41	el-Hawawish	VI.4L-VI end								•	•	
Ḏʿw/Ḏʿw-Šm3j	PM (iv) 244 No12	el-Gebrawi	VI.4M-L	•	•	•	•	•	•	•	•	•	•
DYNASTY VIII													
ʿnḫ-tj.fj	PM (v) 170	Moʿalla	VIII.E	•	•			•					•
W3hj	PM (V) 36-7	el-Hagarsa	VIII		•		•	•	•				•
B3wj	PM n/r BA14	el-Hawawish	VIII		•								•
Mrjj-ʿ3	PM (v) 34 D18	el-Hagarsa	VI.L-VIII.E		•		•	•	•		•	•	•
Rḥw-r-3w.sn	PM n/r BA17	el-Hawawish	VIII		•			•	•				
Ḫwj/Ṯtj-jḳr	PM n/r G1	Geheina	VI.L-VIII.E		•			•	•				•
Ḫw-ns	PM (v) 235	Qubbet el-Hawa	VI.L-VIII.E		•				•				•
Gḥs3/Nbjj	PM n/r GA11	el-Hawawish	VI.L-VIII.E		•		•		•		•	•	•

TABLE 2: MANAGEMENT OF DESERT ANIMALS

MONUMENT			PARADE:	
TOMB OWNER	PM No. FIELD No.	REFERENCE	Held by Horns and Muzzle	Held by Horns
DYNASTY IV.E				
Jtt *	PM (iv) 93 No. 16	Petrie (1892) pl. 27; Harpur (2001) fig. 80, 88.	deer	
Mtn	PM 493 LS 6	LD II 3, 4.		
Nfr-m3ˁt *	PM (iv) 92 No. 16	Petrie (1892) pl. 27.	oryx addax gazelle	
Nfr-m3ˁt	PM 183 G 7060	LD II 17.	oryx	
Rˁ-ḥtp *	PM (iv) 90 No. 6	Petrie (1892) pl. 15; Harpur (2001) fig. 10.	oryx addax ibex	
DYNASTY IV.M-L				
Mr-jb.j/K3-pw-nswt	PM 71 LG 24	LD II 19, 22a.	oryx	ibex
Mrs-ˁnḫ III	PM197 G 7530/40	Dunham–Simpson (1974) fig. 3a, 4.	oryx	
Nj-k3w-Rˁ	PM 232 LG 87	LD II 15.		
Nb.j-m-3ḫtj	PM 230 LG 86	LD II pl. 14.	oryx	gazelle
Nfr	PM 72 G 2110	Reisner (1942) fig. 242.	oryx	oryx
Dbḥnj	PM 235 LG 90	LD II pl. 35.	ibex	
DYNASTY V.E				
Pr-sn	PM 48 LG 20-21	LD II 83b; LD Erg. 8.		
Nswt-nfr	PM143 G 4970	Junker III (1938) Abb. 27.	oryx ibex	
Snb	PM 101	Junker V (1941) Abb. 2, 3.		
Sš3t-ḥtp/Htj	PM 149 LG 36	Junker II (1934) Abb. 28, 30; LD II pl. 24.	oryx addax ibex	
K3-pw-nswt/K3j	PM 135 G 4651	Junker III (1938) Abb. 7[5], 18.	oryx ibex	
DYNASTY V.M				
Jj-mrjj	PM 170 G 6020	Weeks (1994). fig. 29, 31; LD II 50, 54.	ibex oryx	
Jj-nfrt	PM 298	Schürmann, (1983) pl. 21.		
Wr-jr.n-Ptḥ	PM 699	James (1961) pl. 29, 30.	oryx	
Nj-ˁnḫ-Ḫnmw and *Ḫnmw-ḥtp* *	PM 641	Moussa–Altenmüller (1977) Abb. 18, 19, Taf. 38, 40.	oryx ibex	gazelle oryx
Nfr ˁnd K3-ḥ3.j	PM 639	Moussa–Altenmüller, (1971) pl. 24b.	oryx ibex	gazelle
Rˁ-wr I	PM 158 G 5270	Junker III (1938) Abb. 42.	oryx	
Rˁ-m-k3.j *	PM 487 D3=S903	Hayes (1953), fig. 56.		
Ḫntj-k3w.s	PM 148	Junker VII (1944) Abb. 31.	oryx addax	
Sšm-nfr II	PM 146 G5080	Kanawati (2002) pl. .61, 62.	oryx	

DEGREE OF CONTROL			PARADE: ATTITUDE OF ANIMAL			CARE		
Degree of Control	Single Herder at rear	Two Herders	Animal balking	Animal walks without Herder	Animal tethered	Hand feeding	Young carried by Herder	TOMB OWNER
								DYNASTY IV.E
addax								*Jtt* *
							kob gazelle	*M*ṯ*n*Ṯ
								*Nfr-m*ꜣꜥ*t* *
		oryx	oryx		oryx			*Nfr-m*ꜣꜥ*t*
								*R*ꜥ*-ḥtp* *
								DYNASTY IV.M-L
oryx			oryx	oryx			gazelle?	*Mr-jb.j/K*ꜣ*-pw-nswt*
	hyena		oryx					*Mrs-*ꜥ*nḫ* III
hyena								*Nj-k*ꜣ*w-R*ꜥ
gazelle		oryx					gazelle	*Nb.j-m-*ꜣ*ḫtj*
		oryx					ungulate	*Nfr*
	oryx	ibex					hyena?	*Dbḥnj*
								DYNASTY V.E
hyena		hyena						*Pr-sn*
		oryx						*Nswt-nfr*
				ibex				*Snb*
		oryx					ungulate gazelle	*Sš*ꜣ*t-ḥtp/Htj*
			ibex				ungulate gazelle	*K*ꜣ*-pw-nswt/K*ꜣ*j*
								DYNASTY V.M
oryx		oryx ibex	ibex		hyena gazelle oryx			*Jj-mrjj*
								Jj-nfrt
							gazelle	*Wr-jr.n-Ptḥ*
			oryx ibex	addax oryx gazelle B. goat	oryx ibex addax hyena		gazelle	*Nj-*ꜥ*nḫ/Ḫnmw* and *Ḫnmw-ḥtp* *
								Nfr ꜥ*nd K*ꜣ*-ḥ*ꜣ*.j*
		oryx						*R*ꜥ*-wr* I
								*R*ꜥ*-m-k*ꜣ*.j* *
	hyena							*Ḫntj-k*ꜣ*w.s*
oryx		oryx			oryx			*Sšm-nfr* II

TABLE 2: MANAGEMENT OF DESERT ANIMALS

MONUMENT			PARADE:	
TOMB OWNER	PM No. FIELD No.	REFERENCE	Held by Horns and Muzzle	Held by Horns
Kꜣ.j-dwꜣ	PM 244	Hassan (1950) fig.82.	ibex oryx	
DYNASTY V.L				
ꜣhtj-mrw-nswt *	PM 80 G 2184	Smith (1946) fig. 239.		
ꜣḫtj-ḥtp	PM 633 E 17?	Petrie–Murray (1952) pl. 7, 8.	oryx gazelle	ibex addax hartebeest
ꜣḫtj-ḥtp *	PM 599 D 64	Davies, *Ptahhetep II* (1901) pl.19.	oryx ibex	gazelle ibex addax
Jꜣsn	PM 82 G 2196	Simpson (1980) fig. 31.	ibex oryx	gazelle
Jj-nfrt	PM 616	Kanawati–Abder-Raziq (2003) pl. 35, 44.	ibex oryx	
Jntj	PM (iv) 121	Kanawati–McFarlane (1993) pl. 30	addax (by neck)	
Jḥjj (r/u *Sšsšt/Jdwt*)	PM 617	Kanawati–Abder-Raziq (2003) pl. 60, 71.	ibex	gazelle oryx ibex addax hartebeest
Pḥn-wj-kꜣ.j *	PM 491 D 70 = LS 15	LD II 45, 46.		
Ptḥ-ḥtp,	PM 653 LS 31	LD II 102a, 104d.	oryx ibex	
Ptḥ-ḥtp (*ṮI*) (*Ṯfj*) *	PM 600 D 64	Paget–Pirie (1898) pl. 32-33.	oryx gazelle	ibex addax S. gazelle
Ptḥ-špss	PM 340	Verner (1986) 42.	ibex	
Ftk-tꜣṮ	PM 351 LS 1	LD II 96.		
Mꜣ-nfr	PM 575 LS 17 = H 2	LD II 69-70.	addax oryx ibex deer	hartebeest gazelle
Mmj	PM n/r B 12	Kanawati (1987) fig.11.		
Njwtj	PM 133 G 4611	LD *Erg.* 30.	oryx	
*Nj-mꜣʿt-Rʿ**	PM n/r G 2097	Roth (1995) pl.189.		
Nbt	PM 624	Munro (1993), pl 17.	oryx ibex	oryx
Nfr (*I*)	PM 137 G 4761	Junker VI (1943) Abb. 14, 16.	oryx ibex	
Nfr-jrt-n.f	PM 583 D 55	van de Walle (1978) pl.10, 12.	oryx	
Nfr-bꜣw-Ptḥ	PM 169 G 6010=LG 15	Weeks (1994) fig. 18.		
Nfr-ẖw(wj)	PM n/r G 2098	Roth (1995) pl.194.		
Nḫt-kꜣ.j	PM240	Hassan (1953) fig. 22.		
Rʿ-wr II	PM 162 G5470= LG 32	Junker III (1938) Abb. 48; LD Erg. 25.	oryx by muzzle	
Rʿ-špss	PM 494 LS 16 - S 902	LD II 60/2/4a; LD Erg. 41.	ibex oryx	
Ḥm-Mnw	PM n/r M 43	Kanawati (1983) fig. 9.		

DEGREE OF CONTROL			PARADE: ATTITUDE OF ANIMAL			CARE		
Degree of Control	Single Herder at rear	Two Herders	Animal balking	Animal walks without Herder	Animal tethered	Hand feeding	Young carried by Herder	**TOMB OWNER**
hyena		hyena		gazelle feeding				*K3.j-dw3*
								DYNASTY V.L
								3ḫtj-mrw-nswt *
							ungulate addax	*3ḫtj-ḥtp*
gazelle hartebeest								*3ḫtj-ḥtp* *
gazelle hyena			ibex gazelle					*J3sn*
		ibex oryx					gazelle	*Jj-nfrt*
ibex	oryx	ibex addax						*Jntj*
oryx							gazelle ungulate	*Jḥjj* (r/u *Sšsšt/Jdwt*)
	ibex addax gazelle			ibex 2			gazelle hyena	*Pḥn-wj-k3.j* *
						addax gazelle	hartebeest ibex oryx gazelle	*Ptḥ-ḥtp,*
		oryx						*Ptḥ-ḥtp* (*II*) (*Ṯfj*) *
								Ptḥ-špss
								Ftk-t3Ṯ
								M3-nfr
								Mmj
	oryx				oryx ibex			*Njwtj*
							gazelle	*Nj-m3ˁt-Rˁ* *
gazelle		oryx	oryx ibex					*Nbt*
	hyena	oryx						*Nfr* (*I*)
		oryx						*Nfr-jrt-n.f*
oryx								*Nfr-b3w-Ptḥ*
		oryx						*Nfr-ḫw(wj)*
					oryx hyena			*Nḫt-k3.j*
				ibex gazelle	oryx			*Rˁ-wr* II
	gazelle		ibex					*Rˁ-špss*
addax								*Ḥm-Mnw*

TABLE 2: MANAGEMENT OF DESERT ANIMALS

MONUMENT			PARADE:	
TOMB OWNER	PM No. FIELD No.	REFERENCE	Held by Horns and Muzzle	Held by Horns
Ḥzjj-Mnw	PM n/r M22	Kanawati (1983) fig.8, 11.		
Ḥtp-ḥr-3ḫtj	PM 593 D 60	Mohr (1943) fig. 8, 10, 19, 9.	ibex oryx	addax
Ḥtp-k3.j/Tp-k3.j	PM 447 S 3509	Martin (1979) pl. 23.		
Snḏm-jb/Mḥj	PM 87 G 2378 = LG 26	LD II 74b; Brovarski (2001) fig. 110.	oryx	
Sḫm-ˁnḫ-Ptḥ	PM191 G7152	Badawy (1976) fig. 20,23.	oryx ibex addax	
Sḫm-ˁnḫ-Ptḥ	PM 454 D41	Simpson (1976) pl. D.	ibex oryx	addax gazelle hartebeest dama deer goat
Sḫm-k3j	PM 53 G1029	Simpson (1980) fig. 4.	oryx ibex addax gazelle	
Sšm-nfr III	PM 153 G5170	LD II 80; Brunner-Traut (1977), Taf. II.	ibex	
Sšm-nfr IV*	PM 223 LG 53	Junker 11 (1953) Abb. 62, 75, 78, 79; LD II 80.		
Štwj	PM 106	Junker (1950) Abb. 86.	oryx gazelle addax	
K3.j-m-nfrt	PM 208 LG 63	Badawy (1976) fig. 30 (b); LD Erg. 32.	oryx ibex	
K3.j-m-nfrt	PM 467 D 23	Simpson (1992) fig. E.	addax gazelle	
K3.j-m-nfrt II	PM 263	Hassan *Giza* II (1936) fig. 123.		
K3pj	PM 69 G 2091	Roth (1995) pl. 155, 168.	ungulate ibex	
K3-ḥj.f	PM 76	Junker VI (1943) Abb. 40.	oryx gazelle	
Ṯjj *	PM 468 D 22	Wild (1953) pl.110-14, 155, 165-6; Épron–Daumas I (1939) pl. 19, 65.	ibex addax gazelle dama	ibex hartebeest
DYNASTY VI.E				
Jnw-Mnw	PM n/r	Kanawati–Abder-Raziq (2006) pl.		
Wr-nw *	PM 519	Davies et al I (1984) pl. 31.		oryx ibex hartebeest gazelle
Mrjj-Ttj/Mrj *	PM 536	Kanawati–Abder-Raziq (2004) pl. 48.	gazelle ibex oryx damadeer	ibex
Mrw/Bbj	PM (iv) 189 No.20	Davies *Sheikh Saïd* (1901) pl. 19, 20.	oryx ibex	oryx addax
Mrrj	PM 518	Davies et al (1984) pl. 6.		

DEGREE OF CONTROL			PARADE: ATTITUDE OF ANIMAL			CARE		
Degree of Control	Single Herder at rear	Two Herders	Animal balking	Animal walks without Herder	Animal tethered	Hand feeding	Young carried by Herder	**TOMB OWNER**
								Ḥzjj-Mnw
addax oryx		ibex oryx					ungulate	*Ḥtp-ḥr-ꜣḫtj*
	gazelle			ibex				*Ḥtp-kꜣ.j/Tp-kꜣ.j*
		oryx herders 3						*Snḏm-jb/Mḥj*
								Sḫm-ꜥnḫ-Ptḥ
hyena				hyena hare				*Sḫm-ꜥnḫ-Ptḥ*
		gazelle						*Sḫm-kꜣj*
					oryx		gazelle	*Sšm-nfr* III
oryx addax	hyena			gazelle ibex				*Sšm-nfr* IV*
		addax						*Štwj*
								Kꜣ.j-m-nfrt
		gazelle			oryx			*Kꜣ.j-m-nfrt*
					ibex			*Kꜣ.j-m-nfrt* II
hyena?	hyena	hyena		gazelle	ibex			*Kꜣpj*
								Kꜣ-ḥj.f
	hyena	ibex	hartebeest lassoed ibex	hartebeest addax gazelle	addax ibex gazelle	hyena	gazelle ungulate	*Ṯjj* *
								DYNASTY VI.E
						hyena	ungulate	*Jnw-Mnw*
		oryx	oryx	oryx ibex			ungulate gazelle porcupine	*Wr-nw* *
		gazelle ibex oryx damadeer						*Mrjj-Ttj/Mrj* *
hyena oryx hyena		oryx addax						*Mrw/Bbj*
						hyena kob addax		*Mrrj*

TABLE 2: MANAGEMENT OF DESERT ANIMALS

MONUMENT			PARADE:	
TOMB OWNER	**PM No. FIELD No.**	**REFERENCE**	Held by Horns and Muzzle	Held by Horns
Mrrw-k3/Mrj *	PM 525	Duell I (1938) pl. 25, 51; Duell II (1938) pl. 152, 153 Kanawati (2010) pl. 81a, 82a.	oryx ibex addax	
Nj-k3w-Jzzj	PM n/r	Kanawati–Abder-Raziq (2000) pl. 51.	gazelle	oryx gazelle
Nhwt-dšr/Mrj *	PM n/r G95	Kanawati VIII (1988) fig. 3.	oryx	
Rmnj/Mrwj	PM n/r	Kanawati (2009) fig, 45.	oryx	
Ḥnkw/Ḫttj (I)	PM (iv) 242 N39	Kanawati (2005) pl. 39.		
Ḥzj r/u Sšm-nfr	PM n/r	Kanawati–Abder-Raziq (1999) pl.	gazelle	
Sʿnḫ-wj-Ptḥ *	PM n/r	Kanawati–Abder-Raziq (1998)pl. 56, 66.		
St-k3.j and *Ptḥ-ḥtp*	PM 160	Junker VII (1944) Abb. 88.	oryx gazelle	
K3-gmnj/Mnj	PM 521	Harpur (2006), fig. 4.		
DYNASTY VI.M				
Jbj *	PM (iv) 243 No.8	Davies *Gebrâwi* I (1902), pl.11; Kanawati *Gebrawi* II (2007) pl. 52.		
Jn-jt.f/Bḫnt *	BA 63	Kanawati (1987) fig.20-21.		
Ppj-ʿnḫ-ḥrj-jb	PM (iv) 254 D2	Blackman, (1924) pl. 12.		
DYNASTY VI.L				
ʿnḫ-tj.fj *	PM (v) 170	Vandier (1950), pl. 34.		
Ppj-ʿnḫ/Ḥnj-km	PM (iv) 247 A2	Blackman–Apted (1953) pl. 11.	oryx	
Ḥzjj-Mnw/Zzj	PM n/r G 42	Kanawati (1987), fig. 6, 3c.	ungulate	
Ḫnj/Špsj-pw-Mnw	PM n/r H 24	Kanawati (1981) fig. 26.		?gazelle
K3(.j)-ḥp/Ṯtj-jkr	PM (v) 19 H 26	Kanawati (1980) fig.11, 15	gazelle	
Ḏʿw/Ḏʿw-Šm3j	PM (iv) 244 No.12	Davies *Gebrâwi* II (1902) pl.6, 9.		

* Tomb includes scene of desert hunt.

S. gazelle = Soemmerring gazelle

B. goat = Barbary goat

DEGREE OF CONTROL			PARADE: ATTITUDE OF ANIMAL			CARE		
Degree of Control	Single Herder at rear	Two Herders	Animal balking	Animal walks without Herder	Animal tethered	Hand feeding	Young carried by Herder	**TOMB OWNER**
gazelle		addax oryx			oryx addax	hyena addax		*Mrrw-kꜣ/Mrj* *
	oryx	gazelle						*Nj-kꜣw-Jzzj*
		oryx						*Nhwt-dšr/Mrj* *
								Rmnj/Mrwj
oryx					oryx			*Hnkw/Ḫttj* (I)
		gazelle						*Ḥzj r/u Sšm-nfr*
	ibex	gazelle						*Sꜥnḫ-wj-Ptḥ* *
		oryx gazelle						*St-kꜣ.j* and *Ptḥ-ḥtp*
				oryx ibex		hyena		*Kꜣ-gmnj/Mnj*
								DYNASTY VI.M
								Jbj *
								Jn-jt.f/Bḫnt *
	oryx							*Ppj-ꜥnḫ-ḥrj-jb*
								DYNASTY VI.L
								ꜥnḫ-tj.fj *
								Ppj-ꜥnḫ/Ḥnj-km
		ungulate						*Ḥzjj-Mnw/Zzj*
		gazelle oryx						*Ḫnj/Špsj-pw-Mnw*
							ungulate	*Kꜣ(.j)-hp/Ṯtj-jkr*
oryx				oryx				*Ḏꜥw/Ḏꜥw-Šmꜣj*

TABLE 3: FEATURES OF DESERT HUNTING SCENES

TABLE 3(a): Animals attacked by hunting dogs or lion

TOMB OWNER	P/M NO. FIELD NO.	Date/ Dynasty	REFERENCE	ANIMAL ATTACKED	Fenced Enclosure	Attack on Legs/Tail	Attack on Neck/Hea	Packs of 2+ Dogs	Attack by Lion
Rꜥ-ḥtp	PM (iv) 90 No. 6	IV.E	Petrie (1892) pl. 9.	gazelle		•			
				red fox		•			
Nfr-mꜣꜥt	PM (iv) 92 No. 16	IV.E	Petrie (1892) pl. 17.	ungulate?			?		
				Cape hare			•		
				gazelle		•			
Mṯn	PM 493 LS 6	IV.E	LD II pl. 6.	oryx		•			
				gazelle		•			
				kob		•			
Nj-ꜥnḫ/H̱nmw/ H̱nmw-ḥtp	PM 641	V.M	Moussa–Altenmüller (1977) Taf. 40.	ungulate ungulate	•				•
Rꜥ-m-kꜣ.j	PM 487 D3=S903	V.M	Hayes I (1953) fig. 56.	red fox			•		
				gazelle		•			
ꜣḫtj-mrw-nswt	PM 80 G2184	V.L	Smith (1946) fig. 239.	oryx				•	
ꜣḫtj-ḥtp	PM 599 D64	V.L	Davies (1913) pl. 40 [1].	ungulate					•
Pḥn-wj-kꜣ.j	PM 491 D70=LS15	V.L	LD II pl. 46.	golden jackal			•		
Ptḥ-ḥtp (II) (*Ṯfj*)	PM 600 D64	V.L	Paget–Pirie (1898) pl. 32.	oryx			•		
				striped hyena			•		
				ibex		•			
				auroch					•
Ptḥ-ḥtp/Ṯfw	PM 605	V.L	Hassan (1975) fig. 59.	gazelle					•
Ftk-tꜣ	PM 351 LS 1	V.L	LD II pl. 96.	ungulate			•		
Nj-mꜣꜥt-Rꜥ	PM n/r G2097	V.L	Roth (1995) pl. 189	gazelle?		•			
				gazelle			•		
				red fox?		•			
				oryx				•	
				auroch					•
Sšm-nfr IV	PM 223 LG53	V.L	Junker 11 (1953) Abb. 63.	ungulate		•			
				auroch					•
Ṯjj	PM 468 D22	V.L	Wild (1953) pl. 127.	auroch	•				•
Jnw-Mnw	PM n/r	VI.E	Kanawati–Abder-Raziq (2006) pl.47.	ibex			•		
				ibex				10 dogs	
Wr-nw	PM 519	VI.E	Davies et al I (1984) pl. 31.	oryx					•
Mrjj-Ttj/Mrj	PM 536	VI.E	Kanawati–Abder-Raziq (2004) pl.46.	ibex	•			2 dogs	
				ibex				9 dogs	
				addax			•		
				oryx			•		
				gazelle, auroch					•
Mrrw-kꜣj	PM 525	VI.E	Duell I (1938) pl. 25.	ibex	•		•		
				ibex				8 dogs	
				oryx			•		
				auroch					•
Sꜥnḫ-wj-Ptḥ	PM n/r	VI.E	Kanawati–Abder-Raziq (1998) pl. 71.	ibex		•			
Jbj	PM (iv) 243 No. 8	VI.M	Davies *Gebrâwi* I (1902) pl. 11; Kanawati *Gebrawi* II (2007) pl. 52.	gazelle (regs 1, 2)					•
Jn-jt.f/Bḫnt	PM(v) 18 BA63	VI.M	Kanawati (1987) fig. 20-21.	oryx		•			
Ḥzjj-Mnw/Zzj	PM n/r G42	VI.L	Kanawati (1987) fig. 3[c].	oryx		•			
Ḫnj/Špsj-pw-Mnw	PM n/r H24	VI.L	Kanawati (1981) fig. 19.	ibex		•			
				ungulate				2 dogs	
				oryx					•
'Ḏꜥw/Ḏꜥw-Šmꜣj	PM(iv) 244 No.12	VI.L	Davies *Gebrâwi* II (1902) pl. 15.	ungulate					•

TABLE 3: FEATURES OF DESERT HUNTING SCENES

TABLE 3(b): Peaceful activities of animals in desert hunting scenes

TOMB OWNER	PM NO. FIELD NO	Date/ Dynasty	REFERENCE	ANIMAL	Fenced Enclosure	Mating	Female and Young	Calf hiding
Rꜥ-m-kꜣ.j	PM 487 D3=S903	V.L	Hayes I (1953) fig. 56.	gazelle				•
Pḥn-wj-kꜣ.j	PM 491 D70 = LS15	V.L	LD II pl. 46.	gazelle			•	
Ptḥ-ḥtp (II) (*Ṯfj*)	PM 600 D64	V.L	Paget–Pirie (1898) pl. 32.	gazelle			•	
				leopard				
				red fox		•		
				lion		•		
Ftk-tꜣ	PM 351 LS1	V.L	LD II pl. 96.	oryx				•
Nj-ꜥnḫ/Ḫnmw/ Ḫnmw-ḥtp	PM 641	V.L	Moussa–Altenmüller (1977) Taf. 40.	gazelle oryx?	•			•
				red fox		•		
Nj-mꜣꜥt-Rꜥ	PM n/r G2097	V.L	Roth (1995) pl. 189.	lion		•		
				oryx		•		
				wolf?		•		
				donkeys		•		
				red fox?		•		
				weasel		•		
				striped hyena		•		
				addax		•		
				hedgehog		•		
				caracal		•		
				gazelle		•		
				hartebeest		•		
Sšm-nfr IV	PM 223 LG53	V.L	Junker 11 (1953) Abb. 63.	gazelle		•		
				gazelle				•
Jnw-Mnw	PM n/r	VI.E	Kanawati–Abder-Raziq (2006) pl.47.	ungulate				•
Mrjj-Ttj/Mrj	PM 536	VI.E	Kanawati–Abder-Raziq (2004) pl.46.	gazelle	•			•
				ungulate, ibex				•
Mrrw-kꜣj	PM 525	VI.E	Duell I (1938) pl. 25; Kanawati (2010) pl. 73.	ungulate	•			•
Sꜥnḫ-wj-Ptḥ	PM n/r	VI.E	Kanawati–Abder-Raziq (1998) pl. 71.	gazelle				•
Jbj	PM (iv) 243 No.8	VI.M	Davies *Gebrâwi* I (1902) pl. 11; Kanawati *Gebrawi* II (2007) pl. 52.	roan antelope		•		

TABLE 3: FEATURES OF DESERT HUNTING SCENES

TABLE 3(c): Hunters

TOMB OWNER	P/M NO. FIELD NO	Date/ Dynasty	REFERENCE	ANIMAL	Fenced Enclosure	Controls Dogs	Holds Throw-stick	Bow and Arrow	Hunter Lassoing	Handling Animal	Hunter pointing
Nfr-mꜣꜥt	PM (iv) 92 No. 16	IV.E	Petrie (1892) pl. 17.				•				
				Cape hare		•					
				gazelle		•					
Nj-ꜥnḫ/H̱nmw/ H̱nmw-ḥtp	PM 641	V.M	Moussa–Altenmüller (1977) Taf. 40.	oryx	•					•	•
				kob						•	
				n/a		•					
				ibex					•		
Rꜥ-m-kꜣ.j	PM 487 D3=S903	V.M	Hayes I (1953) fig. 56.	ibex					•		
ꜣḫtj-ḥtp	PM 599 D64	V.L	Davies (1913) pl. 40 [2].	ungulate						•	
Pḥn-wj-kꜣ.j	PM 491 D70=LS15	V.L	LD II pl. 46.	ibex			•				
Ptḥ-ḥtp (II) (*Ṯfj*)	PM 600 D64	V.L	Paget–Pirie (1898) pl. 32.	auroch		•			•		•
Ftk-tꜣ	PM 351 LS1	V.L	LD II pl. 96.	ungulate auroch					•		
Nj-mꜣꜥt-Rꜥ	PM n/r G2097	V.L	Roth (1995) pl. 189.	n/a		•					•
Ḥzjj-Mnw	PM n/r M22	V.L	Kanawati (1983) fig.18.	?					•?		
Sšm-nfr IV	PM 223 LG 53	V.L	Junker 11 (1953) Abb. 63.	oryx?					•		
				gazelle					•		
				addax					•		
				ibex						•	
Mrjj-Ttj/Mrj	PM 536	VI.E	Kanawati–Abder-Raziq (2004) pl.46.	oryx	•		•				•
				oryx					•		
Nhwt-dšr/Mrj	PM n/r G95		Kanawati (1988) fig. 3(b).	n/a					•		
Sꜥnḫ-wj-Ptḥ	PM n/r	VI.E	Kanawati–Abder-Raziq (1998) pl. 71.	n/a							•
Jn-jt.f/Bḫnt	PM (v) 18 BA 63	VI.M	Kanawati (1987) fig. 20-21.	oryx		•			•		
Jbj	PM (iv) 243 No.8	VI.M	Davies *Gebrâwi* I (1902) pl. 11; Kanawati *Gebrawi* II (2007) pl. 52.	gazelle				•			•
Mrw/Jjjj	PM n/r N 3737	VI.M	Peck (1959) pl. 7, 10.	ungulate				•			
Ḥzjj-Mnw/Zzj	PM n/r G42	VI.L	Kanawati (1987) fig. 3.	n/a					•		
H̱nj/Špsj-pw-Mnw	PM n/r H24	VI.L	Kanawati (1981) fig. 19.	oryx		•			•		•

TABLE 4: CATTLE CROSSING WATER

MONUMENT			THEMES			HERD	SKIFF	CALF		
TOMB OWNER	**PM No. FIELD No.**	**REFERENCE**	Herd Swimming	Herd Wading	Stepping on to Dry Land	Number of Animals	Number of Craft	Towed in water	Carried by herder	Walking
DYNASTY IV.L										
Nb.j-m-ꜣḫtj	PM 230 LG 86	LD II 12b.		•		9	1			
DYNASTY V.M										
Nj-ʿnḫ/Ḫnmw and *Ḫnmw-ḥtp*	PM 641	Moussa–Altenmüller (1977) pl. 76-77.		•		8	1			
Nfr/Kꜣ-ḥꜣ.j	PM 639	Moussa–Altenmüller (1971) pl. 5.			•	4				•
DYNASTY V.L										
ꜣḫtj-mrw-nswt	PM 80 G2184	Smith (1958) fig. 229.	•				1			
ꜣḫtj-ḥtp	PM 599 D64	Davies *Ptahhetep* II (1901) pl. 14.			•	8			•	•
Jrw-kꜣ-Ptḥ	PM 639	McFarlane (2000) pl. 46.	•			5	2			
Jḫjj (r/u *Jdwt*)	PM 617	Kanawati–Abder-Raziq (2003) pl. 54.	•			13	2	•		
Ptḥ-ḥtp (II) (*Ṯfj*)	PM 600 D64	Davies *Ptahhetep* I (1900) pl. 3.	•			5	1	•		
Nj-mꜣʿt-Rʿ	PM n.r. G2097	Roth (1995) pl.185.	•			5	2	•		
Rʿ-špss (left)	PM 494 LS16	LD II 60.			•	10				
Rʿ-špss (right)	PM 494 LS16	LD II 60.	•			9	2	•		
Ḥzjj-Mnw	PM n.r. M22	Kanawati IV (1983) fig. 18.	•			?	2	•		
Ḥtp-ḥr-ꜣḫtj	PM 593 D60	Mohr (1943) fig. 33.	•			10	1?	•		
Ḫw-ns (left)	PM (4) 134 No.2	LD II 105b.	•			10	1?	?		
Ḫw-ns (right)	PM (4) 134 No.2	LD II 105b.		•	?	8				
Snḏm-jb/Jntj	PM 85 LG 27	Brovarski (2001) fig. 17.	•			3?	1			
Snḏm-jb/Mḥj	PM 87 LG 26	LD Erg. 11.	•			6	2			
Snḏm-jb/Mḥj	PM 87 LG 26	LD Erg. 12.	•			5	2	•		
Kꜣ.j- m- ʿnḫ	PM 192 G4561	Kanawati (2001) pl. 31.		•		4				
Kꜣ.j-m-nfrt	PM 263	Hassan II (1936) fig. 124.	•				2	•		
Kꜣ.j-m-nfrt	PM 263	Hassan II (1936) fig. 140.	•			4	1			
Kꜣ.j-m-rḥw	PM 485	Mogensen (1921) fig. 4.	•			6	2	•		
Ṯjj	PM 468 D22	Wild (1966) pl. 114.			•	7			•	
Ṯjj	PM 468 D22	Wild (1966) pl. 124.	•			8	2	•		
DYNASTY VI.E										
Jttj-Šdw	PM (iv) 122	Kanawati–McFarlane (1993) pl. 46.	•			14	1	•		
ʿnḫ -m- ʿ- Ḥr	PM 512	Kanawati–Hassan (1997) pl. 37.	•			33	2	•		
Mrrw-kꜣ	PM 525	Duell I (1938) pl. 20-1.	•			8	2	•		
Nj-kꜣw-Jzzj	PM n.r.	Kanawati–Abder-Raziq (2000) pl. 49.			•	3				
Rmnj/Mrwj	PM n/r	Kanawati (2009) pl. 45.			•	5+	1		•	
Ḥm-rꜣ/Jzj (I)	PM (iv) 243 N72	Kanawati *Gebrawi* I (2005) pl. 64.	?	?		4?	1	•		
Ḥzj	PM n.r.	Kanawati–Abder-Raziq (1999) pl. 55.	•		•	8	2	•		
Ḥzj	PM n.r.	Kanawati–Abder-Raziq (1999) pl. 55.			•					
Kꜣ-gmnj	PM 521	Harpur (2006) CD 7, p. 493.	•			9	2	•	•	
DYNASTY VI.M										
Jbj	PM (iv) 243 No.8	Davies *Gebrâwi* I (1902) pl. 5-6.	•			10	1			
Kꜣ.j-ḥp/Ṯtj	PM n.r., M8	Kanawati III (1982) fig. 5a.	mostly lost							
DYNASTY VI.L										
Ppj-ʿnḫ/Ḥnj-km	PM (iv) 247 A2	Blackman–Apted MeirV (1953) pl. 30.		•		4	1	•		
Ḫnj/Špsj-pw-Mnw	PM n.r., H24	Kanawati II (1981) fig. 22.	•			6	2	•		
Kꜣ(.j)-ḥp/Ṯtj-jkr	PM (v) 19 H26	Kanawati I (1980) fig. 1.2.	•			5	2	•		
Ḏʿw (right)	PM (iv) 244 no.12	Davies *Gebrâwi* II (1902) pl. 5.	•				?			
Ḏʿw (left)	PM (iv) 244 no.12	Davies *Gebrâwi* II (1902) pl 5.			•	4				

TABLE 4: CATTLE CROSSING WATER

MONUMENT			THEMES			HERD	SKIFF	CALF		
TOMB OWNER	PM No. FIELD No.	REFERENCE	Herd Swimming	Herd Wading	Stepping on to Dry Land	Number of Animals	Number of Craft	Towed in water	Carried by herder	Walking
DYNASTY VIII										
W3hj	PM n.r. D4	Kanawati, (1995) pl. 20.	•			4	1	•		
B3wj	PM n.r. BA14	Kanawati VII (1987) fig. 18.		•(1)		?				
Mrjj-ʿ3	PM (v) 34 D18	Kanawati (1995) pl. 37.	•			15	1	?		
Gḥs3/Nbjj	PM (v) 18 GA11	Kanawati VII (1987) fig. 30.	•			6?	1?	?		

TABLE 5: LIVESTOCK NUMBERS

The following figures giving the size of herds and flocks are cited in tombs.

CATTLE

	Official	Date	Cattle Variety	No of cattle	Presumed total
1.	*Rꜥ-ḥꜥ.f-ꜥnḫ*	V.M	oxen (lyre-shaped horns)	835	
	Rꜥ-ḥꜥ.f-ꜥnḫ	V.M	oxen (hornless)	220	1 055
2.	*Rꜥ-wr* II	V.L	*rn jw3* (oxen lyre-shaped horns)	(x2) 1 000	
	Rꜥ-wr II[1]	V.L	*rn jw3* (oxen hornless)	1 000	3 000
3.	*Nfr-b3w-Ptḥ*	V.L	*rn jw3* (oxen lyre-shaped horns)	(x3) 1000	3 000
4.	*Ḫw-ns*	V.L	oxen (lyre-shaped horns)	1 300	
4.	*Jntj*	V.L	*rn jw3* (oxen lyre-shaped horns)	113[2]	
5.	*Ppj-ꜥnḫ-ḥrj-jb*	VI.M	oxen (lyre-shaped horns)	760 000	
	Ppj-ꜥnḫ-ḥrj-jb	VI.M	cows? (lyre-shaped horns)	7000	
	Ppj-ꜥnḫ-ḥrj-jb	VI.M	bulls (lyre-shaped horns)	700	
	Ppj-ꜥnḫ-ḥrj-jb	VI.M	oxen (lyre-shaped horns)	77	767,777[3]
6.	*Ppj-ꜥnḫ/Ḥnj-km*	VI.L	*rn jw3* (oxen lyre-shaped horns)	300	
	Ppj-ꜥnḫ/Ḥnj-km	VI.L	*ng3w* (oxen lyre-shaped horns)	1 400	
	Ppj-ꜥnḫ/Ḥnj-km	VI.L	*ng3w* (oxen deformed horns)	1 300	
	Ppj-ꜥnḫ/Ḥnj-km	VI.L	cows (oxen lyre-shaped horns)	500	
	Ppj-ꜥnḫ/Ḥnj-km	VI.L	*rn jw3* (oxen hornless)	500	
	Ppj-ꜥnḫ/Ḥnj-km	VI.L	*ng3w* (oxen lyre-shaped horns)	1 400	
	Ppj-ꜥnḫ/Ḥnj-km	VI.L	cows (oxen lyre-shaped horns)	300	5 700
7.	*K3(.j)-ḥp/Ṯtj-jkr*	VI.L	*rn jw3* (oxen lyre-shaped horns)	3 014	
	K3(.j)-ḥp/Ṯtj-jkr	VI.L	*rn jw3* (oxen lyre-shaped horns)	2 012	5 026
8.	*Ḫnj/Špsj-pw-Mnw*	VI.L	*rn jw3* (oxen lyre-shaped horns)	10 000	
	Ḫnj/Špsj-pw-Mnw	VI.L	*rn jw3* (oxen crescent horns)	10 000	
	Ḫnj/Špsj-pw-Mnw	VI.L	*rn jw3* (oxen lyre-shaped horns)	10 000	30 000[4]
9.	*Ḏꜥw*	VI.L	cattle	24 000?[5]	24 000

DESERT ANIMALS

	Official	Date	Species	Herd Numbers
1.	*Rꜥ-wr* II	V.L	oryx	1000
	Rꜥ-wr II	V.L.	oryx	1000
	Rꜥ-wr II	V.L.	ibex	1000
	Rꜥ-wr II	V.L.	gazelle	1000
2.	*K3(.j)-ḥp/Ṯtj-jkr*	VI.M	gazelle	1 023
	K3(.j)-ḥp/Ṯtj-jkr	VI.M	oryx	1 000
3.	*Ḫnj/Špsj-pw-Mnw*	VI.L	gazelle	10 000
	Ḫnj/Špsj-pw-Mnw	VI.L	oryx	10 000

---

1 Desert animals and birds on registers above and below the cattle being presented are also labelled '1000'. LD Erg. 25.
2 Number is branded on the animal. Kanawati, *Deshasha*, pls. 5, 30.
3 The numbers entered on a papyrus sheet by the scribe squatting in front of the lines of animals add up to 760,000 if Blackman's interpretation of the lacunae is accepted. Meir II, p. 43, pl.16.
4 The presumed total may have been 40,000. On the bottom register a lacuna exists following the words *rn m jw3*. This is the only animal whose label does not have the figure for 10,000. Kanawati, *El Hawawish* II, fig. 26.
5 A lacuna after the numbers 20,000 and 4000.

TABLE 5: LIVESTOCK NUMBERS

DOMESTIC GOATS

	Official	Date	Herd Numbers
1.	*Rˁ-ḫˁ.f-ˁnḫ*	V.M	2 235
2.	*Ppj-ˁnḫ/Ḥnj-km*	VI.L	32 400[(1)]
3.	*Mrjj*	VI.L	12 000
4.	*W3hj*	VIII	21 300

(1) The male determinative (Gardiner D52) accompanies the number 32 400.

SHEEP

	Official	Date	Flock Numbers
1.	*Snb*	V.E	M: 10 205 F: 10 103
2.	*Rˁ-ḫˁ.f-ˁnḫ*	V.M	974

DONKEYS

	Official	Date	Herd Numbers
1.	*Snb*	V.E	M: 12 017 F: 10 200
2.	*Rˁ-ḫˁ.f-ˁnḫ*	V.M	760
3.	*Mrjj*	VI.M–L	20 000
4.	*3ḫtj-ḥtp*	V.L	2 500
5.	*Wr-jr.n.j*	V.L	100
6.	*Nfr-jrt-n.f*	V.L	2 300

BIRDS

	Official	Date	Species	Numbers
1.	*Jj-mrjj*	V.M	*d3t* cranes	1 000
			ˁw cranes	1 000
			wdˁ cranes	1 000
2.	*Ptḥ-ḥtp/Ṯfj* (II)	V.L	Geese (*r3*)	121 200
			Geese (*trp*)	121 200
			Geese (Egyptian)	11 110
			Duck (Pintail)	120 000
			Widgeon	121 022
			Pigeon	111 200
			Swan	1 225

TABLE 6: PORTRAYAL OF FLOCK OF SHEEP

MONUMENT			SCENE: Flock of Sheep Trampling								HERDERS			OTHER FLOCK	
NAME	PM No. FIELD No.	REFERENCE	Trampling Seed	Following Plough	Ahead of Plough	Following Hoers	Ahead of Hoers	No Plough or Hoers	Layered Flock	Flock in Disarray	Coaxer/Sower	Beaters with Stick	Beaters with Whip	in Animal Parade	on Threshing Floor
DYNASTY IV.L															
Mrs-ʿnḫ III	PM197 G7530/40	Dunham–Simpson (1974) fig.4.	•					•		•	•	•	•		
Nb.j-m-3ḫtj	PM 230 LG 86	LD II 12(a); Hassan IV (1943) fig. 76.	•	•							•				
DYNASTY V.E															
Wḫm-k3.j	PM 114 D117	Kayser (1954), fig. on p. 36-37.												•	
Snb	PM 101	Junker (1941) Abb. 19.												•	
DYNASTY V.M															
Jj-mrjj	PM 170 G6020	LD II 51; Weeks (1994) fig. 39.	•			•			•			•	•		
Wr-jr.n-Ptḥ	PM 699	Hall–Lambert, (1922) pl. 10-11; James (1961) pl.29.	•					•	•		•	•	•	•	
Nj-ʿnḫ/Ḫnmw and *Ḫnmw-ḥtp*	PM 641	Moussa–Altenmüller (1977) pl. 56, 58.	•	•						•		•	•		
Nfr and *K3-ḥ3.j*	PM 639	Moussa–Altenmüller (1971) pl.6.	•						•		•	•	•		
Rʿ-ḫʿ.f-ʿnḫ	PM 207 G7948=LG 75	LD II 9.												•	
DYNASTY V.L															
3ḫtj-ḥtp (Louvre)	PM 634	Ziegler (1993) 126, 129, 136-7.	•	•							•				
J3sn	PM 82 G2196	Simpson (1980) fig. 30.						•			•	•	•		
Jj-nfrt	PM 298	Schürmann (1983) fig. 11 (a), (b), 21.	•	•					•		•	•	•		
Jrj-n-k3-Ptḥ	PM 644	Moussa–Junge (1975)	•	•							•				•
Wr-jr.n.j	PM (iv) 188 no.25	Davies (1891) pl. 8, 16.	•	•						•	•	•	•		
Ptḥ-ḥtp/Jj-n-ʿnḫ	PM 606	Hassan (1975) fig. 31.						•				•	•		
Nj-m3ʿt-Rʿ	PM 283	Hassan (1936) fig. 240.	•									•	•		
Nbt	PM 624	Munro, (1993) pl.17	•	•					•		•	?	?		
Nfr (1)	PM 137 G4761	Junker VI (1943) Abb. 14, 16.	•					•		•	•	•	•		
Nfr-jrt-n.f	PM 583 D55	van de Walle (1978) pl.12.	•		•					•	•	•	•		
Nfr-b3w-Ptḥ	PM 169 G6010=LG 15	LD II 56 [a, a bis]; Weeks (1994) fig. 9, 11.						•			•	•	•		
Ḥtp-ḥr-3ḫtj	PM 593 D60	Mohr (1943) fig. 44-52.	•			•	•		?		•	•	•		
Ḫw-ns	PM (iv) 134 No.2	LD II 106 (b), 107.	•	•					•			•	•		
Ḫnwt	PM 623	Munro (1993) pl. 43.	•	•											
Snḏm-jb/Jntj	PM 85 G2370=LG 27	LD Erg. 22; Brovarski (2001) fig, 52(b), 53.	•		•				•			•	•		
Snḏm-jb/Mḥj	PM 87 G2378=LG 26	LD Erg. 14; Brovarski (2001) fig, 114-15.		•					?				•		
Sḫm-ʿnḫ-Ptḥ	PM 454 D41	Simpson (1976 pl. D.	•		•				•		•	•	•		
Sḫm-k3j	PM 53 G1029	Simpson (1980) fig. 4.	•	•					•		•	•	•		
Ḥḫntjw and *Nfr-sšm-Ptḥ*	PM 645	Moussa–Junge (1975) pl. 4(b).	?	•					?	?		•	•		

TABLE 6: PORTRAYAL OF FLOCK OF SHEEP

MONUMENT			SCENE: Flock of Sheep Trampling								HERDERS			OTHER FLOCK	
NAME	PM No. FIELD No.	REFERENCE	Trampling Seed	Following Plough	Ahead of Plough	Following Hoers	Ahead of Hoers	No Plough or Hoers	Layered Flock	Flock in Disarray	Coaxer/Sower	Beaters with Stick	Beaters with Whip	in Animal Parade	on Threshing Floor
K3.j-m-nfrt	PM 208 LG 63	LD Erg. 32; Badawy (1976) fig. 30.	•					•	•		•	•	•		
K3.j-m-nfrt	PM 467 D23	Simpson (1992) pl. F, G.	•					•	•		•	•	•		
K3.j-m-ḥzt	PM 542	McFarlane (2003) pl.44.						•		•	•		•		
K3-ḥj.f	PM 76	Junker VI (1943) fig 42.	•					•							
Ṯjj	PM 468 D22	Wild (1953) pl.112-13.	•			•			•		•	•	•		
DYNASTY VI.E															
Mrrw-k3/Mrj	PM 525	Duell II (1938) pl. 168-70; Kanawati (2011) pl. 81-82.													•
Mḫw	PM 619	Altenmüller (1998) pl. 41.	•	?					•			•	•		
Ḥzj r/u *Sšm-nfr*	PM n/r	Kanawati–Abder-Raziq (1999) pl. 52.	•	?				?			?	?	•		
DYNASTY VI.L															
Ppj-ʿnḫ/Ḥnj-km	PM (iv) 247	Blackman–Apted Meir V (1953) pl. 32.												•	
Ḏʿw and *Ḏʿw/Šm3j*	PM (iv) 244 No.12	Davies *Gebrâwi* II (1902) pl.6.												•	

TABLE 7: MANAGEMENT OF DOMESTIC GOATS

MONUMENT			SCENE	
TOMB OWNER	PM No. FIELD No.	REFERENCE	Browsing	Parade
DYNASTY IV.L				
Nb.j-m-3ḫtj	PM 230 LG 86	Hassan (1943), fig. 75.	?	
Sḫm-k3-Rᶜ	PM 233 LG 89	LD Erg. 36; Hassan (1943), fig. 57; Harpur (1985) fig. 10.	?	
DYNASTY V.M				
Nj-ᶜnḫ-H̱nmw and *H̱nmw-ḥtp*	PM 641	Moussa–Altenmüller Taf. 20-1, Abb. 8.	•	
Nfr and *K3-ḥ3.j*	PM 639	Moussa–Altenmüller (1971) pl. 1, 4, 18,19.	•	
Rᶜ-ḫᶜ.f-ᶜnḫ	PM 207 G7948=LG75	LD II 9.		•
DYNASTY V.L				
3ḫtj-mrw-nswt	PM 80 G2184	Smith (1946) fig. 239.	•	
3ḫtj-ḥtp (Louvre)	PM 635	Ziegler (1993) p. 44 (plates); 137-9 (drawing).	•	
Nb-j-k3w-ḥr/Jdw	PM 627	Hassan (1932) fig. 12-13.	•	
Ḥzjj-Mnw	PM n/r M22	Kanawati (1983) fig. 12.		•
Ḥtp-ḥr-3ḫtj	PM 593 D60	Mohr (1943) fig. 21.	•	
H̱w-ns	PM (iv) 134 No.2	LD II, 108.	•	
Sšm-nfr IV	PM 223 LG 53	Junker (1953) Abb. 93.	•	
DYNASTY VI.E				
Jttj-Šdw	PM (iv) 122	Petrie(1898) pl. 15; Kanawati–McFarlane (1993) pl. 14, 46.	•	•
Mrrj	PM 518	Davies et al (1984) pl. 7.	•	
DYNASTY VI.M				
Jbj	PM (iv) 243 No.8	Davies (1902) *Gebrâwi* I, pl. 11; Kanawati *Gebrawi* II pl.71.	•	•
Jḫjj	PM (I) 293 No.186	Saleh (1977) fig. 60, 63, pl. 18.	?	?
Ppj-ᶜnḫ-ḥrj-jb	PM (iv) 254 D2	Blackman (1924) pl. 14.	•	•
DYNASTY VI.L				
Ppj-ᶜnḫ/Ḥnj-km	PM (iv) 247 A2	Blackman–Apted (1953) pl. 32.	•	•
Nj-ᶜnḫ-Ppjj	PM (iv) 137 No 14	LD II, 111 [b]; Varille (1938) pl. 16 [b].	•	
Ḥzjj-Mnw/Zzj	PM n/r G42	Kanawati (1987) fig. 3 [c].	•	
Ḫntj	PM n/r No. 405	Saleh (1977) fig. 36, pl. 6[2].	•	
Ḥnj/Špsj-pw-Mnw	PM n/r H24	Kanawati (1981) fig. 20.		•
K3(.j)-ḥp/Ṯtj-jḳr	PM (v) 19 H26	Kanawati (1980) fig.15.		•
Ḏᶜw and *Ḏᶜw/Šm3j*	PM (iv) 244 No.12	Davies (1902) *Gebrâwi* II, pl. 9, 15.		•
DYNASTY VIII				
W3hj	PM (v) 36-7 D4	Kanawati (1995) pl. 20, 22-3.	•	•
Mrjj-ᶜ3	PM (v) 34 D18	Kanawati (1995) pl. 37, 40.	•	•
Gḥs3/Nbjj	PM n/r GA11	Kanawati (1987) fig. 30.	•	•

GOAT FEATURES							HERDERS/LABOURERS							MONUMENT
on Sub-Register	in Tree Canopy	Fighting	Copulating	Giving Birth	Young	Led by Rope	Felling Trees	Building Boats	Pulling Leaves	Protecting at Birth	Slaughter/Skinning	Cooking Goatmeat	Overseer	**TOMB OWNER**
														DYNASTY IV.L
														Nb.j-m-3ḫtj
							?							*Sḫm-k3-Rˁ*
														DYNASTY V.M
•	•			•	•		•	•	•	•	•	•	•	*Nj-ˁnḫ-Ḫnmw* and *Ḫnmw-ḥtp*
•					•		•		•		•			*Nfr* and *K3-ḥ3.j*
														Rˁ-ḥˁ.f-ˁnḫ
														DYNASTY V.L
	•		•	•							•		•	*3ḫtj-mrw-nswt*
•	•			•						•			•	*3ḫtj-ḥtp* (Louvre)
•	•				•		•				•			*Nb-j-k3w-ḥr/Jdw*
														Ḥzjj-Mnw
•				•	•					?			•	*Ḥtp-ḥr-3ḫtj*
•					•		•	•	•		•			*Ḫw-ns*
			•											*Sšm-nfr* IV
														DYNASTY VI.E
														Jttj-Šdw
														Mrrj
														DYNASTY VI.M
		•	•											*Jbj*
														Jḥjj
•					•				•				•	*Ppj-ˁnḫ-ḥrj-jb*
														DYNASTY VI.L
•					•									*Ppj-ˁnḫ/Ḥnj-km*
	•							•			•	•		*Nj-ˁnḫ-Ppjj*
														Ḥzjj-Mnw/Zzj
														Ḫntj
						•								*Ḥnj/Špsj-pw-Mnw*
		•			•				•					*K3(.j)-ḥp/Ṯtj-jḳr*
														Ḏˁw and *Ḏˁw/Šm3j*
														DYNASTY VIII
					•									*W3hj*
		•												*Mrjj-ˁ3*
														Gḥs3/Nbjj

TABLE 8: REPRESENTATIONS OF DONKEYS

MONUMENT			TRANSPORT OF HARVEST			OTHER THEMES		
TOMB OWNER	**PM No. FIELD No.**	**REFERENCE**	Herd Drivern to Field	Loading Donkeys	Driving laden donkeys	Donkeys Threshing	In Parade of Animals	Other (* see beneath table)
DYNASTY IV.L								
Nj-k3w-Rꜥ	PM 232 LG 87	LD II 15 [b].					•	
DYNASTY V.E								
Wḥm-k3.j	PM 114 D117	Kayser (1954) fig. p. 36-7.			•		•	
Rꜥ-wr	PM 265	Hassan (1932) pl. 3 [3], fig. 21.		•	•	•		
Ḫnmw-ḥtp	PM 578 D49	Petrie–Murray (1952) pl. 17; Sharawi–Harpur (1988) fig. 6.		•	•	•		
Snb	PM 101	Junker (1941) fig. 2, 3.					•	
DYNASTY V.M								
Jj-mrjj	PM 170 G6020	LD II 51; Weeks (1994) fig. 39.	•		•			
Wr-jr.n-Ptḥ	PM 699	Hall–Lambert (1922) pl. 10-11; James (1961) pl. 29.			•		•	
Nj-ꜥnḫ/Ḫnmw and *Ḫnmw-ḥtp*	PM 641	Moussa–Altenmüller (1977) pl 56-9.	•	•	•			*
Rꜥ-m-k3.j	PM 487 D3=S903	Hayes I (1953) fig. 57.		•	•	•		
Rꜥ-ḫꜥ.f-ꜥnḫ	PM 207 G7948=LG75	LD II 9.		•	•	•	•	
K3.j-m-nfrt	PM 208 LG63	LD Erg. 32; Badawy (1976) fig. 30(a).	•	•	•			
DYNASTY V.L								
3ḫtj-mrw-nswt	PM 80 G2184	Smith (1946) fig. 229.				•		
3ḫtj-ḥtp (Louvre)	PM 634	Ziegler (1993) p. 126, 129, 135-7.		•	•	•		
3ḫtj-ḥtp	PM 599 D64	Davies II (1901) pl. 3, 4, 7, 8.	•	•	•	•		
J3sn	PM 82 G2196	Simpson (1980) fig. 30.		•	•			
Jrj-n-k3-Ptḥ	PM 644	Moussa–Junge (1975) pl. 9.	•	•	•	•		
Jḥjj (r/u *Sšsšt/Jdwt*)	PM 617	Macramallah (1935) pl.10 (A); Kanawati–Abder-Raziq (2003) pl. 59(a).	•					
Wr-jr.n.j	PM (iv) 188 No.25	Davies (1901) pl. 8, 16.	•	•	•	•		
Pḥn-wj-k3.j	PM 491 D70=LS15	LD II 47.	•	•	•	•		
Ptḥ-ḥtp I	PM 596 D62	Murray I (1905) pl.11.		•	•	•		
Nj-m3ꜥt-Rꜥ	PM n/r G2097	Roth (1995) pl.189.	•			•		
Nj-ḥtp-Ptḥ	PM 94 G2430=LG 25	LD II 71a; Badawy (1978) pl. 5, 8.	•			•		
Nj-k3w-ḥr	PM 498 S915	Quibell, *1907-08*, pl. 62[2].	•	•	•			
Nfr (1)	PM 137 G4761	Junker (1943) Abb. 14, 17.		•	•			
Nfr-jrt-n.f	PM 583 D55	van de Walle (1978) pl. 11-12.	•	•	•	•		
Nfr-b3w-Ptḥ	PM 169 G6010=LG15	LD II 56; Weeks (1994) fig. 9, 11.	•	•	•			
Nfr-sšm-Ptḥ and *Sḫntjw*	PM 645	Moussa–Junge (1975) pl. 4(b), 5.		•	•			
Rꜥ-wr II	PM 162 G5470=LG32	LD Erg. 25; Junker 3 (1938) fig. 48.	•			•		
Ḥtp-ḥr-3ḫtj	PM 593 D60	Mohr (1943) fig. 44-52.	•		•	•		
Ḥtp-k3.j/Tp-k3.j	PM 447 S3509	Martin (1979) pl. 10, 17.				•		
Ḫww-wr	PM 254 LG95	LD II, 43(a); Hassan (1944) fig. 104.	•	•	•	•		**
Ḫw-ns	PM (iv) 134 No.2	LD II, 106(b), 107.	•	•	•	•		
Ḫnwt	PM 623	Munro (1993) Taf. 41, 42, 43.	•					
S3-jb	PM n/r G2092, G2093	Roth (1995) pl. 74(b), 75(a).	•					
Spd-ḥtp	PM 481 D15	Harpur (1986) fig. 1-7.		•	•	•		

TABLE 8: REPRESENTATIONS OF DONKEYS

MONUMENT			TRANSPORT OF HARVEST			OTHER THEMES		
TOMB OWNER	**PM No. FIELD No.**	**REFERENCE**	Herd Drivern to Field	Loading Donkeys	Driving laden donkeys	Donkeys Threshing	In Parade of Animals	Other (* see beneath table)
Snḏm-jb/Jntj	PM 85 G2370=LG27	LD Erg 22b; Brovarski (2001) fig. 52b, 53.	•	•	•	•		
Snḏm-jb/Mḥj	PM 87 G2378=LG26	LD II, 73; Brovarski (2001) pl. 114.	•		•	•		
Sḫm-ʿnḫ-Ptḥ	PM191 G7152	Badawy (1976) fig. 22.		•	•	•		
Sḫm-ʿnḫ-Ptḥ	PM 454 D41	Simpson (1976) fig. 6-7, 9-10, pl. D.	•	•	•	•		
Sḫm-k3j	PM 53 G1029	Simpson (1980) fig. 4.	•	•				
Sšm-nfr IV	PM 223 LG 53	LD II 80; Junker (1953) fig. 74, 75.			•	•		
K3.j-m-nfrt	PM 467 D23	Simpson (1992) fig. 15, pl. F, G.	•	•	•	•		
K3.j-m-rḥw	PM 485 D2=S905	Mogensen (1921) fig.18-19, 21.				•		
K3-ḥj.f	PM 76	Junker (1943) fig. 42-3, 45-7.	•	•	•	•		
Kd-nfr	PM 56 G1151	Smith (1946) fig. 170.		•	•			
Ṯjj	PM 468 D22	Wild III (1966) pl. 136-39.	•	•	•	•		
DYNASTY VI.E								
Jn-Snfrw-jšt.f	PM 891 No 2	de Morgan (1903) pl. 21; Borchardt (1964) Bl. 102 (CG 1770).		•	•			
Jttj-Šdw	PM (iv) 122	Petrie (1898) pl. 23; Kanawati–McFarlane(1993) pl. 47 [a-c].		•	•	•		
ʿnḫ -m- ʿ-Ḥr	PM 512	Badawy (1978) fig. 24; Kanawati–Hassan (1997) pl. 37(a).				•		
Mrrw-k3/Mrjj	PM 525	Duell II (1938) pl. 168-70; Kanawati (2011) pl. 81-81, 84.	•	•	•	•		
Mḥw	PM 619	Altenmuller (1998) pl. 22[b]-25.	•	•	•	•		
Mṯtj	PM 646	Kaplony (1976) p. 22-4, No. 3.	•					
Nj-ʿnḫ-Ppjj	PM (iv) 37 L14	Varille (1938) pl. 8-9.				•		
Rmnj/Mrwj	PM n/r	Kanawati (2009) pl. 45.			•			
Ḥm-rʿ/Jzj (I)	PM (iv) 243 N72	Davies *Gebrâwi* II (1902) pl. 17; Kanawati *Gebrawi* I (2005) pl. 62.		•	•			
Ḫnmw-ntj	PM 87 G2374	Brovarski (2001) fig. 88.	•	•				
DYNASTY VI.M								
Jbj	PM (iv) 243 No.8	Davies *Gebrâwi* I (1902) pl.12.		•	•	•		
B3wj	PM n/r BA48	Kanawati (1987) fig. 25.		•	•			
Ppj-ʿnḫ-ḥrj-jb	PM (iv) 254 D2	Blackman (1924) pl. 14, 22, 24.		•	•	•		
Mmj	PM n/r B12	Kanawati (1987) fig. 11.		•	•	•		
Snfrw-ḥtp	PM 96 G3008	Fisher (1924) pl. 53.				•		
DYNASTY VI.L								
Ppj-ʿnḫ/Ḥnj-km	PM (iv) 247	Blackman–Apted (1953) pl. 22, 58.		•	•	•	•	
Mḫw and *S3bnj*	PM (v) 231 A1	de Morgan (1894) p. 145 [d].			•			
Ḥzjj-Mnw/Zzj	PM n/r G42	Kanawati (1987) fig. 4(a).		•	•			
Ḫnj/Špsj-pw-Mnw	PM n/r H24	Kanawati (1981) fig. 17.		•	•			
K3(.j)-ḥp/Ṯtj-jkr	PM (v) 19 H26	Kanawati (1980) fig. 5.		•	•			
Ḏʿw and *Ḏʿw/Šm3j*	PM (iv) 244 No.12	Davies (1902) *Gebrâwi* II pl. 6.		•	•		•	
DYNASTY VIII								
ʿnḫ-tj.fj	PM (v) 170	Vandier (1950) pl. 36.			•			
W3ḥj	PM (v) 36-7	Kanawati (1995) pl. 20-23.			?			
Mrjj-ʿ3	PM (v) 34 D18	Kanawati(1995) pl. 35, 37, 39-40.		•	•	•	•	
Rḥw-r-3w.sn	PM n/r BA17	Kanawati (1987) fig. 15, 16.		•	•			

* Carrying young desert animals in container.

** Tomb owner riding.

TABLE 9: PRODUCTION OF FIELD CROPS

MONUMENT			PLANTING			CROP	
TOMB OWNER	PM No. FIELD No.	REFERENCE	Ploughing	Hoeing	Sheep	Crop: (F)=Flax)	Pulling Flax
DYNASTY IV.E							
Jtt	PM (iv) 93 No. 16	Petrie (1892) pl.28; Harpur (2001) fig. 81.	•				
Nfr…w	PM (iv) 94 No. 22	Petrie (1892) pl. 28.					
Nfr-mꜣꜥt	PM (iv) 93	Petrie (1892) pl. 18; Harpur (2001) fig. 75.	•				
Rꜥ-ḥtp	PM (iv) 91 No. 6	Petrie (1892), pl. 12; Harpur (2001) fig. 95.	•				
DYNASTY IV.L							
Mrs-ꜥnḫ III	PM 197 G7530/40	Simpson–Dunham (1974) fig. 4.			•		
Nb.j-m-ꜣḫtj	PM 230 LG 86	LD II 12 [a]; Hassan (1943) fig. 76.	•				
DYNASTY V.E							
Rꜥ-wr	PM 265	Hassan (1932) fig. 21, 26.	•			F	•
H̱nmw-ḥtp	PM 579	Petrie –Murray (1952), pl. 17; Sharawi– Harpur (1988), fig. 6.				F	•
Sḫm-kꜣ-Rꜥ	PM 233 LG 89	LD. Erg. 36; Hassan, (1943) fig. 57.					
DYNASTY V.M							
Jj-mrjj	PM 170 G6020	LD II 51; Weeks (1994) fig. 39.	•	•	•		
Wr-jr.n-Ptḥ	PM 699	Hall–Lambert (1922) pl. 10-11; James (1961) pl. 29.			•	F	•
Ptḥ-špss	PM 340	Vachala (2004) pl. 14-15.				F	•
Nj-ꜥnḫ/H̱nmw and *H̱nmw-ḥtp*	PM 641	Moussa–Altenmüller (1977) Taf. 54 [b], 56-9, Abb. 24.	•		•	F	•
Nfr and *Kꜣ-ḥꜣ.j*	PM 639	Moussa–Altenmüller (1971) pl. 1, 8-9.	•		•		
Rꜥ-ḥtp	PM n/r C 5	El-Khouli–Kanawati (1990) pl. 73 [A].	•				
Rꜥ-ḫꜥ.f-ꜥnḫ	PM 207 G7948	LD II 9 [lower].					
Ḫwfw-ḫꜣ.f II	PM 190 G7150	Simpson (1978) fig. 47.					
Kꜣ.j-m-nfrt	PM 208 LG 63	LD Erg. 32; Badawy (1976) fig. 30 [a].	•		•		
DYNASTY V.L							
ꜣḫtj-mrw-nswt	PM 80 G2184	Smith (1946), fig. 229; Nagoya–BMFA (2001) 58 [36].					
ꜣḫtj-ḥtp (Louvre)	PM 634	Ziegler (1993) pl. 71-4; (drawings) 145-150.			•	F	•
ꜣḫtj-ḥtp	PM 599 D64	Davies (1901) pl. 3, 4, 7-8.					
Jꜣsn	PM 82 G2196	Simpson (1980) fig. 30.			•		
Jj-nfrt	PM 298	Schürmann (1983) Abb. 11 [a, b], 21.	•		•		
Jrj-n-kꜣ-Ptḥ	PM 644	Moussa–Junge (1975) pl. 9.	•			F	•
Jḥjj (r/u *Sšsšt/Jdwt*)	PM 617	Macramallah (1935) pl. 10 [A]; Kanawati–Abder-Raziq (2003) pl. 59 [a].					
Wr-jr.n.j	PM (iv)188 No. 25	Davies *Sheikh Said* (1901) pl. 8, 16.	•		•	F	•
Pḥn-wj-kꜣ.j	PM 491 D70=LS 15	LD II 47.	•			?	•
Ptḥ-ḥtp I	PM 596 D62	Murray (1905) pl. 11.				F	•
Ptḥ-ḥtp	PM 653 LS31	LD II 103.					
Ptḥ-ḥtp/Jj-n-ꜥnḫ	PM 606	Hassan (1975) fig. 31.			•	F	•
Nj-mꜣꜥt-Rꜥ	PM n/r G2097	Roth (1995) pl. 95 [a, b], 189.					
Nj-mꜣꜥt-Rꜥ	PM 282	Hassan (1936) fig. 240.			•	F	•
Nj-ḥtp-Ptḥ	PM 94 G2439=LG 25	LD II 71 [a]; Badawy (1978) pl. 8.					
Nj-kꜣw-ḥr	PM 498 S915	Quibell (1909) pl. 62 [2].					
Nb-j-kꜣw-ḥr/Jdw	PM 627	Hassan (1932) fig. 15.	•		•	F	•
Nbt	PM 624	Munro (1993) pl. 9 [b], 13 .	•		•		
Nfr (1)	PM 137 G4761	Junker (1943) Abb. 14, 17.	•		•	F	•
Nfr-jrt-n.f	PM 583 D55	Van de Walle (1978) pl. 11-12.	•		•	F	•
Nfr-bꜣw-Ptḥ	PM 169 G6010=LG15	LD II 56 [a, a bis]; Weeks (1994) fig. 9, 11.	•	•	•		
Nfr-mḏr-Ḫwfw	PM n/r G2240	Roth (1995) pl. 128 [a, b], 129 [a], 208 [b].					
Nfr-sšm-Ptḥ and *Sḫntjw*	PM 645	Moussa–Junge (1975) pl. 4 [b], 5.	•		•	F	•
Rꜥ-wr II	PM 162 G5470=LG32	LD Erg. 25; Junker (1938) Abb. 48.					

CROP (cont.)				HARVEST								
Binding Flax	Crop: (C)=Cereal)	Crop: (B)=Barley	Crop: E=Emmer Wheat (bdt)	Reaping	Tying Sheaves	Donkeys to Field	Laden Donkeys	Stacks for Threshing	Threshing	Winnowing/ Sweeping	Storage in Granary	**TOMB OWNER**
												DYNASTY IV.E
												Jtt
	C			•								*Nfr…w*
												Nfr-mꜣꜥt
	C		E									*Rꜥ-ḥtp*
												DYNASTY IV.L
										•	•	*Mrs-ꜥnḫ* III
												Nb.j-m-ꜣḫtj
												DYNASTY V.E
	C			•			•	•	•	•		*Rꜥ-wr*
	C			•	•		•		•			*H̱nmw-ḥtp*
	C	B								•		*Sḫm-kꜣ-Rꜥ*
												DYNASTY V.M
	C	B	E?	•	•	•	•	•			B	*Jj-mrjj*
•	C	B		•			•	•			B	*Wr-jr.n-Ptḥ*
	C		E	•			•					*Ptḥ-špss*
•	C	B	E	•		•	•			•	B	*Nj-ꜥnḫ/H̱nmw* and *H̱nmw-ḥtp*
												Nfr and *Kꜣ-ḥꜣ.j*
	C			•								*Rꜥ-ḥtp*
	C			•	•		•	•	•	•		*Rꜥ-ḫꜥ.f-ꜥnḫ*
	C			•	•						B	*Ḫwfw-ḫꜣ.f* II
	C					•	•	•				*Kꜣ.j-m-nfrt*
												DYNASTY V.L
									•	•	•	*ꜣḫtj-mrw-nswt*
•	C	B		•			•	•	•	•		*ꜣḫtj-ḥtp* (Louvre)
	C	B		•		•	•	•	•	•		*ꜣḫtj-ḥtp*
	C			•			•	•				*Jꜣsn*
												Jj-nfrt
	C	B		•		•	•	•	•	•	•	*Jrj-n-kꜣ-Ptḥ*
						•						*Jḫjj* (r/u *Sšsšt/Jdwt*)
•	C	B	E	•	•	•	•	•	•		•	*Wr-jr.n.j*
	C	B		•		•	•	•	•	•	•	*Pḥn-wj-kꜣ.j*
	C	B	E	•			•	•	•	•		*Ptḥ-ḥtp* I
			E								•	*Ptḥ-ḥtp*
												Ptḥ-ḥtp/Jj-n-ꜥnḫ
	C			•		•		•	•			*Nj-mꜣꜥt-Rꜥ*
	C			•								*Nj-mꜣꜥt-Rꜥ*
	C	B				•			•	•		*Nj-ḥtp-Ptḥ*
						•	•	•			•	*Nj-kꜣw-ḥr*
•		B										*Nb-j-kꜣw-ḥr/Jdw*
												Nbt
•	C	B	E	•			•	•				*Nfr* (1)
•	C	B		•		•	•	•	•	•		*Nfr-jrt-n.f*
	C					•	•	•			•	*Nfr-bꜣw-Ptḥ*
	C	B		•						•		*Nfr-mḏr-Ḫwfw*
•	C	B		•			•	•		•		*Nfr-ššm-Ptḥ* and *Sḫntjw*
	C					•			•			*Rꜥ-wr* II

TABLE 9: PRODUCTION OF FIELD CROPS

MONUMENT			PLANTING			CROP	
TOMB OWNER	**PM No. FIELD No.**	**REFERENCE**	Ploughing	Hoeing	Sheep	Crop: (F)=Flax)	Pulling Flax
Rꜥ-m-kꜣ.j	PM 487 D3=S903	Hayes (1953) fig. 57.					
Rꜥ-špss	PM 494 LS16=S902	LD II 60/2/4a; LD Erg. 41; Quibell (1909) 23.	•		•	F	•
Ḥzjj-Mnw	PM n/r M22	Kanawati (1983) fig. 18.				F	•
Ḥtp-ḥr-ꜣḫtj	PM 593 D60	Mohr (1943) fig. 44-52.	•	•	•	F	•
Ḥtp-kꜣ.j/Tp-kꜣ.j	PM 447 S3509	Martin (1979) pl. 10 [7], 17 [c].				F	•
Ḫww-wr	PM 254 LG 96	LD II 43 [a]; Hassan (1944) fig. 104.	•				
Ḫw-ns	PM (iv) 134 No.2	LD II 106 [b], 107.	•		•	F	•
Ḫnwt	PM 623	Munro (1993) pl. 41, 42, 43, 38.	•		•	F	•
Sꜣ-jb	PM 70 G2092, G2093	Roth (1995) pl. 74 [b], 75 [a], 183.			•		
Spd-ḥtp	PM 481 D15	Harpur *MDÄIK* 42 (1986) fig. 1-7.					
Snḏm-jb/Jntj	PM 85 G2370=LG27	Brovarski (2001) fig. 52 [b], 53.	•		•	F	•
Snḏm-jb/Mḥj	PM 87 G2378=LG26	LD Erg. 14; LD II. 73 [left]; Brovarski (2001) fig. 114-15.	•		•	F	•
Sḫm-ꜥnḫ-Ptḥ	PM 191 G7152	Badawy (1976) fig. 22.					
Sḫm-ꜥnḫ-Ptḥ	PM 454 D41	Simpson (1976) fig. 6-7, 9-10, 8, pl. D.	•		•	F	•
Sḫm-kꜣj	PM 53 G1029	Simpson (1980) fig. 4.	•	•	•	F	•
Sšm-nfr IV	PM 223 LG53	LD II 80 [c]; Junker (1953) Abb.74 [a, b].					
Kꜣ.j- m- ꜥnḫ	PM 131 G4561	Junker (1940) Taf. 12; Kanawati (2001) pl. 36.	•				
Kꜣ.j-m-nfrt	PM 467 D23	Simpson (1992) fig. 15, pls. F. G.			•	F	•
Kꜣ.j-m-rḥw	PM 485 D2=S905	Mogensen (1921) figs. 18-19, 21, 24-6.					
Kꜣ.j-m-ḥzt	PM 542	Quibell–Hayter (1927) fig. p. 17; McFarlane (2003) pl. 44.			•		
Kꜣ.j-m-ṯnnt	PM 195 G7411	Smith (1946) p. 310.					
Kꜣpj	PM 69 G2091	Roth (1995) pl. 159.	•			F	•
Kꜣ-ḥj.f	PM 76 G2136	Junker (1943) fig. 42-3, 45-7.	•			F	•
Kd-nfr	PM 56 G1151	Smith (1946) fig. 170.					
Ṯjj	PM 468 D22	Wild, II [1] (1953) pl. 78-79, 112-113; Wild, III [2] (1966) pl. 136-139,151-155.	•	•	•	F	•
Dwꜣ-Mnw	PM n.r. L6	Kanawati (1986) fig. 6.	•				
DYNASTY VI.E							
Jn-Snfrw-jšt.f	PM 891	de Morgan II (1903) pl.21; Borchardt (1964) Bl. 102 [CG 1770].					
Jttj-Šdw	PM (iv) 122	Petrie (1898) pl.23; Kanawati– McFarlane (1993) pl. 47 [a-c].					
ꜥnḫ -m- ꜥ-Ḥr/Zzj	PM 512 D38	Badawy (1978) fig. 24; Kanawati–Hassan (1997) pl. 37[a].					
Mrw/Ttj-snb	PM 520	Lloyd et al (1990) pl. 12.				F	•
Mr.f-nb.f	PM n/r	Mysliwiec (2004) pl. 21, 66.		•			
Mrrw-kꜣ.j/Mrj	PM 533	Duell II (1938) pl.168-70; Kanawati (2011) pl. 81-84.				F	•
Mḥw	PM 620	Altenmüller (1998) pl. 22 [b]-25, 43 (a).			•	F	•
Mṯtj	PM 646	Kaplony (1976) 22-4, No. 3.					
Nj-ꜥnḫ-Ppjj	PM (iv.)137 No. 14	Varille (1938) pl. 8-9.					
Nj-ꜥnḫ-nswt	PM. 694, 755	Sharawi–Harpur (1988) fig. 2.	•			F	•
Ḥm-rꜥ/Jzj (I)	PM (iv) 243 No.72	Davies *Gebrâwi* II (1902) pl. 17.					
Ḥzj	PM n/r	Kanawati–Abder-Raziq (1999) pls. 9-10.			•	F	•
Ḫnmw-ntj	PM 87 G2374/G2385	Brovarski (2001) fig. 88.					
Snfrw-ḥtp	PM 96 G3008	Fisher (1924) pl. 53.			•		
Ṯmrrj	PM n/r N248	Peck (1959) pl. 3.	•				

CROP (cont.)				**HARVEST**								
Binding Flax	Crop: (C)=Cereal)	Crop: (B)=Barley	Crop: E=Emmer Wheat (bdt)	Reaping	Tying Sheaves	Donkeys to Field	Laden Donkeys	Stacks for Threshing	Threshing	Winnowing/ Sweeping	Storage in Granary	**TOMB OWNER**
	C	B	E	•	•		•		•	•		*Rʿ-m-kꜣ.j*
	C		E	•								*Rʿ-špss*
												Ḥzjj-Mnw
•	C		E	•		•	•	•	•			*Ḥtp-ḥr-ꜣḫtj*
	C		E		•				•	•		*Ḥtp-kꜣ.j/Tp-kꜣ.j*
	C			•	•		•		•			*Ḫww-wr*
•	C	B	E	•	•	•	•	•	•		•	*Ḫw-ns*
	C			•		•				•	•	*Ḫnwt*
	C			•		•						*Sꜣ-jb*
	C	B	E	•			•	•	•	•	•	*Spd-ḥtp*
	C	B		•		•		•	•	•		*Snḏm-jb/Jntj*
•	C			•		•	•		•	•		*Snḏm-jb/Mḥj*
							•	•	•	•		*Sḫm-ʿnḫ-Ptḥ*
•	C	B	E	•	•	•	•	•	•	•	•	*Sḫm-ʿnḫ-Ptḥ*
•	C	B	E?	•		•		•				*Sḫm-kꜣj*
	C	B		•	•		•	•	•	•		*Sšm-nfr* IV
	C			•								*Kꜣ.j- m- ʿnḫ*
•	C	B	E	•		•	•	•	•	•	•	*Kꜣ.j-m-nfrt*
									•	•	•	*Kꜣ.j-m-rḥw*
												Kꜣ.j-m-ḥzt
	C			•			•	•				*Kꜣ.j-m-ṯnnt*
												Kꜣpj
•	C	B	E	•		•	•	•	•	•		*Kꜣ-ḥj.f*
	C	B		•			•					*Kd-nfr*
•	C	B	E	•		•	•	•	•	•		*Ṯjj*
												Dwꜣ-Mnw
												DYNASTY VI.E
							•	•		•		*Jn-Snfrw-jšt.f*
	C			•			•		•	•	•	*Jttj-Šdw*
	C	B						•	•	•		*ʿnḫ -m- ʿ-Ḥr/Zzj*
•	C			•								*Mrw/Ttj-snb*
	C			•								*Mr.f-nb.f*
•	C	B		•		•	•	•	•	•		*Mrrw-kꜣ.j/Mrj*
•	C	B		•		•	•	•	•	•	•	*Mḥw*
	C?					•						*Mttj*
	C	B?						•	•			*Nj-ʿnḫ-Ppjj*
•	C			•					•	•		*Nj-ʿnḫ-nswt*
	C			•			•					*Ḥm-rʿ/Jzj* (I)
•	C	B		•						•		*Ḥzj*
				•			•	•				*Ḫnmw-ntj*
	C			•					•	•		*Snfrw-ḥtp*
												Ṯmrrj

TABLE 9: PRODUCTION OF FIELD CROPS

MONUMENT			PLANTING			CROP	
TOMB OWNER	**PM No. FIELD No.**	**REFERENCE**	Ploughing	Hoeing	Sheep	Crop: (F)=Flax)	Pulling Flax
DYNASTY VI.M							
Jbj	PM (iv) 243 No.8	Davies (1902) *Gebrâwi* I pl.12; Kanawati *Gebrawi* II (2007) pl. 51.	•			F	•
Bꜣwj	PM n/r BA48	Kanawati (1987) fig. 25.	•				
Ppj-ʿnḫ-ḥrj-jb	PM (iv) 254 D2	Blackman (1924) pl. 14, 24 [1, 2].	•			F	•
Mmj	PM n/r B12	Kanawati (1987) fig. 11.					
Kꜣ.j-ḥp/Ṯtj	PM n/r M8	Kanawati (1982) fig. 5[a].	•				
DYNASTY VI.L							
Ppj-ʿnḫ/Ḥnj-km	PM (iv) 247 A2	Blackman–Apted (1953) pl. 30, 22, 58.	•			F	•
Mḫw and *Sꜣbnj*	PM (v) 232 A1	de Morgan (1894) p. 145 [d].	•			F	•
Ḥzjj-Mnw/Zzj	PM n/r G42	Kanawati (1987) fig. 4 [a].	•				
Ḫnj/Špsj-pw-Mnw	PM n/r H24	Kanawati (1981) fig. 17.	•				
Kꜣ(.j)-ḥp/Ṯtj-jkr	PM (v) 19 H26	Kanawati (1980) fig. 15.	•				
Ḏʿw/Ḏʿw-Šmꜣj	PM (iv) 244 No.12	Davies *Gebrâwi* II(1902) pl. 6.	•			F	•
DYNASTY VIII							
Wꜣhj	PM (v) 36-7	Kanawati (1995) pl. 20-23, 28, 30-1.	•				
Mrjj-ʿꜣ	PM (v) 34 D18	Kanawati (1995) pl. 35-37, 39-40.					
Rḥw-r-ꜣw.sn	PM n/r BA17	Kanawati (1987) fig. 15-16.	•				
Gḥsꜣ/Nbjj	PM n/r GA11	Kanawati (1987) fig. 26 [c].	•				

CROP (cont.)				HARVEST								
Binding Flax	Crop: (C)=Cereal)	Crop: (B)=Barley	Crop: E=Emmer Wheat (bdt)	Reaping	Tying Sheaves	Donkeys to Field	Laden Donkeys	Stacks for Threshing	Threshing	Winnowing/ Sweeping	Storage in Granary	**TOMB OWNER**
												DYNASTY VI.M
•	C			•	•		•	•	•		•	*Jbj*
	C			•			•					*B3wj*
		B	E	•			•		•			*Ppj-ʿnḫ-ḥrj-jb*
		B		•			•		•			*Mmj*
	C			•								*K3.j-ḥp/Ṯtj*
												DYNASTY VI.L
•	C			•			•		•			*Ppj-ʿnḫ/Ḥnj-km*
	C			•			•					*Mḫw* and *S3bnj*
	C			•			•				•	*Ḥzjj-Mnw/Zzj*
	C			•	•		•				•	*Ḫnj/Špsj-pw-Mnw*
	C			•	•		•					*K3(.j)-ḥp/Ṯtj-jkr*
•	C			•	•		•	•				*Ḏʿw/Ḏʿw-Šm3j*
												DYNASTY VIII
	C			•					•		•	*W3hj*
	C			•			•		•			*Mrjj-ʿ3*
							•					*Rḥw-r-3w.sn*
												Gḥs3/Nbjj

C denotes barley or wheat or cereal of unknown variety.

TABLE 10: GRAIN STORAGE, BREAD AND BEER PRODUCTION

MONUMENT			GRAIN STORAGE					
TOMB OWNER	PM No. FIELD No.	REFERENCE	Grain Taken to Granary	Grain Taken from Stack	Granary Portrayed	Scribes Recording Cereal	Crier Calling Quantities	Bread/Beer Production
DYNASTY IV.L								
Bw-nfr		Hassan (1941) fig. 147 [a] 153.						•
Mrs-ʿnḫ III	PM197 G7530/40	Dunham and Simpson (1974) fig. 11.		•				•
DYNASTY V.E								
Kꜣ.j-ʿpr	PM 501	Bárta (2001) fig. 4.17.						•
DYNASTY V.M								
Jj-mrjj	PM 170 G6020	LD II 51; Weeks (1994) fig. 39.		•		•	•	
		LD Erg. 7; Weeks, (1994) fig. 34.						•
Wr-jr.n-Ptḥ	PM 699	Hall–Lambert (1922) pl. 10; James (1961) pl. 29 [2].				•		
Ptḥ-špss	PM 340	Vachala (2004) pl. 4, 12.						•
Nj-ʿnḫ/Ḫnmw	PM 641	Moussa–Altenmüller (1977) Taf. 54 [a, b] Abb. 24.	•	•	•	•	•	
and *Ḫnmw-ḥtp*		Moussa–Altenmüller (1977) pl. 23, 26, 76.						•
Nfr and *Kꜣ-ḥꜣ.j*	PM 639	Moussa–Altenmüller (1971) pl. 1, 5.						•
Ḫwfw-ḥꜣ.f II	PM 190 G7150	Simpson (1978) fig. 47.		•		•		
DYNASTY V.L								
ꜣḫtj-mrw-nswt	PM 80 G2184	Smith (1946) fig. 229.	•					
Jꜣsn	PM 82 G2196	Simpson (1980) fig. 30, 32.						•
Jj-nfrt	PM 616	Kanawati–Abder-Raziq (2003) pl. 36.						•
Jj-nfrt	PM 298	Wiedemann–Pörtner (1906) Taf. 4-6; Schürmann (1983) Abb. 13 [a-c], 14 [a, b], 15 [a, b], 16 [a-c], 17 [a, b].						•
Jrj-n-kꜣ-Ptḥ	PM 644	Moussa–Junge (1975) pl. 9.				•	•	
		Moussa–Junge (1975) 41, pl. 13.						•
Jtj-sn	PM 252	Hassan (1944) fig. 123.						•
Wpm-nfr/Wp	PM 282	Hassan (1936) fig. 219.						•
Wr-jr.n.j	PM (iv) 188 No.25	Davies *Sheikh Said* (1901) pl. 16.				•	•	
		Davies *Sheikh Said* (1901) pl. 10, 12.						•
Pḥn-wj-kꜣ.j	PM 491 D70=LS15	LD II 47.				•		
Ftk-tꜣ	PM 351 LS 1	LD II 96; Bárta (2001) fig. 3.10, 3.16.						•
Mꜣ-nfr	PM 575 H2=LS17	LD II 66.						•
Nj-mꜣʿt-Rʿ	PM n/r G2097	Roth (1995) pl. 185.						•
Nj-kꜣw-ḥr	PM 498 S915	Quibell (1909) pl. 62 [2], 66 [2].	•	•	•		•	
Nfr-jrt-n.f	PM 583 D55	van de Walle (1978) pl. 13, 12, 20.						•
Nfr-bꜣw-Ptḥ	PM 169 G6010=LG15	LD II 56; Weeks (1994) fig. 11.		•			•	
Rʿ-m-kꜣ.j	PM 487 D3=S903	Hayes (1953) fig. 54.						•
Rʿ-špss	PM 494 LS16=S902	Vachala (2004) Taf. 4, 12.						•
Ḥtp-ḥr-ꜣḫtj	PM 593 D60	Mohr (1943) fig. 11-14, 31.						•
Ḫw-ns	PM (iv) 134 No.2	LD II 105 [b]; Varille (1938) fig. 3.						•
Ḫnwt	PM 623	Munro (1993) pl 36 [a], 38.		•				
Spd-ḥtp	PM 481 D15	Harpur *MDÄIK* 42 (1986) figs. 2, 6.	•	•	•			
Snḏm-jb/Jntj	PM 85 G2370=LG 27	LD II 77; LD Erg. 21; Brovarski (2001) fig. 42, 56-7.						•
Snḏm-jb/Mḥj	PM 87 G2378=LG 26	LD II 74 [a]; Brovarski (2001) fig. 116 [b].						•
Sḫm-ʿnḫ-Ptḥ	PM454 D41	Simpson (1976) fig. 8, pl. D.		•	•	•	•	
Sšm-nfr IV	PM 223 LG53	Junker (1953) Abb. 64.						•
Kꜣ.j- m- ʿnḫ	PM 131 G4561	Junker (1940) Taf. 2, 12;		•	•	•	•	
		Junker (1940) Taf. 8, 13; Kanawati (2001) pl. 35, 37[a].						•
Kꜣ.j-m-nfrt	PM 467 D23	Simpson (1992) pl. F.		•	•		•	
Kꜣ.j-m-rḥw	PM 485 D2=S905	Mogensen (1921) fig. 19, 24-6.		•	•	•	•	
		Mogensen (1921) fig. 29, 31-5; Borchardt (1937) Bl. 48 [CG 1534 A, B].						•
Kꜣ.j-dwꜣ	PM 244	Hassan (1950) fig. 80.						•
Kꜣpj	PM 69 G2091	Roth (1995) pl. 158.						•
Tp-m-ʿnḫ	PM 343	Borchardt (1907) fig. 103 [a].		•	•	•	•	
		Borchardt (1907) fig. 103 [a], 104 [a].						•
Ṯjj	PM 468 D22	Épron–Daumas, Fasc. I (1939), pl. 66-67,70-71.						•

TABLE 10: GRAIN STORAGE, BREAD AND BEER PRODUCTION

MONUMENT			GRAIN STORAGE					
TOMB OWNER	PM No. FIELD No.	REFERENCE	Grain Taken to Granary	Grain Taken from Stack	Granary Portrayed	Scribes Recording Cere	Crier Calling Quantities	Bread/Beer Production
DYNASTY VI.E								
Jttj-Šdw	PM (iv) 122	Petrie (1898) pl. 23i; Kanawati–McFarlane (1993) pl. 47[a].	•		•			
Jdw	PM 185 G7102	Simpson (1976) fig. 38.						•
ʿnḫ -m- ʿ-ḤrZzj	PM 512 D38	Badawy (1978) fig. 29; Kanawati–Hassan (1997) pl. 38.						•
Wnjs-ʿnḫ	PM n/r	Saleh (1977) fig. 6.						•
Mr.f-nb.f		Mysliwiec (2004) pls. 21, 66.						•
Mḫw	PM 619	Altenmüller (1998) [l. 43 [a].		•	•	•		
		Altenmüller (1998) pl 6, 16 [a, b].						•
Nj-k3w-Jzzj	PM n/r	Kanawati–Abder-Raziq (2000) pl. 48.						•
Ḥm-rʿ/Jzj (I)	PM (iv) 243 No.72	Davies (1902) pl. 20.						•
Ḥzj	PM n/r	Kanawati–Abder-Raziq (1999) pl. 61.						•
Ḫntj-k3.j/Jḫhj	PM 508	James–Apted (1953) pl. 9, 42.						•
Ṯmrrj	PM n/r Peck N248	Peck (1959) pl. 3.						•
DYNASTY VI.M								
Jbj	PM (iv) 243 No.8	Davies (1902) pl. 7.				•		
Jḥjj	PM n/r No.186	Saleh (1977) fig. 60-2.						•
Ppj-ʿnḫ-ḥrj-jb	PM (IV) 254 D2	Blackman (1924) pl. 13.						•
Ḥntj	PM n/r No.405	Saleh (1977) fig. 41-3, Colourplate A [c], pl. 11.						•
K3r	PM n/r L 31	Kanawati (1986) fig. 14.						•
DYNASTY VI.L								
Ppj-ʿnḫ/Ḥnj-km	PM (iv) 247 A2	Blackman–Apted (1953) pl. 30.						•
Ḥzjj-Mnw/Zzj	PM n/r G42	Kanawati (1987 fig. 4 [b].						•
Ḥnj/Šszj-pw-Mnw	PM n/r H24	Kanawati (1981) fig. 17.	•	•	•	•		
		Kanawati (1981) fig. 5.						•
K3(.j)-ḥp/Ṯtj-jkr	PM (v) 19 H26	Kanawati (1980) fig. 15.						•
DYNASTY VIII								
W3hj	PM (v) 36-7	Kanawati (1995) pl. 28, 31.	•		•			
Mrjj-ʿ3	PM (v) 34 D18	Kanawati (1995) pl. 36.	•		•			
		Kanawati (1995) pl. 46 [a], 36.						•
Ḥw-ns	PM (v) 235 A6	de Morgan (1894) 160.						•
Ghs3/Nbjj	PM n/r GA11	Kanawati (1987) fig. 26 [c].						•

TABLE 11: TITLES

TITLES RELATING TO FIELD CROPS

Jones [1] Page (number)	Transliteration	Translation
317 (1167)	*jrj mḏ3t*	letter carrier/scribal assistant
325 (1200)	*jrj ḫt*	custodian of property
486 (1818)	*nḫt-ḫrw šnwt*	crier of the granary
670 (2453)	*ḥk3 ḥwt*	estate manager/estate chief
668 (2447)	*ḥk3 njwwt*	chief of the towns/settlements
(no record)	*ḫrj ḏ3 n pr ḏt*	storekeeper of the estate[2]
852 (3112)	*sš n pr-ḏt*	scribe of the estate
875 (3203)	*sš šnwt*	scribe of the granary
992 (3670)	*šmsw n pr-ḏt*	retainer of the estate

TITLES RELATING TO LINEN

Jones Page (number)	Transliteration	Translation
234 (864)	*jmj-r3 sšr*	overseer cloth/linen/clothing distribution
235 (867)	*jmj-r3 sšr nswt* (*jmj-r3 sšr nb nswt*)	overseer of royal cloth/linen
236 (867)	*jmj-r3 sšr nswt mr/š? pr-ʿ3*	overseer of royal linen of the weaving/ weavery of the Great House/overseer of the court weaving of the king's linen
236 (868)	*jmj-r3 sšr pr*	overseer of cloth/linen of the house
237 (870)	*jmj-r3 sšr smsw*	senior? overseer of linen
279 (869)	*jmj-r3 sšr n pr-ʿ3*	overseer of royal cloth/linen of the Great House
336 (1238)	*jrj sšr*	keeper of the linen
336 (1239)	*jrj sšr pr-ʿ3*	keeper of the linen of the Great House
336 (1240)	*jrj sšr n pr-nswt*	keeper of the linen of the royal domain
649 (2377)	*ḥrj-tp sšr*	supervisor of cloth

1 Jones, D. (2001)
2 *Ḫntj-k3.j/Jḫḫj*, James (1953) 45.

TITLES RELATING TO LINEN (continued)

Jones Page (number)	**Transliteration**	**Translation**
649 (2378)	*ḥrj-tp ḏ3t*	supervisor of linen/the wardrobe
873 (3197)	*sš sšr n nswt*	scribe of the royal linen
873 (3198)	*sš sšr nswt ẖkrw nswt*	scribe of the royal linen of the king's regalia
914 (3358)	*sḥḏ jrjw sšr*	inspector of custodians of the *sšr*-linen/cloth
964 (3558)	*sḥḏ sš*(*w*) (*n*) *sšr nswt*	inspector of scribes of the royal linen
965 (3559)	*sḥḏ sš*(*w*) *szp*(*t*) *nswt*	inspector of scribes of the royal light cloth

TABLE 12: INSCRIPTIONS ACCOMPANYING THE TOMB OWNER VIEWING, PLANTING AND HARVESTING FIELD CROPS

MONUMENT			INSCRIPTION	
TOMB OWNER	**PM No. FIELD No.**	**REFERENCE**	**VERTICAL INSCRIPTION**	**TRANSLATION**
DYNASTY V.M				
Jj-mrjj	PM 170 G6020	Weeks (1994) figs. 39-40.	*mꜣꜣ skꜣ m njwwt.f* *nt pr-ḏt*	Viewing the ploughing in his towns of the estate
Nfr and *Kꜣ-ḥꜣ.j*	PM 639	Moussa–Altenmüller (1971) pl. 1, 8-9.	*mꜣꜣ ḥꜣmw ꜥḥw zšw n* *njw(w)t.f nt Tꜣ-mḥw* *Šmꜥw*	Viewing fishermen, fowlers and scribes of his town(s) of Lower and Upper Egypt
DYNASTY V.L				
ꜣḫtj-ḥtp (Louvre)	PM 635	Ziegler (1993) pl. pp. 70-71; fig. pp. 126, 129, 135-7.	*mꜣꜣ skꜣ ꜣzḫ sḫt kꜣt* *nb nt sḫt*	Viewing the ploughing, the reaping, the trapping of birds and all the work of the field
		pl. pp. 70-71; fig. pp. 126, 129, 135-7.	(*mꜣꜣ kꜣt*) *sḫt šdjt nfr* *mꜣꜣ r ḫt nb*	(Viewing the works) of the field, the reaping and transport. A sight more beautiful than anything.
Jꜣsn	PM 82 G2196	Simpson (1980) fig. 30.	*mꜣꜣ skꜣ*	Viewing the ploughing
Jj-nfrt	PM 299	Schürmann (1983) fig. 6-17, 21.	(*m*)*ꜣꜣ kꜣt sḫt jḫt nbt* *nfr.t jrr.t m sḫt*	Viewing the work of the field and all good things which are carried out in the field
Jrj-n-kꜣ-Ptḥ	PM 644	Moussa–Junge (1975) pl. 6-10.	*mꜣꜣ skꜣ wḥꜥ mḥꜥ.w* (*ꜣzḫ*) *šdjt*	Viewing ploughing, gathering flax, (reaping?) and transport
Wr-jr.n.j	PM (iv) 188 No.25	Davies (1901) pl. 7, 16.	*mꜣꜣ skꜣ* (*ḥwj mḥꜥ* *ꜣzḫ šdjt*) *ḥwj spwt* (*j*)*n ḥkꜣw* (*nt*) *pr-ḏt*	Viewing the ploughing (gathering of flax, reaping and transport) and the treading of the threshing floors by the headmen of the estate
Nbt	PM 624	Munro (1993) pl. 13.	*mꜣꜣ jrp rḏ.t m* *stp-zꜣ n* …	Viewing the wine, brought from the palace for ..
Nfr (1)	PM 137 G4761	Junker *Gîza* VI (1943) fig. 14.	*mꜣꜣ skꜣ ꜣzḫ jt bdt* *jn jzwt.f pr.f-ḏt*	Viewing the ploughing, the reaping of barley and emmer wheat by his workers of his estate
Nfr-jrt-n.f	PM 583	van de Walle, (1978) pl. 11.	*mꜣꜣ kꜣt nbt nt sḫt*	Viewing all the work of the field
Ḥtp-ḥr-ꜣḫtj	PM 593 D60	Mohr (1943) figs. 44-52, pls. i, iv-v.	*mꜣ*(*ꜣ*) *skꜣ ḥwj* *mḥ ꜣzḫ*	Viewing the ploughing, the gathering of flax and the reaping
Ḫw-ns	PM (iv) 134 No.2	LD II 106 (b), 107.	*mꜣꜣ skꜣ ḥwj mḥꜥ* *ꜣzḫ šdjt ḥwj spwt* (*j*)*n ḥkꜣw* (*nt*) *pr-ḏt*	Viewing the ploughing, gathering of flax, reaping, transport and the threshing floors by the headmen of the estate
Ḫnwt	PM 623	Munro (1993) pl. 38.	(*mꜣꜣ*) *rr*(*.w*) *m* *ḥwwt.s nwwt.s* *n*(*t*) *pr-ḏt*	(Viewing what is) brought from her estates and her towns of the estate
Sḫm-ꜥnḫ-Ptḥ	PM 454 D41	Simpson (1976) fig. 6-7, 9-10, pl. 9-16, pl. D.	*mꜣꜣ kꜣt sḫt skꜣ* *ꜣzḫ ḥwj mḥꜥw šdjt* *jꜥꜣw ḥwj jꜥꜣw* *spwt ḫꜣḫꜣ*	Viewing the work of the field, ploughing, reaping, gathering flax, donkey transport,donkeys treading the threshing floors and winnowing.
Sḫm-kꜣj	PM 53 G1029	Simpson (1980) fig.4.	*mꜣꜣ skꜣ ꜣzḫ jt jn* *jzwt.f nt njwwt.f* *nwt pr.f ḏ.t*	Viewing the ploughing and the reaping of barley by his crews of his towns of his estate.
Sšm-nfr IV	PM 223 LG53	Junker *Gîza* XI (1953) fig. 74b.	(*m*)*ꜣꜣ nḏt-ḥr jnnt* … *Tꜣ-mḫw Šmꜥw jn* ..	Viewing the presentations which are brought … from Lower and Upper Egypt by …
Kꜣ.j-m-nfrt	PM 208 LG 63	LD *Erg.* pl.32 [lower]; Badawy (1976) fig. 30 [a], pls. 34, 35-6.	(*mꜣꜣ*) *skꜣ*	(Viewing) the ploughing

REGISTERS (n/a = themes other than Agriculture)					
FIRST (HIGHEST)	SECOND	THIRD	FOURTH	FIFTH	**TOMB OWNER**
					DYNASTY V.M
AGRICULTURE Harvest Scribes; reaping	AGRICULTURE Harvest Bringing donkeys	AGRICULTURE Harvest: piling sheaves; transport of harvest	AGRICULTURE Ploughing	AGRICULTURE Hoeing; sheep	*Jj-mrjj*
n/a	n/a	n/a	AGRICULTURE Sowing; sheep		*Nfr* and *Kꜣ-ḥꜣ.j*
					DYNASTY V.L
AGRICULTURE Sheep	AGRICULTURE Harvest Flax	AGRICULTURE Harvest Cereals	AGRICULTURE Harvest Transport	AGRICULTURE Harvest Threshing	*ꜣḫtj-ḥtp* (Louvre)
AGRICULTURE Harvest Reaping; pulling flax	AGRICULTURE Harvest Donkey transport	AGRICULTURE Harvest Winnowing	AGRICULTURE Harvest Threshing	AGRICULTURE Harvest Storage; donkeys return	
AGRICULTURE Harvest Reaping; stacking	AGRICULTURE Harvest Transporting grain;	AGRICULTURE Ploughing sheep treading seed			*Jꜣsn*
n/a	AGRICULTURE Ploughing				*Jj-nfrt*
AGRICULTURE Ploughing	AGRICULTURE Harvest Reaping corn; flax	AGRICULTURE Harvest Transport; threshing	AGRICULTURE Harvest Donkeys; scribes	AGRICULTURE Harvest Scribes; winnowing grain storage	*Jrj-n-kꜣ-Ptḥ*
AGRICULTURE Ploughing; sheep	AGRICULTURE Harvest Scribes; pulling flax; Reaping cereal	AGRICULTURE Harvest Scribes; Piling sheaves; Bringing donkeys; Transport	AGRICULTURE Harvest Reaping; threshing		*Wr-jr.n.j*
AGRICULTURE Ploughing; sheep	n/a	n/a	n/a		*Nbt*
AGRICULTURE Boats; ploughing; flax harvest	n/a	n/a	AGRICULTURE sheep treading seed		*Nfr* (1)
AGRICULTURE Harvest	AGRICULTURE Ploughing; sheep	AGRICULTURE Harvest Flax gathering			*Nfr-jrt-n.f*
AGRICULTURE Ploughing; sheep	AGRICULTURE Harvest Cereals; flax	AGRICULTURE Harvest Transport	AGRICULTURE Harvest Threshing; storage	n/a	*Ḥtp-ḥr-ꜣḫtj*
AGRICULTURE Ploughing; sheep	AGRICULTURE HARVEST flax, corn harvest	AGRICULTURE HARVEST Donkey transport	AGRICULTURE HARVEST Threshing		*Ḫw-ns*
AGRICULTURE Harvest Reaping; storing	n/a	n/a			*Ḫnwt*
n/a	n/a	AGRICULTURE Scribes; sheep; ploughing	AGRICULTURE Harvest Scribe; gathering flax; reaping	AGRICULTURE *(Reg. 5, 6)* Harvest Donkey transport; threshing; storage	*Sḫm-ꜥnḫ-Ptḥ*
AGRICULTURE Scribes; gathering	AGRICULTURE ploughing; sheep treading seed	AGRICULTURE Harvest Gathering flax; reaping; donkeys	n/a	n/a	*Sḫm-kꜣj*
?	?	AGRICULTURE Harvest Reaping	AGRICULTURE Harvest Transport of harvest		*Sšm-nfr* IV
AGRICULTURE Harvest Sctibes; Threshing	AGRICULTURE Harvest Scribe; sheaves; transport	AGRICULTURE Ploughing	AGRICULTURE Sowing; sheep		*Kꜣ.j-m-nfrt*

TABLE 12: INSCRIPTIONS ACCOMPANYING THE TOMB OWNER VIEWING, PLANTING AND HARVESTING FIELD CROPS

MONUMENT			INSCRIPTION	
TOMB OWNER	**PM No. FIELD No.**	**REFERENCE**	**VERTICAL INSCRIPTION**	**TRANSLATION**
Kꜣ.j-m-nfrt	PM 467 D23	Simpson (1992) fig. 15, pls. 19-21, pls. F. G.	*mꜣꜣ skꜣ jt ḥj mḥꜥw ꜣsḫ šd … ḥj ḫꜣḫꜣ jꜥb*	Viewing the ploughing, gathering of flax, reaping, loading (donkeys), treading (of the threshing floors), winnowing, heaping up
Kꜣ-ḥj.f	PM 76	Junker *Gîza* VI (1943) fig. 42-3, 45-7, pl. x [b], 12-14 [a, b].	*(mꜣꜣ skꜣ) ꜣzḫ kꜣt nb nfrt jrrt m sḫt*	Viewing the ploughing and the harvesting and all the beautiful work that is done in the field
Ṯjj	PM 468 D22	Wild (fasc. III) (1966) pl. 136-139 [A, B]; 151-155.	*mꜣꜣ ḥwj mḥꜥ ꜣzḫ jt bdt šdjt wbs m spwt m njwwt nt pr-ḏt*	Viewing the pulling of flax, the reaping of barley and emmer wheat, the transporting, the heaping on the threshing floors from (by?) the towns of the estate
DYNASTY VI.E				
Mrw/Ttj-snb	PM 520	Lloyd (1990) pl. 12.	*(mꜣꜣ …) sḫt jrt (m) njwwt.f m ḥwwt.f nt Tꜣ-mḥw Šmꜥw*	(Viewing …) of the fields which are done in his towns and his estates of Lower and Upper Egypt
Mḥw	PM 619	Altenmüller (1998) pl. 22[A].	*mꜣꜣ spr ḥwꜥw ḥnꜥ kꜣr sḫt jwt ḥwwt.f njwwt.f nt Tꜣ-mḥw Šmꜥw*	Viewing the arrival of the fleet together with the work of the field and the coming of his estates and his towns of Lower and Upper Egypt
		fig. 8, pl. 36, 38.	*mꜣꜣ ḥꜥm mhjt ꜥꜣ wrt kꜣt sḫt sšmt mhjt n ṯzwt nt pr-ḏt*	Viewing the very great fishing and work of the field, the distribution of fish to the teams of the estate
Ḫnmw-ntj	PM 87 G2374	Brovarski (2001) fig. 88.	*mꜣꜣ ḫt nb r jnnt m … kꜣt*	Viewing every good (thing) which is brought from …
DYNASTY VI.M				
Jbj	PM (iv) 243 No.8	Davies *Gebrâwi* I (1902) pl. 12. Kanawati *Gebrawi* II (2007) pl. 51-52, 71.	*(mꜣꜣ) skꜣ ḥwj mḥꜥ ꜣ(zḫ) šdjt ḥb(w) nb(w) nfr(w nw) jt (šmꜥ) n kꜣ …*	Viewing the ploughing, the gathering of flax, the reaping, the grain transport and all good festivals of Upper Egyptian … grain for the ka (of)
DYNASTY VI.L				
Ppj-ꜥnḫ-ḥrj-jb	PM (iv) 254 D2	Blackman (1924) *Meir* 4, pl. 14.	*mꜣꜣ skꜣ m njwwt.f nt Mḥw Šmꜥ*	Viewing the ploughing in his towns of Lower and Upper Egypt
		Meir 4, pl. 14.	*mꜣꜣ ꜣzḫ jt bdt ḥwj mḥꜥ*	Viewing the reaping of barley and emmer wheat and the gathering of flax
Ppj-ꜥnḫ/Ḥnj-km	PM (iv) 247 A2	Blackman–Apted (1953) *Meir* V, pl. 30.	*m(ꜣꜣ) kꜣt nbt nfrt jrrt m sḫwt*	(Viewing) all the good work that is done in the fields of Upper and
Ḥnj/Špsj-pw-Mnw	PM n/r H24	Kanawati (1981) fig. 17.	*mꜣꜣ kꜣt sḫt nb(t) nfrt jn*	Viewing all the good work of the field by …
Kꜣ(.j)-ḥp/Ṯtj-jkr	PM (v) 19 H26	Kanawati (1980) fig. 11.	*(m)ꜣꜣ kꜣt zḫt nb(t) nfrt jn*	Viewing all the good work of the field by …
Ḏꜥw and *Ḏꜥw/Šmꜣj*	PM (iv) 244 No.12	Davies *Gebrâwi* II (1902) pl. 6.	*mꜣ(ꜣ) skꜣ ḥwj mḥꜥ ꜣzḫ šdjt ḥbw nb(w) nfrw nw jt šmꜥ*	Viewing the ploughing, the gathering of flax, the reaping and the transport and every good festival of Upper Egyptian grain
DYNASTY VIII				
Wꜣhj	PM n/r D4	Kanawati (1995) pl. 20, 22.	*mꜣꜣ (jrt jrw?) kꜣw ꜥwt nb m jšt.f ḏs.f*	Viewing (the making of a count?) of all cattle and goats from his own property
Mrjj-ꜥꜣ	PM (v) 34 D18	Kanawati (1995) pl. 38, 39.	*mꜣꜣ jrw jn* (TO) *m jšt.f ḏs.f m wn-mꜣꜥ nn (ꜥb)ꜥ jm*	Viewing the cattle count by (TO) from his own property in reality. There is no boasting therein.
Rḥw-r-ꜣw.sn	PM n/r BA17	Kanawati (1987) fig. 15.	*mꜣꜣ ꜥḥꜣ kꜣw jn*	Viewing the bull fight by …

REGISTERS (n/a = themes other than Agriculture)					
FIRST (HIGHEST)	SECOND	THIRD	FOURTH	FIFTH	**TOMB OWNER**
DESTROYED Sowing?	AGRICULTURE Harvest Gathering flax	AGRICULTURE Harvest Bundling flax	AGRICULTURE Harvest Reaping	AGRICULTURE (+ *regs 6, 7*) Harvest Grain transport; threshing; winnowing	*Kꜣ.j-m-nfrt*
AGRICULTURE Ploughing; sheep FISHING Dragnet	AGRICULTURE Harvest grain; flax; piling sheaves	AGRICULTURE Harvest Transport by donkeys	AGRICULTURE Harvest Threshing; winnowing		*Kꜣ-ḥj.f*
Regs 1, 2 AGRICULTURE Harvest Gathering flax	*Regs 3, 4, 5* AGRICULTURE Harvest Reaping grain	*Regs. 6, 7* AGRICULTURE Harvest Grain transport	*Reg. 8* AGRICULTURE Harvest Piling sheaves	*Regs. 9, 10* AGRICULTURE Harvest Cattle, donkeys threshing; storage	*Ṯjj*
					DYNASTY VI.E
AGRICULTURE Harvest Gathering flax	AGRICULTURE Harvest Reaping grain	n/a	n/a		*Mrw/Ttj-snb*
n/a	AGRICULTURE Harvest Reaping to winnowing	n/a			*Mḥw*
n/a	n/a	AGRICULTURE Ploughing; sheep			
n/a	AGRICULTURE Harvest Reaping, transport	MARSHES Clapenet	MARSHES Returning boatmen	ANIMAL PARADE Cattle	*Ḫnmw-ntj*
					DYNASTY VI.M
DESTROYED	DESTROYED	AGRICULTURE Harvest Gathering flax	AGRICULTURE Harvest Reaping cereal	AGRICULTURE Harvest Transport by donkeys; Threshing	*Jbj*
					DYNASTY VI.L
n/a	AGRICULTURE Ploughing	AGRICULTURE Ploughing	AGRICULTURE Ploughing	AGRICULTURE Ploughing	*Ppj-ʿnḫ-ḥrj-jb*
AGRICULTURE Harvest Gathering flax	Harvest Reaping grain	Harvest Loading donkeys; transport of grain	Harvest Threshing	Harvest Transport; stacking	
AGRICULTURE Ploughing	n/a	n/a	n/a		*Ppj-ʿnḫ/Ḥnj-km*
AGRICULTURE Ploughing; sowing	AGRICULTURE Harvest Reaping; transport	AGRICULTURE granaries; scribe			*Ḫnj/Špsj-pw-Mnw*
AGRICULTURE Ploughing; sowing	n/a	n/a	n/a		*Kꜣ(.j)-ḥp/Ṯtj-jkr*
AGRICULTURE	AGRICULTURE Ploughing; sowing	AGRICULTURE Harvest Flax	AGRICULTURE Harvest Grain		*Ḏʿw* and *Ḏʿw/Šmꜣj*
					DYNASTY VIII
n/a	n/a	n/a	n/a	AGRICULTURE Ploughing; sowing; harvest of grain	*Wꜣhj*
n/a	n/a	n/a	n/a	AGRICULTURE Harvest; Donkey transport; threshing	*Mrjj-ʿꜣ*
n/a	n/a	AGRICULTURE Ploughing			*Rḥw-r-ꜣw.sn*

TABLE 13: RESOURCES, FISH — FISHING METHODS AND MANAGEMENT

MONUMENT		
TOMB OWNER	PM No. FIELD No.	REFERENCE
DYNASTY IV.E		
Nfr…u	PM (iv) 94 No.22	Petrie (1892) pl. 28.
Nfr-m3ʿt	PM (iv) 92 No. 16	Petrie (1892) pl. 18; Harpur (2001) fig. 75.
Rʿ-ḥtp	PM (iv) 90 No. 6	Petrie (1892) pl. 11, 12; Harpur (2001) fig. 94, 95.
DYNASTY IV.L		
Mrt-jt.s	G7650	Smith (1946) fig.66
Nb.j-m-3ḫtj	PM 230 LG 86	LD II 12[b]; Hassan (1943) fig. 77.
Dbḥnj	PM 235 LG 90	LD Erg. 34; Hassan (1943) fig. 121.
DYNASTY V.E		
3ḫtj-ḥtp	PM 2001 G7650	Smith (1946) fig. 66.
Pr-sn	PM 48 LG 20-21	LD Erg. 8.
Sḫm-k3-Rʿ	PM 233 LG 89	LD II 42; Hassan 1943) fig. 401.
K3.j-ʿpr	PM 501	Fischer (1959) fig. 6; Barta (2001) fig. 4.14-16.
DYNASTY V.M		
Jj-mrjj	PM 170 G6020	LD Erg. pl. 6; Weeks (1994). fig.40.
Jj-nfrt	PM 298	Schürmann (1983) Abb. 9, 21.
Wr-jr.n-Ptḥ	PM 699f	Hall–Lambert (1922) pl. 12.
Ptḥ-špss	PM 340	Vachala (2004) p. 130-1; 134-5; 136-7.
Nj-ʿnḫ/Ḫnmw and *Ḫnmw-*	PM 641	Moussa–Altenmüller (1977) Abb. 12.
Nj-k3w-Ḥwt-ḥr	PM 247	Hassan (1950) fig. 168.
Nfr and *K3-ḥ3.j*	PM 639	Moussa–Altenmüller (1971) pl. 1, 4.
Rʿ-m-k3.j	PM 487	Hayes (1953) fig. 55.
Rʿ-ḫʿ.f-ʿnḫ	PM 207 G7948=LG 75	LD II pl. 9.
Ḫwfw-ḥ3.f II	PM 190 G7150	Simpson (1978) fig. 47.
DYNASTY V.L		
3ḫtj-mrw-nswt	PM 80 G2184	*HESPOK* fig. 239.
3ḫtj-ḥtp (Louvre)	PM 634	Ziegler (1993) pp. 133-4, 140, 142-3.
3ḫtj-ḥtp	PM 599 D64	Davies II (1901) pl. 8, 15.
Jj-nfrt	PM 616	Kanawati–Abder-Raziq (2003) pl. 37[b], 39.
Jntj	PM (iv) 121	Petrie (1898) pl. 5; Kanawati–McFarlane (1993) pl. 33.
Jrj-n-k3-Ptḥ	PM 644	Moussa–Junge (1975) pl. 11-12.
Jrw-k3-Ptḥ	PM 639	de Rachewiltz (1960) pl. 16, 17; McFarlane (2000) pl.
Jttj	PM 193 G7391	Badawy (1976) fig. 17.
Jdwt/Sšsšt (*Jḥjj* r/u)	PM 617	Macramallah (1935) pl. 5[D], 6-7; Kanawati– Abder-Raziq (2003) pl. 53 [f], 54.
Wr-jr.n.j	PM (iv) 188 No.25	Davies *Sheikh Saïd* (1901) pl. 12.
Pḥn-wj-k3.j	PM 491 D70 = LS15	LD II 46.
Ptḥ-ḥtp/Jj-n-ʿnḫ	PM 607	Hassan (1975) fig. 36-37
Ptḥ-ḥtp (II) (*Ṯfj*)	PM 600 D64	Paget–Pirie–Griffith (1898) pl. 32; Davies I (1900) pl.
Ptḥ-ḥtp/Ṯfw	PM 605	Hassan (1975) fig. 57.
Ftk-t3	PM 351 LS 1	LD II 40; Barta (2001) fig. 3.11, 2.12.
Nj-m3ʿt-Rʿ	PM n/r G2097	Roth (1995) pl.185.
Nj-m3ʿt-Rʿ	PM 282 G8900	Hassan (1936) fig. 240.
Nj-k3w-ḥr	PM 498 S915	Quibell (1909) pl. 62 [1].
Nb-j-k3w-ḥr/Jdw	PM 627	Hassan (1975) pl. 5, 24.
Nbt	PM 624	Munro (1993) pl. 10, 12.
Nfr (1)	PM137 G4761	Junker *Giza* VI, fig.14, 17.
Nfr-jrt-n.f	PM 583 D55	van de Walle (1978) pl.13.
Ḥzjj-Mnw	PM n/r M22	Kanawati (1983) fig. 12-13, 19.
Ḥtp-ḥr-3ḫtj	PM593 D60	Mohr (1943) fig. 29, 34.
Ḥtp-k3.j	PM 447 S3509	Martin (1979) pl. 16 [18].
Ḫww-wr	PM 254 LG95	LD II 43[a]; Hassan (1944) fig 104.

FISHING METHODS							PRESERVATION OF CATCH					TRANSPORT OF FISH				
Dragnet	Overseer/Dragnet	Weir: large	Weir: small	Basket Trap	Hand Net	Line and Hook	Gutting/Drying	Roe Extraction	Marsh Landscape	Near Dragnet	Brings Fish to Gutting	By Basket/Sack	By Pole or Hand	To Gutting	To Tomb Owner	**TOMB OWNER**
																DYNASTY IV.E
•																*Nfr…u*
•							•									*Nfr-mꜣʿt*
•							•		•				•			*Rʿ-ḥtp*
																DYNASTY IV.L
													•			*Mrt-jt.s*
							•				•		•	•	•	*Nb.j-m-ꜣḫtj*
•																*Dbḥnj*
																DYNASTY V.E
•													•			*ꜣḫtj-ḥtp*
•													•			*Pr-sn*
•																*Sḫm-kꜣ-Rʿ*
•																*Kꜣ.j-ʿpr*
																DYNASTY V.M
•							•	•	•	•			•		•	*Jj-mrjj*
•																*Jj-nfrt*
•							•					•	•			*Wr-jr.n-Ptḥ*
•			•				•									*Ptḥ-špss*
•	•	•	•	•	•	•	•	•	•	•		•	•		•	*Nj-ʿnḫ/Ḫnmw* and *Ḫnmw-*
•																*Nj-kꜣw-Ḥwt-ḥr*
•						•							•		•	*Nfr* and *Kꜣ-ḥꜣ.j*
							•		•		•	•	•	•		*Rʿ-m-kꜣ.j*
•	•						•						•			*Rʿ-ḥʿ.f-ʿnḫ*
•																*Ḫwfw-ḥꜣ.f* II
																DYNASTY V.L
•																*ꜣḫtj-mrw-nswt*
•							•			•		•	•		•	*ꜣḫtj-ḥtp* (Louvre)
					•	•										*ꜣḫtj-ḥtp*
•						•							•		•	*Jj-nfrt*
•							•				•		•	•		*Jntj*
•							•	•		•			•		•	*Jrj-n-kꜣ-Ptḥ*
•																*Jrw-kꜣ-Ptḥ*
•																*Jttj*
•					•	•						•				*Jdwt/Sššt* (*Jḥjj* r/u)
•							•		•							*Wr-jr.n.j*
•	•						•			•			•			*Pḥn-wj-kꜣ.j*
•							•									*Ptḥ-ḥtp/Jj-n-ʿnḫ*
			•		•		•									*Ptḥ-ḥtp* (II) (*Ṯfj*)
					•											*Ptḥ-ḥtp/Ṯfw*
•												•	•			*Ftk-tꜣ*
						•										*Nj-mꜣʿt-Rʿ*
•																*Nj-mꜣʿt-Rʿ*
•							•	•				•	•			*Nj-kꜣw-ḥr*
•							•	•					•			*Nb-j-kꜣw-ḥr/Jdw*
													•			*Nbt*
					•											*Nfr* (1)
•	•						•					•				*Nfr-jrt-n.f*
•						•						•	•			*Ḥzjj-Mnw*
•						•	•					•	•		•	*Ḥtp-ḥr-ꜣḫtj*
•																*Ḥtp-kꜣ.j*
•																*Ḫww-wr*

TABLE 13: RESOURCES, FISH — FISHING METHODS AND MANAGEMENT

MONUMENT		
TOMB OWNER	**PM No. FIELD No.**	**REFERENCE**
Ḫw-ns	PM (iv) 134 No.2	LD II 106; Varille (1938) fig. 4.
Ḫnwt	PM 632	Munro (1993) pl. 33, 34.
Spd-ḥtp	PM 481 D15	Borchardt (1964) Bl. 60 (CG 1671)
Snḏm-jb/Jntj	PM 85 G2370 = LG 27	LD Erg. 17; Brovarski (2001) fig. 24-5.
Snḏm-jb/Mḥj	PM 87 G2378 = LG 26	LD Erg, pl. 12, 14; Brovarski (2001) fig. 100-101, 114-16.
Srf-kꜣ.j	PM (iv) 188	Davies *Sheikh Saïd* (1901) pl. 5.
Sḫm-kꜣj	PM 53 G1029	Simpson (1980) fig. 4.
Sšm-nfr IV	PM 223 LG 53	LD II 80; Junker *Giza* 11 (1963) Abb. 66, 75.
Kꜣ.j- m- ꜥnḫ	PM 131 G4561	Junker (1940) Abb. 8; Kanawati (2001) pl. 31.
Kꜣ.j-m-nfrt II	PM 263	Hassan (1936) fig. 120.
Kꜣ.j-m-nfrt	PM 467 D23	Simpson (1993) pl. A, G.
Kꜣ.j-m-rḥw	PM 485 D2 = S905	Mogensen (1921) fig. 3.
Kꜣ.j-m-ḥzt	PM 542	Quibell–Hayter (1927) p. 17; McFarlane (2003) pl. 44.
Kꜣpj	PM 69 G2091	Roth (1995) pl.159.
Kꜣ-ḥj.f	PM 76	Junker *Giza* VI (1943) Abb. 42.
Tp-m-ꜥnḫ	PM 342-3	Borchardt (1907) p. 122.
Ṯjj	PM 468 D22	Wild, Fasc. II (1963) pl. 85, 90, 91, 117, 119, 123.
DYNASTY VI.E		
Jnw-Mnw	PM n.r.	Kanawati (2006) pl. 48.
Jn-Snfrw-jšt.f	PM 891 No 2	Borchardt (1964) pl. 103, 105.
ꜥnḫ -m- ꜥ-Ḥr	PM 512	Kanawati–Hassan (1997) pl. 43(a).
Wꜥtt-ḫt-ḥr	PM 535	Wreszinski III (1936) pl. 85, 92[D] 95; Kanawati–Abder-Raziq (2008) pl. 56.
Wnjs-ꜥnḫ	PM n.r. no. 413	Saleh (1977) fig. 10.
Wḏꜣ-ḥꜣTtj/Ššj/Nfr-sšm-Ptḥ	PM 515	Wreszinski, (1936) pl. 116[A].
Mrjj-Rꜥ-mrjj-Ptḥ-ꜥnḫ/Nḫbw	PM 89 G2381	Smith (1958) fig. 2.
Mrjj-Rꜥ-nfr/Kꜣr	PM 184 G7101	Simpson (1976) fig. 16, 19.
Mrjj-Ttj/Mrj	PM 536	Kanawati–Abder-Raziq (2004) pl. 45.
Mrw/Ttj-snb	PM 520	Lloyd (1990) pl. 8.
Mr.f-nb.f	PM n.r.	Mysliwiec (2004) pl. 21,66.
Mrrj	PM 518	Davies, (1984) pl. 5.
Mrrw-kꜣ	PM 525	Duell I (1938) pl. 10, 11, 13, 41-43, 45; Duell II (1938)
Mḥw	PM 619	Altenmuller (1998) pl. 20[b], 31, 32, 35.
Nj-ꜥnḫ-Ppjj	PM (iv) 37 L 14	LD II 111a,b; Varille (1938) pl. 5, 6.
Nj-kꜣw-Jzzj	PM n/r	Kanawati–Abder-Raziq (2000) pl. 47, 50.
Hnḳw/Jj …f (II)	PM (iv) 242 N67	Davies *Gebrâwi* II (1902) pl. 26; Kanawati *Gebrawi* I (2005) pl. 55.
Ḥzj	PM n/r	Kanawati–Abder-Raziq (1999) pl. 54, 55.
Kꜣ-gmnj	PM 521	von Bissing I (1905), pl. 17-19, 21; Harpur (2006) 15,
Twꜣw	PM n.r. N359	Peck (1959) pl. 1.
Ṯmrjj	PM (v) 28 N248	Peck (1959) pl. 6.
DYNASTY VI.M		
Jbj	PM (iv) 243 No.8	Davies *Gebrâwi* I (1902) pl 3-4, 5-6, 20, 21; Kanawati *Gebrawi* II (2007) pl. 48, 67.
Jdw I	PM (v) 111	Petrie (1900) pl. 5.
Ppj-ꜥnḫ-ḥrj-jb	PM (iv) 254 D2	Blackman (1924) pl. 8.
Ḫntj	PM n/r No.405	Saleh (1977) fig. 41, 45, 47-9, 54.
DYNASTY VI.L		
Ppj-ꜥnḫ/Ḥnj-km	PM (iv) 247 A2	Blackman–Apted (1953) pl. 24, 27, 30, 40, 57, 60.
Nj-ꜥnḫ-Ppjj-km	PM(iv) 247 A1	Blackman–Apted (1953) pl. 13.

FISHING METHODS							PRESERVATION OF CATCH					TRANSPORT OF FISH				
Dragnet	Overseer/Dragnet	Weir: large	Weir: small	Basket Trap	Hand Net	Line and Hook	Gutting/Drying	Roe Extraction	Marsh Landscape	Near Dragnet	Brings Fish to Gutting	By Basket/Sack	By Pole or Hand	To Gutting	To Tomb Owner	**TOMB OWNER**
•							•	•	•							*Ḫw-ns*
•						•	•			•						*Ḫnwt*
•							•									*Spd-ḥtp*
																Snḏm-jb/Jntj
•	•				•											*Snḏm-jb/Mḥj*
•																*Srf-k3.j*
•																*Sḫm-k3j*
•	•					•										*Sšm-nfr* IV
													•		•	*K3.j- m- ʿnḫ*
•																*K3.j-m-nfrt* II
•						•						•	•			*K3.j-m-nfrt*
•						•										*K3.j-m-rḥw*
•																*K3.j-m-ḥzt*
•							•		•							*K3pj*
•																*K3-ḥj.f*
•					•							•	•			*Tp-m-ʿnḫ*
•		•	•	•	•	•	•	•	•	•		•	•			*Ṯjj*
																DYNASTY VI.E
•	•															*Jnw-Mnw*
•												•				*Jn-Snfrw-jšt.f*
															•	*ʿnḫ -m- ʿ-Ḥr*
•													•			*Wʿtt-ḫt-ḥr*
•						•										*Wnjs-ʿnḫ*
						•						•	•			*Wḏ3-ḥ3Ttj/Ššj/Nfr-sšm-Ptḥ*
						•										*Mrjj-Rʿ-mrjj-Ptḥ-ʿnḫ/Nḫbw*
													•			*Mrjj-Rʿ-nfr/Ḳ3r*
													•		•	*Mrjj-Ttj/Mrj*
•																*Mrw/Ttj-snb*
•																*Mr.f-nb.f*
						•	•									*Mrrj*
•	•	•	•	•	•	•	•		•			•	•		•	*Mrrw-k3*
•		•		•		•						•	•			*Mḥw*
•																*Nj-ʿnḫ-Ppjj*
•	•					•										*Nj-k3w-Jzzj*
•	•															*Hnḳw/Jj …f* (II)
•	•	•			•	•										*Ḥzj*
•		•	•		•	•	•	•	•			•	•			*K3-gmnj*
•																*Tw3w*
•																*Ṯmrjj*
																DYNASTY VI.M
•		•			•	•	•		•			•	•			*Jbj*
•													•			*Jdw* I
•	•															*Ppj-ʿnḫ-ḥrj-jb*
•							•						•		•	*Ḫntj*
																DYNASTY VI.L
•	•				•	•						•	•		•	*Ppj-ʿnḫ/Ḥnj-km*
•	•				•											*Nj-ʿnḫ-Ppjj-km*

TABLE 13: RESOURCES, FISH — FISHING METHODS AND MANAGEMENT

MONUMENT		
TOMB OWNER	**PM No. FIELD No.**	**REFERENCE**
Ḫnj/Špsj-pw-Mnw	PM n/r H24	Kanawati (1981) fig. 22.
K3(.j)-hp/Ṯtj-jkr	PM (v) 19 H26	Kanawati (1980) fig. 12.
Ṯtj-ʿ3	PM n/r H41	Kanawati (1988) fig. 13 [a].
Ḏʿw and *Ḏʿw/Šm3j*	PM (iv) 244 No.12	Davies *Gebrâwi* II (1902) pl. 4, 5.
DYNASTY VIII		
Ghs3/Nbjj	PM (v) 34 D18	Kanawati (1995) pl. 35.
Mrjj-ʿ3	PM n/r GA11	Kanawati (1987) fig. 30.

FISHING METHODS							PRESERVATION OF CATCH					TRANSPORT OF FISH				
Dragnet	Overseer/Dragnet	Weir: large	Weir: small	Basket Trap	Hand Net	Line and Hook	Gutting/Drying	Roe Extraction	Marsh Landscape	Near Dragnet	Brings Fish to Gutting	By Basket/Sack	By Pole or Hand	To Gutting	To Tomb Owner	**TOMB OWNER**
•					•	•										*H̱nj/Špsj-pw-Mnw*
•					•	•										*Kꜣ(.j)-ḥp/Ṯtj-jkr*
•																*Ṯtj-ꜥꜣ*
•					•	•										*Ḏꜥw* and *Ḏꜥw/Šmꜣj*
																DYNASTY VIII
•																*Gḥsꜣ/Nbjj*
•																*Mrjj-ꜥꜣ*

TABLE 14: BIRDS — TRAPPING, POULTRY YARDS, PARADES

MONUMENT		
TOMB OWNER	PM No. FIELD No.	REFERENCE
DYNASTY IV.E		
Jtt	PM (iv) 93	Harpur (2001) fig. 81; Harpur (2001) fig. 82, 86.
Nfr…w	PM (iv) 94	Petrie 1892) pl. 28.
Nfr-m3ʿt	PM (iv) 93	Petrie (1892) pl. 18; Harpur (2001) fig. 75.
Nfr-m3ʿt	PM 183 G7060	LD II 17b.
Rʿ-ḥtp	PM (iv) 90	Petrie (1892) pl. 10; Harpur (2001) fig. 93.
DYNASTY IV.L		
3ḫj	PM 137 G4750	Junker *Gîza* I (1929) Abb. 57.
Mrs-ʿnḫ III	PM 197 G7530/40	Dunham– Simpson (1974) fig. 4.
Nb.j-m-3ḫtj	PM 230 LG86	LD II 12a, 14b.
DYNASTY V.E		
Sḫm-k3-Rʿ	PM 233 LG89	LD II 42a.
DYNASTY V.M		
Jj-mrjj	PM 170 G6020	LD II 50; Weeks (1994) fig.39, 31, 40.
Jj-nfrt	PM 299	Schürmann (1983) 8 [a,b], 21.
Wr-jr.n-Ptḥ	PM 699	Hall–Lambert (1922) pl. 9; James (1961) pl. 29.
Ptḥ-špss	PM 340	Vachala (2004) pp. 130-1, 166; Verner (1986) pl. 38-40.
Nj-ʿnḫ/Ḫnmw and *Ḫnmw-ḥtp*	PM 641	Moussa–Altenmüller (1977) Abb. 12, 24 [a].
Nfr and *K3-ḥ3.j*	PM 639	Moussa–Altenmüller (1971) pl 1, 5-6.
Rʿ-m-k3.j	PM 487	Hayes (1953) fig. 55.
Rʿ-ḥtp	PM n/r C5	El-Khouli–Kanawati (1990) pl. 73a
Rʿ-ḥʿ.f-ʿnḫ	PM 207 G7948	LD II 9, 10b.
Ḫwfw-ḥ3.f II	PM 190 G7150	Simpson (1978) fig. 47.
Sšm-nfr II	PM146 G5080	Kanawati (2002) pl. 62.
K3.j-m-nfrt	PM 208 LG 63	Badawy (1976) fig. 29; LD II, 91[c].
DYNASTY V.L		
3ḫtj-ḥtp (Louvre)	PM 634	Ziegler (1993) 132-4 (drawing).
3ḫtj-ḥtp	PM 599 D64	Davies (1901) pl. 4, 5, 8.
3ḫtj-ḥtp	PM 633 E17	Petrie–Murray (1952) pl. 7.
Jj-nfrt/Š3.n.f	PM 616	Kanawati–Abder-Raziq (2003) pl. 38.
Jj-nfrt	PM 299	Schürmann (1983) Taf. 8 [a, b], 21.
Jntj	PM (iv) 121	Petrie (1898) pl. 5; Kanawati–McFarlane (1993) pl. 33.
Jrj-n-k3-Ptḥ	PM 644	Moussa–Junge (1975) pl. 13.
Jrw-k3-Ptḥ	PM 639	McFarlane (2000) pl. 46.
Jḫjj (r/u *Sšsšt/Jdwt*)	PM 617	Macramallah (1935) pl. 11[B]; Kanawati–Abder-Raziq (2003) pl. 62.
Jtj-sn	PM 252-3	Hassan V (1944) fig. 123-4.
Jttj	PM 193 G7391	Badawy (1976) fig. 17.
Wr-jr.n.j	PM (iv) 188 No.25	Davies (1901) pl. 12.
Pḥn-wj-k3.j	PM 491 D70=LS 15	LD II 46.
Ptḥ-ḥtp I	PM 596 D62	Petrie–Murray (1905) pl. 11.
Ptḥ-ḥtp/Jj-n-ʿnḫ	PM 606	Hassan (1975) fig. 36-7.
Ptḥ-ḥtp (II) (*Ṯfj*)	PM 600 D64	Paget–Pirie–Griffith pl 32; Davies (1900) pl. 21, 25, 26, 31.
M3-nfr	PM 575	LD II pl.69-70.
Nj-m3ʿt-Rʿ	PM 283	Hassan II (1936) fig. 240.
Nj-m3ʿt-Rʿ	PM 70 G2097	Roth (2001) pl.189.
Nfr	PM 259 LG 99	Hassan III (1941) fig. 175.
Nfr (1)	PM 137 G4761	Junker *Gîza* VI (1943), figs. 14, 16, 17.
Nfr-jrt-n.f	PM 583	van der Walle (1978) pls, 13, 21.
Rʿ-wr II	PM 162 G5470=LG 32	Junker *Gîza* III (1938) fig. 48; LD Erg. 25.
Rʿ-špss	PM 494 LS 16	LD II 60/2/4a; LD Erg. 41.

HANDLING CLAPNET			HANDLING TRAPPED BIRDS				FOWL YARD			
No. clapnet scenes	Signaller	Decoy bird	Birds taken to cage	Transport of caged birds	Bird(s) presented to tomb owner	Birds plucked/ cooked by workers	Fowl- yard scene	Force feeding birds	Parade of Birds	TOMB OWNER
										DYNASTY IV.E
1										*Jtt*
1										*Nfr…w*
2					•					*Nfr-m3ʿt*
									•	*Nfr-m3ʿt*
1										*Rʿ-ḥtp*
										DYNASTY IV.L
									•	*3ḫj*
1	•		•			•			•	*Mrs-ʿnḫ* III
2	•		•						•	*Nb.j-m-3ḫtj*
										DYNASTY V.E
1	•		•	•						*Sḫm-k3-Rʿ*
										DYNASTY V.M
1	•		•						•	*Jj-mrjj*
2	•									*Jj-nfrt*
1	•	•								*Wr-jr.n-Ptḥ*
1							•	•	•	*Ptḥ-špss*
1	•				•				•	*Nj-ʿnḫ/H̱nmw* and *H̱nmw-ḥtp*
1	•		•		•	•			•	*Nfr* and *K3-ḥ3.j*
1	•	•							•	*Rʿ-m-k3.j*
			•							*Rʿ-ḥtp*
1	•								•	*Rʿ-ḥʿ.f-ʿnḫ*
1	•									*Ḫwfw-ḥ3.f* II
			•							*Sšm-nfr* II
									•	*K3.j-m-nfrt*
										DYNASTY V.L
1	•	•		•						*3ḫtj-ḥtp* (Louvre)
1	•	?	•							*3ḫtj-ḥtp*
									•	*3ḫtj-ḥtp*
1	•				•					*Jj-nfrt/Š3.n.f*
1	•									*Jj-nfrt*
1	•									*Jntj*
1		•								*Jrj-n-k3-Ptḥ*
1	•									*Jrw-k3-Ptḥ*
			•							*Jḥjj* (r/u *Sšsšt/Jdwt*)
1	•									*Jtj-sn*
									•	*Jttj*
1	•		•							*Wr-jr.n.j*
1			•		•					*Pḥn-wj-k3.j*
1	•	•			•					*Ptḥ-ḥtp* I
1										*Ptḥ-ḥtp/Jj-n-ʿnḫ*
1	•	•	•	•	•				•	*Ptḥ-ḥtp* (II) (*Ṯfj*)
									•	*M3-nfr*
1	•									*Nj-m3ʿt-Rʿ*
									•	*Nj-m3ʿt-Rʿ*
									•	*Nfr*
1	•	?							•	*Nfr* (1)
1		•								*Nfr-jrt-n.f*
									•	*Rʿ-wr* II
									•	*Rʿ-špss*

TABLE 14: BIRDS — TRAPPING, POULTRY YARDS, PARADES

MONUMENT		
TOMB OWNER	PM No. FIELD No.	REFERENCE
Ḥtp-ḥr-ꜣḫtj	PM 593 D60	Mohr (1943) figs. 24-5.
Ḥtp-kꜣ.j/Tp-kꜣ.j	PM 447 S3509	Martin (1979), pls. 16 [18], 17 [19].
Ḫww-wr	PM 254 LG 96	LD II 43a.
Ḫw-ns	PM (iv) 134 No.2	LD II 105b; Varille (1938) fig. 3.
Ḫnwt	PM 623	Munro (1993) pl. 33.
Spd-ḥtp	PM 481-2 D15	Borchardt(1964) pl. 60; Harpur (1986) fig. 1, 5-6.
Snḏm-jb/Jntj	PM 85 G2370=LG 27	LD II 77: Brovarski (2001)50-51, fig. 42.
Snḏm-jb/Mḥj	PM 87 G2378 = LG 26	LD II 74[b]; LD *Erg.* 14; Brovarski (2001) figs. 110, 114.
Sḫm-ꜥnḫ-Ptḥ	PM 191 G7152	Badawy, figs. 20, 23.
Sḫm-ꜥnḫ-Ptḥ	PM 454 D41	Simpson (1976) pl. D.
Sḫm-kꜣj	PM 53 G1029	Simpson (1980) fig. 4.
Sšm-nfr III	PM 153 G5170	Brunner-Traut (1977) Taf. 6.
Sšm-nfr IV	PM 223 LG 53	Junker *Gîza* XI (1953) Abb. 91.
Kꜣ.j- m- ꜥnḫ	PM 131; G4561	Junker *Gîza* IV (1940) Taf. 7; Kanawati (2001) pl. 37.
Kꜣ.j-m-nfrt	PM 467 D23	Simpson (1992) pl. 22, pl. G.
Kꜣ.j-m-rḥw	PM 485 D2=S905	Mogensen (1921) fig. 17.
Kꜣ.j-dwꜣ	PM 244	Hassan VI [3] (1950) fig. 82.
Kꜥpj	PM 69 G2091	Roth (2001) pl. 167.
Kꜣ-pw-Rꜥ	PM 455 D39	Dam, *The Museum Journal, Pennsylvania* 18 [2] (1927) pl. on p. 192.
Kꜣ-ḥj.f	PM 76	Junker *Gîza* VI (1943) Abb. 40.
Kd-nfr	PM 56 G1151	Smith (1946) pl. 59 [b].
Ṯjj	PM 468 D22	Épron–Daumas (1939) pls. 6, 19; Wild II (1953) pl. 87, 120-122; Wild III (1966) pls. 114, 169.
DYNASTY VI.E		
Jnw-Mnw	PM n/r	Kanawati (2006) pl. 48.
Jttj-Šdw	PM (iv) 122	Petrie (1898) pl. 15. Kanawati–McFarlane (1993) pl. 46.
ꜥnḫ -m- ꜥ-Ḥr	PM 512	Badawy (1978) fig. 33; Kanawati–Hassan (1997) pl. 42.
Wḏꜣ-ḥꜣ-Ttj/Ššj/ Nfr-sšm-Ptḥ	PM 515	Wreszinski *Atlas* III (1936) pls. 80-81; Capart (1907) pl. 85.
Mrjj-Ttj/Mrj	PM 536	Kanawati–Abder-Raziq (2004) pl. 28.
Mrw/Ttj-snb	PM 520	Lloyd et al (1990) pl. 8.
Mr.f-nb.f	PM n.r. (2004)	Mysliwiec (2004) pl. 22, 47, 67, 70[a, d].
Mrrj	PM 518	Davies et alI (1984) pl. 6 [A].
Mrrw-kꜣ	PM 525	Duell I (1938), pl. 52, 53 [B].
Mḥw	PM 619	Altenmüller (1998) pls. 6-8, 31, 34.
do		
do		
do		
do		
Mṯtj	PM 646	Ziegler (1990) 128, 143, 144, 151.
Nj-ꜥnḫ-Ppjj	PM (iv) 137 L14	Varille (1938) pls. 5, 6.
Rmnj/Mrwj	PM n/r	Kanawati (2009) pl. 45.
Nj-kꜣw-Jzzj	PM n/r	Kanawati–Abder-Raziq (2000) pl. 49.
Ḥm-rꜣ/Jzj (I)	PM (iv) 243 N72	Davies *Gebrâwi* II (1902) pl. 15, 22; Kanawati *Gebrawi* I (2005) pl. 50.
Ḥzj	PM n/r	Kanawati–Abder-Raziq (1999) pl. 55.
Ḫnmw-ntj	PM 87 G2374	Brovarski (2000) fig. 88.
Sꜥnḫ-wj-Ptḥ	PM n/r	Kanawati–Abder-Raziq (1998) pl. 69, 75.
Kꜣ-gmnj	PM 521	von Bissing I (1905) pls. 8-10.
Twꜣw	PM n/r N 359	Peck (1959) pl. I.

HANDLING CLAPNET			HANDLING TRAPPED BIRDS				FOWL YARD			
No. clapnet scenes	Signaller	Decoy bird	Birds taken to cage	Transport of caged birds	Bird(s) presented to tomb owner	Birds plucked/ cooked by workers	Fowl- yard scene	Force feeding birds	Parade of Birds	**TOMB OWNER**
1			•						•	*Ḥtp-ḥr-3ḫtj*
									•	*Ḥtp-k3.j/Tp-k3.j*
1	•	•								*Ḫww-wr*
1	•		•		•					*Ḫw-ns*
1	•	•								*Ḫnwt*
1							•	•	•	*Spd-ḥtp*
1									•	*Snḏm-jb/Jntj*
1	•		•	•					•	*Snḏm-jb/Mḥj*
									•	*Sḫm-ʿnḫ-Ptḥ*
1	•		•							*Sḫm-ʿnḫ-Ptḥ*
1		•		•						*Sḫm-k3j*
			•							*Sšm-nfr* III
1	•		•	•						*Sšm-nfr* IV
									•	*K3.j- m- ʿnḫ*
1	•	?		•					•	*K3.j-m-nfrt*
				•					•	*K3.j-m-rḥw*
									•	*K3.j-dw3*
									•	*Kʿpj*
									•	*K3-pw-Rʿ*
									•	*K3-ḥj.f*
									•	*Kd-nfr*
1	•	•	•	•			•	•	•	*Ṯjj*
										DYNASTY VI.E
1	•			by boat						*Jnw-Mnw*
1										*Jttj-Šdw*
2	•	•	•		•					*ʿnḫ -m- ʿ-Ḥr*
1	•	•	•				•	•		*Wḏ3-ḥ3-Ttj/Ššj/ Nfr-sšm-Ptḥ*
							•	•		*Mrjj-Ttj/Mrj*
1										*Mrw/Ttj-snb*
									•	*Mr.f-nb.f*
							•	•	•	*Mrrj*
							•	•		*Mrrw-k3*
Reg. 5	•	•	•	•	•					*Mḥw*
Reg. 4		•								do
Reg. 3	•	•								do
Reg. 2	•	•								do
1(top)	•	•								do
2		•	•	•	•		•	•		*Mttj*
1										*Nj-ʿnḫ-Ppjj*
							•			*Rmnj/Mrwj*
							•			*Nj-k3w-Jzzj*
			•						•	*Ḥm-r3/Jzj* (I)
1	•	•					•			*Ḥzj*
1	•									*Ḫnmw-ntj*
3	•	•	•							*Sʿnḫ-wj-Ptḥ*
R	•	•					•	•		*K3-gmnj*
1										*Tw3w*

TABLE 14: BIRDS — TRAPPING, POULTRY YARDS, PARADES

MONUMENT		
TOMB OWNER	PM No. FIELD No.	REFERENCE
DYNASTY VI.M		
Jbj	PM (iv) 243 No.8	Davies *Gebrâwi* I (1902) pl. 6.
Jhjj	PM (i): 291 No.186	Saleh (1977) fig. 61-2.
Jdw I	PM (v): 111 n.r.	Petrie (1900) pl. 5.
Ppj-ˁnḫ-ḥrj-jb	PM (iv) 254 D2	Blackman (1924) pl. 8.
DYNASTY VI.L		
Ppj-ˁnḫ/Ḥnj-km	PM (iv) 247	Blackman–Apted (1953) pl. 22 [2].
Nj-ˁnḫ-Ppjj	PM (iv) 247 A1	Blackman–Apted (1953) pl. 13.
Ḫnj/Špsj-pw-Mnw	PM n/r H24	Kanawati (1981) fig. 22.
K3(.j)-hp/Ṯtj-jkr	PM (v) 19 H26	Kanawati (1980) fig. 12.
Ṯtj-ˁ3	PM n/r H41	Kanawati (1988) fig. 13[a].
Ḏˁw and *Ḏˁw/Šm3j*	PM (iv) 244 No.12	Davies *Gebrâwi* II (1902) pl. 4.
DYNASTY VIII		
Mrjj-ˁ3	PM (v) 34 D18	Kanawati (1995) pl. 35.
Ḥw-ns	PM (v) 235	de Morgan (1894) p. 160.
Gḥs3/Nbjj	PM n/r GA11	Kanawati VII (1987) fig. 30.

HANDLING CLAPNET			HANDLING TRAPPED BIRDS				FOWL YARD			
No. clapnet scenes	Signaller	Decoy bird	Birds taken to cage	Transport of caged birds	Bird(s) presented to tomb owner	Birds plucked/ cooked by workers	Fowl- yard scene	Force feeding birds	Parade of Birds	TOMB OWNER
										DYNASTY VI.M
1	•	?					•			*Jbj*
1										*Jḥjj*
1	•									*Jdw* I
2	•	•	•		•	•				*Ppj-ꜥnḫ-ḥrj-jb*
										DYNASTY VI.L
1	•	•	•		•	•		•		*Ppj-ꜥnḫ/Ḥnj-km*
2	•	•								*Nj-ꜥnḫ-Ppjj*
1	?	•								*Ḫnj/Špsj-pw-Mnw*
1	•	•	•							*Kꜣ(.j)-ḥp/Ṯtj-jkr*
1	•				•					*Ṯtj-ꜥꜣ*
1		•								*Ḏꜥw* and *Ḏꜥw/Šmꜣj*
										DYNASTY VIII
1	•									*Mrjj-ꜥꜣ*
1										*Ḫw-ns*
1			•							*Gḥsꜣ/Nbjj*

TABLE 15: PAPYRUS GATHERING SCENES

MONUMENT		
TOMB OWNER	**PM No. FIELD No.**	**REFERENCE**
DYNASTY IV.E		
Nfr-m3ꜥt	PM 183 G7060	LD II 17.
DYNASTY IV.L		
Mr-jb.j	PM 71 LG 24	LD II 19.
Mrs-ꜥnḫ III	PM 197 G7530/40	Dunham–Simpson (1974) fig. 4.
Nb.j-m-3ḫtj	PM 230 LG 86	LD II 12; Hassan IV (1943) fig. 77.
Dbḥnj	PM 235 LG 90	LD II 35.
DYNASTY V.E		
Pr-sn	PM 48 LG 20-21	Petrie-Murray (1952) pl. 8; Harpur (1985) fig. 7.
DYNASTY V.M		
Jj-mrjj	PM 170 G6020	LD II 50a; LD Erg. pl. 5; Weeks (1994) fig. 31, 33.
Wr-jr.n-Ptḥ	PM 699	James (1961) pl. 30.
Ptḥ-špss	PM 340	Vachala (2004) pp. 132-3.
Nj-ꜥnḫ/Ḫnmw and *Ḫnmw-ḥtp*	PM 641	Moussa–Altenmüller (1977) pl. 76, fig. 17.
Nfr and *K3-ḥ3.j*	PM 639	Moussa-Altenmüller (1971) pl. 1, 5, 6.
Sšm-nfr II	PM 146 G5080	Kanawati (2002) pl. 62.
DYNASTY V.L		
3ḫtj-ḥtp (Louvre)	PM 634	Ziegler (1993) drawing 131.
3ḫtj-ḥtp	PM 599 D64	Davies *Ptahhetep* II (1901) pl. 13-14, 19, 21.
Jj-nfrt	PM 616	Kanawati–Abder-Raziq (2003) pl. 35.
Jntj	PM (iv) 121	Petrie (1898) pl. 5; Kanawati-McFarlane (1993) pl. 33.
Jrj-n-k3-Ptḥ	PM 644	Moussa-Junge (1975) pl. 11, 13.
Jḥjj (r/u *Sšsšt/Jdwt*)	PM 617	Macramallah (1935) pl. 6-7; Kanawati–Abder-Raziq (2003) pl. 54, 71.
Wr-jr.n.j	PM (iv) 188 No.25	Davies *Sheikh Said* (1901) pl. 12.
Pr-nb	PM 497 S913	Hayes (1953) fig. 51.
Ptḥ-ḥtp	PM 653 LS 31	LD II 102, 104a.
Ptḥ-ḥtp (II) (*Ṯfj*)	PM 600 D64	Paget–Pirie (1898) pl.3; Davies *Ptahhetep* I (1900) pl. 3, 21, 23.
M3-nfr	PM 575	LD II 69, 70.
Nb-j-k3w-ḥr/Jdw	PM 627	Hassan I (1975) pl. 24[A].
Nbt	PM 624	Munro (1993) pl. 11, 17, 19.
Nfr-jrt-n.f	PM 583	van de Walle (1978) fig. 10.
Nfr-b3w-Ptḥ	PM 169 G6010	Weeks (1994) fig. 18.
Nṯr-wsr	PM 485	Petrie–Murray (1952) pl. 22.
Ḥzjj-Mnw	n/r M22	Kanawati (1983) fig, 17, 19.
Ḥtp-ḥr-3ḫtj	PM 593 D60	Mohr (1943) fig. 27, 33.
Ḫw-ns	PM (iv) 134 No.2	LD II 105a,b, 106a; Varille (1938) fig. 4.
Ḫnwt	PM 623	Munro (1993) pl. 33.
S3-jb	PM n/r G2092, G2093	Roth (1995) pl. 181.
Snḏm-jb/Jntj	PM 85 G2370=LG 27	LD Erg. pl. 20; Brovarski (2001) fig. 38-8.
Sḫm-k3j	PM 53 G1029	Simpson (1980) fig. 4.
Sšm-nfr IV	PM 223 LG53	Junker *Giza* XI (1953) fig. 78, 79.
Štwj	PM 106	Junker *Giza* XI (1953) fig. 86.
K3.j-m-nfrt	PM 208f LG 63	LD Erg. pl. 32; Badawy, (1976) fig. 30.
K3.j-m-nfrt	PM 467 D23	Simpson (1992) fig. 6, pl. A.
K3pj	PM 69 G2091	Roth (1995) pl. 157, 168.
Tp-m-ꜥnḫ	PM 342	Borchardt (1907) fig. 104b.
Ṯjj	PM 468 D22	Wild II (1953) pl.75[A]; Wild III (1966) pl. 147.

GATHERING				STALKS			USE			AS CATTLE FEED			
Gathering papyrus	Plucking	Tying bundles	Carrying	Stalks shown in bundles	Stalks not shown in bundles	Overseer supervising	Bringing to boat building	Bringing to rope making	Bringing to tomb owner	Papyrus carried by herder	Papyrus brought by skiff	Cattle feeding on papyrus	TOMB OWNER
													DYNASTY IV.E
										•			*Nfr-m3ʿt*
													DYNASTY IV.L
										•			*Mr-jb.j*
										?			*Mrs-ʿnḫ* III
•			•	•			•						*Nb.j-m-3ḫtj*
										•			*Dbḥnj*
													DYNASTY V.E
•			•		•				?				*Pr-sn*
													DYNASTY V.M
•	•		•		•	•			?	•			*Jj-mrjj*
										•			*Wr-jr.n-Ptḥ*
•													*Ptḥ-špss*
										•			*Nj-ʿnḫ/Ḫnmw* and *Ḫnmw-ḥtp*
•										•			*Nfr* and *K3-ḥ3.j*
										•			*Sšm-nfr* II
													DYNASTY V.L
•	•		•	•				•					*3ḫtj-ḥtp* (Louvre)
•	•		•		•				•	•			*3ḫtj-ḥtp*
										•			*Jj-nfrt*
•	•		•	•			•	•					*Jntj*
•			•		•		•						*Jrj-n-k3-Ptḥ*
•	•	•	•		•		•			•			*Jḥjj* (r/u *Sšsšt/Jdwt*)
•	•		•	•			?						*Wr-jr.n.j*
										•			*Pr-nb*
										•			*Ptḥ-ḥtp*
•	•	•	•		•				•	•			*Ptḥ-ḥtp* (II) (*Ṯfj*)
										•			*M3-nfr*
•													*Nb-j-k3w-ḥr/Jdw*
•										•			*Nbt*
										•			*Nfr-jrt-n.f*
										•			*Nfr-b3w-Ptḥ*
										•			*Nṯr-wsr*
•	•						?						*Ḥzjj-Mnw*
•	•						•			•			*Ḥtp-ḥr-3ḫtj*
•			•	?			?			•	•		*Ḫw-ns*
•													*Ḫnwt*
												•	*S3-jb*
•	•		•		•		•						*Snḏm-jb/Jntj*
•			•		•				•		•		*Sḫm-k3j*
										•			*Sšm-nfr* IV
										•			*Štwj*
										•			*K3.j-m-nfrt*
•	•		•		•			?					*K3.j-m-nfrt*
•										•			*K3pj*
•			•	•									*Tp-m-ʿnḫ*
•			•	•					•	•			*Ṯjj*

TABLE 15: PAPYRUS GATHERING SCENES

MONUMENT		
TOMB OWNER	**PM No. FIELD No.**	**REFERENCE**
DYNASTY VI.E		
Jnw-Mnw	PM n/r	Kanawati (2006) pl. 49.
Mrw/Bbj	PM (iv) 189 No.20	Davies *Sheikh Said* (1901) pl. 19.
Mḥw	PM 619	Altenmüller (1998) pl. 14.
Nj-kꜣw-Jzzj	PM n/r	Kanawati–Abder-Raziq (2000) pl. 47, 49.
Ḥm-Rꜥ/Jzj (I)	PM (iv) 243 N72	Kanawati *Gebrawi* I (2005) pl. 64.
Ḥzj	PM n/r	Kanawati–Abder-Raziq (1999) pl. 65b.
Sšm-nfr/Jwfj	PM 614	Barsanti (1900) fig. 9.
DYNASTY VI.M		
Jbj	PM(iv) 243 No.8	Davies Gebrawi I (1902) pl. 6; Kanawati (2007) pl. 48.
Jḥjj	PM n/r No.186	Saleh (1977) fig. 30.
Ḫw.n-wḫ	PM(iv) 243 No. 2	El-Khouli-Kanawati (1989) pl. 37-8.
Ḳꜣr	PM n/r L31	Kanawati (1986) fig. 14.
DYNASTY VI.L		
Ppj-ꜥnḫ/Ḥnj-km	PM(iv) 247	Blackman–Apted (1953) pl. 1, 14, 30, 37-8.
Nj-ꜥnḫ-Ppjj-km	PM (iv) 247 A1	Blackman–Apted (1953) pl. 11.
Nbt	PM n/r H27	Kanawati (1982) fig. 27.
Ḥzjj-Mnw/Zzj	PM n/r G42	Kanawati (1987) fig. 6.
Ḏꜥw and *Ḏꜥw/Šmꜣj*	PM (iv) 244 No.12	Davies *Gebrâwi* II (1902) pl. 5.
DYNASTY VIII		
ꜥnḫ-tj.fj	PM (v) 170	Vandier (1950) fig. 26.

GATHERING				STALKS			USE			AS CATTLE FEED			
Gathering papyrus	Plucking	Tying bundles	Carrying	Stalks shown in bundles	Stalks not shown in bundles	Overseer supervising	Bringing to boat building	Bringing to rope making	Bringing to tomb owner	Papyrus carried by herder	Papyrus brought by skiff	Cattle feeding on papyrus	**TOMB OWNER**
													DYNASTY VI.E
										•			*Jnw-Mnw*
										•			*Mrw/Bbj*
•													*Mḥw*
										•	•		*Nj-kꜣw-Jzzj*
											•		*Ḥm-Rꜥ/Jzj* (I)
										•			*Ḥzj*
										•			*Sšm-nfr/Jwfj*
													DYNASTY VI.M
•			•		•				•				*Jbj*
										•			*Jḥjj*
•													*Ḫw.n-wḫ*
•			•		•				•				*Kꜣr*
													DYNASTY VI.L
•			•		•				•		•		*Ppj-ꜥnḫ/Ḥnj-km*
										•			*Nj-ꜥnḫ-Ppjj-km*
										•			*Nbt*
										•			*Ḥzjj-Mnw/Zzj*
											•		*Ḏꜥw* and *Ḏꜥw/Šmꜣj*
													DYNASTY VIII
										•			*ꜥnḫ-tj.fj*

TABLE 16: PAPYRUS SKIFFS — CONSTRUCTION AND USE

MONUMENT			SKIFF/S Construction			ROPE		USE OF SKIFF/RAFT [R = Raft]					
TOMB OWNER	PM NO. FIELD NO.	REFERENCE	Skiff building	Bringing papyrus to boat	2/3 skiffs being built	Spare rope coils	Rope making	Tomb owner on skiff	Fishing from craft	Hippo hunt	Boatmen return peacefully	Jousting boatmen	Craft with herd crossing water
DYNASTY IV.E													
Jtt	PM (iv) 93 No. 16	Petrie (1892) pl. 23; Harpur (2001) fig. 85.	•										
DYNASTY IV.M-L													
Mrs-ʿnḫ III	PM 197 G7530/40	Dunham–Simpson (1974) fig. 4.						•			•	•	
Nb.j-m-3ḫtj	PM 230 TLG 86	LD II 12 [b]; Hassan, *Gîza* IV (1943) fig. 77.	•					•					•
DYNASTY V.E													
Rʿ-wr	PM 265	Hassan *Gîza* (1932) pl. 3[3] fig. 26.										•	•
Ḫnmw-ḥtp	PM 579	Harpur *JEA* 71 (1985) fig. 8.	?										•
Snb	PM 101	Junker *Gîza* (1941) Abb. 15.						•					
K3(.j)-ḫnt (I)	PM (v) 8 A2	El-Khouli–Kanawati (1990) pl. 35-37.						•					
DYNASTY V.M													
Jj-mrjj	PM 170 G6020	LD Erg. 6; Weeks (1994), fig. 33, 40,	•				•	•				•	
Jj-nfrt	PM 299	Schürmann (1983) 7, 6[a, b], 21.						•		•			•
Ptḥ-špss	PM 340	Vachala (2004) p.128.	•									•	
Nj-ʿnḫ/Ḫnmw and *Ḫnmw-ḥtp*	PM 641	Moussa–Altenmüller (1977) Abb. 5, 6 Taf. 74-5.						•	R•		•	•	•
Nfr and *K3-ḥ3.j*	PM 639	Moussa– Altenmüller, (1971) pl. 1, 5, 10-11, 15..										•	•
Rʿ-ḫʿ.f-ʿnḫ	PM 207 G7948	LD Erg. 38 [b].	•										
Ḫwfw-ḫ3.f II	PM 190 G7150	Simpson (1978) fig. 47.						•			•		
DYNASTY V.L													
3ḫtj-mrw-nswt	PM 80 G2184	Decker–Herb (1994) II, pl. 216; Smith (1946) fig. 229.						•					•
3ḫtj-ḥtp (Louvre)	PM 634	Ziegler (1993) 128, 131, 133-4 (drawing).	•				•				•		•
3ḫtj-ḥtp	PM 599 D64	Davies II (1901) pl. 13, 14.	•									•	•
J3sn	PM 82 G2196	Simpson (1980) fig. 30.						•					
Jj-nfrt/Š3.n.f	PM 616	Kanawati–Abder-Raziq (2003) pl. 37b						•					
Jntj	PM (iv) 121	Petrie (1898) pl. 5; Kanawati –McFarlane (1993) pl. 33.	•		2	•	•						
Jrj-n-k3-Ptḥ	PM 644	Moussa–Junge (1975) pl. 11,12	•					•			•		
Jrw-k3-Ptḥ	PM 639	de Rachewiltz (1960) pls. 16[2], 17 [1]; McFarlane (2000), 41, 46.						•					•
Jḥjj (r/u *Sšsšt/Jdwt*)	PM 617	Macramallah (1935) pl.6 -7, 11, 13; Kanawati–Abder-Raziq (2003) pl. 43, 53, 54.	•					•	R•	•	•		•
Jtj-sn	PM 252	Hassan (1944) fig. 123, pl. 37.						•				•	
Jttj	PM 193 G7391	Badawy (1976) fig. 17						•					
Wr-jr.n.j	PM (iv) 188 No.25	Davies (1901) pl. 11-12.						•					
Ptḥ-ḥtp/Jj-n-ʿnḫ	PM 606	Hassan (1975) fig. 31.									•	•	

TABLE 16: PAPYRUS SKIFFS — CONSTRUCTION AND USE

MONUMENT			SKIFF/S Construction			ROPE		USE OF SKIFF/RAFT [R = Raft]					
TOMB OWNER	PM NO. FIELD NO.	REFERENCE	Skiff building	Bringing papyrus to boat	2/3 skiffs being built	Spare rope coils	Rope making	Tomb owner on skiff	Fishing from craft	Hippo hunt	Boatmen return peacefully	Jousting boatmen	Craft with herd crossing water
Ptḥ-ḥtp (II) (*Ṯfj*)	PM 600 D64	Paget–Pirie (1989) pl. 32, 33; Davies I (1900) pl. 3, 21, 23, 25, 26.	•		3	•	•				•		•
Ftk-tꜣ	PM 351 LS 1	LD Erg. 40d; Bárta (2001) fig. 3.11, 3.12.						•				•	
Nj-kꜣw-ḥr	PM 498 S915	Hassan (1950) fig. 168.								•			
Nj-mꜣʿt-Rʿ	PM n/r G2097	Roth (1995) pl. 185.						•					•
Nj-mꜣʿt-Rʿ	PM 283 G8900	Hassan (1936) fig. 236.						•	•	•			•
Nb-j-kꜣw-ḥr/Jdw	PM 627	Hassan (1975) fig. 16.	•						•		•		
Nbt	PM 624	Munro (1993) pl.10-11, 18, 20.	•								•		
Nfr (1)	PM 137 G4761	Junker VI (1943) fig. 14, 17 facing p. 72.					•						
Nfr-jrt-n.f	PM 583 D55	van de Walle (1978) pl. 1.						•					
Nfr-mḏr-Ḫwfw	PM n/r G2240	Roth (1995) pl. 210.											•
Rʿ-špss	PM 494 LS16	LD II 60.						•			•		•
Ḫm-Mnw	PM n/r M43	Kanawati (1985) fig. 6-7.						•					
Ḥzjj-Mnw	PM n/r M22	Kanawati (1983) fig. 12-13, 17, 19.	•		2			•					•
Ḥtp-ḥr-ꜣḫtj	PM 593 D60	Mohr, (1943) fig. 2, 27, 33, 34.	•					•				•	•
Ḫww-wr	PM 254 LG 95	LD II, 43a (right); Hassan (1944) fig. 104.						•					
Ḫw-ns	PM (iv) 134 No.2	LD II 105 [b]. 106 [a]; Varille (1938) fig. 2-3, 4, 6.	•		2	•		•			•	•	•
Ḫnwt	PM 623	Munro (1993) pl. 33, 34.	•								•		
Sꜣ-jb	G2092, G2093	Roth (1995) pl. 181, 182, 183.					•	•			•		•
Spd-ḥtp	PM 481 D15	Borchardt (1964) pl. 60, 61, 95.											•
Snḏm-jb/Jntj	PM 85 G2370=LG27	LD Erg. 18, 20; Brovarski (2001) fig. 17, 26-7, 38-9, 22-3, 26, 27, 29-30, 42, 43.	•		2	•		•		•	•	•	•
Snḏm-jb/Mḥj	PM 87 G2378=LG26	LD Erg, 11, 12, 14; Brovarski, (2001) fig. 10, 96-8, 103-3, 104-5, 114.						•		•	R•		•
Sḫm-ʿnḫ-Ptḥ	PM 454 D41	Simpson (1976) pl. D.										•	
Sḫm-kꜣj	PM 53 G1029	Simpson (1980) fig. 4.									•	•	
Sšm-nfr IV	PM 223 LG 53	Junker *Giza* (1953) pl. 60, 61, 62, 67.						•	R		•		•
Kꜣ.j- m- ʿnḫ	PM 131 G4561	Junker *Giza* (1940) Abb. 8; Kanawati (2001) pl. 31, 36.						•					•
Kꜣ.j-m-nfrt	PM 467 D23	Simpson (1992), fig. 3, 4, 6, pl. A, fig.16, pl. G.	•				•				•		
Kꜣ.j-m-nfrt II	PM 263	Hassan (1936) fig. 121, 124, 140.								•	•	•	•
Kꜣ.j-m-rḥw	PM 485 D2 S905	Mogensen (1921) fig. 4.											•
Kꜣ.j-m-ḥzt	PM 542	Quibell–Hayter (1927) fig. on p. 17; McFarlane, (2003) pl. 44.							•				
Kꜣpj	PM 69 G2091	Roth (1995) pl. 157-8.	•		2		•						
Kꜣ-ḥj.f	PM 76	Junker (1943) fig. 43.					•						
Ṯjj	PM 468 D22	Épron–Daumas I (1939) pl. 9, 46; Wild, II (1953) pl. 75, 77, 85-6, 93, 111, 114, 119, 124.	•		2			•	•	•			•

TABLE 16: PAPYRUS SKIFFS — CONSTRUCTION AND USE

MONUMENT			SKIFF/S Construction			ROPE		USE OF SKIFF/RAFT [R = Raft]					
TOMB OWNER	PM NO. FIELD NO.	REFERENCE	Skiff building	Bringing papyrus to boat	2/3 skiffs being built	Spare rope coils	Rope making	Tomb owner on skiff	Fishing from craft	Hippo hunt	Boatmen return peacefully	Jousting boatmen	Craft with herd crossing water
DYNASTY VI.E													
Jnw-Mnw	PM n/r	Kanawati–Abder-Raziq (2006) pl. 44, 48.						•				•	
Jn-Snfrw-jšt.f	PM 891 No 2	de Morgan (1903) pl. 24; Borchardt II (1964) pl. 103 [CG 1771].						•			•		
Jttj-Šdw	PM(iv) 122	Petrie (1898) pl. 15; Kanawati–McFarlane (1993) pl.44, 46, 48.						•					•
Jdw	PM 185 G7102	Simpson (1976) fig. 36 [g], 37.									•		
ʿnḫ -m- ʿ-Ḥr	PM 512	Badawy (1978) fig. 24; Kanawati–Hassan (1997) pl.37 [a].											•
Wr-nw	PM 519	Davies et al (1984) pl. 25[A, B].						•			•		
Wḏ3-ḥ3-Ttj/Ššj/ Nfr-sšm-Ptḥ	PM 515	Wreszinski III (1936) fig. 116 [A].									•		
B3wj	PM n/r G 126	Kanawati (1989) fig. 15.						•					
Mrjj-Rʿ-mrjj-Ptḥ-ʿnḫ/Nḫbw	PM 89 G2381	Smith (1958) *BMFA* 56 [304] fig. 2						•					
Mrw/Ttj-snb	PM 520	Lloyd et al (1990) pl. 8, 12.									•		•
Mr.f-nb.f	PM n.r.	Mysliwiec (2004) pls. 21, 22, 63-5, 67, 70 [e-f].						•			•		
Mrrj	PM 518	Davies et al (1984), pl. 5.						•			•		
Mrrw-k3/Mrj	PM 525	Duell I (1938) pl. 9-13, 10, 15.19, 20-1, 43, 45, 127-9; Kanawati (2010) pl. 67-70, 79; Kanawati (2011) 85-86.						•	•	•	•	•	•
Mḥw	PM 619	Altenmüller (1998) pl. 9-13, 14, 35, 39, 40.						•			•		•
Mttj	PM 646	Kaplony (1976) fig. 1-1a; Ziegler (1990) pp. 129, 144.						•					•
Nj-ʿnḫ-Ppjj	PM (iv) 137 No 14	Varille (1938) pl. 5, 6. 9.						•	R•			•	
Nj-k3w-Jzzj	PM n/r	Kanawati–Abder-Raziq (2000) pl. 47, 50.67[b].					•	•			•	•	
Rmnj/Mrwj	PM n/r	Kanawati (2009) pl. 45-47..						•	•				•
Ḥnkw/Jj…f (II)	PM (iv) 242 N67	Kanawati (2005) pl. 23, 54.						•			•		
Ḥnkw/Ḫttj (I)	PM (iv) 242 N39	Davies *Gebrâwi* II (1902) 32, pl. 28, 39.						•					
Ḥm-Rʿ/Jzj (I)	PM (iv) 243 N72	Davies *Gebrâwi* II (1902) pl. 20; Kanawati *Gebrawi* I (2005) pl.64.						•			•		•
Ḥzj r/u *Sšm-nfr*	PM n/r	Kanawati–Abder-Raziq (1999) pl. 53, 54, 55.						•					•
Ḥzj-Mnw /Zzj/Dwdjj	PM n/r F1	Kanawati (1986) fig. 3.						•					
Ḫw.n-wḫ/Ttj	PM (iv)	El-Khouli– Kanawati (1989) pl. 37, 38						•					
Ḫnmw-ntj	PM 87 G2374	Brovarski (2001) fig. 88.									•		
Sʿnḫ-wj-Ptḥ	PM n/r	Kanawati–Abder-Raziq (1998) pl. 69, 76.						•		•			

TABLE 16: PAPYRUS SKIFFS — CONSTRUCTION AND USE

MONUMENT			SKIFF/S Construction			ROPE		USE OF SKIFF/RAFT [R = Raft]					
TOMB OWNER	**PM NO. FIELD NO.**	**REFERENCE**	Skiff building	Bringing papyrus to boat	2/3 skiffs being built	Spare rope coils	Rope making	Tomb owner on skiff	Fishing from craft	Hippo hunt	Boatmen return peacefully	Jousting boatmen	Craft with herd crossing water
K3-gmnj/Mmj	PM 521	Firth–Gunn II (1926) pl. 7, 9 [B], 52; Wreszinski Atlas III (1936) pl. 38, 104[A], 108[A]; von Bissing I (1905) pl 17, 18, 92[A]; Harpur (2006) 6. 7.	•					•	•	•	•		•
DYNASTY VI.M													
Jbj	PM(iv) 243 No.8	Davies *Gebrâwi* I (1902) pl.6-7; Kanawati *Gebrawi* II (2007) pl. 46, 47, 67.						•	R•		•	?	•
Ppj-ʿnḫ-ḥrj-jb	PM(iv) 254 D2	Blackman (1924) pl. 7, 17.						•			•		
Mrw	PM n/r N 3737	Peck (1959) pl.15.						•					
K3.j-ḥp/Ṯtj	PM n/r M8	Kanawati (1982) fig. 5 [a]. 13.						•					•
DYNASTY VI.L													
Ppj-ʿnḫ/Ḥnj-km	PM(iv) 247	Blackman–Apted (1953) pl. 24, 30, 40.						•		•	•		•
Mḫw and *S3bnj*	PM(v) 231 A1	de Morgan (1894) 146.						•					
Ḫnj/Špsj-pw-Mnw	PM n/r H24	Kanawati (1981) fig. 18, 22.						•					•
S3bnj II	PM n/r EQH 35e	Begelsbacher-Fischer–Hottinger (1985) fig. 134; Harpur (1987) l. 24.						•					
K3(.j)-hp/Ṯtj-jkr	PM (v) 19 H26	Kanawati(1980) fig. 8, 11, 12.	•					•	•				•
Ḏʿw and *Ḏʿw/Šm3j*	PM (iv) 244 No.12	Davies *Gebrâwi* II (1902) pl. 3-4, 5-6.						•	•		•	•	•
DYNASTY VIII													
ʿnḫ-tj.fj	PM (v) 170	Vandier (1950) pl. 13.						•					
W3hj	PM n/r	Kanawati (1995) pl. 20, 23.											•
B3wj	PM n/r BA14	Kanawati (1987) fig. 30.											•
Mrjj-ʿ3	PM (v) 34 D18	Petrie 1908) pl. 7, 9; Kanawati (1995) pl. 37, 39-40, 42, 44-5.						•					•
Ḫwj/Ṯtj-jkr	PM n/r G1	El-Masry (2004) *BACE* 15 (2004) 92-94, fig. 2.						•					
Ḫw-ns	PM (v) 235	de Morgan (1894) 159; Decker –Herb I (1994) K 2.82, II, pl. 236						•					
Ghs3/Nbjj	PM n/r GA11	Kanawati (1987) fig. 18.											•

TABLE 17: TOMB OWNER VIEWING THE WORK AND PRODUCE OF HIS ESTATE

MONUMENT			INSCRIPTION	
TOMB OWNER	PM No. FIELD No.	REFERENCE	TRANSLATION	TRANSLITERATION
DYNASTY IV.E				
Nfr-mꜣꜥt	PM (iv) 92 No. 16	Harpur (2001) fig. 70.	Viewing cranes	*mꜣꜣ ḏꜣ<t>*
			Viewing (…) the *pr-ḏt*	*mꜣꜣ … pr-ḏt*
Nfr-mꜣꜥt	PM 183 G7060	LD II 17b.	Viewing the (*ndt-ḥr* offerings) which are brought from the towns of his *pr-ḏt*	*mꜣꜣ n[ḏt-ḥr]jnnt m [njwwt] nt pr.f ḏt*
Rꜥ-ḥtp	PM (iv) 90	Petrie (1892) pl.9.		*mꜣ(ꜣ) ꜥḥt*
		Petrie (1892) pl.11.	Viewing the boat-building in the workshop of the *pr-ḏt*	*mꜣ(ꜣ) šd dšr m wḫrt pr n ḏt*
		Harpur, p. 100.	Viewing the trapping of birds	*mꜣ(ꜣ) sḫt ꜣpdw*
DYNASTY IV.L				
Mrs-ꜥnḫ III	PM 197 G7530/40		She pulls papyrus for Hathor in the marshland, with her mother. They see every good thing which is in the marshland.	*sšš.s wꜣḏ n Ḥwt-ḥr m pḥw mwt.s mꜣꜣ.sn ḫt nbt nfrt ntt m mḥt*
			Viewing the trapping of birds in the marsh	*mꜣꜣ sḫt ꜣpd(w) m mḥt* (Above clapnet)
Nj-kꜣw-Rꜥ	PM 232 LG 87	LD II 15.	Viewing … (honey?) (of) the towns of the *pr-ḏt*	*mꜣꜣ … bjt njwwt pr-ḏt*
Mr-jb.j	PM 71 LG 24	LD II pl. 22b.	Viewing funerary offerings that the royal house has brought: 1000 young cattle, antelope, ibex, gazelles, geese and ducks. (Horizontal inscription above family)	*mꜣꜣ prt-ḫrw jnt pr-nswt rn ḫꜣ kꜣ gsꜣ rn ḫꜣ njꜣw rm ḫꜣ gḥs rn ḫꜣ ꜣpd (?) ḫꜣ ꜣpd (?)*
		LD II pl. 22a.	Viewing the sealed things, the royal house has brought: ?red linen pieces, weaving (1) 100, 5, 4, 1 *pꜣḏ*, a thousand balls incense, a thousand portions of green and black paint and all kinds of the best oil.	*mꜣꜣ ḫtm jnt pr-nswt jdmj ?* 100,5, 4, 2, 1 *ḫꜣ pꜣḏ* *ḫꜣ sntr wꜣḏ smdt ḥꜥtt nb* (horizontal)
DYNASTY V.E				
Wḥm-kꜣ.j	PM 114 D117	Kayser (1954) p. 36-37.	Viewing the record (horizontal)	*mꜣꜣ sš*
			Viewing the small cattle of the *pr-ḏt* and the payment of the cattle tax (horizontal)	*mꜣ(ꜣ) ꜥwt n pr-ḏt n jrt jrw*
Nswt-nfr	PM 143 G4970	Kanawati (2002) pl. 57.	Viewing the record of the reversion offerings	*mꜣꜣ sš n pẖr*
Sḫm-kꜣ-Rꜥ	PM 233 LG 89	Hassan IV (1943) fig. 58; LD II 42.	Viewing the gifts of his chiefs and herdsmen, his fowlers and fishermen, which are brought from the towns of (his *pr-ḏt*) for …	*mꜣ(ꜣ) jnw n ḥḳꜣw(.f) njwwt.f wḥꜥw.f jnnt m njwwt pr(-ḏt.f) n* …
Sšꜣt-ḥtp/Htj	PM 149 LG 36	Kanawati (2002) pl. 45.	Viewing the record of the reversion offerings.	*mꜣ(ꜣ) sš n pẖrt*
		Junker (1953) *Giza* XI, fig, 28.	Bringing the invocation offerings by his towns of Lower Egypt at every feast every day for …	*jnt prt-ḫrw jn njwwt.f nt Tꜣ-mḥw m ḥb nb hrw nb n* …
		Kanawati (2002) pl. 45.	Counting all the property of the *pr-ḏt*	*jp jšt nbt nt pr-ḏt*
Ṯntj	PM 141 LG 47	LD II 30.	Bringing presentations by the herders belonging to the field and herding the *wnḏw* cows of the estate in order to view the count of the herders as it is written down. Giving the document in order to view … *pr-ḏt*	*jnt nḏ(t)-ḥr jn mniww jmj sḫt ḥwy wnḏw.t nt pr-ḏt r mꜣꜣ jp mnjww m jrj mḏꜣwt rdjt zš r mꜣꜣ … pr-ḏt*

REGISTERS					MONUMENT
FIRST (highest)	**SECOND**	**THIRD**	**FOURTH**	**FIFTH/SIXTH**	**TOMB OWNER**
					DYNASTY IV.E
ANIMAL PARADE?					*Nfr-mꜣʿt*
OFFERINGS Domains	OFFERINGS Domains	OFFERINGS Domains	OFFERINGS Domains	OFFERINGS Domains	
RECORDS Presentation	ANIMAL PARADE cattle	ANIMAL PARADE Desert animals	ANIMAL PARADE Birds		*Nfr-mꜣʿt*
DESERT HUNT	DESERT HUNT	ANIMAL PARADE Cattle			*Rʿ-ḥtp*
BOATS Construction	MARSHES Dragnet fishing	ANIMAL PARADE Oryx; calf	OFFERINGS Butchery; Domains	OFFERINGS Domains	
MARSHES Clapnet fowling	MARSHES Returning boats				
					DYNASTY IV.L
OFFERINGS Domains	MARSHES Returning boats	ANIMAL PARADE Birds; cattle	MARSHES Jousting boatmen; trampling sheep		*Mrs-ʿnḫ* III
As above	As above	As above	As above		
ATTENDANTS Presenting record	ANIMAL PARADE cattle	OFFERINGS /PARADE Presenting record Domains; cattle	OFFERINGS/ PARADE Domains; hyena	ANIMAL PARADE Cattle	*Nj-kꜣw-Rʿ*
SINGLE ANIMAL Cattle	SINGLE ANIMAL Cattle	SINGLE ANIMAL Desert animal			*Mr-jb.j*
(Inscription above list) OFFERINGS	ATTENDANTS Presenting record				
					DYNASTY V.E
BOAT	ANIMAL PARADE Scribes/ record cattle; donkeys; small cattle	CHIEFS Presenting chiefs	OFFERINGS	OFFERINGS	*Wḥm-kꜣ.j*
As above	As above	As above	As above	As above	
OFFERINGS Birds	OFFERINGS	OFFERINGS			*Nswt-nfr*
OFFERINGS Family/bearers	MARSHES Fowling	MARSHES Fishing			*Sḫm-kꜣ-Rʿ*
OFFERINGS foods	OFFERINGS	OFFICIAL	OFFERINGS		*Sšꜣt-ḥtp/Ḥtj*
OFFERINGS Scribe; domains	ANIMAL PARADE Desert animals	ANIMAL PARADE Desert animals; cattle			
N/A					
ACCOUNTS Scribes	OFFERINGS	OFFERINGS	OFFERINGS		*Ṯntj*

TABLE 17: TOMB OWNER VIEWING THE WORK AND PRODUCE OF HIS ESTATE

MONUMENT			INSCRIPTION	
TOMB OWNER	PM No. FIELD No.	REFERENCE	TRANSLATION	TRANSLITERATION
DYNASTY V.M				
Jj-mrjj	PM 170 G6020	Weeks (1994) fig. 26.	Viewing the presentations which were brought as invocation offerings from ..	*m33 ndt-ḥr jnnt r prt-ḫrw m*
		Weeks (1994) fig. 41.	Viewing the recording of the presentations which were brought as invocation offerings from the towns of the *pr-dt*	*m33 sš [ndt-ḥr jn]nt r prt-ḫrw m njwwt nt p- dt*
		Weeks (1994) fig. 31.	Bringing invocation offerings from the towns of the *pr-dt* at every feast, every day, forever	*jnt prt-ḫrw m njwwt nt pr-dt m ḥb nb rʿ nb dt*
		Weeks (1994) fig. 39-40.	Viewing the ploughing in his towns of the *pr-dt*	*m33 sk3 m njwwt.f nt pr-dt*
		Weeks (1994) fig. 40.	Viewing (the trapping of) birds from the northern marshes	*m3(3 sḫt) 3pdw m mḥt*
K3(.j)-ḫnt (A2)	PM (v) 8 A2	El-Khouli–Kanawati (1990) pl. 44.	Viewing the cattle count	*m33 jrt*
Wr-jr.n-Ptḥ	PM 699	James, (1961) pl. 29.	(Viewing) the recording of the work of the field	(*m33*) *sš(w) n k3t sḫt*
Nj-ʿnḫ/Ḫnmw and *Ḫnmw-ḥtp*	PM 641	Moussa–Altenmüller (1977) fig. 4-6.	… the marsh waters and viewing all the work …	… *pḥw m33 k3t nb(t)* …
		Moussa–Altenmüller (1977) fig. 8.	Viewing the construction of the Sabt-boat, the papyrus gardens and the possessions of the *pr-dt*.	*m33 šdt m š3bt š n dwjt ʿḥʿt n pr-dt*
		Moussa–Altenmüller (1977) fig. 12 Left/Right.	Viewing the marshland, the marsh thickets and the marsh water, the catching of fish, and the netting of birds. More beautiful (to see) than all things.	*m33 š3w zšw pḥw ḥ3m mḥjt sḫt 3pd nfr* (*m33*) *r jḫt nb*
		Moussa–Altenmüller (1977) fig. 13.	Viewing the presentations which are brought from all his towns and the heaping up of sweet things at the *pr-dšr*	*m33 ndt-ḥr jnnt m njwwt.f nbt tzt jḫt bnrt r pr-dšr*
		Moussa–Altenmüller (1977) figs. 18, 19.	Viewing the desert animals	*m33 ʿwt ḫ3st*
Nfr and *K3-ḥ3.j*	PM 639	Moussa–Altenmüller (1971) pl. 1.	Viewing fishermen, fowlers and scribes of his town(s) of Lower and Upper Egypt.	*m33 ḥ3mw ʿḥw zšw n njw(w)t.f nt T3-mḥw Šmʿw*
		Moussa–Altenmüller (1971) pl. 16-17.	Viewing the records of the *pr-dt* and the presentation of the fowlers and accounting for with the chiefs.	*m33 zšw n pr-dt ndt-ḥr ḥ3mw ḥsb ḥk3w*
Sšm-nfr II	PM 146 G5080	Kanawati (2002) pl. 62.	Viewing the presentation of … from his towns of the *pr-dt*	*m33 n(dt-ḥr)* … *m njwwt.f pr-dt*
DYNASTY V.L				
3ḫtj-ḥtp (Louvre)	PM 634	Ziegler (1993) p. 145.	Viewing the ploughing, the reaping, the trapping of birds and all the work of the field	*m33 sk3 3zḫ sḫt k3t nb nt sḫt*
3ḫtj-ḥtp	PM 633 E17	Petrie–Murray (1952) pl. 5.	Viewing the presentation of all good offerings of the year which are brought from his towns of Lower and Upper Egypt	*m33 ndt-ḥr rnpt nbt nfrt jnnt m njwwt,f n Mḥw Šmʿw*
3ḫtj-ḥtp	PM 599 D64	Davies *Ptahhetep* II (1901) pl. 18, 19.	Viewing the presentation of the animals of the desert	*m33 ndt-ḥr ʿwt ḫ3st*
		Davies *Ptahhetep* II (1901) pl. 20, 21.	Viewing the cattle of the Thoth festival brought from the temples and the towns belonging to the *pr-dt* of Upper and Lower Egypt	*m33 jw3 n dḥwtt inn m ḥwwt-k3 njwwt nt pr-dt m T3-Mḥw Šmʿw*
		Davies *Ptahhetep* II (1901) pl. 14.	Viewing all the works of the fields in the marshlands, whatever is good	*m3(3) k3t nbt sḫt m pḥww ḫt nb nfrt*

REGISTERS					MONUMENT
FIRST (highest)	**SECOND**	**THIRD**	**FOURTH**	**FIFTH/SIXTH**	**TOMB OWNER**
					DYNASTY V.M
Scribes	Animal offerings				*Jj-mrjj*
ACCOUNTS Scribes	ANIMAL PARADE Desert animals	ANIMAL PARADE Cattle	OFFERINGS		
ANIMAL PARADE Cattle	ANIMAL PARADE Cattle	ANIMAL PARADE Desert animals	ANIMAL PARADE Birds; cranes		
HARVEST Scribes; reaping	HARVEST Bringing donkeys	HARVEST Piling sheaves; transport	PLANTING Ploughing	PLANTING Hoeing; sheep	
MARSHES Clapnet fowling	MARSHES Dragnet fishing	MARSHES Dragnet fishing	PAPYRUS SKIFFS Construction; Rope	MARSHES Jousting boatmen	
OFFERINGS Cattle parade	OFFERINGS Cattle parade	BOATS			*K3(.j)-ẖnt* (A2)
HUSBANDRY Cattle	HUSBANDRY Cattle	PLANTING Ploughing Clapnet fowling	HARVEST Scribes; reaping	HARVEST Piling sheaves; transport/BOATS	*Wr-jr.n-Ptḥ*
OFFERINGS	OFFERINGS	Butchery			*Nj-ʿnḫ/H̱nmw* and *H̱nmw-ḥtp*
HUSBANDRY Chopping trees; goats	BOATS Timber boat construction	OFFERINGS Foods; Vineyard	GARDENING		
MARSHES Clapnet fowling	MARSHES Making/mending nets; presentation of birds	MARSHES Dragnet; fishing, presentation of fish	MARSHES Fishing with weirs	MARSHES Drying/gutting fish	
WINE/FRUIT Wine offering; piles of fruit	WINE Record; measuring, wine	ANIMAL PARADE Return from hunt; desert animals	BOATS Bringing gifts		
ANIMAL PARADE Desert animals	ANIMAL PARADE Desert animals	ANIMAL PARADE Desert animals	ANIMAL PARADE Desert animals	Regs 5 - 8 ANIMAL PARADE Desert animals	
HUSBANDRY Cattle; goats	MARSHES Dragnet; fishing	AGRICULTURE Making wine	AGRICULTURE Sowing; sheep		*Nfr* and *K3-ḥ3.j*
MARSHES Scribes; boat building	HUSBANDRY Cattle; fording	MARSHES Clapnet fowling	GAMES Wooden sail boats	MARSHES Jousting boatmen	
ACCOUNTS	ANIMAL PARADE Desert animals	ANIMAL PARADE Cattle	ANIMAL PARADE Birds		*Sšm-nfr* II
					DYNASTY V.L
AGRICULTURE Sheep	HARVEST Flax	HARVEST Cereals	HARVEST Transport	HARVEST Threshing	*3ḫtj-ḥtp* (Louvre)
OFFERINGS	OFFERINGS Offering bearers	OFFERINGS Offering bearers	OFFERINGS Offering bearers		*3ḫtj-ḥtp*
ANIMAL PARADE Desert animals	ANIMAL PARADE Desert animals	OFFERINGS Birds	ANIMAL PARADE Desert animals	ANIMAL PARADE Desert animals; birds	*3ḫtj-ḥtp*
ANIMAL PARADE Cattle	ANIMAL PARADE Cattle	ANIMAL PARADE Cattle	ANIMAL PARADE Cattle	ANIMAL PARADE Cattle	
MARSHES Bringing papyrus	MARSHES Boatmen jousting	MARSHES Boatmen jousting	MARSHES Cattle fording		

TABLE 17: TOMB OWNER VIEWING THE WORK AND PRODUCE OF HIS ESTATE

MONUMENT			INSCRIPTION	
TOMB OWNER	PM No. FIELD No.	REFERENCE	TRANSLATION	TRANSLITERATION
		Davies *Ptahhetep* II (1901) pl. 13.	Viewing the work of the field, everything that is good, done in the Delta marshes.	*(m)ꜣꜣ kꜣt sḫt ḫt nb nfrt jrrt m mḥt*
		Davies *Ptahhetep* II (1901) pl. 6, 10.	Viewing the presentation by …	*mꜣꜣ ndt-ḥr jn*
		Davies *Ptahhetep* II (1901) pl. 8-9.	(Viewing the works) of the field, the reaping and transport. A sight more beautiful than anything.	*(mꜣꜣ kꜣt) sḫt šdjt nfr mꜣꜣ r ḫt nb*
		Davies *Ptahhetep* II (1901) pl. 6, 7.	Viewing the work of the field which is brought by the fishermen of the towns of the *pr-ḏt*	*mꜣ(ꜣ) (kꜣt) sḫt jnnt wḥꜥw m njwwt nt pr-n-ḏt*
Jꜣsn	PM 82 G2196	Simpson (1980) fig. 30.	Viewing the ploughing	*mꜣꜣ skꜣ*
		Simpson (1980) fig. 30.	Viewing the document of the presentations brought (for) an invocation offering every day	*mꜣꜣ sš n ndt-ḥr jnt m pr(t)-ḫrw rꜥ nb*
Jj-nfrt	PM 299	Schürmann (1983) fig. 6-17, 21.	Viewing the work of the field and all good things which are carried out in the field	*(m)ꜣꜣ kꜣt sḫt jḫt nbt nfr.t jrr.t m sḫt*
Jrj-n-kꜣ-Ptḥ	PM 644	Moussa–Junge (1975) pl. 6-10.	Viewing ploughing, gathering flax, (reaping?) and transport	*mꜣꜣ skꜣ wḥꜥ mḥꜥ.w (ꜣzḫ) šdjt*
Jḥjj (r/u *Sššt/Jdwt*)	PM 617	Kanawati–Abder-Raziq (2003) pl. 54.	[Viewing/traversing] the marshlands and ponds by …	*[mꜣꜣ/ḫns] pḥww sšw jn* …
		Kanawati–Abder-Raziq (2003) pl. 71.	Viewing the presentation of all good offerings of the year which are brought for her from the (his) estates and the (his) towns	*mꜣꜣ ndt-ḥr rnpt nb(t) nfrt jnnt n.s m ḥwwt njwwt*
		Kanawati–Abder-Raziq (2003) pl.62.	Viewing all the good work of the field in her estates and her towns of Lower and Upper Egypt	*mꜣꜣ kꜣt nb nfrt nt sšt m ḥwwt.f m njwwt.f m Tꜣ-Mḥw Šmꜥw*
Jtj-sn	PM 252	Hassan V (1944) fig. 122-23.	Viewing the towns of the estate	*mꜣꜣ njwwt nt pr-ḏt*
Wr-jr.n.j	PM (iv) 188 No.25	Davies (1901) *Sheikh Saïd*, pl. 16.	Viewing the ploughing (gathering of flax, reaping and transport) and the treading of the threshing floors by the headmen of the *pr-ḏt*	*mꜣꜣ skꜣ (ḥwj mḥꜥ ꜣzḫ šdjt) ḥwj spwt (j)n ḥkꜣw (nt) pr-ḏt*
Pr-nb	PM 497 S913	Hayes (1953) fig. 51.	Viewing the presentations which are brought from the towns of Upper and Lower Egypt	*mꜣꜣ ndt-ḥr jnnt m njwwt nt Tꜣ-Mḥw Šmꜥw*
Ptḥ-ḥtp	PM 653 LS 31	LD II 104.	(Viewing) … boats, which are travelling to him from all his estates in Upper Egypt	… *ḫꜥww ḥd jnnw n.f m ḥwwt.t.f nb Šmꜥw*
		LD II 102.	(Viewing) … brought to him from his estates and towns of Lower Egypt	… … *jn n.f m ḥwwt.f njwwt.f nt Mḥw*
Ptḥ-ḥtp (II) (*Ṯfj*)	PM 600 D64	Paget–Pirie (1898) pl. 33.	Viewing the offering brought by the estates and towns of Lower and Upper Egypt of the *pr-ḏt* for …	*mꜣꜣ ndt-ḥr jnw n ḥwwt njwwt Mḥw ^mꜥw pr-ḏt n* …
		Paget–Pirie (1898) pl. 31.	Viewing the *jwꜣ* cattle, the contribution of the feast of Thoth from the estates and temples of Upper and Lower Egypt, of the estate. (Above Register 3)	*mꜣꜣ jwꜣ jnw ḏḥwtt m ḥwwt ḥwwt-kꜣ nt Mḥw ^mꜥw pr-n-ḏt*

REGISTERS					MONUMENT
FIRST (highest)	**SECOND**	**THIRD**	**FOURTH**	**FIFTH/SIXTH**	**TOMB OWNER**
HUSBANDRY Cattle	MARSHES Papyrus boat building	MARSHES Gathering papyrus	MARSHES Bringing papyrus Skiff building	Domains	
ACCOUNTS Scribes rendering	OFFERINGS Offering bearers				
HARVEST Reaping, pulling flax	HARVEST Donkey transport	HARVEST Winnowing	HARVEST Threshing	HARVEST Storage; donkeys return	
DESTROYED	OFFERINGS Bearers?	OFFERINGS/ BIRDS Bearers; clapnet	OFFERINGS Bearers with birds	OFFERINGS Marsh products	
HARVEST Reaping , stacking Husbandry; cattle	HARVEST Transporting grain; Animal parade; cattle	PLOUGHING Sheep; Husbandry; cattle			*Jȝsn*
ANIMAL PARADE Scribes; desert animals	ANIMAL PARADE Cattle	ANIMAL PARADE Cattle			
MARSHES Offerings Fishing; fowling	AGRICULTURE Ploughing				*Jj-nfrt*
PLOUGHING Sheep	HARVEST Grain; flax	HARVEST Tying sheaves; transport, threshing	HARVEST Returning donkeys; rendering accounts		*Jrj-n-kȝ-Ptḥ*
HUSBANDRY Cattle	PAPYRUS SKIFFS Construction	MARSH PRODUCTS Brought by boats	MARSH PRODUCTS Brought by boats	MARSHES Cattle fording	*Jḥjj* (r/u *Sšsšt/Jdwt*)
OFFERINGS Foods	OFFERINGS Foods	ANIMAL PARADE Desert animals	ANIMAL PARADE Record; cattle	OFFERINGS Bearers; Butchery	
OFFERINGS Bearers	MARSHES Jousting boatmen	OFFERINGS Bearers	OFFERINGS Brought by boat Return from marshes?	OFFERINGS Bearers	
MARSHES Clapnet fowling BOATS	HUSBANDRY Cattle	HUSBANDRY Cattle	HUSBANDRY Cattle	BOATS	*Jtj-sn*
AGRICULTURE Ploughing; sheep	HARVEST Scribes; pulling flax; reaping cereal	HARVEST Scribes; piling sheaves, bringing donkeys; transport	HARVEST Reaping; threshing		*Wr-jr.n.j*
ANIMAL PARADE Desert animals	ANIMAL PARADE Cattle	FAMILY	SCRIBES	ANIMAL PARADE Offering bearers; cattle	*Pr-nb*
BOATS Cargo; livestock	BOATS Cargo; livestock	BOATS Cargo; livestock	BOATS Cargo; livestock		*Ptḥ-ḥtp*
PART DESTROYED Husbandry: feeding desert animals	ANIMAL PARADE Cattle; feeding desert animals	ANIMAL PARADE Husbandry: feeding cattle	ANIMAL PARADE Husbandry: feeding cattle		
YOUNG MEN Training?	HUNT Returning with desert animals	ANIMAL PARADE Desert animals	HUSBANDRY Cattle	Regs 5, 6, 7 ANIMAL PARADE Cattle; Birds	*Ptḥ-ḥtp* (II) (*Ṯfj*)
HUSBANDRY Cattle	HUSBANDRY Cattle	ANIMAL PARADE Cattle	ANIMAL PARADE Cattle	ANIMAL PARADE Birds	

TABLE 17: TOMB OWNER VIEWING THE WORK AND PRODUCE OF HIS ESTATE

MONUMENT			INSCRIPTION	
TOMB OWNER	PM No. FIELD No.	REFERENCE	TRANSLATION	TRANSLITERATION
		Paget–Pirie (1898) pl. 31.	Viewing the counting of the cattle of the stalls of the estates and temples of the *pr-ḏt* (Above Register 4)	*m33 sjptj n jw3 n mḏwt nt ḥwwt ḥwwt-k3 nt pr-n-ḏt*
		Paget–Pirie (1898) pl. 31.	Viewing the presentation (of) fowl that is brought by the estates and temples of Upper and Lower Egypt, of the *pr-ḏt*.	*m33 nḏt-ḥr 3pdw jnw n ḥwwt ḥwwt-k3 nt Mḥw Šmʿw pr-n-ḏt*
		Paget–Pirie (1898) pl. 33.	Viewing every pleasant amusement that is performed in the whole land	*m33 sḫmḫ-jb nb nfr jrrw m t3 r-ḏr.f*
M3-nfr	PM 575	LD II 68.	Viewing … desert animals?, *jw3* cattle and birds being brought from the towns of the estate	*m33 …ʿwt ḥʿst? jw3 3pdw innw m njwwt nwt pr-ḏt*
		LD II 67.	Viewing the bringing of invocation offerings every day for ... being brought from the towns of the estate	*m33 zḫpt prt-ḫrw rʿ-nb n ... jnnw m njwwt pr-ḏt*
Nbt	PM 624	Munro (1993) pl. 12.	Viewing fowl	*m33 wšnw.w*
		Munro (1993) pl. 16.	Viewing the presentation of offerings brought? …	*m33 nḏt-ḥr jn (…n?).s*
		Munro (1993) pl. 20.	Viewing the work of the field by …	*m33 k3t sḫt jn*
		Munro (1993) pl.17.	(Viewing) the *jw3* cattle, scimitar-horned oryx and the small cattle which are brought from the (estates)	*(m33) jw3 ḥḏ(.w) ʿw.t jnnt m ḥw(wt)*
		Munro (1993) pl. 19.	Viewing all the *jw3* cattle of the western region … from the estates and towns of the (*pr-ḏt*)	*m33 jw3(.w) nb(.w) nw ḫ3s(.wt) … m ḥwwt njwwt n (pr-ḏt)*
		Munro (1993) pl.18.	Viewing the work of the field by …	*m33 k3t sḫt jn*
		Munro (1993) pl. 14.	Viewing the wine, brought from the palace for ..	*(m)33 jrp rḏ.t m stp-s3 n …*
		Munro (1993) pl. 13.	Viewing the wine, brought from the palace for ..	*m33 jrp rḏ.t m stp-s3 n …*
Nfr (1)	PM 137 G4761	Junker *Gîza* VI (1943) fig. 14, 17.	Viewing the ploughing, the reaping of barley and emmer wheat by his workers of his *pr-ḏt*	*m33 sk3 3zḫ jt bdt jn jzwt.f pr.f-ḏt*
Nfr-jrt-n.f	PM 583	van de Walle (1978) pl. 11.	Viewing all the work of the field	*m33 k3t nbt nt sḫt*
Nḫt-k3.j	PM 240	Hassan VII (1953) fig. 22.	Viewing the presentations of the *pr-ḏt*	*(m)33 nḏ(t)-ḥr nt pr-ḏt*
Rʿ-špss	PM 494 LS 16	LD II 61.	Viewing the presentation that is brought from his towns of the *pr-ḏt* of Lower and Upper Egypt	*m33 nḏt-ḥr jnnt m njwwt.f nwt pr-ḏt T3-Mḥw Šmʿw*
		LD II 62.	Viewing the account of the chiefs of his estates and towns of Lower and Upper Egypt	*m33 ḥzb ḥk3w ḥwwt njwwt.f T3-Mḥw Šmʿw*
Ḥtp-ḥr-3ḫtj	PM 593 D60	Mohr (1943) fig. 24.	Viewing the work of the field	*m33 k3t sḫt*
		Mohr (1943) fig. 39-54.	Viewing the ploughing, the gathering of flax and the reaping	*m3(3) sk3 ḥwj mḥ 3zḫ*
Ḫww-wr	PM 254 LG 96	Hassan V (1944) p. 248.	(Viewing) the presentations (which are) brought from the towns of the *pr-ḏt*	*(m33) nḏt-ḥr jn(nt) m njwwt nwt pr-ḏt*
Ḫw-ns	PM (iv) 134 No.2	LD II 105.	Viewing the pools, traversing the marshes by boat, viewing the trapping of birds	*m33 sšw ḫns ḥʿww pḥww m33 sḫt 3pdw*
Ḫnwt	PM 623	Munro (1993) pl. 38.	(Viewing what is) brought from her estates and her towns of the *pr-ḏt*	*(m33) (j)rr(.w) m ḥwwt.s njwwt.s n(t) pr-ḏt*

REGISTERS					MONUMENT
FIRST (highest)	**SECOND**	**THIRD**	**FOURTH**	**FIFTH/SIXTH**	**TOMB OWNER**
HUSBANDRY Cattle	HUSBANDRY Cattle	ANIMAL PARADE Cattle	ANIMAL PARADE Cattle	ANIMAL PARADE Birds	
HUSBANDRY Cattle	HUSBANDRY Cattle	ANIMAL PARADE Cattle	ANIMAL PARADE Cattle	ANIMAL PARADE Birds	
MARSHES Papyrus gathering	YOUNG BOYS Playing	WINE Production	HUNT Desert animals	Regs 5, 6, 7 MARSHES Papyrus boats	
ANIMAL PARADE Desert animals	ANIMAL PARADE Cattle	ANIMAL PARADE Cattle	ANIMAL PARADE Birds		*Mꜣ-nfr*
OFFERINGS Bearers	OFFERINGS Bearers	OFFERINGS Bearers	OFFERINGS Bearers; Butchery		
OFFERINGS Fish	OFFERINGS Birds	OFFERINGS Birds			*Nbt*
OFFERINGS Domains	OFFERINGS Domains	OFFERINGS Offering bearers	OFFERINGS Domains		
MARSHES Returning boatmen	OFFERINGS Bearers	MARSHES Returning boatmen	OFFERINGS Bearers		
ANIMAL PARADE? Records; desert animals; cattle	ANIMAL PARADE? Records; desert animals; cattle	ANIMAL PARADE Cattle	ANIMAL PARADE? Records; desert animals; cattle		
ANIMAL PARADE Cattle	ANIMAL PARADE? Records; desert animals; cattle	ANIMAL PARADE? Records; cattle	ANIMAL PARADE? Records; desert animals; cattle		
MARSHES Returning boatmen	OFFERINGS Bearers	MARSHES Returning boatmen	OFFERINGS Bearers		
BRINGING WINE Bearers; record	BRINGING WINE Bearers; record	BRINGING WINE Bearers; record	BRINGING WINE Bearers; record		
AGRICULTURE Ploughing; sheep	HUSBANDRY Cattle	BRINGING WINE Bearers; record	OFFERINGS Bearers		
AGRICULTURE Boats; ploughing; flax harvest	ANIMAL PARADE Birds; reaping; clapnet	ANIMAL PARADE Cattle; harvest; transport	ANIMAL PARADE Desert animals; sheep treading seed		*Nfr* (1)
HARVEST	AGRICULTURE Ploughing; sheep	HARVEST Flax gathering			*Nfr-jrt-n.f*
OFFERING; BEARERS	ANIMAL PARADE Desert animals	ANIMAL PARADE Cattle	ANIMAL PARADE Cattle	ANIMAL PARADE Cattle; hyena	*Nḫt-kꜣ.j*
ANIMAL PARADE Cattle	ANIMAL PARADE Records; birds; desert animals	ANIMAL PARADE Records; birds	ANIMAL PARADE Records; birds; domains		*Rꜥ-špss*
OFFERING; BEARERS	RECORD Scribes; husbandry; boat with cargo	RECORD Scribes; husbandry; boat with cargo	RECORD Scribes; husbandry; boat with cargo	RECORDS Scribes; chiefs	
OFFERINGS Birds	BOAT (destroyed)				*Ḥtp-ḥr-ꜣḫtj*
AGRICULTURE Ploughing; sheep	HARVEST Cereals; flax	HARVEST Transport	HARVEST Threshing; storage	OFFERINGS Slaughter	
BEARERS?	ANIMAL PARADE Cattle	ANIMAL PARADE Cattle	ANIMAL PARADE Record; cattle	OFFERINGS Butchery	*Ḫww-wr*
OFFERINGS (birds) scribes; trapping fowl	CATTLE PARADE Husbandry	MARSHES Jousting boatmen	MARSHES Cattle fording		*Ḫw-ns*
HARVEST Reaping; storing	BOATS Transport	TRANSPORT Offering cabinets			*Ḫnwt*

TABLE 17: TOMB OWNER VIEWING THE WORK AND PRODUCE OF HIS ESTATE

MONUMENT			INSCRIPTION	
TOMB OWNER	PM No. FIELD No.	REFERENCE	TRANSLATION	TRANSLITERATION
Snḏm-jb/Jntj	PM 85 G2370 =LG 27	LD II 77; Brovarski (2001) fig. 42.	Viewing the marshlands and every good thing consisting of the activity of the fields (by) …	*mꜣꜣ pḥww ḫt nbt nfrt m kꜣt sḫt*
Snḏm-jb/Mḥj	PM 87 G2378 =LG 26	LD Erg,, 12; Brovarski (2001) fig. 97, 98.	(Viewing the work of the fields) and every good thing which is brought from the marshlands (from his fields, his towns and his estates, of Lower Egypt (and Upper Egypt)	*(mꜣꜣ kꜣt sḫt) ḫt nb(t) nfrt jnnt m pḥww (m sḫt.f njwwt.f ḥwwt.f nt) Tꜣ-mḥw (Šmꜥw)*
		LD Erg., 11; Brovarski, (2001) fig. 104-105.	(Viewing the work of the fields and every good thing) which is brought from the marshlands, from his fields, his towns (and his estates of Lower Egypt and Upper Egypt)	*(mꜣꜣ kꜣt sḫt ḫt nb(t) nfrt) jnnt m pḥww m sḫt.f njwwt.f (ḥwwt.f nt Tꜣ-mḥw Šmꜥw)*
Sḫm-ꜥnḫ-Ptḥ	PM 191 G7152	Badawy (1976) fig. 20	Viewing the *jwꜣ* cattle, desert animals and fowl brought by the towns of the *pr-ḏt*	*(mꜣ)ꜣ jwꜣ ꜥwt ḥꜣst ꜣpdw jnw n m njwwt pr-ḏt*
Sḫm-ꜥnḫ-Ptḥ	PM 454 D41	Simpson (1976) pl. D.	Viewing the work of the field, ploughing, reaping, gathering flax, donkey transport, donkeys treading the threshing floors and winnowing.	*mꜣꜣ kꜣt sḫt skꜣ ꜣzḫ ḥwj mḥꜥw šdjt jꜥꜣw ḥwj jꜥꜣw spwt ḫꜣḫꜣ*
Sḫm-kꜣj	PM 53 G1029	Simpson (1980) fig. 4.	Viewing the ploughing and the reaping of barley by his crews of his towns of his *pr.f ḏ.t.*	*mꜣꜣ skꜣ ꜣzḫ jt jn jzwt.f nt njwwt.f nwt pr.f ḏ.t*
Sšm-nfr III	PM 153 G5170	Brunner-Traut (1977) Farbtafel II.	Viewing the document of the presentations which are brought as invocation offerings from the towns of the *pr-ḏt*	*mꜣꜣ zš nḏt-ḥr jnnt r prt-ḫrw njwwt nwt pr-ḏt*
Sšm-nfr IV	PM 223 LG53	Junker *Gîza* XI (1953) fig. 74b.	Viewing the presentations which are brought … from Lower and Upper Egypt by …	*(m)ꜣꜣ nḏt-ḥr jnnt … Tꜣ-mḫw Šmꜥw jn ..*
			(Viewing the work of) the field, the bird pools and the marshlands.	*(mꜣꜣ kꜣt) sḫt sšw pḥww*
Kꜣ.j-m-nfrt	PM 208 LG 63	Badawy (1976) fig. 29.	Viewing the list of presentations which are brought as invocation offerings	*mꜣꜣ zš n nḏt-ḥr nt jn prt-ḥrw*
		LD Erg. pl.32 [lower]; Badawy (1976) fig. 30[a]	(Viewing) the ploughing	*(mꜣꜣ) skꜣ*
Kꜣ.j-m-nfrt	PM 467 D23	Simpson (1992) fig. 15, pls. F. G.	Viewing the ploughing, gathering of flax, reaping, loading (donkeys), treading (of the threshing floors), winnowing, heaping up	*mꜣꜣ skꜣ jt ḥj mḥꜥw ꜣsḫ šd … ḥj ḫꜣḫꜣ jꜥb*
		Simpson (1992) pl. E, pl. 16-18.	Viewing the recording of the presentations brought from his towns of the *pr-ḏt* in Lower and Upper Egypt very abundantly	*mꜣꜣ zš n nḏt-ḥr jnnt m njww.t nt pr-ḏt m Tꜣ-mḥw Šmꜥ ꜥꜣ wrt*
Kꜣ.j-dwꜣ	PM 244	Hassan VI (1946) fig. 82.	Viewing the invocation offerings	*mꜣꜣ prt-ḫrw*
Kꜣ-ḥj.f	PM 76	Junker *Gîza* VI (1941) fig. 41-45.	Viewing the ploughing and the harvesting and all the beautiful work that is done in the field	*(mꜣꜣ skꜣ) ꜣzḫ kꜣt nb nfrt jrrt m sḫt*
		Junker *Gîza* VI (1941) fig. 40.	Viewing of the presentation of the *jwꜣ* cattle, desert animals and the poultry which are brought from his towns	*mꜣꜣ nḏt-ḥr nt jwꜣ wꜥt ḫꜣst ꜥpdw jnnt m njwwt.f*
Ṯjj	PM 468 D22	Épron–Daumas I (1939) pl. 6-8.	Viewing the stalls of (cattle?) and small cattle … the field, in very great (quantity) in order to inspect	*(m)ꜣꜣ mḏt n (jwꜣ?) ꜥwt … …t zḫt ꜥꜣ wrt r mꜣꜣ*
		Épron–Daumas I (1939) pl. 35.	Viewing the feeding of the *zt* geese	*mꜣꜣ znmt zt*
		Épron–Daumas I (1939) pl. 35.	Viewing the feeding (by) the tally-man	*mꜣꜣ snmt nḫt-ḫrw*

REGISTERS					MONUMENT
FIRST (highest)	**SECOND**	**THIRD**	**FOURTH**	**FIFTH/SIXTH**	**TOMB OWNER**
COOKING by herders	HUSBANDRY Cattle	HUSBANDRY Cattle	MAT MAKING Herders	HUSBANDRY Cattle	*Snḏm-jb/Jntj*
MARSHES	?	?	?	Regs 6, 7, 8 OFFERINGS/ HUSBANDRY Cattle fording	*Snḏm-jb/Mḥj*
MARSHES	?	?	?	Regs 6, 7, 8 OFFERINGS/ HUSBANDRY Cattle fording	
ANIMAL PARADE Cattle	ANIMAL PARADE Cattle	ANIMAL PARADE Cattle	ANIMAL PARADE Birds		*Sḫm-ʿnḫ-Ptḥ*
ANIMAL PARADE Scribes; desert animals	MARSHES Scribes; jousting boatmen; clapnet	AGRICULTURE Scribes; sheep; ploughing	HARVEST Scribe; gathering flax; reaping	HARVEST Scribe; bringing donkeys; transport; threshing; storage	*Sḫm-ʿnḫ-Ptḥ*
MARSHES Scribes; gathering plants; clapnet	MARSHES Jousting boatmen; ploughing; sheep	HARVEST Gathering flax; reaping; donkeys	ANIMAL PARADE Desert animals; dragnet fishing	ANIMAL PARADE Cattle; dragnet fishing	*Sḫm-kȝj*
SCRIBES	ANIMAL PARADE Desert animals	ANIMAL PARADE Cattle	OFFERINGS Birds; birds in cages		*Sšm-nfr* III
?	?	HARVEST Reaping	HARVEST Transport of harvest		*Sšm-nfr* IV
?					
ANIMAL PARADE Cattle	ANIMAL PARADE Cattle	ANIMAL PARADE Birds			*Kȝ.j-m-nfrt*
HARVEST Scribes; Threshing	HARVEST Transporting grain; Animal parade; cattle	PLANTING Ploughing	PLANTING Sowing; sheep		
DESTROYED Sowing?	HARVEST Gathering flax	HARVEST Bundling flax	HARVEST Reaping	(Regs 6, 7) HARVEST Grain transport; threshing	*Kȝ.j-m-nfrt*
ATTENDANTS	SCRIBES	ANIMAL PARADE Desert animals	ANIMAL PARADE Desert animals	(Regs 6, 7) ANIMAL PARADE Cattle	
ANIMAL PARADE Birds	ANIMAL PARADE Birds	ANIMAL PARADE Cattle	ANIMAL PARADE Cattle	ANIMAL PARADE Desert animals	*Kȝ.j-dwȝ*
AGRICULTURE Ploughing; sheep FISHING Dragnet	HARVEST grain; flax; piling sheaves	HARVEST Transport by donkeys	HARVEST Threshing; winnowing		*Kȝ-ḥj.f*
ANIMAL PARADE Cattle	ANIMAL PARADE Cattle	ANIMAL PARADE Desert animals	ANIMAL PARADE Birds		
HUSBANDRY Cattle, desert animals	HUSBANDRY Cattle, desert animals	HUSBANDRY Cattle, desert animals	FOWLING Feeding bird	FOWLING Poultry-yard	*Ṯjj*
POULTRY Feeding	POULTRY Feeding	POULTRY Feeding			
as above	as above	as above			

TABLE 17: TOMB OWNER VIEWING THE WORK AND PRODUCE OF HIS ESTATE

MONUMENT			INSCRIPTION	
TOMB OWNER	PM No. FIELD No.	REFERENCE	TRANSLATION	TRANSLITERATION
		Wild III (1966) pl. 136-39.	Viewing the pulling of flax, the reaping of barley and emmer wheat, the transporting, the heaping on the threshing floors from (by?) the towns of the *pr-ḏt*	*mꜣꜣ ḥwj mḥꜥ ꜣzẖ jt bdt šdjt wbz m spwt m njwwt nt pr-ḏt*
		Wild III (1966) pl. 169.	Bringing for viewing and presentation (that) which is brought from the towns of the estate	*jnt r mꜣꜣ nḏ(t)-ḥr jnt m njwwt nt pr-ḏt*
		Wild III (1966) pl. 162.	Viewing the catching of the golden oriole, the treading of grapes and all the work of the field	*mꜣꜣ swḥ gnw ḥwj jꜣrrt kꜣt nb nt zẖt*
		Wild III (1966) pl. 162.	Viewing the presentation which is brought from the towns of the *pr-ḏt*, which is in Lower and Upper Egypt	*mꜣꜣ nḏt-ḥr jnnt m njwwt nt pr-ḏt ntt m Tꜣ-mḥw Šmꜥw*
DYNASTY VI.E				
Jttj-Šdw	PM (iv) 122	Kanawati–McFarlane (1993) pl. 45.	Viewing the work of the marshland and the crossing of the cattle to him	*(mꜣ)ꜣ (kꜣt) zẖt ḏꜥt n.f (kꜣ)w.f*
		Kanawati– McFarlane (1993) pl. 51.	Viewing the bulls and cows of his property and the number of all his bulls, and hearing the speech of the herdsmen of his *pr n ḏt.f*	*mꜣꜣ kꜣ ḥmt nt ẖt.f rẖt n kꜣ.f n sḏm mdw nrw nw pr n ḏt.f*
Ptḥ-špss II	PM 460 E 1,2 H 3	Petrie–Murray (1952) pl. 29.	Viewing presentations which are brought from his estates and his towns of Lower and Upper Egypt	*mꜣꜣ nḏt-ḥr jnnt m ḥwwt.f njwwt.f nt Tꜣ-mḥw Šmꜥw*
			Viewing presentations which are brought from his estates and his towns of Lower and Upper Egypt	*mꜣꜣ nḏt-ḥr jnnt m ḥwwt.f njwwt.f nt Tꜣ-mḥw Šmꜥw*
Mrjj-Ttj/Mrj	PM 536	Kanawati–Abder-Raziq (2004) pl. 47.	Viewing the presentation of offerings which are brought for him from his estates and from his towns of Lower and Upper Egypt	*mꜣꜣ nḏt-ḥr jnnt n.f m ḥwwt.f m njwwt.f nt Tꜣ-mḥw Šmꜥw*
Mrw/Ttj-snb	PM 520	Lloyd (1990) pl. 8.	Viewing the work (of the field and what is done) in the marshland in (trapping) birds and catching fish	*mꜣꜣ kꜣt sẖt … jrt m pḥw m ꜣpdw m ḥꜥm*
		Lloyd (1990) pl. 12.	(Viewing …) of the fields which are done in his towns and his estates of Lower and Upper Egypt	*(mꜣꜣ …) sẖt jrt (m) njwwt.f m ḥwwt.f nt Tꜣ-mḥw Šmꜥw*
Mrrj	PM 518	Davies et al (1984) pl. 9-10.	Viewing all the good things that are brought for him from his towns and his estates of Lower and Upper Egypt	*mꜣꜣ ẖt nbt nfrt jnnt n.f m njwwt.f m ḥwwt.f nt Tꜣ-mḥw Šmꜥw*
Mrrw-kꜣ	PM 525	Duell I (1938) pl. 14.	(… …) which are brought to him from his estates and his towns of Lower Egypt …	*(… j)nnt n.f m ḥwwt m njwwt.f nt Tꜣ-mḥw*
		Duell I (1938) pl. 96-99.	Viewing the sealed (items) and clothing brought from the Residence as a boon which the king gives for …	*mꜣꜣ ẖtmt mnẖt jnt m ẖnw m ḥtp dj nswt n …*
		Duell II (1938) pl. 113-116.	Viewing the sweet offerings which are brought to him from the Residence as a boon that the king gives there for …	*mꜣꜣ ẖt bnrt jnnt n.f m ẖnw m ḥtp dj nswt jm n …*
Mḥw	PM 619	Altenmüller (1998) pl. 22[A].	Viewing the arrival of the fleet together with the work of the field and the coming of his estates and his towns of Lower and Upper Egypt	*mꜣꜣ spr ḥwꜥw ḥnꜥ kꜣr sẖt jwt ḥwwt.f njwwt.f nt Tꜣ-mḥw Šmꜥw*
		Altenmüller (1998) fig. 8, pl. 36, 38.	Viewing the very great fishing and work of the field, the distribution of fish to the teams of the *pr-ḏt*	*mꜣꜣ ḥꜥm mḥjt ꜥꜣ wrt kꜣt sẖt sšmt mḥjt n ṯzwt nt pr-ḏt*
Nj-kꜣw-Jzzj	PM n/r	Kanawati–Abder-Raziq (2000) pl. 49.	Viewing the feeding on grain by the cranes and the drawing of milk for …	*mꜣꜣ wnm jt jn ḏwꜣt sšr jrṯt n*

REGISTERS					MONUMENT
FIRST (highest)	**SECOND**	**THIRD**	**FOURTH**	**FIFTH/SIXTH**	**TOMB OWNER**
Regs 1, 2 HARVEST Reaping, stacking HUSBANDRY: cattle	Regs 3, 4, 5 HARVEST Reaping grain	Reg. 6, 7 HARVEST Grain transport	Reg. 8 HARVEST Piling sheaves	Reg. 9, 10 HARVEST Grain transport, threshing, storage	
Regs 1-4 ANIMAL PARADE Desert animals	Reg 5 ANIMAL PARADE Scribe; cattle	Reg 6 ANIMAL PARADE cattle	Reg 7 ANIMAL PARADE Scribes; cattle	Reg 8, 9, 10 ANIMAL PARADE cattle, birds	
Regs 1-6 OFFERINGS	Reg. 7 Offering bearers	Reg. 8 Offering bearers	Reg. 9 Offering bearers; music	Reg. 10 , 11 Offering bearers; butchery	
as above	as above	as above	as above	as above	
					DYNASTY VI.E
BOATS Destroyed	MARSHES Fowling; clapnet	GOATS Browsing	HUSBANDRY Cattle fording		*Jttj-Šdw*
OFFERINGS Scribe; foods	HUSBANDRY Cattle mating; birth of calf; milking	ANIMAL PARADE Cattle	BUTCHERY Oxen; cattle being led		
OFFERINGS Butchery	OFFERINGS	OFFERINGS	OFFERINGS		*Ptḥ-špss* II
OFFERINGS Butchery	OFFERINGS	OFFERINGS	OFFERINGS		
ATTENDANTS	Regs 2-4 ATTENDANTS Offering bearers	Regs 5-9 OFFERING BEARERS			*Mrjj-Ttj/Mrj*
HUSBANDRY Cattle	HUSBANDRY Cattle fording	MARSHES Fowling: clapnet	MARSHES Fishing: dragnet		*Mrw/Ttj-snb*
HARVEST Gathering flax	HARVEST Reaping grain	MARSHES Returning boatmen	OFFERINGS Butchery		
OFFERING BEARERS	OFFERING BEARERS	OFFERING BEARERS	OFFERING BEARERS	OFFERING BEARERS	*Mrrj*
ATTENDANTS	ATTENDANTS	OFFERING BEARERS	OFFERING BEARERS	OFFERING BEARERS	*Mrrw-kꜣ*
OFFERINGS	OFFERINGS	OFFERINGS	OFFERINGS		
OFFERINGS	OFFERINGS	OFFERINGS	OFFERINGS		
OFFERINGS/BOATS Offering bearers	HARVEST Reaping to winnowing	OFFERINGS Domains			*Mḥw*
OFFERINGS FOWLING: Clapnet	OFFERINGS FISHING: dragnet; Scribes	AGRICULTURE Ploughing; sheep; FISHING			
POULTRY Feeding birds	HUSBANDRY Milking				*Nj-kꜣw-Jzzj*

TABLE 17: TOMB OWNER VIEWING THE WORK AND PRODUCE OF HIS ESTATE

MONUMENT			INSCRIPTION	
TOMB OWNER	**PM No. FIELD No.**	**REFERENCE**	**TRANSLATION**	**TRANSLITERATION**
Ḫnmw-ntj	PM 87 G2374	Brovarski (2001) fig. 88.	Viewing every good (thing) which is brought from …	*m33 ḫt nb r jnnt m … k3t*
S3bw/Jbbj	PM 460 E 1,2 H 3	Borchardt (1964) pl. 21.	Viewing the *jw3* cattle which are brought from the estates and his towns of the *pr-ḏt*	*m33 jw3w jnnw m ḥwwt njwwt.f n pr-ḏt*
Sšm-nfr/Jwfj	PM 614	Barsanti, ASAE (1900) fig. 9.	Viewing the work of the field by …	*m33 k3t zḫt jn*
K3-gmnj	PM 521	Firth–Gunn II (1926) pl. 9b.	Viewing the marshlands	*m33 pḥww*
		von Bissing I (1905) pl. 8-10.	Viewing the presentation of cattle	*(m)33 sjʿ n wnḏw*
		von Bissing I (1905) pl. 8-16.	(Viewing) the presentation of poultry, the stalls of the *jw3* cattle, the stalls of the small cattle of the desert, the bird-pools of the *pr-ḏt*	*(m33) s(ḫp) 3pd mḏwt nt jw3 mḏwt nt ʿwt ḫ3st štbw ḥrt-mw nt pr-ḏt*
		von Bissing I (1905) pl. 17-21.	(Viewing the work) of every marshland and the conduct of the fish of the gang of the *pr-ḏt*	*(m33 … k3t) sḫt nb sšmt mḥjt nt ṯzt nt pr-n-ḏt*
		von Bissing I (1905) pl. 2-3.	(Viewing) the presentation of all offerings of the year which are brought from his estates and his towns of Lower and Upper Egypt by the inspectors of priests, assistant priests and priests and all people of the *pr-ḏt*	*(m33) nḏt-ḥr rnpt nb(t) jnnt m ḥwwt.f m njwwt.f nt T3-mḥw Šmʿw jn sḥd ḥmw-k3 jmj-ḫt ḥmw-k3 ḥmw-k3 rmṯ nb nw pr-ḏt*
		von Bissing II (1911) pl. 4.	(Viewing) the presentation of all offerings of the year which are brought from his estates and his towns of Lower and Upper Egypt and his chapels of supplies of the two lakes? of his *pr-ḏt*	*(m33) nḏt-ḥr rnpt nb(t) jnnt m ḥwwt.f m njwwt.f nt T3-mḥw Šmʿw ḥwwt-k3.t.f nt jmjt w3ḏt nwt pr-ḏt*
		von Bissing II (1911) pl. 37-39.	(Viewing) … from the Residence as an offering for	*(m33) … m ḫnw m ḥtp-dj-nzwt n*
		von Bissing II (1911) pl. 41.	(Viewing) the royal clothing and the oils which come from the court as a royal offering for …	*(m33) ḥbs nswt mrḥt m ḫnw m ḥtp-dj-nswt*
DYNASTY VI.M				
Jbj	PM(iv) 243 No.8	Kanawati *Gebrawi* II (2007) pl. 71.	Viewing the ploughing, the gathering of flax, the reaping, the grain transport and all good festivals of Upper Egyptian grain for the ka (of) …	*(m33) sk3 ḥwj mḥʿ 3(zḫ) šdjt ḥb(w) nb(w) nfr(w nw) jt (šmʿ) n k3 …*
			Viewing the desert animals and the beautiful herd of cattle for the ka of …	*(m)33 ʿwt ḫ3st z3 nfr n k3w n k3 n*
Jn-jt.f/Bḫnt	PM (v) 18 BA63	Kanawati (1987) fig. 20-21.	Viewing the lassoing of wild game by the hunters	*m33 spḥ ʿwt ḫ3st jn nww*
B3wj	BA48	Kanawati (1987) fig. 43.	Viewing the cattle which are brought from the towns of …	*(m)33 jw3w jnnw m njwwt nt*
Ppj-ʿnḫ-ḥrj-jb	PM (iv) 254 D2	Blackman *Meir* IV (1924) pl. 14.	Viewing the ploughing in his towns of Lower and Upper Egypt	*m33 sk3 m njwwt.f nt Mḥw Šmʿ*
		Blackman *Meir* IV (1924) pl. 14.	Viewing the reaping of barley and emmer wheat and the gathering of flax	*m33 3zḫ jt bdt ḥwj mḥʿ*

REGISTERS					MONUMENT
FIRST (highest)	**SECOND**	**THIRD**	**FOURTH**	**FIFTH/SIXTH**	**TOMB OWNER**
ANIMAL PARADE Cattle	HARVEST Reaping, transport	MARSHES Clapnet	MARSHES Returning boatmen	ANIMAL PARADE Cattle	*H̱nmw-ntj*
ANIMAL PARADE Scribes; cattle					*Sꜣbw/Jbbj*
OFFERINGS Domains; games	MARSHES Jousting boatmen	HUSBANDRY Cattle; fowl	MARSHES Clapnet; offerings		*Sšm-nfr/Jwfj*
ANIMAL PARADE Scribes; cattle	MARSHES Papyrus boat building	DESTROYED	HUSBANDRY Cattle	HUSBANDRY Cattle fording	*Kꜣ-gmnj*
ANIMAL PARADE Cattle	HUSBANDRY Scribes; cattle	HUSBANDRY Scribes; cattle	FOWLYARD Force-feeding birds	FOWLYARD Offerings; fowlyard MARSHES clapnets	
as above	as above	as above	as above	as above	
BOATS	MARSHES Processing fish	MARSHES Dragnet; dragnet	MARSHES Scribes; transport of fish		
OFFERINGS	OFFERINGS	OFFERINGS	OFFERINGS	OFFERINGS	
OFFERINGS	OFFERINGS	OFFERINGS	OFFERINGS	OFFERINGS	
OFFERINGS	OFFERINGS	OFFERINGS	OFFERINGS	OFFERINGS	
OFFERINGS	OFFERINGS	OFFERINGS	OFFERINGS	OFFERINGS	
					DYNASTY VI.M
DESTROYED	DESTROYED	HARVEST Gathering flax	HARVEST Reaping cereal	HARVEST Transport by donkeys; threshing	*Jbj*
HUNT Desert animals	HUNT Desert animals	HUSBANDRY Scribe; Birth of calf	HUSBANDRY Bull fight; cattle mating	HUSBANDRY Milking; goats browsing	
PARADE Cattle; slaughter	BUTCHERY	HUNT Desert animls			*Jn-jt.f/Bḫnt*
ANIMAL PARADE Cattle	ANIMAL PARADE Cattle				*Bꜣwj*
HUSBANDRY Goats; chopping trees	AGRICULTURE Ploughing	AGRICULTURE Ploughing	AGRICULTURE Ploughing	AGRICULTURE Ploughing HUSBANDRY Cattle	*Ppj-ꜥnḫ-ḥrj-jb*
HARVEST Gathering flax	HARVEST Reaping grain	HARVEST Loading donkeys; transport of grain	HARVEST Threshing	HARVEST Transport; stacking	

TABLE 17: TOMB OWNER VIEWING THE WORK AND PRODUCE OF HIS ESTATE

MONUMENT			INSCRIPTION	
TOMB OWNER	**PM No. FIELD No.**	**REFERENCE**	**TRANSLATION**	**TRANSLITERATION**
		Blackman *Meir* IV (1924) pl. 8.	Viewing the work of the field, the netting of waterfowl, the catching of very great fish	*(m33) k3t sḫt 3pdw ḥˁm mḥjt ˁ3 wrt*
		Blackman *Meir* IV (1924) pl. 16.	Exaction of the impost of cattle and goats due from the middle nomes.	*jrt jrw n k3w ˁwt n sp3wt ḥrjw-jb*
			Viewing *jw3* cattle and goats,	*m33 jw3 wnḏw*
DYNASTY VI.L				
Ppj-ˁnḫ/Ḥnj-km	PM (iv) 247	Blackman–Apted *Meir* V (1953) pl. 30.	(Viewing) all the good work that is done in the fields of Upper and Lower Egypt	*m(33) k3t nbt nfrt jrrt m sḫwt nt Šmˁ T3-mḥw*
		Blackman–Apted *Meir* V (1953) pl. 29.	Viewing the presentation brought to him from (his) towns … his headmen	*m33 nḏt-ḥr jnnt n.f m njwwt … n ḥk3w.f*
		Blackman–Apted *Meir* V (1953) pl. 32.	Viewing the exaction of the impost of cattle and all small cattle	*m33 jrt jrw m k3w ˁwt nb(w)t*
		Blackman–Apted *Meir* V (1953) pl. 41.	Viewing the cattle and small cattle from his towns	*m33 k3w ˁwt nb(wt) m njwwt.f*
Mrjj	PM (v) 35	Kanawati (1993) pl. 43.	Viewing the cattle, goats and donkeys (of the town of his estate?) by …	*m33 k3w ˁwt ˁ3w [njwt ḏt].f jn …*
Nj-ˁnḫ-Ppjj-km	PM (iv) 247 A1	Blackman–Apted *Meir* V (1953) pl. 13.	Viewing the work of the field, the netting of waterfowl, the catching of very great fish by …	*(m33) k3t sḫt sḫt 3pdw ḥˁm mḥjt ˁ3 wrt jn …*
Ḥnj/ Špsj-pw-Mnw	PM n/r H24	Kanawati II (1981) fig. 17.	Viewing all the good work of the field by …	*m33 k3t sḫt nb(t) nfrt jn*
		Kanawati (1981) fig. 20.	Viewing the bull fight by …	*m33 ˁḥ3 k3w jn*
		Kanawati (1981) fig. 26.	Viewing the *jw3* cattle …	*m33 jw3w jn …*
K3(.j)-ḥp/Ṯtj-jkr	PM (v) 19 H26	Kanawati (1980) fig. 9.	(Viewing the exaction) of the cattle tax of bulls …	*… jrw m k3w*
		Kanawati (1980) fig. 10.	Viewing the bull fight by …	*m33 ˁḥ3 k3w jn …*
		Kanawati (1980) fig. 11.	Viewing all the good work of the field by …	*(m)33 k3t sḫt nb(t) nfrt jn …*
Ḏˁw and *Ḏˁw/Šm3j*	PM (iv) 244 No.12	Davies *Gebrâwi* II (1902) pl. 6.	Viewing the ploughing, the gathering of flax, the reaping and the transport and every good festival of Upper Egyptian grain	*m3(3) sk3 ḥwj mḥˁ 3zḫ šdjt ḥbw nb(w) nfrw nwjt šmˁ*
DYNASTY VIII				
W3hj	PM (v) 36 D4	Kanawati (1995) pl. 20, 22.	Viewing (the making of a count?) of all cattle and goats from his own property	*m33 (jrt jrw?) k3w ˁwt nb m jšt.f ḏs.f*
Mrjj-ˁ3	PM (v) 34 D18	Kanawati, (1995) pl. 38, 39.	Viewing the cattle count by (tomb owner) from his own property in reality. There is no boasting therein.	*m33 jrw jn (...) m jšt.f ḏs.f m wn-m3ˁ nn (ˁb)ˁ jm*
Rḥw-r-3w.sn	PM n/r BA17	Kanawati (1987) fig 15.	Viewing the bull fight by …	*m33 ˁḥ3 k3w jn*
Gḥs3/Nbjj	PM n/r GA11	Kanawati (1987) fig 30.	Viewing the work of…	*(m)33 k3t …*

REGISTERS					MONUMENT
FIRST (highest)	**SECOND**	**THIRD**	**FOURTH**	**FIFTH/SIXTH**	**TOMB OWNER**
MARSHES Fowling; clapnet	OFFERINGS Bearers with fowl; cooking fowl	MARSHES Fowling; clapnet	MARSHES Fishing; dragnet		
ANIMAL PARADE	ANIMAL PARADE	BOATS			
CATTLE COUNT	CATTLE COUNT Goat count				
					DYNASTY VI.L
AGRICULTURE Ploughing	MARSHES Fowling; clapnet	MARSHES Fishing; dragnet	MARSHES /HUSBANDRY Cattle; fording; papyrus gathering; spearing hippo		*Ppj-ʿnḫ/Ḥnj-km*
OFFERINGS	DESTROYED				
ANIMAL PARADE Scribes; cattle	HUSBANDRY Goats feeding	ANIMAL PARADE Cattle	ANIMAL PARADE Cattle	HUSBANDRY Bulls fighting	
ANIMAL PARADE Cattle	ANIMAL PARADE Oryx	ANIMAL PARADE Oryx	ANIMAL PARADE Oryx		
OFFERINGS	PARADE Goats; donkeys	HUSBANDRY Bulls fighting			*Mrjj*
MARSHES Fowling; clapnet	BIRDS Cooking, presenting	MARSHES Fowling; clapnet	MARSHES Fishing: dragnet		*Nj-ʿnḫ-Ppjj-km*
AGRICULTURE Ploughing; sowing	HARVEST Reaping	MARSHES/ GRANARIES Jousting boatmen			*H̱nj/ Špsj-pw-Mnw*
HUSBANDRY Bull fights	HUSBANDRY Bull fights	HUSBANDRY Mating, birth	HUSBANDRY Milking	ANIMAL PARADE Goats	
BUTCHERY	BUTCHERY	ANIMAL PARADE Cattle; gazelle	ANIMAL PARADE Cattle; oryx		
MANUFACTURE	MANUFACTURE	ANIMAL PARADE Cattle; goats			*Kꜣ(.j)-ḥp/Ṯtj-jkr*
HUSBANDRY Bull fight	HUSBANDRY Bull fight	HUSBANDRY Bull fight	HUSBANDRY Bull fight		
AGRICULTURE Ploughing; sowing	MARSHES Jousting boatmen	MARSHES Papyrus boat building	MARSHES Jousting boatmen		
AGRICULTURE Ploughing; sowing	AGRICULTURE Ploughing; sowing	HARVEST Flax	HARVEST Reaping cereal		*Ḏʿw* and *Ḏʿw/Šmꜣj*
					DYNASTY VIII
ANIMAL PARADE Bulls fighting	BEARERS/ HUSBANDRY Cattle fording	BEARERS	ANIMAL PARADE Goats		*Wꜣhj*
ANIMAL PARADE Cattle, bearers	HUSBANDRY Bulls fighting; fording	ANIMAL PARADE/ HUSBANDRY Goats	ANIMAL PARADE Donkeys	HARVEST Donkeys; transporting, threshing	*Mrjj-ʿꜣ*
ANIMAL PARADE Cattle	HUSBANDRY Bulls fighting	AGRICULTURE Ploughing			*Rḥw-r-ꜣw.sn*
HUSBANDRY Cattle fording	MARSHES Bird trapping	MARSHES Fishing			*Gḥsꜣ/Nbjj*

www.ingramcontent.com/pod-product-compliance
Ingram Content Group UK Ltd.
Pitfield, Milton Keynes, MK11 3LW, UK
UKHW061846041026
15591UKWH00015B/943

9 781407 309842